VEHICLE ROUTING: METHODS AND STUDIES

Studies in Management Science and Systems

Volume 16

Series Editor

Burton V. Dean

Department of Organization and Management
San Jose State University
San Jose, California, U.S.A.

Previous volumes in the series:

Y. T. Haimes (ed.), *Large Scale Systems*
T. H. Naylor (ed.), *Corporate Strategy*
B. Lev (ed.), *Energy Models and Studies*
T. H. Naylor / C. Thomas (eds.), *Optimization Models of Strategic Planning*
B. V. Dean (ed.), *Project Management: Methods and Studies*
A. Kusiak (ed.), *Flexible Manufacturing Systems: Methods and Studies*
B. Lev (ed.), *Production Management: Methods and Studies*
D. O. Gray, T. Solomon, and W. Hetzner (eds.), *Technological Innovation: Strategies for a New Partnership*
B. Lev et al. (eds.), *Strategic Planning in Energy and Natural Resources*

NORTH-HOLLAND
AMSTERDAM • NEW YORK • OXFORD • TOKYO

Vehicle Routing: Methods and Studies

Edited by:

Bruce L. GOLDEN

and

Arjang A. ASSAD
College of Business and Management
University of Maryland
College Park, Maryland 20742
U.S.A.

1988

NORTH-HOLLAND
AMSTERDAM • NEW YORK • OXFORD • TOKYO

ISBN: 0 444 70407 8

Published by:
Elsevier Science Publishers B.V.
P.O. Box 1991
1000 BZ Amsterdam
The Netherlands

Sole distributors for the U.S.A. and Canada:
Elsevier Science Publishing Company, Inc.
52 Vanderbilt Avenue
New York, N.Y. 10017
U.S.A.

PRINTED IN THE NETHERLANDS

PREFACE

The vehicle routing problem (VRP), like its well-known traveling salesman cousin, is a fascinating problem. It is easy to describe, but difficult to solve. One might say that such problems belong to the class of easy-NP-hard!

A further fascinating feature of vehicle routing is that the basic problem can be extended into an untold number of variations that are not just mathematical diversions; they do occur in the real-world problem-solving arena. Thus, it is no wonder that researchers from the fields of operations research, mathematics, transportation, and computer science, such as those who have contributed papers to this volume, have found this problem most challenging.

Although difficult to solve in an optimizing sense, VRPs are solved in an operational sense. The world's economies could not operate except for the fact that VRPs and their extensions have readily available "practical" solutions. Things are delivered and picked up, and customer demands are more or less satisfied on time without too much pain. But competition and the desire to improve profits call for better solutions. To their credit, VRP researchers have not been lured to the rocks by the siren of optimality. They recognized early that important improvements in vehicle routing could be made by non-optimal, directed investigations into the mathematical and computational structures that describe VRPs. Such efforts have helped to form the theoretical field of heuristics, as well as aiding in the development of heuristic solution procedures. The resulting heuristic algorithms go way beyond the replication of how the experts run their operational systems; these algorithms seek and find improved solutions that can be implemented.

The papers in this volume not only summarize the past developments in VRP technology, but also describe many new advances. One can only be impressed by the abilities of the authors as they attack and resolve a diverse set of important problems. Researchers and practitioners will find much here that is new and rewarding.

A closing thought. The original and classical VRP can be traced back in time many hundreds of years. It is a problem that arises every year at about this time; that it gets solved each time has always, I am sure, amazed us all. Our wonderment in how the solution is obtained is an unconscious force (in a psychological sense), stemming from our childhood, that motivates our search for algorithms to solve the VRP. The problem: How does Santa Claus do it? Santa has a single vehicle with finite capacity that leaves from a single depot; millions of stochastic demands having tight time windows must be satisfied within a 24-hour period. I believe that the charm and challenge of VRPs are reflected in our wondering about Santa's problem, and in our desire to help Santa out and get into his good graces. He knows whether we (our solutions) are good or bad.

Saul I. Gass
College Park, Maryland
December 18, 1987

CONTENTS

LIST OF CONTRIBUTORS

Arjang A. Assad
College of Business and Management
University of Maryland
College Park, Maryland 20742

Edward K. Baker
Department of Management Science
University of Miami
Coral Gables, Florida 33124

Michael O. Ball
College of Business and Management
University of Maryland
College Park, Maryland 20742

Lawrence D. Bodin
College of Business and Management
University of Maryland
College Park, Maryland 20742

Waverly E. Bolkan
The Boeing Company
Seattle, Washington 98124

Daniel O. Casco
General Electric Information Services
Rockville, Maryland 20850

Martin Desrochers
Centre de Recherche sur les Transports
University de Montreal
Case Postale 6128
Succursale A
Montreal, Quebec
Canada H3C 3J7

Matteo Fischetti
D.E.I.S.
University of Bologna
Viale Risorgimento 2
40136 Bologna
Italy

Saul I. Gass
College of Business and Management
University of Maryland
College Park, Maryland 20742

Bruce L. Golden
College of Business and Management
University of Maryland
College Park, Maryland 20742

Peter Greenberg
Centro de Investigacion y de Estudios Avanzados del IPN
Mexico 14
DF, Mexico

Paul Gregory
Office of Resource Recovery
Department of Sanitation
New York, New York 10007

Mordecai Haimovich
Department of Statistics
Tel-Aviv University
Ramat-Aviv 69978
Tel-Aviv, Israel

John N. Holt
Department of Mathematics
University of Queensland
St. Lucia, Queensland 4067
Australia

J. Michael Hooban
MicroAnalytics, Inc.
2054 N. 14th Street
Suite 307
Arlington, Virginia 22201

Patrick Jaillet
CERMA - Ecole Nationale des Ponts et Chaussees
La Courtine, B.P. 105
93194 Noisy-le-grand
France

Gilbert Laporte
Centre de Recherche Sur Les Transports
Universite de Montreal
C.P. 6128
Succursale "A"
Montreal, Quebec
Canada H3C 3J7

Richard C. Larson
Operations Research Center
Massachusetts Institute of Technology
Cambridge, Massachusetts 02139

Jan Karel Lenstra
Centre for Mathematics and Computer Science
Postbus 4079
1009 AB Amsterdam
The Netherlands

Laurence Levy
DISTINCT Management Consultants
10705 Charter Drive
Suite 440
Columbia, Maryland 21044

Alan Minkoff
IBM Corporation
P.O. Box 950
Poughkeepsie, New York 12602

Barindra Nag
School of Business and Economics
Towson State University
Towson, Maryland 21204

Kendall E. Nygard
Department of Computer Science and Operations Research
North Dakota State University
Fargo, North Dakota 58105

Amedeo R. Odoni
Operations Research Center
Massachusetts Institute of Technology
Room 33-404
Cambridge, Massachusetts 02139

Uwe Pape
Technische Universitaet
Franklinstrasse 28-29
1000 Berlin 10
West Germany

Warren B. Powell
Department of Civil Engineering and Operations Research
Princeton University
Princeton, New Jersey 08544

Harilaos N. Psaraftis
Massachusetts Institute of Technology
Room 5-211
Cambridge, Massachusetts 02139

Alexander H.G. Rinnooy Kan
Department of Economics
Econometrics Institute
Erasmus Universiteit Rotterdam
Postbus 1738
3000 DR Rotterdam
The Netherlands

Jean-Marc Rousseau
Centre de Recherche sur les Transport
University de Montreal
Case Postale 6128
Succursale A
Montreal, Quebec
Canada H3C 3J7

Martin W.P. Savelsbergh
Center for Mathematics and Computer Science
Kruislaan 413
1098 SJ Amsterdam
The Netherlands

Joanne R. Schaffer
Department of Management Science
University of Miami
Coral Gables, Florida 33124

Marius M. Solomon
College of Business Administration
Northeastern University
Boston, Massachusetts 02115

Francois Soumis
Ecole Polytechnique
P.O. Box 6079, Station A
Montreal, Quebec
Canada H3C 3A7

Leen Stougie
University of Amsterdam
Jodenbreestraat 23
1011 NH Amsterdam
The Netherlands

Elizabeth J. Swenson
AT&T Bell Labs
6200 East Broad Street
Columbus, Ohio 43213

Paolo Toth
D.E.I.S.
University of Bologna
Viale Risorgimento 2
40136 Bologna
Italy

Edward A. Wasil
Kogod College of Business Administration
American University
Washington, D.C. 20016

A.M. Watts
Department of Mathematics
University of Queensland
St. Lucia, Queensland 4067
Australia

OVERVIEW

Vehicle Routing: Methods and Studies
B.L. Golden and A.A. Assad (Editors)

INTRODUCTION

Arjang A. Assad and Bruce L. Golden

College of Business and Management, University of Maryland, College Park, Maryland 20742

ABSTRACT

In these opening remarks, we introduce and preview the contents of this volume. Our discussion highlights the five basic clusters into which the contributed papers are divided, and points out the balance between the methodology of routing and its applications.

DISCUSSION

In the last decade, enormous advances have been made in the field of vehicle routing. In fact, vehicle routing, as an area of both research and practice, stands out as one of the great success stories of operations research. Innovative algorithmic research has played a major role in aiding the cost-effective movement of goods and delivery of products within a wide variety of firms and organizations. Mainframe as well as microcomputer-based systems are now widely available.

With this as background, the purpose of this volume is to collect and synthesize much of the current state of vehicle routing knowledge and to convey some of the excitement that still pervades this fertile area of research and practice. Possible directions for future research are indicated by numerous authors.

In planning the contents of this book, our intention was to represent methodological advances and recent algorithmic developments, as well as the rich applications of vehicle routing in practice. We believe that it is the fruitful interplay between these two areas of investigation that is responsible for the successful implementation of numerous computer-assisted vehicle routing systems. We now briefly review the contents of this text.

The first section of the book contains two articles addressing the theory and practice of vehicle routing. The first article by Assad focuses on modeling and implementation issues and adopts a practitioner's viewpoint. Since this paper provides an overview of modeling issues and refers to other papers in this collection, it can serve as a useful introduction to the text as a whole. The article by Haimovich et al. reviews some recent work in the analysis of VRP heuristics. The authors show that under a Euclidean metric, elegant geometrical techniques can be used to provide bounds on the performance of intuitively appealing heuristics. While the paper is not meant to be a comprehensive review of the area, it does bring out the richness of the research questions that may be posed for VRP heuristics.

Section II of the book contains five papers that address variants of the VRP. Papers 4 through 8 focus on vehicle routing algorithms in the presence of complicating constraints. In particular, Desrochers et al. review recent research on vehicle routing with time window constraints (a good part of which is less than 6 years old), while Solomon et al. describe how certain well-known VRP heuristics can be modified to incorporate time window restrictions. Nygard et al. and Nag et al. both describe algorithms based on the Generalized Assignment Problem structure of the VRP. Casco et al. discuss the backhaul problem, addressing both modeling and implementation issues. This problem is a good example of how various hard-to-quantify constraints impact the design of heuristics and the quality of the solutions they generate.

The third section contains six papers that address routing in complex environments. In the first two papers, the routing component is integrated with other decisions: location decisions in the case of Laporte and a multi-period allocation problem in Ball's paper. The latter considers the allocation of service to different periods in problems where delivery time itself is a decision variable. This problem arises in inventory/routing where customer inventory levels determine the timing of deliveries and also in the design of periodic routes. Psaraftis considers dynamic routing problems where information on demands becomes available in real-time. The paper by Powell considers the vehicle allocation problem for truckload transportation. In this problem, the routing component is simplified due to transportation of full loads. On the other hand, the information on future demands complicates the allocation of vehicles. The problem is multi-period in nature and can be modeled in a variety of ways. Powell brings out the role of different model formulations for this problem.

The last two papers of Section III have to do with the design of routes where a subset of customers can be left out. In the paper by Jaillet and Odoni, the subset served is determined by a probability distribution, so that the route must be optimized over its various realizations. In the work of Fischetti and Toth, the model is deterministic; the performance measure includes rewards associated with serving customers and costs for leaving them out. Both of these models can be applied to the design of fixed routes.

Section IV focuses on applications. The papers in this section cover a diverse set of application scenarios and they all reflect practical work done by the various authors. The applications include the design of routes for the daily distribution of newspapers (Holt and Watts), the determination of workloads and routes for postal carriers (Levy and Bodin), the distribution of cars by trucks from manufacturers to dealerships (Pape), and the scheduling of tugboats and barges for the transport of refuse (Larson et al.). This cluster ends with an application drawn from the Coast Guard in which ships must visit, inspect, and service buoys on a regular basis to enhance waterway safety.

The last section of the text (Section V) focuses on commercial VRP systems. The article by Hooban presents the reflections of a developer and marketer of a commercial VRP package on the nature of the market, its potential, and its needs. Naturally, the paper draws heavily upon Hooban's personal experiences. In soliciting this article, we did not set out to endorse a particular routing package among the various systems commercially available. Rather, we thought that observations from some developers would be interesting and useful in mapping out some of the trends of this market, trends that academic researchers should not ignore. Hooban's account of the need for general purpose packages is counterbalanced by Rousseau's paper, which argues for the development of customized algorithms by referring to certain applications he has been exposed to. Together, these two articles exhibit the diversity of system requirements for practical routing and a divergence of opinion as to the future of computer-assisted vehicle routing systems.

As editors, we are delighted with the international perspective of this book. We thank the individual contributors, who are affiliated with a diversity of institutions all around the globe, for their dedication and cooperation. In addition, we thank Burton V. Dean and Gerard Wanrooy for their encouragement and support.

Vehicle Routing: Methods and Studies
B.L. Golden and A.A. Assad (Editors)

MODELING AND IMPLEMENTATION ISSUES IN VEHICLE ROUTING

Arjang A. Assad
College of Business and Management
University of Maryland,
College Park, Maryland 20742
U.S.A.

The area of vehicle routing is distinguished by a highly successful interplay between algorithmic techniques and the development of effective routing systems for industry. Consequently, a large body of practical experience with routing problems has accumulated over the last decade. This paper focuses on modeling and implementation issues that arise in practice. We discuss the challenges of model formulation and implementation and illustrate them with examples drawn from practice. Our discussion includes an overview of recent developments in geographical data bases and commercial vehicle routing systems, as well as comments on future directions in routing practice.

1. INTRODUCTION

Vehicle routing has been called "one of the great success stories of operations research in the last decade" in a recent issue of Operations Research. Most observers would agree that the success of vehicle routing must be attributed to the close interplay between theory and practice: On the one hand, operations researchers with academic affiliations have gone beyond the design and development of algorithms to play an important role in the implementation of routing systems. On the other hand, developments in computer hardware and software and their growing integration into the operational activities of commercial concerns have created a high degree of awareness of the potential benefits of vehicle routing, leading to a generally receptive climate for its use in the industry. As successful implementations continue to be reported or advertised, and as the distribution industry starts to develop an "experience curve" with routing systems, one may expect to see an even greater appreciation of the particularly effective conjunction of theory and practice exhibited by vehicle routing.

If vehicle routing does in fact constitute a major success story, a large share in the success must be attributed to effective modeling and implementation. The key to the success has been to bring the power of algorithms and the capabilities of the computer to bear on the routing problem at hand, while preserving a high degree of model realism. From the standpoint of the underlying methodologies of Mathematical Programming and Combinatorial Optimization, one could argue that existing vehicle routing algorithms are no more technically involved or sophisticated than, say, solution techniques for the classical traveling salesman problem. However, the major advance in vehicle routing has been to capture enough characteristics of the "real-world" distribution environment to enable the solution procedures to obtain a useful answer, without thereby precluding their computational tractability. In most successful applications, this desirable state of affairs has resulted from a combination of careful modeling, the design of clever heuristics, and an appropriate interactive user interface.

In this paper, we focus on the role of modeling and implementation issues in vehicle routing from an applied perspective. Our vantage point is that of the "routing expert" interested and involved in applications and the challenges they present. Indeed, we find it useful to view routing using a consultant's paradigm for structuring and defining problems. This perspective leads us to draw upon the experiences of the developers and users of routing systems over the past decade, including our own. In keeping with this focus, we avoid detailed descriptions or discussions of the algorithms. This paper does not purport to be a comprehensive review of the state of the art. Such a survey would probably require a full-length book. The interested reader may consult a number of useful surveys of the field including Golden and Assad [39], Bodin et al. [14], Bott and Ballou [16], Christofides [21], [22], and Mole [59]. Moreover, the contents of the present text cover or review a substantial part of the research performed in routing. In fact, throughout this paper, we refer to the other papers of this collection in an attempt to relate their contents to the issues raised here. Finally, we apologize for any omissions the reader may detect in our citations: To control the length of our list of references, we were forced to cite only a fraction of the publications on routing, biasing our selections in favor of the more recent work.

The plan of this paper is as follows: We begin Section 2 with an examination of how the "routing expert" might approach a given application environment with the goal of structuring it into a manageable model. Section 3 proceeds to discuss model formulation and solution techniques, using examples to point out some of the difficulties involved. The implementation of routing systems is discussed in Section 4, followed by a look at the role of geographical databases and their role in routing (Section 5). Section 6 presents a brief summation of certain areas of research and our views on some useful directions for further investigations.

2. PROBLEM DEFINITION

Consider the scenario where an expert is asked to examine the routing requirements of a commercial firm. We endow the expert with knowledge of the current research on routing techniques and algorithms and a reasonable understanding of the present computer-based technology and capabilities. In this section, we focus on the process of defining the problem, clearly the first step in any application of routing techniques.

2.1. Problem Definition

In defining the problem to be solved, the expert must consider both the nature of the routing problem and the larger environment of the firm being studied. A sample of the information relevant to the distribution activities follows:

- Size of the fleet used by the firm
- Number of drivers
- Number of routes run daily and average stops/route
- Inter-city or intra-city operations
- Total annual cost of distribution activities
- Crew and vehicle costs as percentages of total cost
- Future demand and service area predictions
- Current computer capabilities and support for routing
- Integration of routing with other activities
- Stand-alone or fully integrated routing system?

The preceding macro-level information often provides an initial assessment of the magnitude of resources the firm can allocate to the distribution function. This, in turn, allows the expert to estimate the range of system costs that the firm could justify economically. We mention it here since this step draws upon the knowledge-base of the expert in interesting ways: The macro characteristics of the distribution activity and the current practice of the firm can be taken as rough guides for estimating the potential benefits of installing a more elaborate routing or dispatching system. In a practical context, the use of this information to narrow down the options saves time and is important in focusing subsequent investigations. In many cases, this initial estimate is simply drawn from similar applications known to the expert. In dealing with a larger variety of applications settings and a broader spectrum of problem characteristics, one clearly feels the need for a more explicit model of the reasoning involved at this step. Among other things, such a model would require a classification of routing problems according to the nature of the delivery activities involved.

2.2. Classes of Routing Problems

The basic activity at customer sites of most commercial distribution settings falls into one of the following three categories:

i) pure pick up or pure delivery,

ii) the preceding with the backhaul option, and

iii) combined pick up and delivery.

In the backhaul option, discussed in some detail by Casco et al. [20] in this volume, can be viewed as converting a pure delivery (or pure pick up) problem into a more general pick up and delivery problem. However, it is usually modeled and solved differently to take advantage of the fact that the activities on a single route are predominantly of the same type, say deliveries, with a much smaller number of backhaul activities (usually just one) occurring towards the end of the route. The pickup and delivery problem is characterized by the simultaneous presence and mixing of pickup and delivery activities. These problems form a rich class due to the possible variations in the ratio and distribution of pickup and delivery tasks. For instance, it is possible to distinguish few-to-many (a few pickup locations with loads destined for a much larger number of delivery locations) or many-to-few problems from many-to-many problems.

If we go beyond the context of commercial distribution, we encounter a large number of other routing problems. In the service industry, the routing problem may not involve distribution as such, but rather arise from required visits of skilled workers at customer sites. This is the case with servicemen, postal carriers, or nurses (see [12], [53], [58]). In another category of problems, classified as arc routing problems, the basic activity occurs on the links of a road network rather than its nodes. Examples include street sweeping, waste collection, and postal carrier problems similar to the one described by Levy and Bodin [53] in this volume. In service activities such as these, the timing of services generally assumes a larger importance, causing routing models to be time-driven.

The wide variety of routing problems calls for an appropriate taxonomy or classification scheme, even when the nature of the activity is limited to (i)-(iii) above. In their lengthy review of routing and scheduling problems, Bodin et al. [14] divided problems into routing, scheduling, and combined routing and scheduling problems based on the relative importance of the spatial and temporal aspects of a problem. However, as they point out, all these problems finally result in a set of vehicle schedules so that the distinction cannot be made based on the final output of the system. One difficulty with finding a suitable classification scheme is whether the constraints of the problem or the solution technique should be used as the basis for classification. For example, one may start with a problem that is classified as a vehicle routing problem with time windows. For certain distributions of the time windows, it may be feasible to solve the problem by an algorithm that seeks good spatial configurations for the routes. In effect, the use of a pure routing algorithm reflects a decision for reclassifying the problem based on the flexibility of the timing constraints.

Instead of addressing the classification question in general terms (as in [13], say), we return to the procedure that an expert might use to elicit the problem characteristics in a practical setting. One could envision the expert asking a number of questions relating to the problem characteristics listed in Table 1. The questions specify the nature of the demand, the vehicle fleet, the crew availability, working rules, and pay structure. One can use such a "questionnaire" as a checklist to record various features of a particular problem.

To illustrate some possible complexities of routing problems, it may be useful to take a closer look at one of the categories in Table 2. We examine the category of demand characteristics with respect to the following issues:

a) selection of customers to be served,

b) uncertainty in the sizes of deliveries to customers,

c) the option of partial filling of customer demand, and

d) the time at which demand information becomes available.

a) The demand for distribution can be distinguished by its service requirements. In the classical vehicle routing problem (VRP), all customer demand points must be visited and served. In inventory/routing problems, however, the set of customer locations can be divided into three classes--those that must be served in this period, those that can be served, and those that should be delayed until a future period (see Ball [4]). Thus the multi-period nature of the problem, and the different run-out times for the customer inventories, introduce priority classes within the customers, so that a customer selection step is needed to identify the set of customers to be served in the current period. Another example of customer selection occurs when a subset of the customers can be assigned to a common carrier, a problem discussed by Ball et al. [5] and also by Fisher and Rosenwein [37] in the context of ship scheduling for bulk cargo movements.

b) Even when the subset of customers to be visited is specified, the amount of goods delivered to (or picked up from) a customer site may not be known with certainty. A good example in the beverage industry is the driver-sell problem described by Golden and Wasil [43]. In driver-sell, the exact demands of customers visited on a given route are not known in advance. Instead, the driver sells as much product as he/she can at each location. The problem in which the demand of a customer is not known with certainty is called the stochastic vehicle routing problem.

c) In some cases, the customer needs may be predicted exactly, but the amount actually delivered may fall below the full customer need. This occurs in the distribution of industrial gases when the tank levels of customers is known via meter readings, but the distribution system may decide to satisfy the customer need only partially due to routing economies (see Bell et al. [11]). Thus, the size of a delivery becomes a decision variable of the model. Similarly, certain locations may be chosen as "dump sites" where the vehicle empties its remaining load prior to returning to the depot.

TABLE 1. General Characteristics of Routing Problems

NATURE OF DEMAND	Pure pickups or pure deliveries
	Pickups (Deliveries) with backhaul option
	Single or multiple commodities
	Must serve all demand?
	Common carrier option
	Priorities for customers
INFORMATION ON DEMAND	All demand known in advance?
	Many repeat demands?
	Fixed frequencies for visits?
	Uncertain demands?
	Real-time inflow of demands
VEHICLE FLEET	Homogeneous fleet or multiple vehicle types
	Weight and capacity restrictions
	Compartments
	Loading restrictions/equipment
	Vehicle type/site dependencies
	Vehicle type/commodity compatibility
	Fixed or variable fleet size
	Fleet based at single depot or multiple terminals
CREW REQUIREMENTS	Pay structure:
	length of workday
	minimum and maximum on duty times
	overtime option
	Fixed or variable number of drivers
	Driver start times and locations
	Lunch or other breaks
	Multiple-day trips allowed
SCHEDULING REQUIREMENTS	Assignment of customers to day of the week
	Time windows for pickup/delivery (soft, hard)
	Open and close times
	Load/unload (dwell) times
DATA REQUIREMENTS	Geographic database, road networks
	Customer addresses and locations
	Travel times
	Vehicle location information
	Customer credit and billing information

d) One classification of routing problems is based on the time at which information on the demand becomes available. In the classical VRP, all demands are assumed to be known in advance. In dynamic routing, however, demands occur in real time as the routing activity proceeds. In practice, the expert would need to know how substantial the dynamic component of the problem is. In one common distribution scenario, most of the demand for a given day is known in advance, while other demands are specified dynamically as the day unfolds. In most messenger services and certain Dial-a-Ride operations, most of the demand is dynamic (see Psaraftis [61]).

2.3. Characteristics of Operational Routes

Once the problem characteristics are defined, the expert would question the firm on the operational characteristics of its routing activity. One goal of this step is to become familiar with the actual process of devising routes as it is currently done in the firm. A more important benefit is to specify the constraints that govern the route configurations. In an operational setting, the routes driven on any given day are subject to a number of constraints reflecting restrictions imposed by the vehicles, the driver schedules, customers' preferences, or the geography of the delivery region. A partial list of such constraints appears in Table 2. We now briefly discuss certain constraints listed in the table.

The start and end locations of a route are normally taken to be the depot. However, the choice becomes more complicated in multiple depot problems or when certain vehicles originate or terminate their trips at the driver's home. In location routing problems, the choice of the base for a set of routes may be a decision variable that is optimized at the same time that the routes are designed. Laporte [51] discusses a number of applications settings for this problem. Levy and Bodin [53] describe a mail delivery application where the routes are based at parking locations where the carrier's vehicle is parked at different points of the day. Their model selects the parking location in such a way as to make the design of a desirable set of routes more likley.

To schedule a driver's route, one must determine its start time and duration while observing customer time windows and breaks for the driver. In some problems, the determination of start times requires special attention. For example, start times may have to be staggered due to limits on the number of vehicles that can be loaded simultaneously. The route start times in the newspaper delivery problem described by Holt and Watts [44] are linked to the production rate of the presses, which places constraints on the loading of trucks.

The scheduling task for routes that extend beyond a single work day involve an additional set of constraints that reflect driver safety regulations. These regulations do not allow more than 10 hours of over-the-road driving per day

TABLE 2. Constraints on Route Configuration

Duration of the route
Start and end locations for the route/driver
Start times of routes
Over-night trips and associated work rules
Intermediate breaks

Weight and volume restrictions on vehicle load
Loading constraints and restrictions

Driver territories or delivery regions
Natural or legal region boundaries
Grouping of customers
Rules for split deliveries (if any)

Importance of balanced routes
Multiple routes per vehicle
Use of fixed routes and fixed stops on route
Modification of routes based on new incoming orders

Special rules for certain road segments
One-way streets
Avoiding U-turns or other safety rules

and place upper limits of 15 hours per day and 60 hours per week on working time. Moreover, the schedule must include a minimum of 8 hours of sleeping time per day. These constraints are operative in the applications described in [33], [67], and Chapter 4 of [14].

A second group of constraints on the routes have to do with the vehicle load. Classical VRP formulations allow restrictions on the total load carried by the vehicle. In practice, this simple capacity may have to be replaced by both weight and volume restrictions which are suitably scaled for different commodities. In the case of petroleum products, the weight and volume relations depend on the temperature as well [17]. For some weight restrictions, one may have to track the weight of the vehicle load over the entire itinerary of the route as the weight restrictions on the vehicle may vary according to the territory traversed by the route. For example, certain road segments and bridges can impose particular weight limits on vehicles; just as the legal vehicle weight limit may also vary by state [11]. Of course, when the customer demands are sizeable in comparison to the maximum vehicle load, and the travel times among customer sites are not large, vehicle capacity restrictions lead to multiple trips per day for a single vehicle or possible split deliveries.

In addition to the total vehicle load, loading considerations may also constrain route configurations. As seen in the case of delivery trucks for household goods and furniture, the sequence of stops along the route requires a compatible loading of the vehicle (the goods for the final stops are loaded first). Indeed, loading restrictions could require backhaul stops to be placed at the very end of the vehicle route, even when enough capacity for the backhaul load is available on the vehicle earlier on the route [20], [67].

An interesting example of the interaction between the loading issue and the route itinerary occurs in the distribution of trucks shipped from Navistar to its customers. As described by Clavey [24], individual truck units fit into "deck-sets" that are transported together. The loading of trucks depends on the technology of deck formation and hence may not correspond to the sequence of deliveries made. The costs incurred when a driver "undecks" and "re-decks" at a delivery site form part of the distribution costs and are affected by the sequencing decision for the route.

The configuration of routes can be strongly linked to certain geographical or driver territories defined in advance. The simplest example is when the driver is confined to a fixed geographical territory. For instance, one service operation serving the Washington DC, Maryland, and Virginia area was unwilling to let its servicemen cross state (and district) boundaries for cost accounting reasons. Littich [54] mentions another instance of geographical territories in the context of beer distribution where region boundaries are defined for the delivery area. Routes crossing these boundaries incur surcharges. In certain applications, the customer's expectation to see the same driver from one visit to the next motivates stable assignments of customers to drivers to be defined, so that driver territories form in effect. In the driver-sell problem, territories are first defined for the drivers in a tactical planning step based on the size, the geography, and sales potential of the defined area (see Golden and Wasil [43]). Once the territories are set, the servicing of customers within a single territory is an allocation routing problem described by Ball [4]. Apart from territories, other groupings of customers may be defined: In a fuel delivery problem described by Golden et al. [40], customers in certain areas such as trailer parks required simultaneous delivery. It is also possible for service times of different customers to be linked. For example, in the postal carrier arc routing problem described by Levy and Bodin [53], the times for covering the two sides of a street cannot be more than an hour apart.

Some constraints in Table 2 have to do with the details of the route configurations. Naturally, a long list of such constraints can be prepared for any rich application, reflecting in some way the multitude of desirable characteristics that a dispatcher (or the driver) would like to see in a route. Together, these constraints define the nebulous notion of "solution quality," to which we return in Section 4. For now, we only mention two such constraints as examples. In arc routing problems, where a vehicle has to traverse and serve streets of a road network, one must pay close attention to the actual path the vehicle follows on the network to discourage U-turns and other unsafe driving practices. Thus if the output of the routing algorithm contains such U-turns, these must be replaced by detour that may increase the deadhead costs (see Bodin et al. [15] for detailed modifications of the output). The next example is a constraint that occurs in beer distribution according to Littich [54]. In that application, union contracts attached a cost to "covering a stop," that is, passing a stop and returning to serve it at a future time. A similar situation can easily be envisioned within a service environment: Most people would not enjoy seeing the mailman or the garbage truck drive by their domicile, delaying service until two hours later when the vehicle returns to their location. We mention these two issues as examples of desirable route characteristics that influence the "acceptability" of generated routes and that also have certain cost implications. It is also evident that these considerations require information at the level of the detailed geographical network database.

3. SOLVING THE ROUTING PROBLEM

3.1. Model Formulation and Objectives

The last section focused on the process of eliciting the characteristics of the routing environment and the constraints on route design. While this process paves the way for further analysis, the expert's modeling task really lies beyond this initial inquiry: An appropriate model of the routing problem at hand must be formulated, followed by the choice of the effective algorithm, the performance of which must then be tested. In this section, we focus on the two issues of model formulation and choice of a solution technique.

It is a commonplace observation that problem formulation is colored by the arsenal of existing models and techniques. Currently, the expert can draw upon the substantial literature on vehicle routing and related distribution models, as well as the capabilities of commercially available routing software, if appropriate. Once the expert decides that a commercial package may not be directly applicable, an appropriate model has to be constructed for the problem. Two key issues are the level of detail incorporated into the model and the nature of the objective function driving the model. The first issue usually involves a tradeoff between accuracy and tractability. Naturally, the model must reflect the basic underlying economies of the distribution process, the nature of which the expert (modeler) must determine early on.

To take a simple example, the expert must determine if the real objective is to minimize distance-related costs over the fleet of vehicles or just the number of vehicles used. In applications where drivers are guaranteed a certain minimum number of hours of work each day, the real economies may be associated with eliminating drivers. Savings in the number of drivers (or vehicles) may be accomplished by using a certain amount of overtime, which introduces the need for tracking the length-of-the-day constraints within the model. The bulk pickup and delivery problem Fisher et al. [34] encountered in Shanghai provides another example of the reasoning underlying the choise of an objective function. The objective used is to minimize empty miles (deadheading) traveled by all vehicles in the fleet. This is equivalent to minimizing the total travel distance since all shipments involve full loads between pickup and delivery locations and all demand must be met. That this is an appropriate objective hinges on the following problem characteristics:

a) it is more important to save vehicle-related variable costs (fuel and maintenance) than crew costs,

b) all drivers work seven hours and overtime is not considered, and

c) both vehicle and driver costs are highly correlated with distance.

The authors contrast assumption (a) to distribution in the United States, where "driver costs account for roughly two-thirds of most US vehicle fleet budgets." We add that as crew costs become more important, the crew schedules may have to be considered and optimized closely. In commercial vehicle routing, the scheduling of crews is generally derived from vehicle schedules as determined by the routing model. In mass transit operations, however, vehicle and crew schedules interact heavily and must ideally be developed together. As minimizing the deadheading time for vehicles is often a reasonable surrogate objective for reducing crew costs, the problem is often solved sequentially; vehicle schedules are formed first and crew schedules second. Note further that in mass transit problems, the driver does not have to stay with the vehicle for the duration of the vehicle schedule as this may be longer than a normal work day and also because the mass transit system provides transportation to the driver between relief points.

The question of what problem characteristic "drives the problem" is naturally at the very heart of the modeling process and must be faced by the modeler early on. In certain distribution problems, the routing costs per se may account for a small fraction of the total distribution costs. In the distribution of food products to supermarkets, the total drive time may be small compared to dwell times (for unloading) at the customer sites. This would suggest that the packing of trucks (loading) may be the driving factor in the model. In school bus scheduling (see Chapter 4 of [14] or [27]), the choice of school start times is a key driving factor; the variable of interest is the maximum number of vehicles needed in any given time period--a minimax objective which seeks to reduce peak load requirements. Since the various time window requirements are crucial to the design of bus schedules, the time periods used in the model must also be short to allow a fuller optimization of the schedules. For example, the time windows are in the range of 10-30 minutes in the TRANSCOL school bus routing system described by Desrosiers et al. [27]. Correspondingly, their model uses a time period 10 minutes in duration.

Our next example is drawn from the more complex routing environment of inventory/routing. In this problem, the timing of the deliveries is governed by the inventory levels of the distributed commodities at the client site. Deliveries for both industrial gases and fuel follow a "push system," that is, the timing of most deliveries is controlled centrally by the dispatcher, rather than the individual customers. Given this flexibility in the timing of deliveries, the problem involves a multi-period model for planning future deliveries, with the option of assigning a delivery to several different time periods. Research on this problem indicated early on that the temporal degree of freedom is more important than purely spatial routing considerations. Fisher and Jaikumar [36] have noted that dispatchers are good in optimizing the spatial characteristics of a route, but not the temporal ones. Bell et al. [11] provide an interesting example of how a solution with a long haul joining relatively distant customers may be preferable due to the usage rates of the customers involved and argue that such "non-local" clusterings of customers may easily escape the dispatcher's attention. Assad et al. [3] and Golden et al. [40] also emphasize the relative importance of the temporal aspects. Their results indicate that if the assignment of customers to days is fixed to be the same as the firm's, the use of standard vehicle routing to design the routes offers little savings. Thus, most of the economies are associated with sliding the customer delivery time, even when the routing component is non-trivial (60-120 customers daily). For this reason, the system implemented based on this work generated three days of routes each day, allowing a post-processor to exchange customers among the three days.

Another important modeling consideration in the inventory/routing problem is the choice of an objective function. A favorite aggregate performance measure in the industry is gallons/mile, which reflects a long-run average efficiency measure, but does not fully capture savings in vehicle and crew fixed costs. This measure is unwieldy and difficult to use directly within an optimization model, especially in view of the truncation effects caused by converting the long run problem into an operational model with a short planning horizon. For this reason, it is important to introduce surrogate terms in the objective function designed to induce gallons/mile reductions. As Fisher and Jaikumar [36] have noted for the Air Products application, the more a delivery is delayed, the larger the size of the delivery. Thus, given a 120-hour week, delaying delivery by 1 hour yields a savings of one percent roughly. This observation guides the choice of the level of detail in the model of Bell et al. [11], where one-hour increments for the route start times are allowed. Moreover, there are terms in the objective function that encourage larger deliveries to the customers. In the case of fuel delivery, Dror and Ball [30] discuss the inclusion of terms into the objective that reflect the long run advantages of longer cycle times for customer deliveries (see also Ball [4]).

The preceding example shows that defining the appropriate model and objective function for a realistic but complicated model may constitute a difficult exercise in itself. In many routing problems, the objective function has a multi-attribute nature: In a vehicle routing problem formulated as a distance minimization problem, it may also be very important to ensure that balanced routes are generated (vehicle schedules of equal duration are highly desirable). Naturally, the objective function can be expanded to penalize large imbalances, reflecting a weighting of multiple objectives (distance, balance, etc.). Indeed, Levy and Bodin [53] describe an arc routing problem where the main objective is route balance. Similar considerations apply to Dial-a-Ride problems where the "excess ride time"--time a passenger spends on the vehicle in excess of the travel time--is the main driving term in the objective function (see the article by Psaraftis [61]). In general, as the problem constraints become more complicated, and to the extent that the solution technique is composed of different interacting sub-models, the objective function starts to assume the role of a suitably chosen control on the quality of the solution, rather than an exact representation of the total operating costs.

3.2. Solution Strategies

In keeping with the applied focus of this paper, we begin our discussion of solution techniques with the option of using a commercial routing system. The expert investigating a firm's distribution problem is often asked the standard question whether the problem can be solved by a commercial system already offered in the market. If so, the substantial costs and time expenditures associated with developing a customized system can be avoided. While many considerations impact the firm's choice between commercial general purpose packages and customized systems, it should be acknowledged that the availability of the former places a burden of economically justifying the use of the latter.

Commercial routing systems currently offer a wide range of features geared, for the most part, to the commercial distribution of goods. Indeed, most packages solve a version of the standard vehicle routing problem involving deliveries only (or pickup only) with a variety of side constraints and possibly the backhaul option. An exception is a well-known package that handles the pickup and delivery problem with time windows. Table 3 provides some features of commercially available packages. We should remark that these features are pooled from different packages and do not simultaneously exist in a single package as a full discussion of packages on an individual basis is beyond the scope of this paper. A review of selected commercial packages with more package-specific details can be found in Golden et al. [42]. Hooban [45] describes the status of commercial routing systems from the perspective of a developer and marketer of such software.

TABLE 3. Typical Features of Commercial Vehicle Routing Software

Features	Objectives
Different vehicle types	Minimize distance
Pickups & deliveries, backhauls	Minimize travel time
Time windows	Minimize no. of vehicles
Load/unload times	Minimize total cost
Variable speeds	
Common carrier option	**Output**
	Route itinerary & schedule
Route Information	Vehicle utilization report
Capacity & duration limits	Driver utilization report
Use of overtime	Graphics for routes
Start and end times	Road network overlays
Layovers, 2-driver teams	Address matching
Fixed stops	
Multiple routes/vehicle	

Commercial packages are naturally designed for the "standard" vehicle routing problem faced by most firms in the distribution of goods, although the needs of certain industries such as the beverage industry have led to the development of special features for specific problems such as the driver-sell (see Golden and Wasil [43]). Most developers of packages, however, are not interested in customized designs and can not afford, or do not have the resources and expertise, to add special purpose functionalities unless the client firm agrees to pay substantial additional costs. In our experience, when a problem has nonstandard features, some modeling expertise is needed to decide whether a fully customized system is warranted, or whether the problem can be suitably modified or decomposed to take advantage of the available standard algorithms. This can be illustrated with the following examples of "nonstandard" problems we have encountered [2].

Consider the operations of a messenger service operating in a major metropolitan area, offering same-day pick up and delivery of packages between designated origin-destination pairs. The delivery area is divided into a number of regions or territories, each assigned to a driver. The work day is divided into four "time buckets": 8-10 AM, 10 AM - 1 PM, 1-3 PM, and 3-5 PM. During each of the four time buckets, a vehicle must cover its territory with a tour based at the depot, carrying out pickup and delivery activities as needed. Some parcels are picked up and delivered within the same time bucket, while others are picked up in one time interval, brought into the depot, and delivered in the next time bucket. Note that this operating scheme allows the vehicle routes to remain relatively localized, while the depot acts as a transshipment point for parcels traveling between distinct regions.

At first sight, this appears to be a complicated routing problem beyond the capabilities of the average commercial routing package. However, the following observations and assumptions can help us to simplify the problem considerably.

1) Vehicle capacity is not binding (parcels are small relative to the vehicle capacity, e.g., letters, legal documents, etc.).
2) Route durations are limited to the length of the time bucket in which they occur.
3) Real-time changes to the routes are not very frequent (demand is generally known at least one time bucket in advance).
4) Territories must remain stable, allowing only limited overlaps between the operating regions of different drivers.

Given these characteristics, one can decompose the problem by time bucket into separate routing problems. The routing problem in each interval has no capacity constraints, but a strict limit on the duration of the route. However, it deviates from the classical VRP in that both pickups and deliveries occur on the same route. Since vehicle capacity is not an issue, the main complication introduced is precedence relations among the tasks--pickups must precede their associated deliveries. The dispatcher must decide which parcels are picked up and delivered on the same route based on customer needs and route duration restrictions. If the number of such packages is not large, the standard VRP may be a very good starting solution for the construction of the routes associated with each time bucket. We should also remark that the definition of regions is also key in this problem since it effectively defines the options open to a customer: By ensuring sufficiently early pickups, the customer can estimate the delivery time of the parcel.

The next example involves the operations of an armored car service. In this problem, armored cars deliver money from businesses to various banks. There are two types of services--same day and overnight. In the former, both pickup and delivery occur on the same route; whereas in the latter overnight service, the money is stored in the depot vault overnight for delivery early the next day. The problem is thus somewhat similar to the preceding one in using the depot as a transshipment point, however the length of the "time bucket" has increased to a full day. It is useful to list some of the characteristics of this problem:

1) The vehicle capacity is not binding.
2) The length of the routes (work day) must be controlled. Overtime (work beyond 8 hours) results in a 50% surcharge.
3) Most customers are serviced each day or on fixed days of the week; the repeat business is therefore high.
4) Time windows apply. Deliveries or pick ups should not occur during busy times. Open and close times of businesses must be respected.
5) Routes for the entire week must be formed one week in advance.

It is clear that this problem is more difficult to transform into a standard (pure delivery) VRP. Viewed as a pickup and delivery problem, an important characteristic for routing is the ratio of pickup points to delivery locations. If the number of banks is much smaller than the number of businesses, the problem is a many-to-few pickup and delivery problem, and it may be possible to use the banks as "depot" locations and form routes that terminate at the depot. This would suggest an approximate decomposition of the problem by bank and time period of the day. The standard VRP model (with time windows) is more appropriate for the initial and final portions of the daytime due to the overnight service option: Two VRP problems can be solved--one to route the pickups destined for the depot at the end of the day, and the other for routes making the bank deliveries at the beginning of the next day.

The last two examples show some of the complications present in realistic routing problems and indicate how the expert might try to decompose the problem into simpler components to take advantage of existing algorithms or packages.

Note that both problems can be viewed as pickup and delivery problems, but that the different mix of pickups versus deliveries and the nature of operative time windows causes the problems to have very different flavors. It is our impression that once the domain of applications extends beyond commercial distribution of goods out of a single depot, most applications have special complicating constraints that can not be directly handled by general purpose codes. Rousseau [63] also doubts the widespread applicability of general purpose routing software and relates some experiences that point to the need for customized systems.

To conclude, the expert must ultimately decide whether a general purpose code can be suitably used or controlled to give acceptable solutions, possibly with the aid of other heuristics or appropriate post-processing procedures. If this approach fails to give an acceptable solution strategy, a special purpose algorithm must be designed for the problem at hand.

3.3. Algorithmic Techniques

As mentioned in the introduction, a review or classification of existing routing algorithms lies outside the scope of this paper. Consequently, our comments on algorithmic techniques in this section are from a practitioner's viewpoint.

From an applied perspective, the salient feature of algorithmic development in vehicle routing over the past 5-8 years has been the ability to enrich previous models to handle more realistic and versatile operating environments. In fact, one could argue that the main impetus for algorithmic developments originates from practical applications. To take only one example, the incorporation of time window constraints alone into routing problems has already resulted in a significant body of research (see Savelsberg [64], Kolen et al. [50], the collection of articles in Golden and Assad [38], as well as the papers by Solomon [65] and Solomon et al. [66] and Desrochers et al. [26] in this volume). At the risk of some oversimplification, we can identify the following three trends towards greater realism in routing models.

a) One basic thrust in the design of algorithms is to enrich existing routing heuristics to handle some of the realistic constraints on route configurations discussed in Section 2. A traditional approach to this is to select a well-known vehicle routing heuristic (which may be called the base heuristic), such as the Clarke and Wright savings algorithm, and modify it to account for the added complication. A simple example is the early work on the backhaul option described by Casco et al. [20], where the savings and insertion heuristics were suitably modified to assign backhaul demand points to routes. The early work of Solomon [65] on time windows reflected the same basic approach. This approach has its advantages since it allows the developer to maintain the structure of the base heuristic and simply build upon it, thereby avoiding a large marginal development cost. The modification often involves simple checks on the conditions for insertions or route mergers within the base heuristic. Other examples of such heuristic modifications can be found in the routing literature; for example Beasley [9] handles time-dependent travel times in this manner. While the preceding approach marks the early stages of attacking a VRP problem with side constraints, subsequent work usually goes beyond standard VRP heuristics to devise more customized methods based on the specific problem structure.

b) A second basic direction of research involves the use of mathematical programming techniques to construct frameworks for the solution of complicated vehicle routing problems. The mathematical programming structures that inform algorithm development in routing are carefully reviewed in Magnanti [57]. From the practitioner's point of view, however, two classes of models have proved especially successful in applications--those based on the Generalized Assignment Method and those using Set Partitioning formulations.

The Generalized Assignment Heuristic of Fisher and Jaikumar [35] has proved to be a powerful technique when the packing aspects of the problem are important and can incorporate realistic vehicle type/route/customer compatibility constraints. But it also provides a general framework for assigning delivery patterns or frequencies to customers in more complicated problems such as allocation routing (see the article by Ball [4]). The recent work of Lee and Shapiro [52] suggests that further theoretical insights into the nature of the Benders Decomposition underlying this approach may be forthcoming.

Set partitioning approaches are most useful when the number of candidate routes can be limited due to the restrictive constraints imposed on routes (tight time windows, limited route duration, or capacity limitations). In practice, this implies routes with only a few stops. The work of Desrosiers and coworkers [27], [28] demonstrates how column generation can be an effective solution technique for such models. Finally, in addition to these techniques, one should mention the state space relaxation method of Christofides et al. [23] that is described as "one of the few effective methods for the exact solution of the VRP" [22]. It appears that this approach underlies a powerful package for solving VRP's that is commercially available in Europe.

c) A third area of research has focused on the design of what may be termed "sequential heuristics" for routing problems. In these approaches, the problem is decomposed into a sequence of subproblems that are sequentially solved with different algorithms. Examples include well-known "route first-cluster second" and "cluster first-route second" approaches to the VRP and its variants. For the simpler problems, such as the classical VRP, the more powerful approaches mentioned in (b) above have to some extent supplanted these approaches. However, these approaches are convenient ways of attacking complex routing problems such as location/routing and allocation/routing problems where the simultaneous optimization of the all decisions is computationally impractical. The practitioner's interest in such models is to develop some insights into appropriate sequential decompositions for various problems when different sets of complicating constraints are operative. One may ask, for example, what an appropriate clustering scheme is in the case of simultaneous pickups and deliveries with relatively tight time windows or in the presence of multiple vehicle types. Computational studies could also indicate the accuracy of such methods and the extent of suboptimality resulting from the decomposition or the possibility of including feedback loops within the decomposition. Examples of sequential approaches include Beasley [7], Raft [62], and Perl and Daskin [60]. An account highlighting practical issues is available in the form of a case study in algorithm design in Golden et al. [41].

The research efforts just mentioned show that the VRP in the presence of a single set of complicating constraints has been studied in some detail and the literature now offers a choice of techniques. Further work and insights are needed for designing practical routing systems when a number of such complicating constraints are simultaneously present. Short of devising a completely problem-specific algorithm, routing systems must still rely on a set of interacting algorithms. Modular design of these algorithmic components has been popular in practice and has obvious advantages in the context of interactive systems; since individual modules may be called separately by the dispatcher. However, a better understanding of the information flow between the various modules and the logic of their interactions should be obtained in the future. The issue of modular design also brings up the idea of a common "shell" to allow various modules to communicate through a common database, and also to standardize the basic routing "operators." A brief outline of such a system is provided by Rousseau [63].

4. IMPLEMENTATION OF VEHICLE ROUTING SYSTEMS

The implementation and installation of vehicle routing systems in industry have resulted in a wide exposure to practical implementation issues. In many instances, the commercial software developer or consultant has had the largest share of interaction with the ultimate client and accumulated much experience as a result. Usually, this type of "hands-on" experience is not publicly disseminated. To be sure, both success stories and "war stories" are told, but the full process of implementing a vehicle routing system often goes unrecorded due to lack of time or incentives, as well as restrictions on proprietary information. Fortunately, the involvement of academic researchers in vehicle routing, and the interest of professional societies in operations research and management science have resulted in a number of papers documenting the implementation process, at least in part.

We believe that lessons drawn from practice deserve fuller documentation and greater attention. In this section, we select certain topics from these accounts (and our own experience) in order to impart some sense of the challenges presented by system implementation.

4.1. Some Pervasive Implementation Issues

a) The dispatcher/computer division of labor.

Most routing systems ultimately communicate with the outside world through the dispatcher. A fundamental question in system design is the nature of the interaction between the dispatcher and the computer: On the one hand, the system may be nothing more than a convenient "scratch-pad" for the dispatcher who is responsible for all routing decisions. Consider, for example, the dispatching of servicemen to customer locations in a utilities company. The "dispatch room" has several maps of the service areas, and two or three dispatchers who assign the standing and incoming customer service requests to the available drivers on the road. It is possible to "computerize" this operation without in any way changing the nature of decision-making for the dispatchers. In this case, the "dispatching system" merely captures the available information, maps, driver availabilities, and so on electronically, eliminating the flow of paper through the dispatch room.

On the other hand, one can envision routing systems that heavily rely on the power of routing algorithms to generate complete driver schedules obeying a complicated set of constraints. The dispatcher still checks and approves the generated routes, but expects the output to be nearly acceptable. In this case, the normal planning is relegated to the system, and the dispatcher assumes the role of a supervisor or troubleshooter. Naturally, the nature of the division is intimately linked to the overall system design. Brown et al. [17] refer to this decision in the context of their computer assisted dispatching system (CAD). They mention that since "many crucial aspects of the dispatching process are not quantifiable," the CAD system allows the dispatcher "to concentrate on the qualitative aspects of the specific dispatching situation and to quickly see the economic impact of manually overriding CAD's recommended solutions." Thus, an algorithm is relied upon to specify the loading of products in a given order onto truck compartments. This makes sense as the problem is combinatorial in nature and possesses a quantifiable objective. We should note that the nature of the final division of work is reflected in the response times set for the algorithmic component of the overall system: The dispatcher is expected to spend a total of 5-10 minutes on a dispatch, which may typically involve 30 trucks delivering 116 orders out of 7 terminals. The automated processing for CAD takes only 5-10 seconds per call. This 60 to

1 ratio in elapsed time indicates the importance of providing quick interaction times to the dispatcher.

To summarize, the division of the routing responsibility between the dispatcher and the system obviously varies for different applications and depends on the following factors:

- the "quality" of the solutions desired,
- the ability of the algorithms to generate high-quality solutions and the associated computing time,
- how capable and how fast the dispatchers are,
- how much time do dispatchers have to interactively construct and modify solutions, and
- the stability of the operating environment.

Different choices for this division are reflected in commercial routing packages as well. Some systems, similar to the interactive system developed by Belardo et al. [10], do not rely on high-powered algorithms for route construction. Instead, they allow the user to construct routes by using simple route editing functions (insert or delete stop, merge, split, etc.). What an appropriate level of interaction is for a given market segment is an issue that software developers still disagree on.

b) The quality of solutions.

An important issue that impacts all phases of system design, model formulation and solution, and system implementation is that of "quality of solutions." One way to define this concept is to examine what the dispatcher expects to see before he/she checks off on a solution. To start with a simple example, a dispatcher looking at a graphical display of a route would tend to question a route that crosses itself. Such a route could be improved with simple exchange routines such as the 3-opt. In this case, the objective of minimizing distance can preclude the generation of such inefficient routes. Consider the same criterion in the presence of time windows, however. In this case, the optimal (minimum distance) route may in fact cross itself. As more complicated constraints, such as those listed in Table 2 are added, finding an intuitive performance measure that captures the notion of a "good" route becomes more difficult. Correspondingly, an algorithm that takes all desirable route characteristics into account may become very difficult to design. An example where complex rules governing route design may be necessary is the backhaul option as discussed by Casco et al. [20]. If the notion of solution quality is linked to how the dispatcher evaluates the output, then empirical evidence on the nature of this process must be gathered in practice. Some software developers view the dispatcher as a satisficer: The route will be accepted if it stays within the gross boundaries of feasibility. Such developers argue for producing output quickly even if it contains some inefficiencies (including, say, the use of overtime or an extra vehicle). Others relate their installation experiences with dispatchers that examined the routes in great detail and were eager to point out its inefficiencies or infeasibilities. Naturally, repeated identification of such inadequacies could cause the dispatcher to lose faith in the system. Such varied experiences point to a need for studying the evaluation criterion, preferably in advance of final system specification: What characteristics cause the dispatcher to reject a route (when the dispatcher claims that the route is not "driveable")? Is the dispatcher rejecting the route for the right reasons? What is the effect of learning on this process? Finally, we should point out that the notion of solution quality sometimes requires detail information from the geographic data base, as discussed in Section 5.

c) Incorporating constraints into the model.

The issues of solution quality and the split between human/machine decision-making raised above are crucial considerations in model formulation. What set of constraints should the model include explicitly and in what form? How should the algorithms be designed to drive the solutions towards feasibility with respect to these constraints? The answers clearly depend on the nature of the constraints and the intractibility caused by their inclusion into the solution technique. For example, in complicated problems such as crew scheduling where many constraints exist, the approach has been to capture only the basic economies in an optimization problem and to rely on a large amount of human input.

To illustrate these issues in the context of routing, consider the case of time window constraints. Bell et al. [11] recount a situation where the algorithm suppressed a natural and efficient solution (in favor of an inferior solution) because it violated a delivery time constraint. The dispatcher then "pointed out in rather direct language that delaying delivery by one hour to customers receiving deliveries once a week is no big deal." The authors therefore argue in favor of "soft constraints" and mention that when the optimization algorithm exhibited high sensitivity to certain constraint limits, these values were set loosely. In our experience, we have found that dispatchers often violate stated time windows in their own solutions. The analyst then discovers that the time windows were set conservatively as inputs to the model, compared to the leeway the dispatcher allows in creating actual solutions.

Similar practical considerations in other instances have led to a soft time window formulation. One standard approach is to attach penalties to time window violations so that they can be included in the objective. A case study reported in Golden et al. [41] follows this approach and describes how the solution quality can be evaluated with the dual objectives of total travel time and total penalty incurred. Briefly stated, if one uses these measures as the two axes and plots the solutions generated for various values of the penalty term, the familiar "exchange curve" emerges, providing a visual representation of the trade-off between meeting time windows and reducing travel time. While the use of penalty values for soft constraints is an option that can quickly win the user's approval in principle, it creates the problem of fixing penalty values. In our practice, the dispatcher would need initial help in this process and also has to develop "hands-on" experience with the choice of the values used. Brown et al. [17] also mention the need to familiarize the dispatcher with penalty terms for a loading model within their system.

d) Installation and learning effects.

It is a commonplace observation that most of the implementation challenges are first seen when the system is actually installed. Thus, the "burn-in" period is crucial to the success of the system and may necessitate some changes in the system design, output, and interactive capabilities. When designing the algorithms, the analyst hopes to foresee most needed changes and allow them to be handled with parameter settings. However, unforeseen problems do occur. For instance, even though the required response time is specified in advance, the running time of the algorithm may be unacceptable for certain scenarios run by the dispatcher. Alternatively, the dispatcher may not be able to modify the generated solutions to obtain an acceptable solution quality within a reasonable time allotted for interaction.

An important component of the "burn-in" process has to do with learning effects. If the dispatcher is rejecting certain routes because they are unfamiliar, the system should revert to routes that mimic the dispatcher's choices to a greater extent, possibly by requiring the user to fix more decisions. In the work of Bell et al. [11], "neighborhoods" were defined for each customer to control the clustering of customers onto the same route. By shrinking the neighborhoods, the system precluded the clustering of distant customers to conform with the initial expectations of the dispatcher. As the users gained in experience, however, these neighborhoods were expanded to allow the system to optimize more fully.

Another aspect of learning is the choice of candidate solutions, that is, solution patterns that are known to be effective from experience. This issue is especially important for set partitioning or set covering based approaches such as [11] and [67]. In such models, the system optimizes over a set of candidate routes available to it so that the inclusion of effective patterns enhances the system solution. This motivates saving effective route configurations for future use. The CAD system of [17] stores the results of each dispatch run off-line for later use. Note that the idea of storing good routes appears in elementary form in the notion of fixed routes (see [8]) since these are routes that can be used repeatedly with minor modifications to accommodate the daily changes in demand.

e) System integration.

The integration of routing systems into the overall information processing system of the firm is an important practical consideration. In distribution of goods, the order entry and billing systems are usually in place when a routing system is implemented. Moreover, information from components such as credit checking or a forecasting based on past customer history (in fuel delivery) may affect the routing decision. The firm must then decide to what extent the routing system is a "stand-alone" system. This decision also depends on hardware constraints, naturally. The systems described in [11] and [17] are good examples of fully integrated systems. System integration issues are bound to become more complex as detailed geographical databases and vehicle location/communications technology starts to play a larger role in routing.

4.2. Benefits and Savings of Routing Systems

The measurement of the savings produced by implementing a routing system or other associated benefits is a much-needed, but often neglected, exercise. In principle, the reporting of actual savings involves a careful measurement of the company practice before and after the installation. One approach is to run the system on a set of past delivery data and compare the results with the routes actually run in the past. This approach is taken in certain tests studies, such as the one in Evans and Norback [31], usually prior to full system installation. The skeptic could argue that in such a test the system has the advantage of hindsight and a completely static routing environment. A fuller test would involve running a newly installed system in parallel with the old and recording the results of both. Of course, only one set of routes is actually driven, making a full over-the-road comparison problematic. Moreover, such parallel runs at the early stages of system installation probably underestimate the full system potential. Another issue in reporting savings is how a particular dollar figure is to be computed. Does one include only operating costs, or does one also look at capital savings associated with the lower need for equipment due to higher utilization? In the latter case, how should one extrapolate the capital savings into the future? Moreover, how does one factor in major strategic decisions that arise as a result of using the system in planning mode (such as depot relocation)?

Despite the difficulties mentioned above, a number of implementation efforts have reported savings figures, some of which are summarized for seven different studies in Table 4. The information on savings is at times vague and is available at different levels of detail, due in part to proprietary constraints imposed by the firms releasing the data. The table also briefly mentions some of the other benefits reported for the study, where applicable. In contrast to these studies, not much information is available in the open literature from the developers of commercial routing software. Naturally, developers quote some figures on potential savings based on the distribution environment of the client firm. For example, a rough savings may be estimated from various pieces of information such as the fleet size, the number of routes per day and stops per route, as well as the total distribution-related operating costs. It would be interesting to construct a simple "macro-model" that provides such savings estimates and to test its validity. One could take the preceding factors as explanatory variables and attempt to empirically derive the structure of a savings model. Currently, however, we doubt if a sufficient database for calibrating such a model exists. In fact, some developers claim that their clients do little in the way of measuring savings systematically. In many cases, it appears that the major appeal of the system does not primarily lie in the savings it offers, but in its user interface, graphics, and database capabilities, link with geographic database, and the ease with which the dispatcher can incorporate it into the operational activities of the firm. Of course, the relative importance of obtaining good data to justify the economics of system implementation depends on the size of the firm and the system's price tag.

Our final remark on the measurement issue has to do with how different sources of savings are identified and separated. In some implementation studies cited in Table 4, one can attempt to identify and measure the different components of savings. For example, in the study by Fisher et al. [33], the system was used in three stages: First, the routes were revamped by the system, keeping the assignment of customers to terminals and delivery modes fixed. These assignments were relaxed in the second stage, while the third stage used the system to evaluate the effects of different terminal location configurations. If the study reported the various savings associated with each stage, one could examine the contribution of decisions of each stage to the total savings. As another example, Evans and Norback [31] report a 10.7% improvement by using the system for the construction of tours one at a time and estimate another 3-5% savings if customers were allowed to move between tours. Similarly, a series of runs in [58] showed the savings to increase from 2% to 31% if only one visit to the depot is allowed instead of two. For the propane distribution problem reported in [3] and [40], tests showed that the main savings was realized by saving a vehicle, while fixing the number of vehicles would result in much smaller savings. In the same way, tests tried to separate the effect of forecast errors (for customer usage) on the model.

The preceding examples argue for more detailed reporting of savings and benefits. In particular, suitably designed runs can attempt to separate the effects of different components of the routing environment on the quality and

TABLE 4. Savings and Benefits in Selected Applications

REFERENCE	NATURE OF ACTIVITY	SIZE OF PROBLEM	SAVINGS AND BENEFITS
Brown et al. [17]	Distribution of petroleum products. System aids dispatching of vehicles.	430 vehicles 50,000 orders/month 120 bulk terminals	Savings of \$2-3 million in operating expenses. Helped formalize delivery policy decisions. Standardized policy across dispatchers. Allowed replacement of 3 manual centers with one automatic one.
Bell et al. [11]	Distribution of industrial gases nationwide.	340 vehicles 3500 customers 23 depots	Savings of 6-7.5% on the average or \$1.5-1.7 million for 16 depots. Better utilization of equipment. 9% increase in miles/gallon. Increase of 1.3% in weighted delivery radius.
Fisher et al. [33]	Distribution of clinical equipment and supplies.	50 routes 1500 customers (1000 cities) two plants and 5 depots	15% reduction in delivery costs. Model used for strategic facility location.
Belardo et al. [10]	Distribution of goods to convenience stores.	7000 stores	Test site with 35 routes serving 1500 locations showed savings of one route. Cost of a route estimated at \$1000.
Yano et al. [67]	Delivery of goods to retail stores with backhaul option.	11 trucks 40 stores single depot	Estimated savings of \$450,000 in 1986. Allowed use of backhaul option instead of common carrier. Showed need for trailers. Reduced amount of store-to-warehouse flow.
Bartholdi et al. [6]	Delivery of meals to senior citizens	4 vehicles 200 customers	Reduction of travel times by 13%. Low cost/minimal technology system. Allowed construction of balanced routes.
Evans and Norback [31]	Delivery operations for a foodservices firm.	13-18 vehicles 150-250 customers/day single depot	Variable cost savings of 10.7%. More savings possible if customers are re-assigned among routes.

cost effectiveness of the routes. Naturally, such studies are complicated by measuring the effects of man/machine interaction. In view of the emphasis on operational systems, our discussion in this section has not expanded on the use of routing systems for planning studies, such as those reported in [5], [12], [58], [27]. Clearly, routing models can be important components of medium or long range planning models for distribution.

5. GEOGRAPHIC DATABASES AND VEHICLE LOCATION IN ROUTING

The use of accurate geographic information on customer location and road networks has been a major concern in distribution studies. In recent years, major developments have occurred in the two areas of geographic databases and vehicle location and tracking, both of which are expected to impact routing systems significantly. This section attempts to outline some of these developments and discuss their integration into routing systems. Some of the material in this section is based on discussions with R. Dial [29] who has had first hand experience with the new technologies in this area.

5.1. Geographic Databases

Firms engaged in routing use maps very heavily. If one walks into any "dispatching station," one is apt to see the walls covered with maps; drivers, too, regularly rely on road and city maps to make deliveries to customers that are not visited routinely.

Given this heavy reliance on traditional maps, it is interesting to observe how static some of the technology on mapping has been in the past. Maps have been prepared in certain standard display formats for years, and few efforts were made to have the information conveniently available in "computer readable" (digitized) form until recently. Any user of road maps has at some point been bothered by their monolithic organization. A road map showing all streets at the same time, together with street names and other legends, presents the reader with a labyrinthine wealth of detail that is often hard to read, use, or decipher. Naturally, the user may find most of this detail irrelevant but has no control on what is displayed. Thus, even at the level of mere presentation (or what Dial calls "dumb maps"), there is much room for improvement. A good map must be hierarchically organized and selective in the information it displays.

An interesting example of geographic network database design is the system marketed by ETAK, a California-based firm that develops tracking devices for vehicles. This system provides the driver with a map of the road network that matches the view through the windshield. The vehicle always appears at the center of the map, and the display is frequently updated to reflect the changing "view" as the vehicle covers ground or makes turns. The small (4.5 or 7 inch) screen installed in the vehicle displays an area of 1/4 square miles at the highest level of detail. The driver can zoom out from this level of detail to see a nested sequence of increasingly more aggregate maps. This hierarchical structure is one way of selecting the level of detail presented. Note that since the ETAK system is used in the process of driving, a special effort must be made to design the display so that key information would be quickly grasped--the driver can not look away for long.

Clearly, the preceding system requires a very detailed geographic database to be stored. In recent years, a number of efforts have been made to make detailed geographic information electronically available. The starting point of most efforts is the Bureau of the Census DIME files where information is available on street segments at the block face level. This results in large networks: A metropolitan area such as Atlanta or Washington DC can easily involve 100,000 links (road segments), while Southern California may require 500,000 records. While these are not large numbers for a mainframe machine the handling of such networks on a microcomputer presents a number of challenges:

First, there is the issue of cheap storage. The ETAK system uses cassettes that contain the geographic data base for a given area. Thus, the driver has to switch cassettes when moving to another area. In the future, a compact disc (CD ROM) may be the favored storage device. The second issue has to do with fast algorithms, designed to observe the selective and hierarchical organization of the network. Good map drawing algorithms require some new thinking in this context, as do path finding algorithms. Currently, fast algorithms can construct shortest path trees on a large metropolitan network in only a few seconds. Dial [29] believes that the fastest algorithms for this purpose currently go beyond the published state of the art and will likely remain unpublished, as in the case of certain sorting algorithms.

Once a detailed geographic database is viewed as a general database rather than just a map, it is easy to envision a wide range of features this database may offer. Clearly, address matching is the main feature. In the ETAK product, a customer may be identified by a street address or an intersection in the road network. This is sufficient for the system to locate the customer on the map and save the driver the trouble of paging through many pages of a city map. From this basic feature, it is a conceptually short leap to have the database include "yellow pages" features, a listing of the closest restaurants, hotels, repair facilities, and so on.

5.2. Vehicle Location and Tracking

Vehicle drivers often have to call the base (depot) to inform the dispatcher about their current location. In the absence of any other means of communication, the position of the vehicle can only be estimated (based on the route to its next stop) in the interval between two such calls. The new vehicle location technology allows the central station to track the vehicle location in real time, without burdening the driver to radio information in. Clearly, such information is of great interest in dynamic routing, e.g., for messenger services or dial-a-ride problems. It is also useful to know the vehicle location more precisely where security is an issue such as the routing of patrol cars, or in the transportation of valuable or hazardous loads.

There are three major techniques for vehicle location which we now briefly describe following Zygmont [68]:

a) Dead Reckoning -- Given a highly accurate geographic database, it is possible to track the position of the vehicle on the road network based on detailed information on its movement. Navigational sensors can combine information on vehicle velocity and changes in direction with the data available on the road network being traversed. Using a modem between the navigator and the mobile radio in the vehicle, the position of the vehicle can be transmitted back to the base. This is the approach used by ETAK and is reported to be accurate to within 50 feet.

b) Loran C -- In this system, low frequency radio pulses are sent from pairs of stations to the vehicle, which then relays them to the base station. This allows the computer to locate the vehicle on a hyperbolic path based on the times the two signals were received by the vehicle. Signals from another pair of stations provide another path; and the vehicle location is fixed at the intersection of the two paths calculated. The coverage of this system is clearly tied to the availability of Loran beacons.

c) Two-Satellite Radio Determination System -- In this approach, a signal from the base triggers the vehicle to send a signal to a pair of satellites, which then relay it back to the base. Given the known locations of the satellites, the vehicle location can be determined by triangulation. This system, used by Geostar, is expected to have an eventual accuracy of 20-30 feet and provide nationwide coverage. However, the satellite facilities are not in place at this point.

At this writing, it is unclear which of the preceding three approaches will predominate. It is also possible for combinations to emerge, since one approach may correct for the "blind spots" of another. Major distribution firms are evidently interested in using vehicle tracking technology, although the cost is still viewed as being prohibitive. The recent (Fall 1986) acquisition of II Morrow, a leading producer of vehicle tracking systems based on Loran C, by United Parcel Service of America is one indication of the interest on the part of a major carrier. It should be clear that once the appropriate communications link between the vehicle and the base is established, the driver can also transmit other information, such as the current driver status or appropriate codes to signal the completion of the individual orders.

5.3. The Use of Geographic Information in Routing

We now turn to the use of geographical information within distribution systems. Here, it is useful to distinguish between two major types of routing applications based on the area of service [2]:

a) inter-city routing, and

b) intra-city distribution.

Inter-city problems involve long hauls between cities and can therefore rely on generally accessible data on state and federal highway networks. Intra-city problems involve the distribution of goods within urban areas and hence operate on the more detailed network of streets and road segments. Databases at this level of detail are not universally available in all parts of the United States.

The basic information required from a geographic database for the purpose of routing is as follows:

- the precise customer locations,
- site-to-site distances,
- site-to-site travel times,
- the geography of the delivery region and the road network,
- detailed route information.

In the remainder of this section, we discuss these needs and outline how they are met in practice.

1) Customer locations. The customer location is usually available in the form of an address that must be located on a map of the region. One simple way of doing this is to attach (x,y) coordinates to the customer. In the past, this was often done manually by plotting the customer on the map, a process currently simplified with digitizers. Given an accurate geographic database, the customer location can be derived directly from the address, a procedure called address matching. The vehicle location technology can also aid this process since one can simply track the vehicle that makes a stop at the customer location and use the initial visit to record the customer's latitude and longitude in the database. Indeed, Littich [54] has described the use of such a procedure in beer distribution.

Address matching is especially convenient to use when the customer base varies a lot and changes over time. One example is school bus routing where the location of the students is needed to develop suitable bus stops. Another example is given by the "Meals-on-Wheels" problem of Bartholdi et al. [6] who report 14% monthly turnover in the customer base. Naturally, address matching would reduce the burden of keeping customer locations (and associated travel times) current. However, its cost is prohibitive in the context of "minimal technology" applications such as [6].

2) Inter-site distances. The distances between customers can be computed by finding the shortest paths from each site to all others. In principle, this is immediately available from the geographic database network. It is important to note that the networks involved may be large. Even if the number of customers (sites) is moderate, the number of nodes in the underlying road network remains large. This distinction between the nodes of the road network and the nodes of the "delivery network" is important in the design of path finding algorithms. This distinction has been noted by Kleindorfer et al. [49]; but their procedure seems most suitable for cases when the number of sites constitutes a large percentage of the total number of nodes. This is not usually the case in intra-city delivery operations, so that it is useful to construct a suitable network abstraction of the underlying road network.

In discussing inter-site distances, we should consider the option of using (x,y) coordinates as a basis for developing the road distance. While the true network distances are of course more accurate, their use requires a detailed geographic database which may not be available in some cases and also impacts the hardware requirements.

Given two points i and j and their (x,y) coordinates, one can define the usual ℓ_p norm distance between them as

$$d_p(i,j) = [|x_i - x_j|^p + |y_i - y_j|^p]^{1/p}. \tag{1}$$

This more general expression yields the usual Euclidean distance for the choice p = 2, and the rectangular (Manhattan) distance for p = 1. Using observations of actual road distances, one can fit a linear model relating the true distance to various explanatory variables such as d_1 and d_2, as well as other possible zonal factors.

Let DISTR be the required true road distance. A relation for estimating DISTR that is frequently cited and utilized is

$$\text{DISTR} = k\ (\text{Straight-line distance}). \tag{2}$$

Love and Morris [55] estimated the value k = 1.16 or 1.18, using data from the Wisconsin region and the U.S. respectively; while Fildes and Westwood [32] obtained k = 1.17 using data from the U.K. In the Love and Morris study, the relations

$$\text{DISTR}\ (i,j) = 1.15\ d_{1.78}(i,j), \quad \text{and}$$

$$\text{DISTR}\ (i,j) = 1.01\ [d_{1.74}(i,j)]^{1.74/1.71}$$

are considered more accurate than (2). Note, however, that these are more complicated expressions. Since the simple replacement of the Euclidean distance by a linear approximation that avoids the computation of the square root can speed up running times of microcomputer implementations significantly (factor of 3-4, say); it is important to use easily computed explanatory variables. More importantly, it is important to reconsider the estimation and statistical issues addressed in both studies [32] and [55]. For example, the measure of error used naturally influences the choice of the parameters. Moreover, one may expect the results to depend on the underlying distribution of the distances traveled (relative frequency of trips with distances in a given interval). Finally, it is useful to recall that the data used to calibrate the above models reflected inter-city trips.

The use of straight-line distances as the main explanatory variable can be refined by partitioning the area of interest into zones as in [32], or by correcting for barriers. Naturally, all such modifications are costly. For instance, Madsen [56] cites one study where the introduction of 17 barriers increased the computation times by a factor of 30. The simple model in (2) has formed the basis for distance computations in a number of implementations (see [31], [44], [58]); one of which reports accuracy to within 2% for inter-city route distances with such a model [33]. We refer the reader to the survey by Madsen [56] for further discussion and references on the use of geographic information in routing.

3) Travel times. Road distances can easily be converted to travel time using an average speed function. The determination of the appropriate average speed must take the following factors into account.

- nonlinearity of the average speed as a function of distance (speed is smaller on shorter segments),
- dependency on time of the day and congestion effects,
- the classification of the road link,
- zonal information on region of travel (downtown vs. suburban areas).

Some other possible factors are discussed in [19]. We cannot enter a discussion of the techniques for accurate travel time measurement using sensors and on-the-road data capture. Naturally, the vehicle tracking technology can serve as an important source of such data in the future.

4) and 5) The road network and route information. In some applications, the information required on the geography of the region or the road network is not limited to the computation of travel times. One example is the use of zip code areas or other geographic units to define driver areas and territories as in the driver-sell problems [43]. Another natural use is the need for displaying the underlying road network in interactive systems. As humans are good processors of visual maps, a simple overlay of the road network on top of the display of routes greatly aids the dispatcher. In the system implementation described by Bell et al [11], the shortest paths between pairs of customers reflected a generalized cost composed of travel

time, distance, and tolls. For this purpose, a network database of the U.S. highway network of about 40,000 nodes and 65,000 links was used. For each road segment, the length and toll cost were specified, together with other road characteristics used in the assessment of an average speed over the segment.

An interesting example of detailed use of the geographic database is the system designed by Bodin et al. [12] for vehicle routing to perform solid waste collection in the Town of Oyster Bay, New York. In this application, collection service is provided to over 64,000 residential and commercial stops. One curious aspect of this problem is the role of the geobased information system in the estimation of service times at demand locations (rather than just travel time); as the total collection time depends on the total amount of refuse generated along each street segment. Detailed network information is also used to evaluate deadhead times and the effect of eliminating certain undesirable turns.

5.4. The Role of Accurate Geographic Information in Routing.

Briefly stated, there are two basic reasons why accurate geographic information is needed. First, for the accurate estimation of travel times; and second, for improved solution quality. We do not dwell on the former issue. Clearly, there is a loss function (to use the words of [32]) associated with the estimation of travel times. Underestimates of travel time lead to missed deadlines whereas overestimates can lower the utilization of drivers and vehicles, and may create unproductive wait times as well. Note that if the estimation method is unbiased, the total duration of the tour may still be estimated reasonably accurately since errors over route segments "average out." However, this "aggregate" accuracy cannot counteract the effect of errors on hard time windows associated with individual stops.

The issue of using geographic databases for reasons of solution quality is more complex. In Section 2.3, we cited the constraint on "covering stops" as an example of how the solution quality may be linked to the route configuration on the road network. To take another example we have seen in practice, consider a driver that uses a major highway to drive out to a suburb where most of his/her daily deliveries are located. Such a driver is usually reluctant to get off the highway during the early part of the drive to deliver goods to one or two customers close to the highway. Thus while from the viewpoint of distance minimization, the low cost insertion of the customer onto the route seems justified, the resulting route configuration would be unacceptable to the driver.

The preceding example illustrates the use of the geographic database in fine-tuning the quality of solutions. The main point is that the detailed geographic information goes beyond what is captured in travel times alone. This brings up the related question of the interplay between geographic database and routing algorithms. Most routing algorithms use an inter-site travel time matrix as their basic input. To avoid the $0(n^2)$ storage required for such a matrix (n = number of sites), a number of other possible forms of input information have been considered. One approach is to store only the p nearest neighbors of each site, thereby reducing the requirements to $0(n\ p)$. Jacobsen and Madsen [46] used a value of p = 14 in an application with over 4500 customers, but sensitivity with respect to the vale of p has not been reported in any detail. In general, the following options for presenting travel time information have been used:

a) storage of the full travel time matrix,

b) computation of travel time information upon demand directly from the network,

c) storage of a "reduced" travel time matrix, and

d) computing travel time information based on (x,y) coordinates.

Option (b) requires frequent calls to a path finding algorithm, while (c) includes the use of site neighborhoods as controlled by the value of p just mentioned, or through other partitions imposed on the delivery area. Option (d) was discussed in Section 5.3 and is generally less accurate. Given these options, the design and storage of travel time inputs to the algorithm require some thought in the presence of limited storage and required short response times.

The preceding options all relate to how accurate travel time and path information can be passed to an algorithm. As detailed road network databases become available, a more basic question emerges: How can routing algorithms be re-designed to make good use of the significant topological information available in the network? It appears that algorithms can be much smarter if they use inputs that go beyond pure travel times by drawing upon the underlying topology of the network. Some initial steps in this new direction have been taken by Dial [29], and the area of "network-based routing algorithms" appears to be a promising and fertile one.

6. CONCLUDING REMARKS

In this paper, we have attempted to examine the field of vehicle routing from an applied vantage point. Naturally, a full survey of modeling and implementation issues in routing would take us beyond the contents and scope of the present paper. For instance, we have emphasized the routing of vehicles and have not entered a discussion of other modes of transportation such as airline scheduling, ship scheduling, or rail transport where routing plays a significant part.

In this final section, we wish to make a few observations on fruitful directions of research and applications focusing, once again, on issues related to the implementation and use of routing models in practice. The following thoughts reflect some of the challenges and concerns raised in the course of the preceding five sections:

1) Theory versus Practice in Vehicle Routing. Most interested readers are already well aware of the rich interaction between theory and practice in vehicle routing. The traditional Traveling Salesman Problem, which was once cited sardonically as a total abstraction from real-life complications, has now evolved into functional VRP systems that are used on an operational basis. We believe that the use of VRP systems will grow as distribution managers become increasingly aware of their capabilities. Stated otherwise, we believe that the process of increasing this awareness among practitioners has just begun and that the market potential is significantly greater than what the present number of installations might indicate.

   Together with this growth in commercial applications come problems associated with the dissemination of information and proprietary restrictions. It is conceivable that the inner workings of certain highly successful commercial systems will remain unpublished or at least partially veiled. In the worst case, this may result in "two cultures"--the academically-based research or publications, and the developers or marketers of routing software--a situation that tends to impede the sharing of experiences and precludes scientific testing of algorithm performance. Naturally this possibility is not unique to the routing field and can be expected to arise whenever an area in operations research begins to assume a sizeable market potential (recall certain concerns relating to Karmarkar's Linear Programming Algorithm). In any case, it appears to us that the rewards of routing applications outweigh the risks associsted with such a bifurcation.

2) The Integration of New Technology into Routing. In Section 5 we briefly described the new technologies emerging in geographic databases and vehicle location and tracking. We predict that the integration of these advances with routing systems will constitute a major area of research and activity. As it is now possible to install powerful microcomputers in vehicles, drivers may soon have access to both routing and powerful geographic databases directly in the vehicle. We should also note that the incorporation of routing functions into geographic databases allows the construction of "intelligent maps" (to use a term of Dial's [29]), the market for which extends considerably beyond commercial routing for the distribution of goods, with many potential applications in the service or leisure industries.

3) Design of Interactive Systems. The experience with installations of routing systems to date indicate that interaction plays a major role in the design

of acceptable routes of high quality. Algorithms, however useful, cannot be rigidly designed and must in any event count on human input and modifications. The need for further work on interactive systems has often been remarked and is clearly felt in the context of routing. Different research strands relate to this issue: For example, graphical presentation of information on temporal characteristics of a route (e.g., time windows) should be designed to facilitate interaction with the human. Further insights are needed into interactive functions that the human can perform "well." Finally, the design of modular algorithms (discussed in Section 3.3) and work station functions must also receive careful attention in the context of human interaction.

4) Policy Issues in Vehicle Routing. We have remarked how a salient feature of the recent development of routing systems is model realism, as seen in the incorporation of many realistic constraints and objectives into the traditional model (see Section 2). While work will and should continue in this direction, it appears that routing models are now powerful enough to address a different set of questions that we call "policy questions." In policy questions, insights into the design of routing operations are of key interest; the following examples illustrate the nature of policy issues: Consider the question of dividing the demand between the available fleet and the common carrier. Are there any rules that can guide this decision and work in a majority of cases? How should fixed routes (see [8]) be used in operational planning, and what principles should inform the design of such routes? The recent work of Jaillet and Odoni [47] shows that novel modes of analysis can be applied to this problem. How should a dynamic real-time routing system deploy and use its vehicles and what are good measures of utilization for such systems? How can routing models provide useful information for strategic planning, as in location decisions? We are not suggesting that these questions are easily answered or that a single answer may be found to work in all contexts. Rather, we simply believe that the power and richness of current routing algorithms have reached a stage that allows methodical investigations of such questions.

In conclusion, one can state with confidence that the field of vehicle routing is alive and well with an exciting future before it. It is sobering to recall that less than a quarter of a century has passed since the appearance of the often-cited Clarke and Wright paper (1964). In vehicle routing, the advances made in practice over this period are truly impressive, especially because they have also resulted in a rich area for algorithmic development and analysis. Researchers and practitioners in vehicle routing should be pleased if during the next decade, their field shows the same level of growth and activity that it did during the last decade.

ACKNOWLEDGMENTS

The author wishes to thank Michael Ball for his comments on an early version of the manuscript. Many observations in this paper derive from my joint work and discussions with Michael Ball, Larry Bodin, and Bruce Golden. I am grateful for their generous sharing of information. Finally, certain sections of the paper draw upon two earlier presentations at professional meetings [1], [2].

REFERENCES

[1] Assad, A., "Modeling and Implementation Issues in Vehicle Routing," presented at the ORSA/TIMS Meeting, Los Angeles (1986).

[2] Assad, A., Ball, M., Bodin, L. and Golden, B., "Vehicle Routing and Scheduling with Special Emphasis on Commercial Vehicle Routing Systems," Workshop presented at the ORSA/TIMS Meeting, Atlanta (1985).

[3] Assad, A., Golden, B., Dahl, R., and Dror, M., "Evaluating the Effectiveness of an Integrated System for Fuel Delivery," Proceedings of the 1983 S.E. TIMS Conference, J. Eatman, editor, 153-160 (1983).

[4] Ball, M., "Allocation/Routing: Models and Algorithms," this volume.

[5] Ball, M., Golden, B., Assad, A. and Bodin, L., "Planning for Truck Fleet Size in the Presence of a Common Carrier Option," Decision Sciences, 14, 103-120 (1983).

[6] Bartholdi, J., Platzman, L., Collins, R. and Warden, W., "A Minimal Technology Routing System for Meals on Wheels," Interfaces, 13 (3), 1-8 (1983).

[7] Beasley, J., "Fixed Routes,: J. Operational Research Society, 35, 49-55 (1984).

[8] Beasley, J., "Route First-Cluster Second Methods for Vehicle Routing," Omega, 11, 403-408 (1983).

[9] Beasley, J., "Adapting the Savings Algorithm for Varying Inter-customer Travel Times," Omega, 9, 658-659 (1981).

[10] Belardo, S., Duchessi, P. and Seagle, J., "Microcomputer Graphics in Support of Vehicle Fleet Routing," Interfaces, 15 (6), 84-92 (1985).

[11] Bell, W., Dalberto, L., Fisher, M., Greenfield, A., Jaikumar, R., Mack, R. and Prutzman, P., "Improving the Distribution of Industrial Gases with an On-Line Computerized Routing and Scheduling System," Interfaces, 13 (6), 4-23 (1983).

[12] Bodin, L., Fagan, G., Welebney, R. and Greenberg, J., "The Design of a Computerized Sanitation Vehicle Routing and Scheduling for the Town of Oyster Bay, New York," unpublished manuscript (1987).

[13] Bodin, L. and Golden, B., "Classification in Vehicle Routing and Scheduling," Networks, 11, 97-108 (1981).

[14] Bodin, L., Golden, B., Assad, A. and Ball, M., "Routing and Scheduling of Vehicles and Crews: The State of the Art," Computers & Operations Research, 10, 63-211 (1983).

[15] Bodin, L. and Kursh, S., "A Computer-Assisted System for the Routing and Scheduling of Street Sweepers," Operations Research, 26, 525-537 (1978).

[16] Bott, K. and Ballou, R., "Research Perspectives in Vehicle Routing and Scheduling," Transportation Research, 20A, 239-243 (1986).

[17] Brown, G., Ellis, C., Graves, G. and Ronen, D., "Real-Time Wide Area Dispatch of Mobil Tank Trucks," Interfaces, 17 (1), 107-120 (1987).

[18] Brown, G. and Graves, G., "Real-Time Dispatch of Petroleum Tank Trucks," Management Science, 27, 19-32 (1981).

[19] Camp, R. and DeHayes, D., "A Computer-based Method for Predicting Transit Time Parameters Using Grid Systems," Decision Sciences, 5, 339-346 (1974).

[20] Casco, D., Golden, B. and Wasil, E., "Vehicle Routing with Backhauls: Models, Algorithms, and Case Studies," this volume.

[21] Christofides, N., "Vehicle Routing," in The Traveling Salesman Problem, edited by Lawler, E., Lenstra, J., Rinnooy Kan, A. and Shmoys, D., Wiley, Chichester, 1985, pp. 431-448.

[22] Christofides, N., "Vehicle Routing," in Combinatorial Optimization Annotated Bibliographies, edited by O'hEigeartaigh, M., Lenstra, J. and Rinnooy Kan, A., Wiley, Chichester, 1985, pp. 148-163.
[23] Christofides, N., Mingozzi, A. and Toth, P., State Space Relaxation Procedures for the Computation of Bounds to Routing Problems," Networks, 11, 145-164 (1981).
[24] Clavey, W., "Development of Distribution System for Navistar Unites Routing/Scheduling with Production/Inventory Control," presented at the ORSA/TIMS Meeting, St. Louis (1987).
[25] Cullen, F., Jarvis, J. and Ratliff, D., "Set Partitioning Based Heuristics for Interactive Routing," Networks, 11, 125-144 (1981).
[26] Desrochers, M., Lenstra, J., Savelsberg, M. and Soumis, F., Vehicle Routing with Time Windows: Optimization and Approximation," this volume.
[27] Desrosiers, J., Ferland, J., Rousseau, J-M., Lapalme, G. and Chapleau, L., "TRANSCOL: A Multi-Period School Bus Routing and Scheduling System," TIMS Studies in Management Sciences, 22, 47-71 (1986).
[28] Desrosiers, J., Soumis, F. and Desrochers, M., "Routing with Time Windows by Column Generation," Networks, 14, 545-565 (1984).
[29] Dial, R., private communication (1987).
[30] Dror, M. and Ball, M., "Inventory/Routing: Reduction from an Annual to a Short-Period Problem," Naval Research Logistics, 34, 891-905 (1987).
[31] Evans, S. and Norback, J., "The Impact of a Decision-Support System for Vehicle Routing in a Food Service Supply Situation," J. Operational Research Society, 36, 467-472 (1985).
[32] Fildes, R. and Westwood, J., "The Development of Linear Distance Functions for Distribution Analysis," J. Operational Research Society, 29, 585-592 (1979).
[33] Fisher, M., Greenfield, R., Jaikumar, R. and Lester, J., "A Computerized Vehicle Routing Application," Interfaces, 12 (4), 42-52 (1982).
[34] Fisher, M., Huang, J. and Tang, B., "Scheduling Bulk-Pickup-Delivery Vehicles in Shanghai," Interfaces, 16 (2), 18-23 (1986).
[35] Fisher, M. and Jaikumar, R., "A Generalized Assignment Heuristic for Vehicle Routing," Networks, 11, 109-124 (1981).
[36] Fisher, M. and Jaikumar, R., "Experiences with Real World Vehicle Routing Problems," presented at the TIMS XXVI International Meeting, Copenhagen (1984).
[37] Fisher, M. and Rosenwein, M., "Scheduling a Fleet of Bulk Transport Ships," presented at the ORSA/TIMS Meeting, San Francisco (1984).
[38] Golden, B. and Assad, A., "Vehicle Routing with Time-Window Constraints," American J. Mathematical and Management Sciences, 6, 251-260 (1986).
[39] Golden, B. and Assad, A., "Perspectives on Vehicle Routing: Exciting New Developments," Operations Research, 34, 803-810 (1986).
[40] Golden, B., Assad, A. and Dahl, R., "Analysis of a Large-Scale Vehicle Routing Problem with an Inventory Component," Large Scale Systems, 7, 181-190 (1984).
[41] Golden, B., Assad, A., Wasil, E. and Baker, E., "Experimentation in Optimization," European J. Operational Research, 27, 1-16 (1986).
[42] Golden, B., Bodin, L. and Goodwin, T., "Microcomputer-Based Vehicle Routing and Scheduling Software," Computers & Operations Research, 13, 277-285 (1986).
[43] Golden, B. and Wasil, E., "Computerized Vehicle Routing in the Soft Drink Industry," Operations Research, 35, 6-17 (1987).
[44] Holt, J. and Watts, A., "Vehicle Routing and Scheduling in the Newspaper Industry," this volume.
[45] Hooban, J., "Marketing a Vehicle Routing Package," this volume.
[46] Jacobsen, S. and Madsen, O., "A Comparative Study of Heuristics for a Two-Level Routing-Location Problem," European J. Operational Research, 5, 378-387 (1980).

[47] Jaillet, P. and Odoni, A., "The Probabilistic Vehicle Routing Problem," this volume.
[48] Jaw, J., Odoni, A., Psaraftis, H. and Wilson, N., "A Heuristic Algorithm for the Multi-Vehicle Advance-Request Dial-a-Ride Problem with Time Windows," Transportation Research, 20B, 243-257 (1986).
[49] Kleindorfer, G., Kochenberger, G. and Reutzel, E., "Computing Inter-site Distances for Routing and Scheduling Problems," Operations Research Letters, 1, 31-38 (1981).
[50] Kolen, A., Rinnooy Kan, A. and Trienekens, H., "Vehicle Routing with Time Windows," Operations Research, 35, 266-273 (1987).
[51] Laporte, G., "Location-Routing Problems," this volume.
[52] Lee, A. and Shapiro, J., "A Nested Decomposition Method for Vehicle Routing and Scheduling Models," presented at the ORSA/TIMS Meeting, St. Louis (1987).
[53] Levy, L. and Bodin, L., "Scheduling the Postal Carriers for the U.S. Postal Service: An Application of Arc Partitioning and Routing," this volume.
[54] Littich, J., "Routing and Scheduling at Anheuser-Busch," presented at the ORSA/TIMS Meeting, St. Louis (1987).
[55] Love, R. and Morris, J., "Modelling Inter-city Road Distances by Mathematical Functions," Operational Research Quarterly, 23, 61-71 (1972).
[56] Madsen, O., "On the Use of Road Databases in the Planning of Locational Decisions and Distribution Systems," presented at the TIMS XXVI International Meeting, Copenhagen (1984).
[57] Magnanti, T., "Combinatorial Optimization and Vehicle Fleet Planning: Perspectives and Prospects," Networks, 11, 179-214 (1981).
[58] Mathews, B. and Waters, C., "Computerized Routing for Community Nurses--A Pilot Study," J. Operational Research Society, 37, 677-683 (1986).
[59] Mole, R., "A Survey of Local Delivery Vehicle Routing Methodology," J. Operational Research Society, 30, 245-252 (1979).
[60] Perl, J. and Daskin, M., "A Warehouse Location-Routing Problem," Transportation Research, 19B, 381-396 (1985).
[61] Psaraftis, H., "Dynamic Vehicle Routing Problems," this volume.
[62] Raft, O., "A Modular Algorithm for an Extended Vehicle Scheduling Problem," European J. Operational Research, 11, 67-76 (1982).
[63] Rousseau, J-M., "Customization Versus a General Purpose Code for Routing and Scheduling: A Point of View," this volume.
[64] Savelsberg, M., "Local Search in Routing Problems with Time Windows," Annals of Operations Research, 4, 285-305 (1985).
[65] Solomon, M., "Algorithms for the Vehicle Routing and Scheduling Problem with Time Window Constraints," Operations Research, 35, 254-265 (1987).
[66] Solomon, M., Baker, E. and Schaffer, J., "Vehicle Routing and Scheduling Problems with Time Window Constraints," this volume.
[67] Yano, C., Chan, T., Richter, L., Cutler, T., Murty, K. and McGettigan, D., "Vehicle Routing at Quality Stores," Interfaces, 17 (2), 52-63 (1987).
[68] Zygmont, J., "Keeping Tabs on Cars and Trucks," High Technology (September 1986).

Vehicle Routing: Methods and Studies
B.L. Golden and A.A. Assad (Editors)

Analysis of heuristics for vehicle routing problems

M. Haimovich
Tel-Aviv University

A.H.G. Rinnooy Kan
Erasmus University Rotterdam

L. Stougie
University of Amsterdam

Almost all vehicle routing and scheduling problems are *NP*-hard, and most of the algorithms designed for their solution are of a heuristic nature. Due to complications present in most vehicle routing problems e.g. time-windows, conflicting objectives, precedence constraints, these heuristics are often equipped with problem specific features that turn the theoretical analysis of their performance quality into a far from trivial task. Consequently, the number of results in this direction is still very limited. We present examples from the recent literature, in which mathematical analysis has led to rigorous statements on the performance of vehicle routing heuristics. We shall describe more or less extensively examples of worst-case and probabilistic analysis of heuristics for capacitated vehicle routing problems. Notes on selected references have been added.

1985 Mathematics Subject Classification: 90B05, 90B35, 90C27, 60B10
Key Words and Phrases: vehicle routing problem, capacitated routing, travelling salesman problem, route selection, worst-case analysis, probabilistic analysis, almost sure convergence.

1. Introduction

Almost all problems of vehicle routing and scheduling are *NP-hard* [Lenstra & Rinnooy Kan 1981]. Computational complexity theory has provided strong evidence that any optimization algorithm for their solution is likely to perform very poorly on some occasions: more formally, its *worst case running time* is likely to grow exponentially with *problem size*.

One possible attitude towards this unpleasant feature which has been adopted by many algorithm designers is to abolish the ideal of optimization and settle for an approximation of the optimal solution. This has resulted in a variety of *approximation algorithms* or *heuristics* for the solution of vehicle routing and scheduling problems. For an overview and a classification of these heuristics, we refer to [Bodin & Golden 1983]. The question how well the heuristic solution approximates the optimal one is a very natural one to pose.

Empirical analysis is one way to arrive at some insight into the behavior of heuristics, but it involves the evaluation of not necessarily representative computational experiments. In some cases it is also possible to perform a *theoretical analysis* of the *absolute* or *relative error* of the heuristic, i.e. the absolute or relative difference between the heuristic solution value and the optimal one. For example, the error incurred on the least favorable problem instance can sometimes be computed. Such a *worst-case performance analysis* provides an absolute guarantee on the quality of the heuristic solution.

This approach yields a pessimistic view, in the sense that even though a heuristic may behave poorly on some instances, its performance may be much better in general. Therefore, an attractive alternative to a worst-case analysis is an *average-case analysis*. For such an analysis, probability theory provides a natural setting. One first defines a probability distribution over the space of all possible problem instances. The heuristic solution value can then be viewed as a random variable whose behavior can be studied and evaluated so as to arrive at a probabilistic estimate of the error.

Due to complications present in most vehicle routing problems, e.g. time-windows, conflicting objectives, precedence constraints, heuristics for their solution are often equipped with problem specific features that turn the theoretical analysis of their performance quality into a far from trivial task. Consequently, the number of results in this direction is still very limited.

In this chapter we will present examples from the recent literature, in which mathematical analysis has led to rigorous statements on the performance of vehicle routing heuristics. We shall describe more or less extensively examples of, respectively, worst-case and probabilistic analyses of heuristics for *capacitated vehicle routing problems*. In these problems *customers* have to be served by *vehicles* of limited capacity from a common *depot*. The objective is to find a set of *routes*, where each route starts at the depot and returns there after visiting a subset of the customers, such that each customer is visited exactly once, the capacity of the vehicles is not exceeded and the *total length* of the routes is as small as possible. We assume that each customer corresponds to a point in the Euclidean plane and that the distances between customers are given by the Euclidean distance between the corresponding points. Of course, this is only a stylized version of real life routing problems (see [BODIN & GOLDEN 1983] for a survey). We concentrate on this simple problem, because it is the only one that has been studied in any depth.

In Section 2 we illustrate the use of worst-case analysis; in Section 3 examples are provided on probabilistic analysis. Section 4 contains bibliographical notes, including selected references to related work.

Notation

Let $X = (x_1, x_2, \ldots, x_n)$ be the set of customers (points in the Euclidean plane) where customer x_i is at distance r_i from the origin O (the depot). We denote the maximum distance $\max_i\{r_i\}$ by $r_{\max}$ and the average distance $(\Sigma_{i=1}^n r_i)/n$

by $\bar{r}$. The demand of customer x_i is denoted by d_i ($i = 1,2,...,n$).

Each customer has to be visited by a vehicle. The jth vehicle starts from the depot and returns there after visiting a subset $X_j \subseteq X$; the total set of points visited by this vehicle is denoted by $X_j^O = X_j \cup \{O\}$. The capacity of each vehicle is equal to q units and we assume that there is an infinite supply of vehicles.

We shall be interested in finding a feasible set of routes such that their total length is as small as possible; this optimal solution value will be denoted by $R^*(X)$. The value produced by a heuristic H will be denoted by $R^H(X)$. Finally, an important role in the analysis will be played by the *shortest travelling salesman tour* visiting each customer exactly once. The length of this shortest tour will be denoted by $T^*(X)$.

2. Worst case analysis

2.1. *Bounds on the optimal solution value*

In this subsection and the following one we will assume that the customers have unit demand; i.e., $d_1 = d_2 = \cdots = d_n = 1$. We shall be interested in worst case bounds on the performance ratio $R^H(X)/R^*(X)$, i.e. bounds that are valid for all possible X. In general such bounds are obtained by comparing an *upper bound* on $R^H(X)$ to a *lower bound* on $R^*(X)$. To derive the latter will be our first concern.

THEOREM 2.1.

$$R^*(X) \geqslant \max\{2\frac{n}{q}\bar{r},\ T^*(X^O)\}.$$

PROOF. Consider the subset X_j of customers visited by the j-th vehicle in the optimal solution. Clearly

$$T^*(X_j^O) \geqslant 2\max_{x_i \in X_j}\{r_i\} \geqslant 2\frac{\Sigma_{x_i \in X_j} r_i}{|X_j|} \geqslant \frac{2}{q}\Sigma_{x_i \in X_j} r_i$$

and hence

$$R^*(X) = \Sigma_j T^*(X_j^O) \geqslant \frac{2}{q}\Sigma_{i=1}^{n} r_i = 2\frac{n}{q}\bar{r}.$$

The inequality $R^*(X) \geqslant T^*(X^O)$ is an immediate consequence of the triangle inequality of the Euclidean metric. □

How sharp is this lower bound? A partial answer is provided by the next result which essentially shows that $R^*(X)$ is bounded from above by the *sum* of the two terms whose *maximum* appears in Theorem 2.1. The usual technique to obtain an upper bound on $R^*(X)$ is to establish such a bound on a particular feasible solution. Here, we obtain such a feasible solution by means of a *tour partitioning* heuristic.

The heuristic starts from the shortest travelling's salesman tour through all customers and the depot. In an arbitrary orientation of the tour, the customers are renumbered $x^{(1)},x^{(2)},...,x^{(n)}$, in order of appearance. We partition the path from $x^{(1)}$ to $x^{(n)}$ into $\lceil n/q \rceil$ disjoint segments so that each segment contains no more than q customers, and connect the endpoints of each segment to the depot. (Note that $\lceil n/q \rceil - 1$ arcs from the travelling salesman's tour are excluded from the construction.)

This simple heuristic is known as *Optimal Tour Partitioning* (OTP). To improve on its result, we repeat the above construction q times. The first time, we start with $x^{(1)}$ as a single element segment, and partition the path into further segments starting from $x^{(2)}$ so that each segment contains exactly q customers (except possibly the last one). In consecutive iterations, the endpoints of these segments are moved up by $1,2,...,q-1$ positions in the direction of the orientation (see Figure 1). We then choose the best of the resulting solutions.

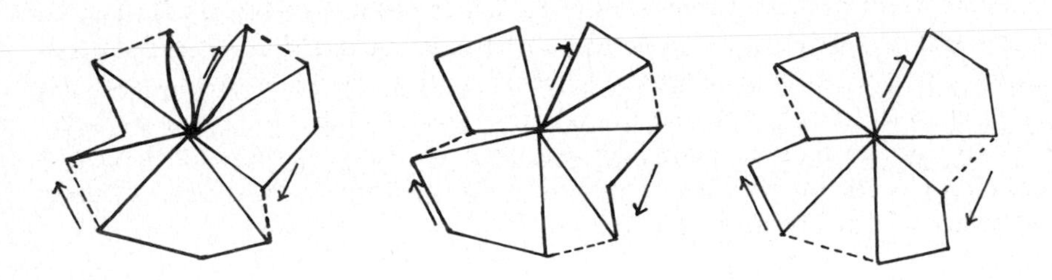

FIGURE 1. IOTP with $q=3$ and $n=11$

To analyze the value of the heuristic (known as *Iterated Optimal Tour Partitioning* (IOTP)), we observe that the cumulative length of the q solutions is equal to

$$2\Sigma_i r_i + (q-1)T^*(X^O). \tag{2.1}$$

To understand the first term, we notice that all customers except $x^{(1)}$ and $x^{(n)}$ occur once as first points and once as last points of a segment. Therefore each of these customers is connected twice to the depot. Customer $x^{(1)}$ is first point of a segment in all q solutions, and in the first solution it is also last point of a segment. Thus, $x^{(1)}$ is connected $q+1$ times to the depot. In the same way, $x^{(n)}$ is always a last point and once a first point of a segment, which implies that $x^{(n)}$ is also connected to the depot $q+1$ times. Two connections of $x^{(1)}$ and $x^{(n)}$ to the depot are included in the first term of (2.1), so that $(q-1)$ connections are still to be accounted for in the second term. But these connections are arcs of the optimal travelling salesman's tour. This, together with the fact that each of the other arcs in the tour is excluded in one solution and therefore present in $q-1$ solutions, justifies the second term of (2.1).

Clearly, the value $R^{IOTP}(X)$ of the best solution must be less than the average value found. Thus we arrive at the following theorem.

THEOREM 2.2.

$$R^*(X) \leqslant R^{IOTP}(X) \leqslant 2(\frac{n}{q})\bar{r} + (1-\frac{1}{q})T^*(X^O). \quad \square$$

We note that the bounds in Theorem 2.1 and 2.2 consist of two terms. One term, $2\frac{n}{q}\bar{r}$, reflects the *radial collection cost*, the cost of setting out to the customers from the depot. The second term, $T^*(X^O)$, reflects the total *local collection cost* that is incurred as soon as the subset of customers has been reached. In combination with Theorem 2.1, Theorem 2.2 shows that the dominant one of these two cost factors ultimately determines $R^*(X)$.

2.2. Worst-case analysis of tour partitioning heuristics for the unit demand case

In this section we study the performance behavior of tour partitioning heuristics for the solution of capacitated vehicle routing problems from a worst-case point of view. The first result is a bound on the *performance ratio* $R^{IOTP}(X)/R^*(X)$ of the iterated optimal tour partitioning heuristic described in Section 2.1 for the problem with unit customer demand. This result follows immediately from Theorems 2.1 and 2.2.

THEOREM 2.3.

$$R^{IOTP}(X)/R^*(X) \leqslant 2-\frac{1}{q}. \quad \square$$

We notice that IOTP presupposes the availability of an optimal travelling salesman tour. It is well known that finding such a tour is an *NP*-hard problem, and hence the heuristic is of little practical value. Any reasonable tour partitioning heuristic will be based on an approximation of this tour. Let us denote a tour partitioning heuristic that starts from an approximation of the optimal travelling salesman tour, whose length is bounded by $\alpha T^*(X)$ for a constant $\alpha > 1$, by αTP.

An upper bound on the solution value produced by $I\alpha TP$ can be obtained in the same way as the bound in Theorem 2.2.

$$R^{I\alpha TP}(X) \leqslant 2\frac{n}{q}\bar{r} + (1-\frac{1}{q})\alpha T^*(X^O). \tag{2.2}$$

This bound, together with the lower bound on $R^*(X)$ in Theorem 2.1, yields the following bound on the performance ratio of $I\alpha TP$.

THEOREM 2.4.

$$R^{I\alpha TP}(X)/R^*(X) \leqslant 1 + (1-\frac{1}{q})\alpha. \quad \square$$

The best value for α known to be achievable in polynomial time is 3/2 [CHRISTOFIDES].

2.3. Worst-case analysis of tour partitioning heuristics for the unequal demand case

Let us now consider the problem with unequal customer demands. We assume that customer i has integer demand d_i ($i = 1,2,...,n$). It is not allowed to split the demand of a customer over several vehicles. Any tour partitioning heuristic for this problem would divide a travelling salesman tour into segments, each containing a set of customers with total demand no more than q.

The tour partitioning heuristic that we study consists of two phases. In the first phase, we construct a tour partition by neglecting the requirement that the demand of a customer may not be split. Customer i is then represented by d_i copies all having unit demand and interdistance zero ($i = 1,2,...,n$). Then a tour partitioning heuristic as above for the unit demand problem is applied to this problem with $m = \Sigma_{i=1}^{n} d_i$ customers, that have average distance $\overline{r}_d = (1/m)\Sigma_{i=1}^{n} d_i r_i$ to the depot, and where, the vehicle capacity is set equal to $q/2$, so that segments are obtained containing $q/2$ customers each. Without loss of generality we may assume that q is even. If we again use the IOTP heuristic here, then an upper bound on the solution value produced is easily obtained from the bound in Theorem 2.2 by substitution of m for n,$\overline{r}_d$ for $\overline{r}$ and $q/2$ for q. The index s indicates that demand splitting is allowed; we find that

$$R_s^{IOTP}(X) \leqslant 2\frac{2m}{q}\overline{r}_d + (1-\frac{2}{q})T^*(X^0).$$

In the second phase the tour partition obtained above is converted into a feasible partition for the original problem with total length $R^{IOTP}(X)$. The following lemma shows that the solution thus obtained has length no more than $R_s^{IOTP}(X)$.

LEMMA 2.5. $R^{IOTP}(X) \leqslant R_s^{IOTP}(X)$.

PROOF. The first-phase partition yields $\lceil 2m/q \rceil$ segments. Starting from a customer that is an endpoint of any of the segments and following the orientation of the travelling salesman tour previously used, the customers are renumbered $x^{(1)},x^{(2)},...,x^{(n)}$. Then there are indices i_j, $j = 1,2,..., \lceil 2m/q \rceil$, such that customer $x^{(i_j)}$ is the last customer (partially) belonging to the j-th segment, denoted by s_j. Let $x^{(i_0)}$ be $x^{(1)}$. Hence, $s_j = [x^{(i_{j-1})},...,x^{(i_j)}]$ or $s_j = [x^{(i_{j-1}+1)},...,x^{(i_j)}]$, depending on whether or not the demand of customer $x^{(i_{j-1})}$ has been split ($j = 1,2,...,\lceil 2m/q \rceil$). We note that, similarly, $x^{(i_{\lceil 2m/q \rceil})} = x^{(1)}$ or $x^{(i_{\lceil 2m/q \rceil})} = x^{(n)}$.

For the first new segment s'_1 we choose $x^{(1)}$ as its starting point and $x^{(i_1)}$ or $x^{(i_1-1)}$ as its endpoint, depending on whether or not inclusion of $x^{(i_1)}$ in s'_1 is feasible with respect to the capacity q. In the former case s'_2 starts with $x^{(i_1+1)}$, in the latter it starts with $x^{(i_1)}$. The procedure is continued, each time transforming segment s_j into segment s'_j, $j = 1,2,...,\lceil 2m/q \rceil$. In this way s'_j starts either with $x^{(i_{j-1})}$ or $x^{(i_{j-1}+1)}$ and ends with either $x^{(i_j)}$ or $x^{(i_{j-1})}$. The capacity $q/2$ of s_j ensures that all customers that were served completely in s_j, i.e. that were inner points of s_j, will be served completely in s'_j. The only possible differences between s_j and s'_j are in their endpoints. Therefore, for a comparison between $R_s^{IOTP}(X)$ and the total length of the routes induced by the new partition, we only have to consider pairs of segments s_j and s'_j that have different endpoints. We know that if s_j and s'_j differ in their starting points then s_j must start with $x^{(i_{j-1})}$ and s'_j with $x^{(i_{j-1}+1)}$. This means that the arc $(x^{(i_{j-1})},x^{(i_{j-1}+1)})$ that is part of the old j-th route does not appear in the new j-th route. The triangle inequality ensures that the sum of the length of $(O,x^{(i_{j-1})})$ and $(x^{(i_{j-1})},x^{(i_{j-1}+1)})$ is greater than the length of $(O,x^{(i_{j-1}+1)})$. The same argument can be applied if s_j and s'_j differ in their other endpoints. Hence, for every $j = 1,2,...,\lceil 2m/q \rceil$ the length of the j-th route according to the new partition is smaller than or equal to the length of the j-th route in the old partition. □

Thus, we have

$$R^{IOTP}(X) \leqslant R_s^{IOTP}(X) \leqslant 2\frac{2m}{q}\bar{r}_d + (1-\frac{2}{q})T^*(X^O).$$

It is obvious that the optimal solution value of the problem allowing demand to be split is smaller than the one in which demand may not be split. Substitution of m for n and $\bar{r}_d$ for $\bar{r}$ in Theorem 2.1 yields the following lower bound

$$R^*(X) \geqslant \max\{2\frac{m}{q}\bar{r}_d, T^*(X^O)\}.$$

It is now a simple matter of calculation to establish a bound on the performance ratio of the above heuristic.

THEOREM 2.6.

$$R^{IOTP}(X)/R^*(X) \leqslant 3-\frac{2}{q}. \quad \square$$

2.4. A bound on the travelling salesman tour

For use in the next section, we finally derive a sharp upper bound on $T^*(X^O)$.

THEOREM 2.7.

$$T^*(X^O) \leqslant 2\sqrt{2n\pi r_{\max}\bar{r}} + (2+\pi)r_{\max}.$$

PROOF. First the circle with radius $r_{\max}$ and the depot as its centre is partitioned into $2h$ equal sectors. The boundaries and the circumference of the circle are used to construct a flower shaped closed path traversing the circle as in Figure 2. The path is converted into a travelling salesman tour by a double connection of minimal length from each point x_i; the sum of these double connections for x_i is less than $2(2\pi/2h)r_i/2$. Hence the length of the tour is less than

$$\pi n\bar{r}/h + 2hr_{\max} + \pi r_{\max},$$

which is the sum of the double connections plus the length of the flower shaped path.
By taking h equal to the value minimizing the above expression and rounding up, i.e., $h = \lceil \sqrt{\pi n\bar{r}/2r_{\max}} \rceil$, we obtain the desired result. □

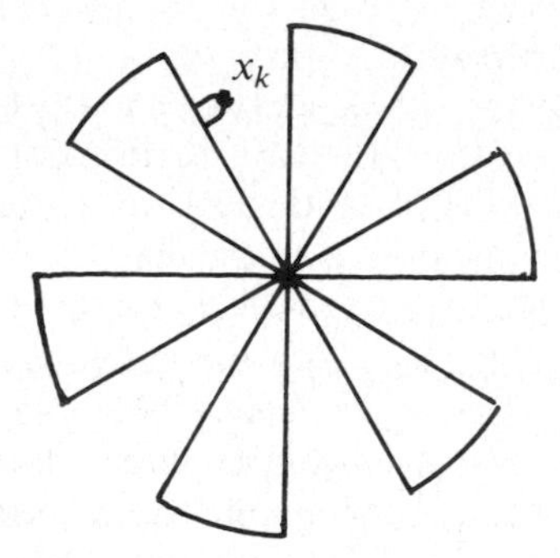

FIGURE 2.

3. PROBABILISTIC ANALYSIS

3.1. Asymptotic analysis of the optimal solution value

In this section we again study the case of unit customer demand. The following theorem gives an asymptotic characterization of the optimal solution value of the capacitated vehicle routing problem. If we combine the bounds in Theorems 2.2 and 2.9 and ignore for the time being the influence of the $r_{\max}$ and $\bar{r}$ terms, we see that the radial collection cost is $O(n/q)$ and the local collection cost is $O(\sqrt{n})$. It is therefore intuitively clear that the former cost will dominate the latter as long as $q = o(\sqrt{n})$ (hence, in particular, if q is constant as assumed here).

THEOREM 3.1. *Let $x_1, x_2, \ldots, x_n$ be a sequence of independent, identically distributed random points with finite expected distance Er_1 from the origin, and let $X^{(n)}$ denote the first n points of the sequence. For constant q,*

$$\lim_{n\to\infty} \frac{1}{n} R^*(X^{(n)}) = \frac{2Er_i}{q}$$

almost surely (a.s.).

PROOF. By Theorems 2.1, 2.2 and 2.7.

$$\frac{2}{q}\bar{r} \leqslant \frac{1}{n} R^*(X^{(n)}) \leqslant \frac{2}{n}(\frac{n}{q})\bar{r} + O(\sqrt{\bar{r}\frac{r_{\max}}{n}} + \frac{r_{\max}}{n}).$$

The theorem follows from the fact that $Er_1 < \infty$ implies that $\bar{r} \to Er_1$ (a.s) by the strong law of large numbers and $r_{\max}/n \to 0$ (a.s). (Since $\Sigma_n \epsilon Pr\{r_n \geqslant \epsilon n\} = \Sigma_n \epsilon Pr\{r_1 \geqslant \epsilon n\} \leqslant Er_1$, the latter convergence follows from the Borel-Cantelli lemma and the fact that $r_n/n \to 0$ implies that $r_{\max}/n \to 0$.) □

3.2. Probabilistic analysis of heuristics

Two types of heuristics may be considered for solving the capacitated vehicle routing problem. The first type are the tour partitioning heuristics introduced in Section 2. A travelling salesman tour through all customers was cut into segments and the segments were connected to the depot.

These tour partitioning heuristics, however, hardly exploit the topological structure of the Euclidean plane, in which the points are located. therefore, as a second type of heuristics, *region partitioning* heuristics, is considered. These heuristics first partition the set of customers into subsets by partitioning the planar region containing the customers into connected subregions, each of which contains no more than q customers. Subsequently, a travelling salesman tour is constructed through the customers in each subregion and the depot.

We shall consider various partitioning schemes of this type. In *rectangular region partitioning* (RRP), onc starts with a rectangle containing X and cuts it into smaller rectangles in much the same way as Karp proposes in [KARP 1977]. In *polar region partitioning* (PRP), one starts with a circle centered at the origin and partitions it in a similar way, with circular concentric arcs substituting for horizontal cuts and radial lines for vertical cuts as in [MARCHETTI SPACCAMELA ET AL. 1984]. If only radial cuts are used, we have a *sectorial region partitioning* (SRP) scheme. These schemes are illustrated in Figure 3.

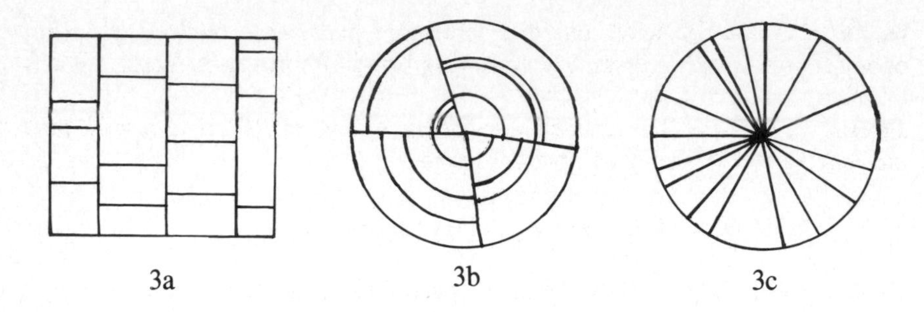

FIGURE 3a. RRP, 3b PRP, 3c SRP.

All these heuristics construct subsets X_j, the points in which are subsequently linked by a travelling salesman tour. In the error analysis of the heuristics the value of $\Sigma_j T^*(X_j)$ plays an important role. This value depends heavily on the sum P^H of the perimeters of the subregions generated by heuristic H. Indeed, Karp proves in [KARP 1977] that

$$\Sigma_j T^*(X_j) \leqslant T^*(X) + \frac{3}{2}P^H \leqslant T^*(X^O) + \frac{3}{2}P^H. \tag{3.1}$$

Hence, as a first step we analyze P^H for the three partitioning schemes proposed above.

LEMMA 3.2. *The three partitioning schemes RRP, PRP and SRP can be implemented to run in $O(n \log n)$ time so that*
(i) $P^{RRP} \leqslant 16\sqrt{(n/q)}r_{\max}$;
(ii) $P^{PRP} \leqslant 10\pi\sqrt{(n/q)}r_{\max}$;
(iii) $P^{SRP} \leqslant (2\lceil n/q \rceil + 2\pi)r_{\max}$.

PROOF. (i), (ii) The proofs follow the original one in [KARP 1977] for RRP or the simplified version in [MARCHETTI SPACCAMELA ET AL. 1984] for PRP. The total number of partitioning cuts k is proportional to $\log_2(n/q)$. In the case of RRP, P^{RRP} is bounded by $2^{k/2}p \leqslant 2\sqrt{n/q}p$, where p is the perimeter of the original rectangle. In the case of PRP, a bound of $5\sqrt{n/q}p$, where p is now the perimeter of the original circle, can be obtained similarly. Using the implementation described in [KARP 1977] and [MARCHETTI SPACCAMELA ET AL. 1984], the bounds follow immediately.
(iii) The circle with radius $r_{\max}$ is partitioned into $\lceil n/q \rceil$ sectors. Therefore the total perimeter of the sectors consists of $2\lceil n/q \rceil$ radii of length $r_{\max}$ and the circle itself. □

Note that the results in Lemma 3.2 already bode ill for the quality of SRP: it will usually produce solutions not shorter than $2(n/q)r_{\max}$. The $O(\sqrt{n}\, r_{\max})$

results obtained for RRP and PRP are more promising, but not quite good enough for all our purposes. The following lemma improves on Lemma 3.2.

LEMMA 3.3. *There is a circular partitioning scheme (CRP) that can be implemented to run in $O(n \log n)$ time, for which*

$$P^{CRP} \leqslant 4\sqrt{3\pi \frac{n}{q} r_{\max} \bar{r}} + (3+2\pi) r_{\max}.$$

PROOF. We start the CRP scheme by partitioning the circle with radius $r_{\max}$ into h equal sectors. Each sector is then partitioned into several subregions by circular arcs, such that all of them contain q points except possibly the one closest to the origin which may contain less than q. The latter group of at most h subregions is repartitioned by radial cuts into at most $h-1$ subregions with q points and at most one containing less than q points.

The total length of the initial radial cuts is $hr_{\max}$. The length of an inner circular arc cut bounding a subregion containing X_j is no more than

$$\frac{2\pi}{h} \min_{x_i \in X_j}\{r_i\} \leqslant \frac{2\pi}{h} \cdot \frac{\Sigma_{x_i \in X_j} r_i}{|X_j|} = \frac{2\pi \Sigma_{x_i \in X_j} r_i}{hq}.$$

Finally, the excess length added by the repartitioning of the central subregions is no more than $hr_{\max}/2$. (It is no more than half the total radial distance along the external boundary of the subregions.) It follows that

$$P^{CRP} \leqslant 2(hr_{\max} + \frac{2\pi n \bar{r}}{hq} + \frac{hr_{\max}}{2}) + 2\pi r_{\max}.$$

Taking $h = \lceil \sqrt{(4\pi n \bar{r})/(3qr_{\max})} \rceil$, we obtain the desired result. □

After any region partitioning scheme, RP, the tours through the resulting sets X_j^O may be optimal ones with value $T^*(X_j^O)$ or heuristic ones with value $T^H(X_j^O)$. If, in the latter case there exist a $\alpha > 1$ such that $T^H(X) \leqslant \alpha T^*(X)$ for all X, we denote the partitioning heuristic as a whole by $RP\alpha T$; the former case is denoted by RPT.

We now have all the necessary ingredients for an asymptotic analysis of the performance of partitioning heuristics under assumptions similar to those of Theorem 3.1. The following lemma contains the essential inequalities.

LEMMA 3.4.
(i) $R^{\alpha TP}(X) \leqslant \min\{2\lceil n/q \rceil r_{\max} + \alpha T^*(X), 2(n/q)\bar{r} + 2r_{\max} + 2\alpha T^*(X)\}$,
(ii) $R^{I\alpha TP}(X) \leqslant 2(n/q)\bar{r} + \alpha T^*(X^O)$,
(iii) $R^{RP\alpha T}(X) \leqslant 2(n/q)\bar{r} + 2r_{\max} + \alpha T^*(X^O) + \alpha(3/2)P^{RP}$.

PROOF. (i) The first bound is easily justified by the observation that the travelling salesman tour is partitioned into $\lceil n/q \rceil$ segments, each of which is connected twice to the depot. To establish the second one, note that the radial distances from O to the endpoints x'_j and x''_j of the jth tour segment X_j is less than twice the distance from O to the closest point on X_j, say x'''_j, plus the sum of the distances from x'''_j to x'_j and from x'''_j to x''_j. If $|X_j| = q$, the first term is less than $2(\Sigma_{x_i \in X_j} r_i)/q$; it is smaller than $2r_{\max}$ for the single segment with $|X_j| \leq q$. The second terms add up to correspond to a tour through x'_j, x'''_j and $x''_j (j = 1, ..., \lceil n/q \rceil)$ the length of which is less than the original tour length by virtue of the triangle inequality. Hence,

$$R^{\alpha TP}(X) \leq 2\frac{\Sigma_i r_i}{q} + 2r_{\max} + 2\alpha T^*(X).$$

(ii) The proof is immediate from Theorem 2.2.

(iii) Let us assume that the subsets X_j created by the RP scheme are numbered in such a way that $|X_j| = q$ with the possible exception of $|X_1|$. We have that

$$\begin{aligned} R^{RP\alpha T}(X) &\leq \Sigma_j (2\min_{x_i \in X_j}\{r_i\} + \alpha T^*(X_j)) \\ &\leq 2r_{\max} + \Sigma_{j=2}^{\lceil n/q \rceil}(\frac{2}{q}\Sigma_{x_i \in X_j} r_i) + \alpha\Sigma_j T^*(X_j) \\ &\leq 2\frac{n}{q}\bar{r} + 2r_{\max} + \alpha\Sigma_j T^*(X_j). \end{aligned}$$

This, together with (3.1), yields the desired result. □

Let us now define, for any capacitated routing heuristic H, the relative error by

$$e^H(X) = \frac{R^H(X) - R^*(X)}{R^*(X)}.$$

If, as before, $X^{(n)}$ denotes the first n points of a randomly generated sequence, we call H *asymptotically optimal* if $\lim_{n\to\infty} e^H(X^{(n)}) = 0$ (a.s).

Theorem 3.5. *Let $x_1, x_2, ...$ be a sequence of independent, identically distributed random points with finite expected distance Er_1 from the origin.*

(i) *For any fixed α, αTP, $I\alpha TP$ and $CRP\alpha T$ are asymptotically optimal.*

(ii) *Suppose that in addition $x_1, x_2, ...$ are uniformly bounded. Then, for any fixed α, αTP, $I\alpha TP$ and all the $RP\alpha T$ heuristics except $SRP\alpha T$ are asymptotically optimal.*

Proof. We consider each of the proposed heuristics in turn.

$I\alpha TP$. The results for this heuristic are immediate from the proof of Theorem 3.1, i.e. from the fact that $T^*(X)/2(n/q)\bar{r}$ converges almost surely to 0 for (i) and (ii).

Keeping that fact in mind, we restrict our attention through the rest of the proof to upper bound terms (in Lemma 3.4) other than $2(n/q)\bar{r}$ or multiples of $T^*(X)$.

αTP. We use the second upper bound in Lemma 3.4(i) combined with the facts that $r_{max}/n \to 0$ (a.s.) under the conditions of (i) and that r_{max} is bounded by a constant under the conditions of (ii), to obtain the results under (i) and (ii).

$CRP\alpha T$. From Lemma 3.3 P^{CRP} is $O(r_{max} + \sqrt{nr_{max}\bar{r}/q})$. As in the proof of Theorem 3.1, this and the fact that $r_{max}/n \to 0$ (a.s.) are sufficient to establish the result under (i). The one under (ii) is implied by the results for the other *RP* heuristics.

$RRP\alpha T$, $PRP\alpha T$. From Lemma 3.2(i),(ii) P^{RRP} and P^{PRP} are both $O(\sqrt{n/q}r_{max})$ which for bounded r_{max} is equivalent to $O(\sqrt{n/q})$. Since $R^*(X) = \theta(max(n/q, \sqrt{n}))$ (a.s.) the results under (ii) follow for both heuristics. □

A noteworthy aspect of the previous theorem is the poor performance of the sectorial partitioning scheme, the more so since it is very similar to the well-known commercially available *sweep heuristic* [GILLET & MILLER 1974]. Note, for example, that if the points x_i are uniformly distributed over the unit circle around the origin, then an elementary calculation shows that $\lim_{n\to\infty} R^*(X^{(n)}) = 4/3q$ (a.s.) whereas $\lim_{n\to\infty} R^{SRPOT}(X^{(n)}) = 2/q$ (a.s.).

On the positive side, the performance of the tour partitioning heuristics as well as of the circular partitioning scheme is uniformly good. These heuristics have real practical potential.

4. BIBLIOGRAPHICAL NOTES

In the recent literature only a limited number of other results in the field of analysis of heuristics for vehicle routing and scheduling problems have appeared. The results from this paper are extracted from [HAIMOVICH & RINNOOY KAN 1985], [ALTINKEMER & GAVISH 1985] and [ALTINKEMER and GAVISH 1985a]. Specifically, the results in Sections 2.1, 2.4 and 3 come from [HAIMOVICH & RINNOOY KAN 1985]. A minor error in Theorem 3 of that paper has been corrected in Theorem 2.7. The bound in Theorem 2.2 and Sections 2.2 and 2.3 come from [ALTINKEMER & GAVISH 1985] and [ALTINKEMER & GAVISH 1985a]. The argumentation that leads to Theorem 2.6 is a simplification of the proof of Lemma 4 in [ALTINKEMER & GAVISH 1985a].

In the rest of this section we will briefly review some selected references to related work.

In [MARCHETTI SPACCAMELA ET AL. 1984] a probabilistic analysis of a tour partitioning heuristic for the multiple travelling salesman problem in the Euclidean plane is presented. The objective is to minimize the longest distance to be travelled by any of the salesmen. A spatial distribution for the location of customers is specified. First, an asymptotic characterization of the optimal solution value is derived in the spirit of Theorem 3.1. Then almost sure asymptotic optimality of the heuristic is proved.

In [STOUGIE 1985, Section 3.3] similar results as above are presented for a combined location and routing problem. A number of depots are to be located and vehicles with capacity q are to be routed from these depots so as to minimize the total length of the routes. A heuristic composed of the honeycomb heuristic for the location of the depots (see [PAPADIMITRIOU 1981]) and a tour partitioning heuristic for the routing of the vehicles is shown to be asymptotically optimal almost surely under appropriate assumptions.

A remarkable result is presented in [FRIEZE 1986]. A randomized algorithm for the solution of the travelling salesman is given. Integer weights (distances) for the edges of a complete graph are drawn randomly. The mass of the distribution in zero is high enough to ensure a high expected number of edges with length 0. The algorithm then yields the optimal solution in $O(n \log n)$ steps with probability tending to 1. Stated otherwise, the probability that the algorithm fails to find the optimal solution within $O(n \log n)$ steps tends to zero when n tends to infinity.

Further results on worst case analyses are found in [HAIMOVICH & RINNOOY KAN 1985, Section 6] and [SOLOMON 1986]. In the former paper strong results on the worst case performance of the partitioning heuristics are established. The authors show how an *ϵ-approximation scheme* for the solution of the capacitated vehicle routing problem can be obtained. I.e., a sequence of heuristics is provided, depending on ϵ and q, such that for any given ϵ and q the corresponding heuristic in the sequence yields a relative error no greater than ϵ. The running time of the heuristic is polynomial in n but superpolynomial in q/ϵ, which is to be expected since a *polynomial* ϵ-approximation scheme that runs in time polynomial in q/ϵ would imply $P = NP$ [GAREY & JOHNSON 1979].

Solomon studied the worst-case performance ratios of a variety of heuristics proposed for the solution of vehicle routing and scheduling problems with time window constraints [SOLOMON 1986]. He showed that, for all heuristics considered, this ratio is growing faster than linear in the number of customers, indicating that introduction of time window constraints increases the difficulty of routing problems substantially.

Given the current state of the art in vehicle routing and scheduling, the results described in this paper represent the first steps towards the derivation of much needed theoretical statements about performance evaluations, which complement the empirical evaluations presented in the literature. Hopefully, researchers in the field, who are frequently involved in the solution of real life problems, will thereby be motivated to consider the use of mathematical analysis to arrive at rigorous performance evaluations of their solution methods.

REFERENCES

1. K. ALTINKEMER, B. GAVISH, *Heuristics for equal weight delivery problems with constant error guarantees.* Report Graduate School of Management, University of Rochester, 1985.
2. K. ALTINKEMER, B. GAVISH, *Heuristics for unequal weight delivery problems with a fixed error guarantee.* Report Graduate School of Management, University of Rochester, 1985a.
3. J. BEARDWOOD, J.L. HALTON, J.M. HAMMERSLEY, The shortest path through many points. *Proc. Cambridge Phil. Soc.* 55, 1959, pp. 299-327.
4. L. BODIN, B.L. GOLDEN, A. ASSAD, M. BALL, The state of the art in the routing and scheduling of vehicles and crews. *Computers and Oper. Res.* 10, 1983, pp. 63-212.
5. N. CHRISTOFIDES, Worst case analysis of a new heuristic for the travelling salesman problem. *Math Programming* (to appear).
6. A.M. FRIEZE, *On the exact solution of random travelling salesman problems with medium size coefficients.* Report Queen Mary College, London, 1986.
7. M.R. GAREY, D.S. JOHNSON, *Computers and Intractibility; a guide to the theory of NP-Completeness.* Freeman, San Francisco, 1979.
8. B.E. GILLET, L.R. MILLER, A heuristic algorithm for the vehicle dispatch problem. *Oper. Res.* 22, 1974, pp. 340-350.
9. M. HAIMOVICH, A.H.G. RINNOOY KAN, Bounds and heuristics for capacitated routing problems. *Math. Oper. Res.* 10, 1985, pp. 527-542.
10. R.M. KARP, Probabilistic analysis of partitioning algorithms for the travelling salesman problem in the plane. *Math. Oper. Res.* 2, 1977, pp. 204-244.
11. J.K. LENSTRA, A.H.G. RINNOOY KAN, Complexity of vehicle routing and scheduling problems. *Networks* 11, 1981, pp. 221-227.
12. A. MARCHETTI SPACCAMELA, A.H.G. RINNOOY KAN, L. STOUGIE, Hierarchical vehicle routing problems. *Networks*, 1984, pp.
13. C.H. PAPADIMITRIOU, Worst-case and probabilistic analysis of a geometric location problem. *SIAM J. Comput.* 10, 1981, pp. 542-557.
14. M.M. SOLOMON, On the worst-case performance of some heuristics for the vehicle routing and scheduling problem with time window constraints. *Networks* 16, 1986, pp. 161-174.
15. L. STOUGIE, *Design and analysis of algorithms for stochastic integer programming.* CWI Tract 37, Stichting Mathematisch Centrum, Amsterdam, 1987.

ALGORITHMIC TECHNIQUES FOR VEHICLE ROUTING

Vehicle Routing: Methods and Studies
B.L. Golden and A.A. Assad (Editors)

Vehicle Routing with Time Windows: Optimization and Approximation

M. Desrochers
Centre for Mathematics and Computer Science, Amsterdam
Ecole des Hautes Etudes Commerciales, Montréal

J.K. Lenstra, M.W.P. Savelsbergh
Centre for Mathematics and Computer Science, Amsterdam
Erasmus University, Rotterdam

F. Soumis
Ecole Polytechnique, Montréal

This is a survey of solution methods for routing problems with time window constraints. Among the problems considered are the traveling salesman problem, the vehicle routing problem, the pickup and delivery problem, and the dial-a-ride problem. We present optimization algorithms that use branch and bound, dynamic programming and set partitioning, and approximation algorithms based on construction, iterative improvement and incomplete optimization.

1. Introduction

Over the past ten years, operations researchers interested in vehicle routing and scheduling have emphasized the development of algorithms for real-life problems. The size of the problems solved has increased and practical side constraints are no longer ignored.

Most of the existing algorithms have been designed to solve pure routing problems and hence only deal with spatial aspects. They are not capable of handling all kinds of constraints that frequently occur in practice. One such constraint is the specification of time windows at customers, i.e., time intervals during which they must be served. These lead to mixed routing and scheduling problems and ask for algorithms that also take temporal aspects into account.

In this survey, we consider two types of problems. One is the *vehicle routing problem with time windows* (VRPTW), which is defined as follows. A number of vehicles is located at a single depot and must serve a number of geographically dispersed customers. Each vehicle has a given capacity. Each customer has a given demand and must be served within a specified time window. The objective is to minimize the total cost of travel.

The special case in which the vehicle capacities are infinite is called the *multiple traveling salesman problem with time windows* (*m*-TSPTW). It arises in school bus routing problems. The problem here is to determine routes that start at a single depot and cover a set of trips, each of which starts within a time window. Trips are considered as customers. There are no capacity constraints, since each trip satisfies those by definition and vehicles moving between trips are empty.

The second problem type is the *pickup and delivery problem with time windows* (PDPTW). Again, there is a single depot, a number of vehicles with given capacities, and a number of customers with given demands. Each customer must now be picked up at his origin during a specified time window, and delivered to his destination during another specified time window. The objective is to minimize total travel cost.

The special case in which all customer demands are equal is called the *dial-a-ride problem* (DARP). It arises in transportation systems for the handicapped and the elderly. In these situations, the temporal constraints imposed by the customers strongly restrict the total vehicle load at any point in time, and the capacity constraints are of secondary importance. The cost of a route is a combination of travel time and customer dissatisfaction.

We will denote the time window of an address i (whether it be a customer in the VRPTW or an origin or destination in the PDPTW) by $[e_i, l_i]$, the time of arrival at i by A_i, and the time of departure at i by D_i. It is assumed throughout this paper that the service time at i is included in the travel time t_{ij} from address i to address j. Since service must take place within the time windows, we require that $e_i \leqslant D_i \leqslant l_i$ for all i. If $A_i < e_i$, then a waiting time $W_i = e_i - A_i$ occurs before the opening of the window at i.

There are several ways to define the tightness of the time windows. One could say that the windows are tight when the underlying network with addresses as vertices contains no time-feasible cycles. This guarantees that all feasible routes are elementary paths. However, this condition is difficult to verify, and we do not get much information if it does not hold. The following two definitions may be more useful:

$$T_1 = \frac{\overline{(l_i - e_i)}}{\overline{t_{ij}}}, \quad T_2 = \frac{\overline{(l_i - e_i)}}{\max_i\{l_i\} - \min_i\{e_i\}}.$$

T_1 is the ratio between the average window width and the average travel time. If T_1 is at its minimum value 0, we have a pure scheduling problem. If T_1 is inbetween 0 and 2, we can expect that there are not many time-feasible cycles, and the temporal aspects are likely to dominate the spatial aspects. If T_1 is large, we have almost a pure routing problem. These are, of course, only rough indications.

T_2 is the ratio between the average window width and the time horizon. The value of T_2 is between 0 and 1, with 0 indicating a pure scheduling problem and 1 a problem with identical time windows.

In the following, VRP denotes the VRPTW without time windows. TSPTW is the m-TSPTW with a single salesman, and TSP is the TSPTW without time windows. Since the TSP is already NP-hard, one has to obtain solutions to the VRPTW and PDPTW by fast approximation or enumerative optimization. In

Section 2, we present mathematical programming formulations for these problems and some of their extensions. In Section 3, we survey optimization algorithms based on dynamic programming and set partitioning. In Section 4, we review various types of approximation algorithms.

There are more time-constrained routing problems and more solution approaches than we can cover in this survey. The interested reader is referred to a recent collection of papers on this topic [Golden and Assad 1986].

2. FORMULATION

In this section, the VRPTW and the PDPTW are defined and formulated as mathematical programs. We concentrate on the basic problems, with a single depot and a single vehicle type. We indicate generalizations involving multiple depots, multiple vehicle types, and constraints on the travel time of the vehicles.

2.1. *The vehicle routing problem with time windows*

Given is a graph $G = (V,A)$ with a set V of vertices and a set A of arcs. We have $V = \{0\} \cup N$, where 0 indicates the depot and $N = \{1,...,n\}$ is the set of customers, and $A = (\{0\} \times N) \cup I \cup (N \times \{0\})$, where $I \subset N \times N$ is the set of arcs connecting the customers, $\{0\} \times N$ contains the arcs from the depot to the customers, and $N \times \{0\}$ contains the arcs from the customers back to the depot. For each customer $i \in N$, there is a demand q_i and a time window $[e_i, l_i]$. For each arc $(i,j) \in A$, there is a cost c_{ij} and a travel time t_{ij}. Finally, the vehicle capacity is given by Q; we note that the number of vehicles is unbounded in the present formulation. We also note that an arc $(i,j) \in I$ may be eliminated by temporal constraints $(e_i + t_{ij} > l_j)$, by capacity constraints $(q_i + q_j > Q)$, or by other considerations. The objective is to minimize total travel cost.

The mathematical programming formulation has three types of variables: x_{ij} $((i,j) \in A)$, equal to 1 if arc (i,j) is used by a vehicle and 0 otherwise; D_i $(i \in N)$, specifying the departure time at customer i; and y_i $(i \in N)$, specifying the load of the vehicle arriving at i. The problem is now to minimize

$$\sum_{(i,j)\in A} c_{ij} x_{ij} \tag{1}$$

subject to

$$\sum_{j\in N} x_{ij} = 1 \qquad \text{for } i \in N, \tag{2}$$

$$\sum_{j\in N} x_{ij} - \sum_{j\in N} x_{ji} = 0 \qquad \text{for } i \in N, \tag{3}$$

$$x_{ij} = 1 \Rightarrow D_i + t_{ij} \leqslant D_j \qquad \text{for } (i,j) \in I, \tag{4}$$

$$e_i \leqslant D_i \leqslant l_i \qquad \text{for } i \in N, \tag{5}$$

$$x_{ij} = 1 \Rightarrow y_i + q_i \leqslant y_j \qquad \text{for } (i,j) \in I, \tag{6}$$

$$0 \leqslant y_i \leqslant Q \qquad \text{for } i \in N, \tag{7}$$

$$x_{ij} \in \{0,1\} \qquad \text{for } (i,j) \in A. \tag{8}$$

The objective function (1) represents total travel cost; it is possible to include the fixed charge of using a vehicle by adding it to all c_{0j}. Minimizing (1) subject to (2),

(3) and (8) is a minimum cost flow problem, which has an integral solution. Constraints (4) and (5) ensure feasibility of the schedule, and constraints (6) and (7) guarantee feasibility of the loads.

This VRPTW formulation is more compact than the VRP formulation due to Bodin and Golden [1981]. The latter formulation has $O(n^3)$ variables and an exponential number of subtour elimination constraints. The above formulation has $O(n^2)$ variables, while the subtours are eliminated by (4), as well as by (6). These constraints can be rewritten as follows, where M is a large constant:

$$D_i + t_{ij} - D_j \leqslant (1-x_{ij})M \qquad \text{for } (i,j) \in I, \tag{4a}$$

$$y_i + q_i - y_j \leqslant (1-x_{ij})M \qquad \text{for } (i,j) \in I. \tag{6a}$$

In their TSP formulation, Miller, Tucker and Zemlin [1960] propose the following subtour elimination constraints:

$$D_i - D_j + nx_{ij} \leqslant n - 1 \qquad \text{for } (i,j) \in I.$$

These appear as a special case of (4a) when all $t_{ij} = 1$ and $M = n$, and as a special case of (6a) when all $q_i = 1$ and $M = n$.

The above single-depot formulation is based on a single-commodity flow. There is no explicit flow conservation constraint for the depot, as this is implied by the flow conservation constraints (3) for the customers. Let us now consider the multi-depot case. The single depot 0 is replaced by a set M of depots. In the graph $G=(V,A)$, we now have $V = M \cup N$ and $A = (M \times N) \cup I \cup (N \times M)$, where N and I are as before. There are two variants. In case each vehicle must return to its home depot, we need a multi-commodity flow formulation with a separate commodity for each depot. Each variable x_{ij} is replaced by variables x_{ij}^k $(k \in M)$, where $x_{ij}^k = 1$ if arc (i,j) is used by a vehicle from depot k, and 0 otherwise. In case vehicles do not have to return to their points of origin, all we have to do is to add a flow conservation constraint for each depot.

The case of multiple vehicle types is modeled with fictitious depots. For each type of vehicle at a given depot, we create a fictitious depot with a separate commodity to ensure that the number of vehicles of each type at each depot is balanced. The case that the vehicles have upper bounds on their total travel time is handled by the specification of a time window for the depot. The case that the vehicles have different periods of availability is obviously dealt with by the introduction of fictitious depots with time windows.

2.2. *The pickup and delivery problem with time windows*

As in the previous section, there is a set N of customers. In the current situation, however, each customer $i \in N$ requests the transportation from an origin i^+ to a destination i^-. We write $N^+ = \{i^+ \mid i \in N\}$ for the set of origins and $N^- = \{i^- \mid i \in N\}$ for the set of destinations. The graph $G=(V,A)$ is now defined as follows. The vertex set is given by $V = \{0\} \cup N^+ \cup N^-$, where 0 denotes the depot. The arc set is given by $A = (\{0\} \times N^+) \cup I \cup (N^- \times \{0\})$, where $I \subset (N^+ \cup N^-) \times (N^+ \cup N^-)$ is the set of arcs corresponding to feasible trips between origins and destinations. For each customer $i \in N$, there are a demand q_i and two time windows $[e_{i^+}, l_{i^+}]$ and $[e_{i^-}, l_{i^-}]$. For each arc $(i,j) \in A$, there is a cost c_{ij} and a travel time t_{ij}. Finally, there is a set M of vehicles, each with capacity Q. The objective is to minimize total travel cost.

The mathematical programming formulation has the same three types of variables as in the case of the VRPTW: x_{ij}^k ($(i,j) \in A$, $k \in M$), equal to 1 if arc (i,j) is used by vehicle k and 0 otherwise; D_i ($i \in N^+ \cup N^-$), specifying the departure time at vertex i; and y_i ($i \in N^+ \cup N^-$), specifying the load of the vehicle arriving at i. We note that the flow variables have now a third index in order to ensure that the pickup at i^+ and delivery to i^- are done by the same vehicle. The problem is to minimize

$$\sum_{(i,j) \in A, k \in M} c_{ij} x_{ij}^k \tag{9}$$

subject to

$$\sum_{k \in M} \sum_{j \in V} x_{ij}^k = 1 \qquad \text{for } i \in N^+, \tag{10}$$

$$\sum_{j \in V} x_{ij}^k - \sum_{j \in V} x_{ji}^k = 0 \qquad \text{for } i \in N^+ \cup N^-, \ k \in M, \tag{11}$$

$$\sum_{j \in V} x_{i^+ j}^k - \sum_{j \in V} x_{j i^-}^k = 0 \qquad \text{for } i \in N, \ k \in M, \tag{12}$$

$$D_{i^+} + t_{i^+ i^-} \leqslant D_{i^-} \qquad \text{for } i \in N, \tag{13}$$

$$x_{ij}^k = 1 \Rightarrow D_i + t_{ij} \leqslant D_j \qquad \text{for } (i,j) \in I, \ k \in M, \tag{14}$$

$$e_i \leqslant D_i \leqslant l_i \qquad \text{for } i \in N^+ \cup N^-, \tag{15}$$

$$x_{ij}^k = 1 \Rightarrow y_i + q_i \leqslant y_j \qquad \text{for } (i,j) \in I, \ k \in M, \tag{16}$$

$$0 \leqslant y_i \leqslant Q \qquad \text{for } i \in N^+, \tag{17}$$

$$x_{ij}^k \in \{0,1\} \qquad \text{for } (i,j) \in A, \ k \in M. \tag{18}$$

Minimizing (9) subject to (10), (11) and (18) is a multi-commodity minimum cost flow problem of a more complex structure than in the case of the VRPTW. Constraints (12) ensure that each i^+ and i^- are visited by the same vehicle. Constraints (13) represent the precedence relation between pickup and delivery points. Constraints (14) and (15) ensure feasibility of the schedule, and constraints (16) and (17) guarantee feasibility of the loads; we note that capacity constraints are only specified for origins because a vehicle reaches its maximum load after a pickup. We also note that all model extensions presented for the VRPTW can be applied to the PDPTW.

3. Optimization

Optimization algorithms for routing problems with time windows employ the two standard principles of implicit enumeration: dynamic programming and branch and bound. Among the branch and bound methods, two approaches stand out. One is the set partitioning approach, which uses column generation to solve a continuous relaxation of the problem and branch and bound to obtain integrality. The other approach uses state space relaxation to compute lower bounds.

Dynamic programming is mainly applied to solve single-vehicle problems. Those problems arise in the context of column generation and state space relaxation, so that dynamic programming algorithms appear as subroutines in branch and bound methods.

In Section 3.1, we collect the applications of dynamic programming, including state space relaxation. In Section 3.2, we discuss the set partitioning approach. A variety of other branch and bound methods is reviewed below.

Baker [1983] presents a branch and bound method for the TSPTW, in which bounds are derived from longest path problems. He solves small problems with this method.

The most widely studied routing problem with time windows is the school bus routing problem [Orloff 1976], which is essentially an m-TSPTW. Two objectives are distinguished: minimizing fleet size and minimizing a weighted combination of fleet size and total travel time.

As to the first objective, Swersey and Ballard [1984] discretize the time windows and solve the linear programming relaxation of the resulting integer programming problem. For most instances, the solution is integral; otherwise, they are able to modify the solution so as to obtain integrality without increasing the fleet size. Desrosiers, Sauvé and Soumis [1985] study the Lagrangean relaxation which is obtained by relaxing constraints (2). As one visit to each customer is no longer required, the Lagrangean problem is a shortest path problem with time windows. Although the lower bound is often equal to the optimal fleet size, this dual method does not necessarily produce a feasible solution, in which case branch and bound has to be applied.

For the m-TSPTW with the second objective function, Desrosiers, Soumis, Desrochers and Sauvé [1985] study the network relaxation which is obtained by removing the scheduling constraints (4) and (5). If $e_i = l_i$ for all $i \in N$, then this relaxation produces an optimal solution in view of the definition of I. The quality of the bounds deteriorates with an increasing number of customers and an increasing width of the time windows. Two branching rules are proposed: branching on the flow variables and branching by splitting time windows. In the case of very tight time windows, Soumis, Desrosiers and Desrochers [1985] apply the first rule to solve problems with up to 150 customers; as the time windows become wider, the tree grows rapidly in size. The second branching rule can handle wider time windows, but it is concluded that the network relaxation is inferior to the set partitioning relaxation considered in Section 3.2.

Sörensen [1986] suggests the use of Lagrangean decomposition [Guignard 1984; Jörnsten, Nasberg and Smeds 1985] for the VRPTW. The two resulting subproblems are the shortest path problem with time windows and the generalized assignment problem. No computational results have been reported.

3.1. *Dynamic programming*

Dynamic programming is a traditional solution method for constrained shortest path problems. The constituents of a dynamic programming algorithm are states, transitions between states, and recurrence equations that determine the value of the objective function at each state. Let us consider the standard shortest path problem on a graph $G=(V,A)$ with vertex set V, arc set A, a source $0 \in V$, and a travel time t_{ij} for each $(i,j) \in A$. Each vertex represents a state, each arc represents a transition between two states, and the value $d(j)$ associated with state j is the shortest path duration from the source 0 to vertex j. The recurrence equations to compute these values are:

$$d(0)=0,$$

$$d(j)=\min_{(i,j)\in A}\{d(i)+t_{ij}\} \text{ for } j \in V \setminus \{0\}.$$

This algorithm has a running time that is polynomially bounded in the size of G.

Constraints are treated by expansion of the state space and modification of the recurrence equations. Such a dynamic programming approach can be useful for several NP-hard routing problems. However, the cardinality of the state space is usually exponential in the problem size. The practical use of dynamic programming in this context is restriced to state spaces of at most pseudopolynomial size and relatively small problem instances.

3.1.1. *Single-vehicle problems with time windows*

We will consider four problems in this section: the traveling salesman problem with time windows, the single-vehicle dial-a-ride problem, and two constrained shortest path problems.

The TSPTW can be viewed as the problem of finding a shortest path from an origin 0 to a destination $n+1$ that visits all vertices in the set N and respects the time window of each vertex. Christofides, Mingozzi and Toth [1981c] propose the following dynamic programming algorithm. There are states of the form (S,j) with $S \subset N$ and $j \in S$, and $d(S,j)$ denotes the shortest duration of a feasible path starting at 0, visiting all vertices in S, and finishing at j. The optimal solution value $d(N \cup \{n+1\}, n+1)$ is determined by the following recurrence equations:

$$d(\{0\},0)=e_0,$$

$$d(S,j)=\min_{i\in S-\{j\},(i,j)\in A}\{d(S-\{j\},i)+t_{ij}\} \text{ for } j \in N \cup \{n+1\},$$

where we redefine $d(S,j)=e_j$ in case $d(S,j)<e_j$ and $d(S,j)=\infty$ in case $d(S,j)>l_j$.

Psaraftis [1983a] uses dynamic programming to solve the single-vehicle DARP. The states are of the form $(j,k_1,\ldots,k_n)$, where j is the vertex presently visited and each k_i can assume three values that denote the status of customer i: not yet picked up, picked up but not yet delivered, and delivered. It is now straightforward to define the feasible transitions between states. The algorithm has $2n$ stages, each of which extends the paths constructed so far with one arc. The total time requirement is $O(n^2 3^n)$. Psaraftis estimates that this approach is able to solve problems with up to ten customers.

Desrosiers, Dumas and Soumis [1986b] give a similar $2n$-stage algorithm for the capacitated single-vehicle PDPTW. They propose a number of state elimination rules to reduce the computational effort. In addition to Psaraftis' feasibility tests which eliminate states on the basis of information about customers picked up so far, they also have feasibility tests which use information about customers not yet delivered. The algorithm can solve real-life problems with up to 40 customers.

Two types of constrained shortest path problems have been considered: the shortest path problem with time windows (SPPTW) and the capacitated shortest path problem with pickups, deliveries and time windows (SPPPDTW). The main difference between these problems and the single-vehicle DARP is that the path is no longer required to visit all customers. For the SPPTW, which is defined by (1), (3)-(5) and (8), Desrosiers, Pelletier and Soumis [1984] propose a label correcting method. Desrochers and Soumis [1985a, 1985b] give two pseudopolynomial algorithms. One is a label setting method, the other a primal-dual method. Desrochers [1986] generalizes the latter algorithm to the case of multidimensional time windows. For the SPPPDTW ((9) and (11)-(18)), Dumas [1985] and Dumas and Desrosiers [1986] present an algorithm which is similar to the one for the capacitated single-vehicle PDPTW.

As we have mentioned before, dynamic programming algorithms are mostly used as subroutines in other solution methods. This is because the problems considered in this section occur as subproblems in multi-vehicle problems. The TSPTW and the single-vehicle DARP arise in the second phase of cluster-first route-second approaches, where the first phase allocates customers to vehicles and the second phase asks for single-vehicle routes. The SPPTW occurs as a subproblem in the set partitioning algorithm for the m-TSPTW due to Desrosiers, Soumis and Desrochers [1984], in the Lagrangean relaxation algorithm for the fleet size problem due to Desrosiers, Sauvé and Soumis [1985], and in the Lagrangean decomposition algorithm for the VRPTW due to Sörensen [1986]. The SPPPDTW is a subproblem in the set partitioning algorithm for the PDPTW due to Desrosiers, Dumas and Soumis [1987].

3.1.2. *State space relaxation*

For a number of problems, Christofides, Mingozzi and Toth [1981a, 1981b, 1981c] have developed branch and bound algorithms that obtain lower bounds by dynamic programming on a relaxed state space. They take a dynamic programming algorithm for the problem under consideration as a starting point and replace its state space by a smaller space in such a way that the recursion over the new state space requires only polynomial time and yields a lower bound on the optimal solution value of the original problem.

State space relaxation is based on a mapping g from the original state space to a space of smaller cardinality. If there is a transition from S_1 to S_2 in the original state space, then there must be a transition from $g(S_1)$ to $g(S_2)$ in the new state space. We illustrate this idea on the TSPTW [Christofides, Mingozzi and Toth 1981c].

With each vertex i, an arbitrary integer β_i is associated, with $\beta_0 = \beta_{n+1} = 0$.

The mapping is defined by $g(S,j)=(k,\beta,j)$, where $k=|S|$ and $\beta=\sum_{i\in S}\beta_i$. The new recurrence equations are:

$$d(0,\beta,0)=\begin{cases}0 & \text{if } \beta=0,\\ \infty & \text{if } \beta\neq 0,\end{cases}$$

$$d(k,\beta,j)=\min_{i\neq j,(i,j)\in A}\{d(k-1,\beta-\beta_j,i)+t_{ij}\} \text{ for } j\in N\cup\{n+1\},$$

where we redefine $d(k,\beta,j)=e_j$ in case $d(k,\beta,j)<e_j$ and $d(k,\beta,j)=\infty$ in case $d(k,\beta,j)>l_j$. The lower bound is now given by

$$\min_{j\in N,(j,n+1)\in A}\{d(n,\textstyle\sum_{i\in N}\beta_i,j)+t_{j,n+1}\}.$$

This lower bound can be improved by the use of vertex penalties and state space modifications. Vertex penalties serve to decrease the travel times of arcs incident to undercovered vertices and to increase the travel times of arcs incident to overcovered vertices; these penalties are adjusted by subgradient optimization. Similarly, the weights β_i can be modified by subgradient optimization. The resulting branch and bound method is able to solve problems with up to 50 vertices.

Kolen, Rinnooy Kan and Trienekens [1987] extend this approach to the VRPTW. They use a two-level state space relaxation. At the first level, a lower bound on the costs of a time-constrained path from the depot to vertex j with load q is computed. This is done with an adaptation of the above method for the TSPTW. The states are of the form (t,q,j), where q is the load of a shortest path arriving at vertex j no later than time t. We have $0\leqslant t\leqslant T$ where T is the scheduling horizon, $0\leqslant q\leqslant Q$ where Q is the vehicle capacity, and $j\in N$. At the second level, a lower bound on the costs of m routes with total load $\Sigma_{i\in N}q_i$ and different destination vertices is computed. The states are now of the form (k,q,j), where q is the total load of the first k routes and j is the destination vertex of route k. Vertex penalties are used to improve the lower bounds. Problems with up to fifteen customers are solved.

3.2. *Set partitioning*

Vehicle routing problems and in particular the VRPTW and the PDPTW can be reformulated as set partitioning problems, with variables (columns) corresponding to feasible routes.

Let R be the set of feasible routes of the problem under consideration. For each route $r\in R$, we define γ_r as the sum of the costs of its arcs and δ_{ri} $(i\in N)$ as a binary constant, equal to 1 if route r visits customer i and 0 otherwise. If x_r $(r\in R)$ is equal to 1 if route r is used and 0 otherwise, the set partitioning problem is to minimize

$$\sum_{r\in R}\gamma_r x_r \tag{19}$$

subject to

$$\sum_{r \in R} \delta_{ri} x_r = 1 \qquad \text{for } i \in N, \tag{20}$$

$$x_r \in \{0,1\} \qquad \text{for } r \in R. \tag{21}$$

Although problems (9)-(18) and (19)-(21) are equivalent, their continuous relaxations are not. This is because the variables in the latter problem are restricted to feasible paths in which each customer is included or not. Any solution to the relaxed version of (19)-(21) is a feasible solution to the relaxation of (9)-(18), but not vice versa. We can therefore expect to obtain better lower bounds on the basis of the set partitioning formulation.

Because of the cardinality of R, the relaxed set partitioning problem cannot be solved directly and column generation is used. That is, a new column of minimum marginal cost is generated by solving an appropriate subproblem. If its marginal cost is negative, then it is added to the linear program, the problem is reoptimized and column generation is applied again; otherwise, the current solution to the linear program is optimal. Before discussing results for specific vehicle routing problems, we first describe some general aspects of this approach.

3.2.1. *The subproblem*

The objective function of the subproblem has coefficients that depend on the values of the dual variables π_i ($i \in N$) of the continuous relaxation of the set partitioning problem. The constraints define a path subject to side constraints but not necessarily visiting all customers. They include (3)-(8) for the VRPTW, (3)-(5) and (8) for the m-TSPTW, and (11)-(18) for the PDPTW.

As we have seen in Section 3.1.1, dynamic programming is a suitable method to solve these subproblems to optimality, because the state spaces are relatively small.

3.2.2. *The master problem*

The continuous relaxation of the set partitioning problem is solved by the simplex algorithm. This method produces the dual values π_i that are needed for column generation and enables easy reoptimization each time new columns are generated.

To obtain an integral solution to the master problem, we add cutting planes or we use branch and bound. Each time a new constraint is added, another round of column generation is applied in order to solve the modified master problem. We must restrict ourselves to types of constraints that are compatible with the column generation method. For any cutting plane, the method must be able to compute its coefficients in order to evaluate the marginal cost of new columns. For any branching rule, the method must be able to exclude the columns that have become infeasible by branching

In case the c_{ij} are integral, a compatible type of cut is the one that rounds the objective up to the next integer. In the particular case that we minimize fleet size, this cut has the same coefficient 1 in each column; if it has a dual value π, a new column is generated by minimizing the reduced cost

$$\sum_{(i,j)\in A}(c_{ij}-\pi_i)x_{ij}-\pi.$$

However, we cannot use Gomory cuts or other types of cuts whose coefficients are not known before the new column is generated.

As to branching, the usual rule to fix a fractional variable x_r to 0 or 1 is not compatible. We can fix $x_r = 1$ by simply deleting the customers on route r from the subproblem. But we cannot fix $x_r = 0$: there is no way to prevent route r from being generated again. Four types of compatible branching rules have been proposed: branching on the flow variables of route r; branching on the position of a customer in route r; branching by splitting time windows; and branching on the number of vehicles of a given type in problems with multiple vehicle types. These rules have been listed here in order of increasing effectiveness.

3.2.3. *Acceleration techniques*

There are various ways to improve the performance of the set partitioning approach.

First of all, the set partitioning problems that arise in the context of vehicle routing are highly degenerate. It is an obvious idea to improve the convergence of the simplex method by a perturbation strategy.

Secondly, the solution of the relaxed master problem can be accelerated by the simultaneous generation of columns. The solution of a subproblem by dynamic programming produces not only a column of minimum reduced cost, but also many other columns of negative reduced cost. Several of these can be added.

In the third place, the solution of most of the subproblems can be greatly sped up by the heuristic elimination of vertices, arcs, and states. The first columns are generated in subnetworks, which only consist of customers with large dual values and inexpensive arcs; in addition, less promising states are ruled out during the recursion. At later stages, the elimination rules are gradually relaxed, until at the final stage the full network and state space are used in order to prove optimality.

3.2.4. *The multi-salesman and vehicle routing problem with time windows*

Desrosiers, Soumis and Desrochers [1984] propose a set partitioning approach to the m-TSPTW. The column generation problem is the SPPTW, which was reviewed in Section 3.1.1. In their algorithm, two cuts are added to the master problem: one to round up the number of vehicles and one to round up the total costs. After that, branching on flow variables is applied. With this rule, it is time consuming to achieve optimality, even if the integrality gap is small. They solve problems with up to 151 customers; the solution time on a CDC Cyber 173 ranges from 100 to 1000 seconds, depending on the width of the time windows. A recent improvement of the algorithm is able to solve problems with 223 customers within 600 seconds. A branching rule based on time window splitting is under development.

Desrosiers, Dumas and Soumis [1986a] extend this algorithm to the case of multiple vehicle types. Several SPPTW's are now to be solved, one for each type of vehicle. Branching is first done on the number of vehicles of a given type; when this number is integral for each type, the usual branching on flow variables is applied.

No set partitioning algorithm for the VRPTW has been proposed so far. However, Desrochers [1986] presents a dynamic programming algorithm for the shortest path problem with a variety of constraints. This method is suitable for solving the subproblems that occur in this context.

3.2.5. *The pickup and delivery problem with time windows*

Dumas [1985] develops a set partitioning approach for the PDPTW. He solves problems with 30 customers (60 vertices) within 100 seconds on a CDC Cyber 173. These problems have tight capacity constraints ($q_i \geqslant Q/3$) and loose time window constraints. Narrowing the time windows significantly decreases the cardinality of the state space and thereby the computation time.

The subproblem in this case is the SPPPDTW, which was reviewed in Section 3.1.1. The algorithm of Dumas [1985] first branches on the number of vehicles per type and then on flow variables. Desrosiers, Dumas and Soumis [1987] replace the latter branching rule by branching on the position of customers in routes and obtain improved results.

4. Approximation

In spite of the recent success of optimization algorithms for vehicle routing with time windows, it is unlikely that they will be able to solve large-scale problems. In many situations one has to settle for algorithms that run fast but may produce suboptimal solutions. In this section, we review three types of approximation algorithms. *Construction* methods try to build a feasible solution starting from the raw data. *Iterative improvement* methods start from a feasible solution and seek to improve it through a sequence of local modifications. *Incomplete optimization* methods use a combination of enumeration of the solution space and heuristic rules to truncate the search. These types of methods have been widely applied to unconstrained routing problems. Their extension to constrained problems has only recently become a subject of investigation. In presenting this work, we will concentrate on feasibility rather than optimality aspects.

As in Section 3.1, we split the depot (vertex 0) in an 'origin' (vertex 0) and a 'destination' (vertex $n+1$). In the sequel, when we refer to a route, we assume that it is given by $(0, 1, ..., i, ..., n, n+1)$, where i is the ith customer visited by the vehicle. There are two quantities associated with a subpath $(h, ..., k)$ that play a dominant role in the algorithms below. The *possible forward shift* S_{hk}^{+} is the largest increase in the departure time D_h at h which causes no violation of the time windows along the path $(h, ..., k)$:

$$S_{hk}^{+} = \min_{h \leqslant j \leqslant k} \{l_j - (D_h + \textstyle\sum_{h \leqslant i < j} t_{i,i+1})\}.$$

The *possible backward shift* S_{hk}^{-} is the largest decrease in the departure time D_h at h which causes no waiting time along the path $(h, ..., k)$:

$$S_{hk}^{-} = \min_{h \leqslant j \leqslant k} \{D_j - e_j\}.$$

These quantities express the flexibility we have when we want to push customers forward or backward in time. It is not hard to see that all values S_{jk}^{+} and S_{jk}^{-} for $j = h,...,k$ can be computed in $O(n)$ time [Savelsbergh 1985].

4.1. *Construction*

In the design of construction methods, there are two key questions:

(1) Selection criterion: *which* customer is selected next to be inserted into the current solution?

(2) Insertion criterion: *where* will it be inserted?

While such decisions may be made at the same time, several of the algorithms in this section employ different criteria for selection and insertion.

4.1.1. *The vehicle routing problem with time windows*

Solomon [1983] was one of the first who attempted to adapt the existing approximation algorithms for the VRP to the VRPTW. Much of the material in this section is based on his work.

Savings. The savings method of Clarke and Wright [1964] is probably the first and certainly the best known heuristic proposed to solve the VRP. It is a sequential procedure. Initially, each customer has its own route. At each iteration, an arc is selected so as to combine two routes into one, on the basis of some measure of cost savings and subject to vehicle capacity constraints. Note that in this case the selection criterion applies to arcs rather than customers and that the insertion question does not occur.

In order to adapt this procedure for the VRPTW, we must be able to test the time feasibility of an arc. While in pure routing problems the direction in which a route is traversed is usually immaterial, this is not the case anymore in the presence of time windows. Hence, we only consider arcs from the last customer on one route to the first customer on another.

If two routes are combined, the departure times on the first route do not change. As to the second route, one necessary condition for feasibility is that the departure time at the first customer is no more than his latest service time, but that is not all. The other departure times on the route could be pushed forward, and one of them could become infeasible. This is where the possible forward shift enters the picture. For any path $(1,...,n+1)$, a change in the departure time at 1 is feasible if and only if it is no more than $S_{1,n+1}^{+}$.

By selecting of a cost effective and time feasible arc, the modified heuristic could link two customers whose windows are far apart in time. This suggests a further modification which selects arcs on the basis of both spatial and temporal closeness of customers, e.g., by adding a waiting time penalty to the cost savings.

Nearest neighbor. Initially, a route consists of the depot only. At each iteration, an unvisited customer who is closest to the current end point of the route is selected and added to the route to become its new end point. The selection is restricted to those customers whose addition is feasible with respect to capacity and time window constraints. A new route is started any time the search fails, unless there are no more customers to schedule.

The measure of closeness should include spatial as well as temporal aspects. Solomon [1983] proposes the following:

$$\alpha_1 t_{ij} + \alpha_2(\max\{e_j, D_i + t_{ij}\} - D_i) + \alpha_3(l_j - (D_i + t_{ij})), \text{ with } \alpha_1 + \alpha_2 + \alpha_3 = 1.$$

This measures the travel time between customers i and j, the difference between their respective delivery times, and the urgency of a delivery to j.

Insertion. Insertion methods treat the selection and insertion decisions seperately. We distinguish sequential and parallel insertion rules. The former construct the routes one by one, whereas the latter build them up simultaneously. All methods considered here are of the sequential type.

The general scheme of an insertion method is simple. Let U be the set of unrouted customers. For each customer $u \in U$, we first determine the best feasible point i_u after which it could be inserted into the emerging route:

$$\iota(u, i_u) = \min_{0 \leqslant i \leqslant n}\{\iota(u,i)\} \text{ for } u \in U.$$

We next select the customer u^* to be inserted into the route:

$$\sigma(u^*, i_{u^*}) = \min_{u \in U}\{\sigma(u, i_u)\}.$$

The insertion criterion ι and the selection criterion σ are still to be specified; we refer to Solomon [1983] and Savelsbergh [1985] for a number of possible definitions which take both spatial and temporal aspects into account. When no more customers can be inserted, a new route is started, unless all customers have been routed.

Insertion of u between i and $i+1$ could change all departure times on the path $(i+1,...,n+1)$. Again, the insertion is feasible if and only if the change in departure time at $i+1$ is no more than $S^+_{i+1,n+1}$.

Solomon [1983] concludes on the basis of extensive computational experiments that insertion methods outperform other types of construction methods.

4.1.2. *The pickup and delivery problem with time windows*

Jaw, Odoni, Psaraftis and Wilson [1986] consider a variant of the DARP. Their approach seems to be applicable to the proper DARP as well.

The customers that are to be picked up and delivered have the following types of service constraints. Each customer i specifies either a desired pickup time $\bar{D}_{i^+}$ or a desired delivery time $\bar{A}_{i^-}$, and a maximum travel time $\bar{T}_i$; in addition, there is a tolerance $\bar{U}$. If customer i has specified a desired pickup time, the actual pickup time D_{i^+} should fall within the time window $[\bar{D}_{i^+}, \bar{D}_{i^+} + \bar{U}]$; if he has specified a desired delivery time, the actual delivery time A_{i^-} should fall within the window $[\bar{A}_{i^-} - \bar{U}, \bar{A}_{i^-}]$. Moreover, his actual travel time should not exceed his maximum travel time: $A_{i^-} - D_{i^+} \leqslant \bar{T}_i$. Note that this information suffices to determine two time windows $[e_{i^+}, l_{i^+}]$ and $[e_{i^-}, l_{i^-}]$ for each customer i. Finally, waiting time is not allowed when the vehicle is carrying passengers.

The selection criterion is simple: customers are selected for insertion in order of increasing e_{i^+}. The insertion criterion is as follows: among all feasible points of insertion of the customer into the vehicle schedules, choose the cheapest; if no feasible point exists, introduce an additional vehicle.

For the identification of feasible insertions, the notion of an *active period* is introduced. This is a period of time a vehicle is active between two successive periods of slack time. For convenience, we drop the superscript indicating pickup or delivery. For each visit to an address i during an active period, we define the following variants of possible backward and forward shifts:

$$\Sigma_i^- = \min\{\min_{j \leqslant i}\{A_i - e_i\}, \Lambda\},$$

$$\Sigma_i^+ = \min_{j \leqslant i}\{l_i - A_i\},$$

$$S_i^- = \min_{j \geqslant i}\{A_i - e_i\},$$

$$S_i^+ = \min\{\min_{j \geqslant i}\{l_i - A_i\}, L\},$$

where Λ and L are the durations of the slack periods immediately preceding and following the active period in question. Σ_i^- (Σ_i^+) denotes the maximum amount of time by which every stop preceding but not including i can be advanced (delayed) without violating the time windows, and S_i^- (S_i^+) denotes the maximum amount of time by which every stop following but not including i can be advanced (delayed). These quantities thus indicate how much each segment of an active period can be displaced to accommodate an additional customer. Once it is established that some way of inserting the pickup and delivery of customer i satisfies the time window constraints, it must be ascertained that it satisfies the maximum travel time constraints.

The cost measure that is used to choose among feasible insertions is a weighted combination of customer dissatisfaction and resource usage.

Sexton and Bodin [1985a, 1985b] consider a variant of the single-vehicle DARP in which only deadlines for the deliveries are specified. Their solution approach is to apply Benders decomposition to a mixed 0-1 nonlinear programming formulation, which separates the routing and scheduling component.

4.2. *Iterative improvement*

Croes [1958] and Lin [1965] introduced the notion of k-exchanges to improve solutions to the TSP. Lin and Kernighan [1973] generalized the approach, and many authors reported on its application to related problems. In the context of vehicle routing, Christofides and Eilon [1969] and Russell [1977] adapted the approach to the basic VRP, and Psaraftis [1983b] used it for the DARP.

In this section, we will show how time windows can be handled in k-exchange procedures for the TSPTW [Savelsbergh 1985]. It is not hard, however, to extend the techniques to multi-vehicle problems with various types of side constraints [Savelsbergh 1987, 1988]. The issue is also addressed in another contribution to this volume [Solomon, Baker and Schaffer 1988].

A k-exchange is a substitution of k arcs of a route with k other arcs. In the TSP, the processing of a single k-exchange takes constant time for any fixed value of k. One only has to test whether the exchange is profitable and does not have to bother about feasibility. In the case of the TSPTW, the processing of a k-exchange may take $O(n)$ time. This is because a modification at one point may affect the departure times on the entire route, so that feasibility questions arise. It will be indicated below that, even in the presence of time windows, constant time suffices for the processing of a single k-exchange.

4.2.1. *The traveling salesman problem with time windows*

The number of possible k-exchanges in a given route is $O(n^k)$. The computational requirement of k-exchange procedures thus increases rapidly with k, and one usually only considers the cases $k = 2$ and $k = 3$. A 2-exchange replaces two arcs $(i,i+1)$ and $(j,j+1)$ by (i,j) and $(i+1,j+1)$, thereby reversing the path $(i+1,...,j)$. In a 3-exchange, three arcs are deleted and there are seven possibilities to construct a new route from the remaining segments. Or [1976] proposes to restrict attention to those 3-exchanges in which a string of one, two or three consecutive customers is relocated between two others. Note that no paths are reversed in this case and that there are only $O(n^2)$ exchanges of this kind. We will illustrate the combination of the k-exchange concept and time windows on those Or-exchanges in which one customer is relocated.

The basic idea of the approach is the use of a search strategy and of a number of global variables such that, for each considered exchange, testing its feasibility and updating the global variables require no more than constant time.

The search strategy is as follows. Suppose that customer i is relocated between j and $j+1$; this means that the arcs $(i-1,i)$, $(i,i+1)$ and $(j,j+1)$ are substituted by $(i-1,i+1)$, (j,i) and $(i,j+1)$. The cases of backward relocation ($j<i$) and forward relocation ($j>i$) are handled separately. In the former case, j is successively chosen to be equal to $i-1$, $i-2$, ..., 0; note that in each exchange the path $(j+1,...,i-1)$ of the previous exchange is expanded with the arc $(j,j+1)$. In the latter case, j assumes the values $i+1$, $i+2$, ..., n in that order; in each exchange the path $(i+1,...,j-1)$ of the previous exchange is expanded with the arc $(j-1,j)$.

The global variables we need are:

(1) the possible forward shift S^+, which is equal to $S^+_{j+1,i-1}$ as defined above;
(2) the possible backward shift S^-, which is equal to $S^-_{i+1,j-1}$ as defined above;
(3) the gain G made by going directly from $i-1$ to $i+1$:

$$G = A_{i+1} - (D_{i-1} + t_{i-1,i+1});$$

(4) the loss L incurred by going from j through i to $j+1$:

$$L = \max\{D_j + t_{ji}, e_i\} + t_{i,j+1} - A_{j+1};$$

(5) the waiting time W on the path $(j+1,...,i-1)$:

$$W = \textstyle\sum_{j+1 \leqslant k \leqslant i-1} W_k.$$

During the backward search, an exchange is feasible if $D_k^{\text{new}} \leqslant l_k$ for $k = j+1, \ldots, i-1$, and potentially profitable if $D_{i+1}^{\text{new}} < D_{i+1}$. The superscript 'new' indicates the value if the exchange were carried out. Note that a decrease in the departure time at $i+1$ does not guarantee an earlier arrival at the depot, but 'potential profitability' is still a suitable criterion for accepting an exchange. In terms of global variables, feasibility and potential profitability are equivalent to

$$L < \min\{S^+, G + W\}.$$

The global variables are updated by

$$S^+ := W_{j+1} + \min\{l_{j+1} - D_{j+1}, S^+\};$$

$$W := W + W_{j+1}.$$

During the forward search, an exchange feasible if $D_i^{\text{new}} \leqslant l_i$ and potentially profitable if $D_{j+1}^{\text{new}} < D_{j+1}$. This equivalent to

$$L < \min\{S^-, G\}.$$

The only update is

$$S^- := \min\{D_j - e_j, S^-\}.$$

It follows that a single exchange of this type can be handled in constant time. The adaptation of other types of exchange procedures to time window constraints is conceptually similar but technically more complicated.

4.3. *Incomplete optimization*

Fast approximation algorithms can also be derived from the optimization algorithms presented in Section 3.2. The two principal ideas are the heuristic generation of columns and the partial exploration of the branch and bound tree.

Heuristic generation of columns is based on the third type of acceleration technique mentioned in Section 3.2.3. While solving the relaxed master problem, we eliminate vertices, arcs and states in a heuristic fashion. The elimination rules are not relaxed, so that an approximate solution to the linear program is obtained.

Partial exploration of the search tree can take place in several ways. One is to obtain an integral solution by depth-first search and then to explore the tree for the remaining available time. Another way is to use an invalid branching rule, i.e., to eliminate branches on heuristic grounds.

A combination of these ideas has been used to obtain feasible integral solutions within two percent from the optimum with highly reduced running times.

5. Conclusion

If one conclusion emerges from the preceding survey, it is that algorithm designers have turned their attention to the development of efficient methods that are capable of solving large-scale routing problems subject to real-life constraints.

A striking example is the set partitioning approach, which appears to be particularly efficient for strongly constrained problems. The continuous relaxation of the set partitioning formulation can be solved by the use of a column generation scheme and provides for better bounds than the relaxation of other formulations. Dynamic programming turns out to be a powerful tool to generate columns. This family of algorithms is well designed to produce approximate solutions to problems of a realistic size. Optimization algorithms of this type are being used for school bus scheduling.

The construction and iterative improvement algorithms that have received so much attention in the context of the TSP and the VRP have now been adapted to incorporate time windows and other constraints, such as precedence constraints and mixed collections and deliveries. These types of algorithms are all familiar, but their modification to handle practical problems is nontrivial. Although the worst-case performance of these methods is very bad [Solomon 1986], they have been successfully incorporated in distribution management software [Anthonisse, Lenstra and Savelsbergh 1987].

Our survey is no more than an interim report. The developments in the area of constrained routing problems have just started. The practical need for effective routing algorithms will continue to stimulate further advances.

References

J.M. Anthonisse, J.K. Lenstra, M.W.P. Savelsbergh (1987). *Functional Description of CAR, an Interactive System for Computer Aided Routing*, Report OS-R8716, Centre for Mathematics and Computer Science, Amsterdam.

E.K. Baker (1983). An exact algorithm for the time-constrained traveling salesman problem. *Oper. Res. 31*, 65-73.

L. Bodin, B. Golden (1981). Classification in vehicle routing and scheduling. *Networks 11*, 97-108.

N. Christofides, S. Eilon (1969). An algorithm for the vehicle dispatching problem. *Oper. Res. Quart. 20*, 309-318.

N. Christofides, A. Mingozzi, P. Toth (1981a). Exact algorithms for the vehicle routing problem, based on spanning tree and shortest path relaxations. *Math. Programming 20*, 255-282.

N. Christofides, A. Mingozzi, P. Toth (1981b). State space relaxation procedures for the computation of bounds to routing problems. *Networks 11*, 145-164.

N. Christofides, A. Mingozzi, P. Toth (1981c). *Exact Algorithms for the Travelling Salesman Problem with Time Constraints, Based on State-Space Relaxation*, Unpublished manuscript.

G. Clarke, J.W. Wright (1964). Scheduling of vehicles from a central depot to a number of delivery points. *Oper. Res. 12*, 568-581.

A. Croes (1958). A method for solving traveling salesman problems. *Oper. Res. 5*, 791-812.

M. Desrochers (1986). *An Algorithm for the Shortest Path Problem with Resource Constraints*, Publication 421A, Centre de recherche sur les transports, Université de Montréal.

M. Desrochers, F. Soumis (1985a). *A Generalized Permanent Labelling Algorithm for the Shortest Path Problem with Time Windows*, Publication 394A, Centre de recherche sur les transports, Université de Montréal.

M. DESROCHERS, F. SOUMIS (1985b). *A Reoptimization Algorithm for the Shortest Path Problem with Time Windows*, Publication 397A, Centre de recherche sur les transports, Université de Montréal.

J. DESROSIERS, Y. DUMAS, F. SOUMIS (1986a). *The Multiple Vehicle Many to Many Routing and Scheduling Problem with Time Windows*, Cahiers du GERAD G-84-13, Ecole des Hautes Etudes Commerciales de Montréal.

J. DESROSIERS, Y. DUMAS, F. SOUMIS (1986b). A dynamic programming solution of the large-scale single-vehicle dial-a-ride problem with time windows. *Amer. J. Math. Management Sci. 6*, 301-326.

J. DESROSIERS, Y. DUMAS, F. SOUMIS (1987). *Vehicle Routing Problem with Pick-up, Delivery and Time Windows*, Cahiers du GERAD, Ecole des Hautes Etudes Commerciales de Montréal.

J. DESROSIERS, P. PELLETIER, F. SOUMIS (1984). Plus court chemin avec contraintes d'horaires. *RAIRO Rech. Opér. 17(4)*, 357-377.

J. DESROSIERS, M. SAUVÉ, F. SOUMIS (1985). *Lagrangian Relaxation Methods for Solving the Minimum Fleet Size Multiple Traveling Salesman Problem with Time Windows*, Publication 396, Centre de recherche sur les transports, Université de Montréal.

J. DESROSIERS, F. SOUMIS, M. DESROCHERS (1984). Routing with time windows by column generation. *Networks 14*, 545-565.

J. DESROSIERS, F. SOUMIS, M. DESROCHERS, M. SAUVÉ (1985). Routing and scheduling with time windows solved by network relaxation and branch-and-bound on time variables. J.-M. Rousseau (ed.). *Computer Scheduling of Public Transport 2*, North-Holland, Amsterdam, 451-469

Y. DUMAS (1985). *Confection d'itinéraires de véhicules en vue du transport de plusieurs origines à plusieurs destinations*, Publication 434, Centre de recherche sur les transports, Université de Montréal.

Y. DUMAS, J. DESROSIERS (1986). *A Shortest Path Problem for Vehicle Routing with Pick-up, Delivery and Time Windows*, Cahiers du GERAD G-86-09, Ecole des Hautes Etudes Commerciales de Montréal.

B.L. GOLDEN, A.A. ASSAD (1986). *Vehicle Routing with Time-Window Constraints: Algorithmic Solutions, Amer. J. Math. Management Sci. 6*, 251-428 (special issue).

M. GUIGNARD (1984). *Lagrangean Decomposition: an Improvement over Lagrangean and Surrogate Duals*, Report 62, Department of Statistics, University of Pennsylvania, Philadelphia.

J.-J. JAW, A.R. ODONI, H.N. PSARAFTIS, N.H.M. WILSON (1986). A heuristic algorithm for the multi-vehicle advance request dial-a-ride problem with time windows. *Transportation Res. Part B 20B*, 243-257.

K.O. JÖRNSTEN, M. NASBERG, P.A. SMEDS (1985). *Variable Splitting - a New Lagrangean Relaxation Approach to Some Mathematical Programming Models*, Report LITH-MAT-R-85-04, Department of Mathematics, Linköping Institute of Technology.

A.W.J. KOLEN, A.H.G. RINNOOY KAN, H.W.J.M. TRIENEKENS (1987). Vehicle routing with time windows. *Oper. Res.*, to appear.

S. LIN (1965). Computer solutions to the traveling salesman problem. *Bell System Tech. J. 44*, 2245-2269.

S. LIN, B.W. KERNIGHAN (1973). An effective heuristic algorithm for the traveling salesman problem. *Oper. Res. 21*, 498-516.

C. MILLER, A. TUCKER, R. ZEMLIN (1960). Integer programming formulation of travelling salesman problems. *J. Assoc. Comput. Mach. 7*, 326-329.

I. OR (1976). *Traveling Salesman-Type Combinatorial Problems and Their Relation to the Logistics of Blood Banking*, Ph.D. thesis, Department of Industrial Engineering and Management Sciences, Northwestern University, Evanston, IL.

C.S. ORLOFF (1976). Route constrained fleet scheduling. *Transportation Sci. 10*, 149-168.

H. PSARAFTIS (1983a). An exact algorithm for the single vehicle many-to-many dial-a-ride problem with time windows. *Transportation Sci. 17*, 351-360.

H. PSARAFTIS (1983b). *k*-Interchange procedures for local search in a precedence-constrained routing problem. *European J. Oper. Res. 13*, 391-402.

R.A. RUSSELL (1977). An effective heuristic for the *m*-tour traveling salesman problem with some side constraints. *Oper. Res. 25*, 517-524.

M.W.P. SAVELSBERGH (1985). Local search for routing problems with time windows. *Ann. Oper. Res. 4*, 285-305.

M.W.P. SAVELSBERGH (1987). *Local Search for Constrained Routing Problems*, Report OS-R8711, Centre for Mathematics and Computer Science, Amsterdam.

M.W.P. SAVELSBERGH (1988). *Computer Aided Routing*, Ph.D. thesis, Centre for Mathematics and Computer Science, Amsterdam.

T.R. SEXTON, L.D. BODIN (1985a). Optimizing single vehicle many-to-many operations with desired delivery times: I. Scheduling. *Transportation Sci. 19*, 378-410.

T.R. SEXTON, L.D. BODIN (1985b). Optimizing single vehicle many-to-many operations with desired delivery times: II. Routing. *Transportation Sci. 19*, 411-435.

M.M. SOLOMON (1983). *Vehicle Routing and Scheduling with Time Window Constraints: Models and Algorithms*, Report 83-02-01, Department of Decision Sciences, The Wharton School, University of Pennsylvania, Philadelphia.

M.M. SOLOMON (1986). On the worst-case performance of some heuristics for the vehicle routing and scheduling problem with time window constraints. *Networks 16*, 161-174.

M.M. SOLOMON, E.K. BAKER, J.R. SCHAFFER (1988). Vehicle routing and scheduling problems with time window constraints: implementations of solution improvement procedures. This volume.

B. SÖRENSEN (1986). *Interactive Distribution Planning*, Ph.D. thesis, Technical University of Denmark, Lyngby.

F. SOUMIS, J. DESROSIERS, M. DESROCHERS (1985). Optimal urban bus routing with scheduling flexibilities. *Lecture Notes in Control and Information Sciences 59*, Springer, Berlin, 155-165.

A.J. SWERSEY, W. BALLARD (1984). Scheduling school buses. *Management Sci. 30*, 844-853.

Vehicle Routing: Methods and Studies
B.L. Golden and A.A. Assad (Editors)

VEHICLE ROUTING AND SCHEDULING PROBLEMS WITH TIME WINDOW CONSTRAINTS: EFFICIENT IMPLEMENTATIONS OF SOLUTION IMPROVEMENT PROCEDURES

Marius M. Solomon

College of Business Administration, Northeastern University
Boston, Massachusetts

Edward K. Baker and Joanne R. Schaffer

Department of Management Science, University of Miami
Coral Gables, Florida 33124

A number of heuristic algorithms have been proposed for the vehicle routing and scheduling problem with time window constraints. These algorithms include both route construction and route improvement procedures. This paper extends branch exchange solution improvement procedures, well known from the standard vehicle routing literature, to vehicle routing and scheduling problems with time window constraints. We focus on efficient implementations of these procedures and present extensive computational results. The methods presented are completely robust in that significant reductions in running time are achieved without any degradation in the quality of the solution.

1. INTRODUCTION

The vehicle routing problem (VRP) involves the design of a set of minimum cost vehicle routes for a fleet of vehicles of known capacity which services a set of customers with known demands. All routes must originate and terminate at a common depot. Each customer is serviced exactly once and, in addition, all customers must be assigned to vehicles such that the total demand on any route does not exceed the capacity of the vehicle assigned to that route. We refer to Bodin et al. [1983] for a comprehensive survey.

The vehicle routing and scheduling problem with time window constraints (VRSPTW) is an important generalization of the VRP. In the VRSPTW, a number of customers have one or more time windows during which service must be scheduled. These allowable service times stem from the fact that customers impose service deadlines and/or earliest service time constraints. A restaurant, for example, may require deliveries before 11:00 am to avoid conflicts with luncheon customers. Baker and Schaffer [1986] and Solomon [1987] describe other examples and many important aspects of this problem.

Although the VRP has been studied extensively, the VRSPTW has only recently received the attention it deserves. Part of the literature has been directed at special structures such as the single and multiple traveling salesman problem (Christofides et al. [1981], Baker [1983], Desrosiers et al. [1986]), the dial-a-ride problem (Psaraftis [1983a], Sexton and Bodin [1985a], [1985b], Jaw et al. [1986]), the school bus scheduling problem (Orloff [1976], Swersey and Ballard [1984], Desrosiers et al. [1984], [1985]), and ship routing and scheduling problems (Psaraftis et al. [1986]). The recent paper of Kolen et al.

[1985] considers a branch and bound algorithm for the multiple vehicle time window constrained routing problem, but presents computational results only for problems with less than 15 customers.

More typically, the work on the general VRSPTW has been focused on heuristic methods. This is due to the computational complexity of this problem class. The findings of Savelsbergh [1984] and Solomon [1986] indicate that the VRSPTW is fundamentally more difficult than the VRP. Not only is the VRSPTW NP-hard, but even finding a feasible solution when the fleet size is fixed is itself an NP-complete problem (Savelsbergh [1984]).

A number of VRSPTW route construction methods have been proposed and analyzed by Solomon [1987]. In related work, Baker and Schaffer [1986] have conducted a computational study of route improvement procedures which were applied to heuristically generated initial solutions. The methods considered were extensions of the 2-opt and 3-opt branch exchange procedures of Lin [1965], adapted to include the vehicle capacity and time window constraints of the VRSPTW.

The computational experience reported in this earlier work suggests that while these methods were effective in improving the quality of the solutions generated by the route construction heuristics, the processing time required to obtain these results was large. A very large computational requirement was also reported by Cook and Russell [1978], in early work on the VRSPTW for a k-optimal improvement heuristic, M-Tour (Russell [1977]), which was effective in solving a real-world problem with a few time-constrained customers.

These findings, and similar ones for different VRP variants, indicate that one major drawback of the k-optimal branch exchange heuristics is their large processing time requirement. In addition, a k-optimal solution is rarely optimal. This is due to the fact that the number of locally optimal solutions grows exponentially with the problem size (Tovey [1985]). Furthermore, Solomon [1986] has shown that the worst-case behavior for k-optimal improvement heuristics for the VRSPTW on problems with 'n' customers is at least of order 'n'.

In this paper, we develop and implement methods for the acceleration of branch exchange procedures for the VRSPTW. The algorithms presented in this computational study are extensions of known VRP improvement heuristics. The novelty of our approach consists of improved implementations of these procedures that streamline the time window feasibility checks necessary to guarantee the feasibility of the vehicle route. These improvements have led to significant increases in algorithmic efficiency with no deterioration in solution quality. The methods developed here are used to accelerate branch exchanges by exploiting the inherent structure of the VRSPTW to eliminate only unnecessary feasibility checks, hence the quality of the solutions produced is not degraded. While we will focus on within-route methods, these techniques can be easily embedded within an overall improvement procedure for the VRSPTW where exchanges between routes are also examined.

The remainder of this paper follows in four sections. In the next section, Section 2, the methods for the acceleration of the branch exchange procedures are developed. This development focuses on three types of branch exchanges: the exchange of two branches, three branch exchanges, and the special case of the three branch exchange where the orientation of the route segments is preserved. In Section 3, the implementation of the acceleration procedures is discussed. Each of the implementations is coded in the C programming language for the IBM-PC AT microcomputer. In Section 4, computational results for the accelerated branch exchange procedures are presented. Finally, a summary of the results and the conclusions that are drawn are presented in Section 5.

2. ACCELERATION OF BRANCH EXCHANGE HEURISTICS FOR THE VRSPTW

In this section, we develop acceleration methods for branch exchange heuristics for the vehicle routing and scheduling problem with time window constraints. The analysis presented in Sections 2.1 and 2.2 is an adaptation of the recent work of Psaraftis [1983b] on the dial-a-ride problem to the VRSPTW. In the following discussion we show that with the use of a preprocessor, the addition of time window constraints to the VRP can, under certain conditions, be handled without an increase in running time of the algorithm. In Section 2.3 we show how to streamline feasibility checks even further for orientation preserving branch exchanges.

2.1 Exchanges Involving Two Branches, 2-opt

Before the discussion of the 2-opt case, we introduce the notation to be used throughout the remainder of this paper. Let

$e(i)$ = the earliest arrival time at customer i
$f(i)$ = the latest arrival time at customer i
$s(i)$ = the service time for customer i.
$t(i)$ = the time at which service begins for customer i
$a(i,j)$ = the travel time from i to j
$c(i,j)$ = the cost of traversing arc (i,j)

In this study, the service time for each customer was considered to include any minimum dwell time. Additionally, we note that the cost of traveling from i to j, $c(i,j)$, was assumed to be equal to a function of travel time and distance. In the VRSPTW, each customer is assumed to have one or more time windows of the form:

$$e(i) \leq t(i) \leq f(i) .$$

We then define the time at which service begins at customer i, $t(i)$, as follows:

$$t(i) = \max \{e(i),\ t(j) + s(j) + a(j,i)\} ,$$

where j is the immediate predecessor of i and where the arrival time at node i may be determined as

$$t(j) + s(j) + a(j,i) .$$

If a vehicle arrives at a customer before the time window opens, that is before e(i), then the vehicle must wait. We define the wait at node i, w(i), as:

$$w(i) = \max\{0, e(i) - (t(j) + s(j) + a(j,i))\}$$

Although we will consider only a single time window per customer in order to simplify the exposition, multiple time windows could be considered with only minor additional effort. The VRSPTW is also assumed to have N customers to be serviced by M vehicles. Additionally, we assume that the distances and the travel times of the problem are symmetric and are scalar multiples of each other. In this discussion, we consider only exchanges made within a single vehicle route of the VRSPTW.

An example VRSPTW vehicle route is presented in Figure 1. The indicated orientation of the route is clockwise from the depot. It is noted that time window constrained vehicle routing problems typically cannot be traversed in either direction. The time window constraints usually impart an orientation to the route that must be followed. When the time window constraints are very loose, however, parts or all of the route may be traversed in either direction without causing an infeasibility. It is the investigation of this aspect of the problem's feasibility that we shall pursue in greater detail below.

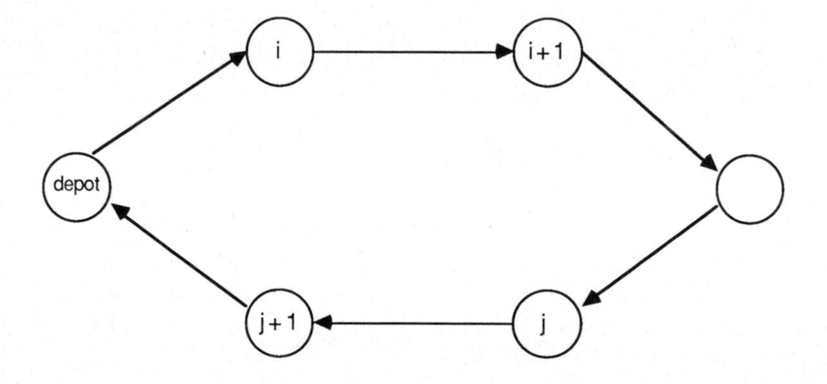

Figure 1. An Example VRSPTW Route

Consider the 2-interchange depicted in Figure 2. The proposed change would substitute arcs (i,j) and (i+1, j+1) for arcs (i, i+1) and (j, j+1), respectively. Using total route distance as the performance measure, this 2-interchange would result in a local route improvement if and only if:

$$c(i,j) + c(i+1, j+1) < c(i, i+1) + c(j, j+1) .$$

Additionally, the time window constraints of the customers affected by the exchange would also need to be satisfied. This will include all customers from i+1 to the end of the route, but particularly those customers on the route between j and i+1 since this section of the route will now be traversed in the reverse direction. Since the number of customers that may be contained in the section of the route between customers j and i+1 may vary, the computational effort required to check the feasibility of each customer's time window is of $O(N_k)$, where N_k is the number of customers on route k. This additional level of complexity in the 2-opt procedure would result in O(N**3) computations for the VRSPTW as opposed to the O(N**2) computations required in the standard TSP. In the following discussion, we consider a method of pre-processing the vehicle route which will streamline the feasibility check required of the 2-opt exchanges.

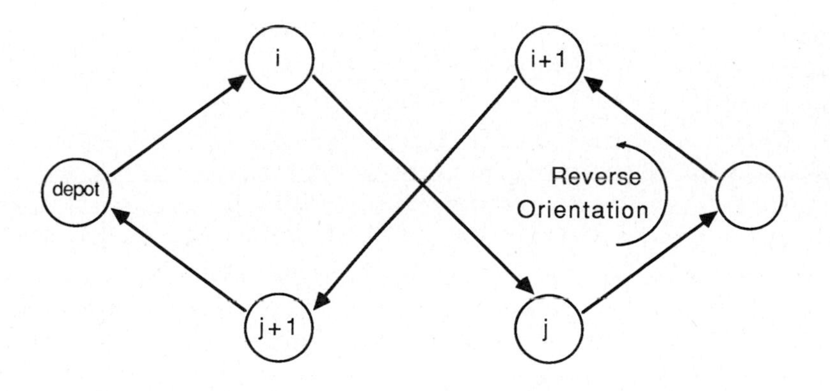

Figure 2. A 2-opt Exchange

We would now like to determine if precedence relationships exist between any pair of customers on the vehicle route. By testing if:

$$t(i) + s(i) + a(i,j) > f(j) ,$$

we may determine whether it is necessary for customer i to precede customer j. For any vehicle route, an examination of all such precedence relationships may be performed in O(N**2) time. The resultant matrix may then be defined:

$$VP(i,j) = \begin{cases} +1 \text{ if customer } i \text{ must precede customer } j \,; \\ \ \ 0 \text{ if no precedence relationship exists} \,; \\ -1 \text{ if customer } j \text{ must precede customer } i \,. \end{cases}$$

From the vehicle precedence matrix it is now possible to define a value for each customer on the route which reflects the precedence dependence at that point with regard to customers to be visited later on the route. For each customer i on the route, we define a node precedence value, NP(i), equal to the number of the first customer beyond customer i+1, that also has a predecessor which occurs after customer i on the route. (It is noted that the customer numbers referred to in this and subsequent analyses refer to the customer's sequence number on a specific route, not the customer's unique identification number). More formally, we define:

NP(i) = the smallest value of k, k > i+1, such that VP(j,k) = +1, j ≥ i+1. If no such k exists, then set NP(i) = N_k+1 .

It is noted that the NP(i) array may be obtained in O(N**2) time from the VP(i,j) array. Additionally, the following feasibility condition may be stated:

Condition 1. A necessary condition for the feasibility of the 2-exchange of arcs (i, i+1) and (j, j+1) with arcs (i,j) and (i+1, j+1) is that j < NP(i) .

The proof of the condition is easily seen by contradiction. If, for some 2-exchange, a j ≥ NP(i) is selected, then the route segment from j to i+1, which will be traversed in the reverse direction, contains a precedence relationship that will be violated during the operation of the proposed improved route.

Consider the example given in Figure 3. We assume that the route distance between all customers on the route is five units. Additionally, two precedence relationships are specified: 1 must precede 2, and 3 must precede 4. The NP(i) array must be computed as given below:

i	NP(i)
0	2
1	4
2	4
3	6
4	6
5	6
0	6

The interpretation of the NP(i) values implies that arc (1,2), for example, could be used in a 2-exchange with arc (3,4), but not with arc (4,5), which

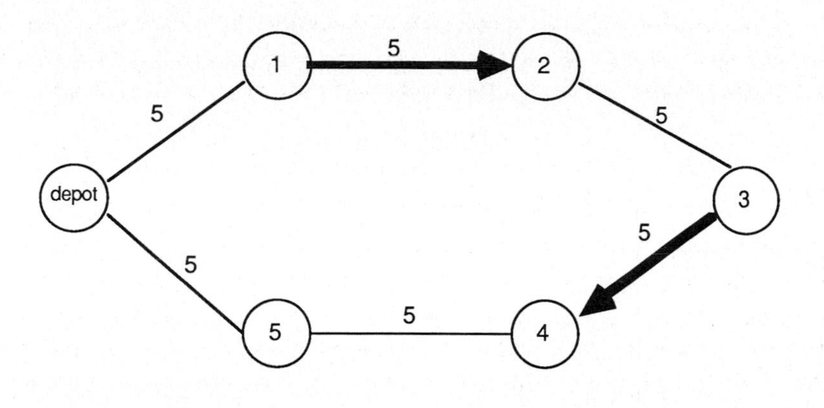

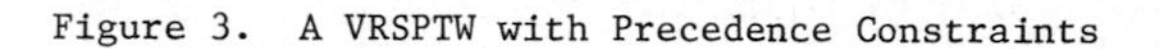

Figure 3. A VRSPTW with Precedence Constraints

would require the 3 before 4 precedence relationship to be violated. This situation is illustrated in Figure 4.

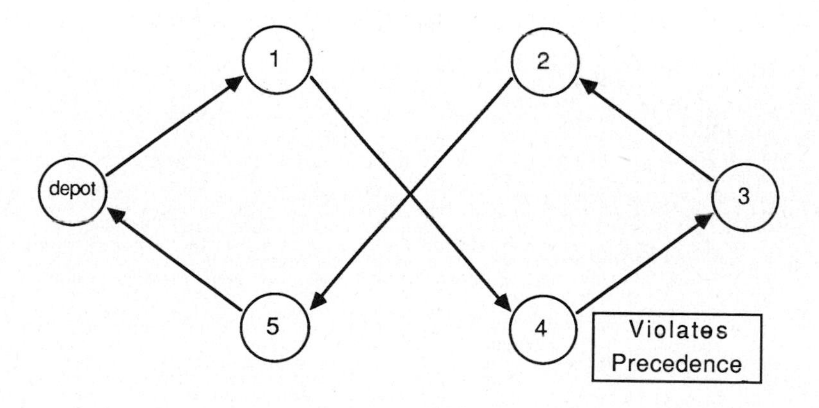

Figure 4. An Example of an Infeasible 2-opt Exchange

Unlike the result shown by Psaraftis [1983b] for the dial-a-ride problem, Condition 1 for the VRSPTW remains only a necessary condition for feasibility.

The presence of the time window constraints prohibits the sufficiency result. Consider, for example, the VRSPTW of Figure 3 with the additional time window constraint on the visitation time of customer two, i.e.,

$$e(2) = 8$$

and $$f(2) = 12$$

so that $$8 \leq t(2) \leq 12 .$$

The current route of Figure 3 is feasible since $t(2) = 10$. Note however, that the 2-exchange of arce (1,3) and (2,4) for arcs (1,2) and (3,4), which satisfies Condition 1, will force $t(2) = 15$, and become infeasible. <u>Although the use of the NP(i) array does not eliminate the need for further checking of the feasibility of a 2-exchange, it may be used as an effective filter to reduce the number of complete feasibility checks required to obtain a 2-optimal solution for the VRSPTW.</u>

2.2. Exchanges Involving Three Branches, 3-opt

The techniques discussed above for the 2-exchange heuristic may also be extended to the 3-exchange procedure. Generally, there are four ways in which a 3-exchange may be used to uniquely reconfigure a route. Using Figure 5 to illustrate the first of these four cases, the arcs (i, i+1), (j, j+1), and (k, k+1) may be replaced by:

Case 1. arcs (i, j+1), (k, i+1), and (j, k+1),
Case 2. arcs (i,k), (j+1, i+1), and (j, k+1),
Case 3. arcs (i,j), (i+1, k), and (j+1, k+1),
Case 4. arcs (i, j+1), (k,j), and (i+1, k+1) .

It is noted that in a 3-exchange, the order in which two segments of the route are traversed may be exchanged. Additionally, the segments may or may not be traversed in the same direction. In Case 1, for example, the segments are reversed, but the orientation of the segments is preserved. In Case 2, the order is reversed, and the orientation of the k to j+1 segment is reversed.

When segments of the route are to be performed in reverse order, the precedence relationships defined in the VP(i,j) array may be examined to determine whether or not the necessary conditions for a feasible exchange are met. If, for example, the segments from i+1 to j and from j+1 to k are to be performed in reverse order, then it is possible to check the VP(i,j) array to determine if there are any predecessors of any customers of the second segment, j+1 to k, on the first, i+1 to j. Operationally, we define a segment predecessor array SP(i,j), as follows:

SP(i,j) = the smallest customer number r, $j+1 \leq r$, such that $VP(q,r) = +1$, $i+1 \leq q \leq j$. If no such r exists, then set $SP(i,j) = N_k + 1$.

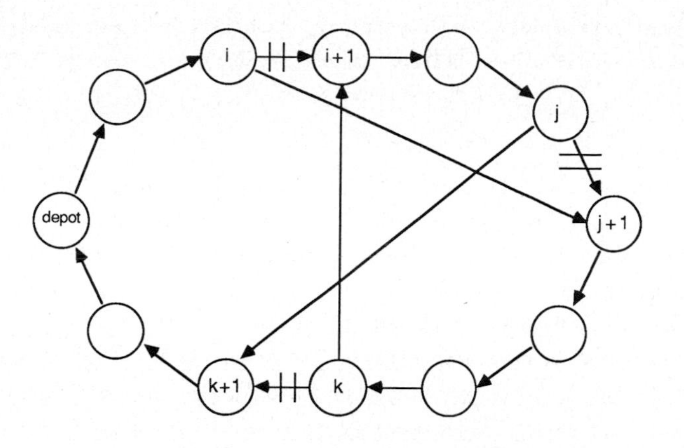

Figure 5. An Example of a 3-opt Exchange

It is noted that the SP(i,j) array may be obtained from the VP(i,j) array in O(N**3) time.

To illustrate this concept, consider the example shown in Figure 6. In the figure, the vehicle route, displayed linearly, is to be reconfigured by a Case 1, 3-opt exchange. This implies that the route segment from i+1 to j (node 2 to node 3) will be preceded by route segment j+1 to k (node 4 to node 5). If, for example, a precedence relationship is specified such that node 3 must precede node 4, then, by the above definition, SP(1,3) = 4. Since node 4 is contained within the route segment j+1 to k, this would cause the proposed 3-exchange to be infeasible.

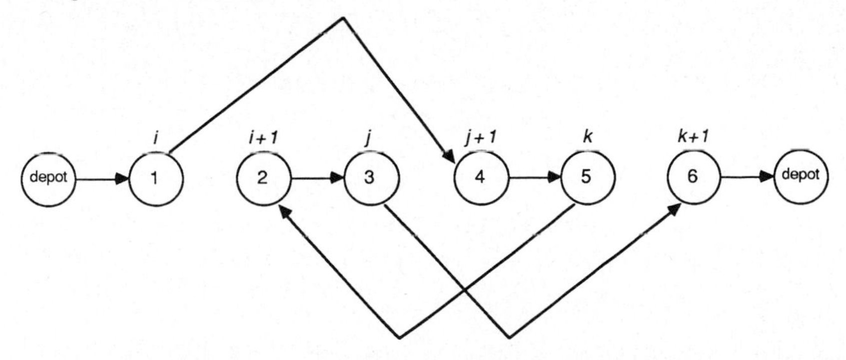

Figure 6. An Example of a 3-Exchange (Case 1)

The segment precedence array may be used to establish a necessary condition on the feasibility of a 3-exchange. We may now state:

Condition 2. A necessary condition for the 3-exchange of arcs (i, i+1), (j, j+1), and (k, k+1) is that $k < SP(i,j)$.

The proof of Condition 2 is obvious by contradiction. It is noted that in Case 2 above, the reverse orientation of segment k, j+1 may also be tested for feasibility using a minor modification of Condition 1. Similar proofs apply for Case 3 and Case 4.

2.3. Orientation Preserving Branch Exchanges

Of primary importance to the efficiency of the VRSPTW k-optimal improvement heuristics is the way in which the time window violation checks are incorporated in the solution process when an exchange is performed. For a 3-opt procedure, this involves all the customers from i+1 to the end of the route (see Figure 5). Hence the computational effort required to check time feasibility is O(N). This added level of complexity would require O(N**4) computations in the VRSPTW 3-opt procedure as opposed to only O(N**3) in its VRP counterpart.

In the following discussion we show that by taking advantage of orientation preserving exchanges, e.g., Case 1 above, one can significantly reduce the computational effort required. Note that this type of 3-exchange may be viewed as an insertion of a string of customers between two currently adjacent customers. Additionally, in this type of exchange, the direction in which each route segment is traversed remains the same as in the original vehicle route. In Figure 6, for example, the customers on the route segment from i+1 to j are inserted between k and k+1 and the orientation on each segment is preserved.

Without loss of generality, we assume that the triangle inequality holds for both travel distances and times. Then, this insertion defines a push backward in the schedule at j+1,

$$PB(j+1) = t(j+1) - t'(j+1)$$

where $t'(j+1)$ is the new time to begin service at j+1.
Furthermore,

$$PB(r+1) = \min\{PB(r),\ t(r+1) - e(r+1)\},\quad j+1 \le r \le k-1 .$$

It is easy to see that, while all the customers on the route segment from j+1 to k will remain feasible, the times to begin service, t(i), need to be adjusted sequentially, starting with t(j+1) up to the customer r, $j < r < k$, where $PB(r) = 0$.

Similarly, inserting the route segment from i+1 to j between k and k+1 produces a push forward in the schedule at i+1:

$$PF(i+1) = t'(i+1) - t(i+1) .$$

Moreover,

$$PF(r+1) = \max\{0,\ PF(r) - w(r+1)\},\quad i+1 \le r \le j-1,$$

where w(i) is the waiting time at customer i.

If PF(i+1) > 0, some customers from i+1 to j could become infeasible. One simply needs to sequentially examine time feasibility on the route segment from i+1 to j up to a customer, say r, where PF(r) = 0, or r is time infeasible, or, in the worst case, r = j.

To illustrate these concepts, consider the proposed 3-opt exchange of Figure 6. We assume that the initial vehicle route visits the customers in numerical order and that the travel time between sequential customers is five units. With these assumptions, t(0) = 0, t(1) = 5, t(2) = 10, and so on. We also assume that the branches to be exchanged in the 3-opt all have a travel time equal to four, hence, a(1,4) = 4, a(5,2) = 4, and a(3,6) = 4. The push forward defined at the j+1st node, node 4, is then,

$$PB(j+1) = 20 - 9 = 11 .$$

The push backward at the i+1st node, node 2, is defined as:

$$PF(i+1) = 18 - 10 = 8 .$$

Although the push backward will not cause an infeasibility on the segment of the route between j+1 and k, the push forward in the schedule at node i+1 may affect the time window feasibility of any customer from i+1 to the end of the new vehicle route. The use of the PF(i) vector, however, may obviate the need to examine all of these nodes. If, for example, the original vehicle schedule was such that the vehicle was required to wait at node i+1 for some time equal to or greater than eight time units, the entire push forward in the schedule could be absorbed and no further feasibility checks would be required.

In conclusion, it can be seen that the push forward and the push backward concepts can lead to substantial reductions in the number of customers being examined for time feasibility. Note that these concepts can also be incorporated in any type of exchange, such as Case 2 above, for the orientation preserving portions of the insertions.

3. IMPLEMENTATION

In the previous section, various procedures were introduced that allow conditions to be placed upon the feasibility of 2- and 3-exchange procedures for the VRSPTW without increasing the theoretical worst-case computational complexity of the original algorithm. (Note, the complexity of the actual coding of the algorithm has increased, however, as we will see, running time is decreased.) The 2-opt solution improvement procedure was implemented using the precedence testing conditions described in Section 2.1. Since the VP(i,j) matrix is symmetric, only half of this matrix was actually stored. Then for each vehicle route, the NP array was updated as each route improvement was made.

The 3-opt solution improvement procedure was implemented using the combined precedence testing conditions of Sections 2.1 and 2.2. When the order of the segments is preserved, but the direction of traversal is reversed, the conditions of Section 2.1 apply. When the order of the segments is reversed, but direction preserved, the conditions of Section 2.2 apply. When both order and direction are reversed, both sets of conditions are employed. Additionally, the NP and SP arrays are updated after each exchange is made.

The final improvement technique considered is an extension of the accelerated 3-opt method proposed by Or [1976] for the single vehicle VRP. The motivation for considering this approach stems from its success for the VRP. Golden and Stewart [1985] report that the OROPT produced improvements on already good starting solutions which were considerably better than 2-opt improvements and on a par with the 3-opt improvements. The processing time required was, however, substantially less than that needed for the full 3-opt heuristic.

The OROPT heuristic is a restricted branch exchange procedure which examines only exchanges where a string of one, two, or three currently adjacent nodes between j+1 and k are inserted between i and i+1. Hence, the OROPT considers only approximately 3*N**2 ways to remove arcs from the current route out of the possible N**3. However, since the heuristic accounts for both the geographical and temporal closeness of customers, most of the arcs it does not consider are probably superfluous.

Additional improvements in efficiency can be made by only considering orientation preserving cases (e.g., Case 1 above). In an earlier study, Baker and Schaffer [1986] have observed that less than ten percent of the improvements found involved the reversal of the orientation of a sequence of two or more customers.

We have implemented an OROPT procedure that only considers orientation preserving branch exchanges, utilizes the push forward and push backward ideas (see Section 2.3.) for time window violation checks, and the lexicographic ordering for processing suggested by Savelsbergh [1984]. Additionally, a version of the code has been created which has the option of using the pre-processing of Section 2.2.

4. COMPUTATIONAL RESULTS

Each of the branch exchange solution improvement procedures, the 2-opt, the 3-opt, and the OROPT, was applied to two sets of test problems. The test problems were generated using the procedures discussed in an earlier paper by Solomon [1987]. The problems were specifically created to allow a large number of customers per vehicle route so as to test the effectiveness of the within-route improvement procedures investigated in this paper. Each test problem has 100 customers. Due to the vehicle capacity and the time window constraints, three to five vehicles are needed to service all customers. The average number of customers per route is generally 33, while occasionally as many as 50 customers may be on a single vehicle route.

The test data set includes two distinct problem types. The first set, designated NC201 - NC208, contains problems with a structured configuration of customers. In these problems, three clusters of customers were created. A single vehicle route was then created for each cluster using a standard 3-opt procedure. Time windows for the customers on the route were then created by using the 3-opt arrival time as the center of the window. The second group of test problems, designated R201 - R2011, has customer configurations that have been obtained by random placement of customers on a unit area. In both problem sets, the width of the time windows and the quantity demanded by each customer are randomly obtained from user specified probability distributions.

The results of the computational tests are presented in Tables 1 through 4. Each test problem was initially solved using an insertion heuristic described in Solomon [1987]. The measure of performance in this and the subsequent route improvement procedures was arbitrarily chosen as a 50-50 weighting of route time and distance to allow a general form of a total cost function to be considered. The results reported for each test problem include total schedule time, total distance, and number of routes. The total schedule time includes service time, travel time, and waiting time.

Since the route improvement procedures considered here are restricted to within route improvements, the number of routes is determined for each test problem by the route construction heuristic.

The procedures were applied as follows. First, each problem was solved by the route construction heuristic. This initial solution was then improved by the 2-opt procedure. The 2-opt solution was then used as input to the various implementations of the 3-opt and the OROPT procedures. The total time to produce the OROPT solution reported in Table 1 for the test problem NC201 is: 22.30 seconds to produce the initial insertion solution, 0.82 seconds to produce the most efficient 2-opt solution, and 1.32 seconds to produce the most efficient OROPT solution.

Each of the algorithmic implementations was coded in the C programming language and compiled for an IBM-PC AT using the DeSmet compiler. The programs used all-integer arithmetic. The AT did not have a math co-processor. Timings of the various implementations were made using a user generated function based on the microcomputer's internal clock. The timings are reported in minutes, seconds, and hundredths of seconds.

A summary of the computational results is presented in Table 5. The overall average reduction in processing time obtained for all procedures across all problem types was 57 percent. The performance of the procedures on the structured problems, NC201 - NC208, being slightly superior, 61 percent reduction

Table 1. Computational Results of Solution Improvement Procedures
Test Problems NC201 - NC205

	Test Problems				
	NC201	NC202	NC203	NC204	NC205
Initial Insertion Solutions					
Total Schedule Time	9604	10090	12923	10756	9717
Total Distance	604	796	937	1134	670
# Routes	3	4	4	4	4
Running Time	22.30	23.50	26.15	29.11	24.77
Within Route 2-Opt Solutions					
Total Schedule Time	9604	10049	12893	10706	9676
Total Distance	604	797	907	1055	655
Running Time	1.26	1.93	3.24	7.53	3.13
/with preprocessor					
Running Time	0.82	1.21	1.86	4.94	1.65
% Reduction	35	37	43	35	47
Within Route 3-Opt Solutions					
Total Schedule Time	9604	9971	12566	10429	9655
Total Distance	604	852	844	1048	655
Running Time	1:32.77	3:10.21	5:17.08	3:53.55	2:11.60
/with preprocessor					
Running Time	2.74	14.27	1:01.41	1:32.71	9.17
% Reduction	97	92	81	61	94
Within Route OROPT Solutions					
Total Schedule Time	9604	9971	12573	10511	9655
Total Distance	604	852	887	1075	655
Running Time	13.02	21.42	34.65	47.78	21.48
/with preprocessor					
Running Time	7.91	11.20	19.94	25.60	11.59
% Reduction	39	48	43	47	46
/with preprocessor and push					
Running Time	1.32	5.22	15.93	30.37	3.63
% Reduction	90	76	54	36	83

Running Time is total elapsed time on an IBM-PC AT.

Table 2. Computational Results of Solution Improvement Procedures Test Problems NC206 - NC208, R201 - R202

	Test Problems				
	NC206	NC207	NC208	R201	R202
Initial Insertion Solutions					
Total Schedule Time	9722	9733	9765	3482	3328
Total Distance	659	746	720	1809	1573
# Routes	3	3	3	5	4
Running Time	17.95	26.59	18.94	23.12	28.73
Within Route 2-Opt Solutions					
Total Schedule Time	9640	9722	9700	3479	3298
Total Distance	640	719	700	1800	1538
Running Time	4.89	6.26	7.86	2.25	11.21
/with preprocessor					
Running Time	2.64	3.02	4.56	1.21	3.07
% Reduction	46	52	42	46	73
Within Route 3-Opt Solutions					
Total Schedule Time	9633	9717	9694	3479	3305
Total Distance	633	714	694	1773	1486
Running Time	3:12.90	2:58.90	3:14.98	2:32.97	6:16.79
/with preprocessor					
Running Time	21.14	32.01	34.54	7.85	33.61
% Reduction	89	82	82	95	89
Within Route OROPT Solutions					
Total Schedule Time	9633	9717	9695	3481	3305
Total Distance	633	714	695	1774	1488
Running Time	29.94	31.97	32.85	14.50	51.90
/with preprocessor					
Running Time	15.15	9.61	17.09	9.39	24.99
% Reduction	50	70	48	35	52
/with preprocessor and push					
Running Time	7.25	9.50	10.82	3.07	15.38
% Reduction	76	70	67	79	71

Running Time is total elapsed time on an IBM-PC AT.

Table 3. Computational Results of Solution Improvement Procedures
Test Problems R203 - R207

	Test Problems				
	R203	R204	R205	R206	R207
Initial Insertion Solutions					
Total Schedule Time	3300	2844	2772	2576	2630
Total Distance	1539	1191	1386	1308	1156
# Routes	4	3	4	3	3
Running Time	28.44	55.64	29.50	23.33	48.11
Within Route 2-Opt Solutions					
Total Schedule Time	3270	2764	2687	2545	2633
Total Distance	1364	1120	1314	1265	1120
Running Time	17.96	29.00	10.98	10.33	10.87
/with preprocessor					
Running Time	7.69	16.14	5.71	4.67	5.11
% Reduction	57	44	48	55	53
Within Route 3-Opt Solutions					
Total Schedule Time	3172	2616	2631	2474	2491
Total Distance	1334	1022	1279	1217	1046
Running Time	21:20.09	93:20.37	14:10.52	18:58.38	35:09.25
/with preprocessor					
Running Time	4:16.38	42:43.21	2:44.66	3:06.85	8:20.64
% Reduction	80	55	81	84	76
Within Route OROPT Solutions					
Total Schedule Time	3172	2630	2631	2479	2438
Total Distance	1361	1033	1270	1224	1008
Running Time	1:42.83	2:56.08	45.64	1:22.61	3:46.02
/with preprocessor					
Running Time	50.91	1:11.62	25.49	35.76	1:27.22
% Reduction	50	59	44	57	61
/with preprocessor and push					
Running Time	50.97	1:37.66	16.48	22.96	1:38.15
% Reduction	50	45	64	72	57

Running Time is total elapsed time on an IBM-PC AT.

Table 4. Computational Results of Solution Improvement Procedures
Test Problems R208 - R211

	R208	R209	R210	R211
Initial Insertion Solutions				
Total Schedule Time	2755	2942	3096	2170
Total Distance	1066	1278	1408	1017
# Routes	3	4	4	3
Running Time	55.26	35.32	32.63	59.49
Within Route 2-Opt Solutions				
Total Schedule Time	2705	2886	3016	2126
Total Distance	1049	1236	1363	1013
Running Time	17.74	11.42	15.16	17.57
/with preprocessor				
Running Time	8.40	6.43	7.96	13.90
% Reduction	53	44	47	21
Within Route 3-Opt Solutions				
Total Schedule Time	2580	2865	2933	2091
Total Distance	962	1183	1285	1027
Running Time	80:31.02	45:51.71	37:12.00	5:04.84
/with preprocessor				
Running Time	30:45.77	17:16.85	13:24.66	2:16.92
% Reduction	62	62	64	55
Within Route OROPT Solutions				
Total Schedule Time	2592	2861	2960	2113
Total Distance	981	1187	1289	1020
Running Time	2:58.24	1:10.91	1:55.28	57.67
/with preprocessor				
Running Time	1:07.56	31.03	50.53	27.25
% Reduction	60	56	56	53
/with preprocessor and push				
Running Time	1:32.61	27.13	46.09	41.98
% Reduction	48	62	60	27

Running Time is total elapsed time on an IBM-PC AT.

Table 5. A Summary of Average Percent Reduction in Execution Time by Procedure and Problem Type

Procedure	Test Problem Type	
	Structured	Random
Pre-processed 2-opt	42%	49%
Pre-processed 3-opt	85%	73%
Push Oriented OROPT	49%	53%
Pre- and Push OROPT	69%	42%
Average Reduction	61%	54%

overall, to that on the random problems, 54 percent. The most dramatic reduction in processing time, 85 percent reduction, was obtained for the pre-processed 3-opt procedure on the structured problems. It would appear from the results that pre-processing is definitely advantageous when using a 3-opt procedure whether working with structured or unstructured problems. The effect of using both the pre-processor and the push concepts produced mixed results. In a number of cases, the additional computational effort required for a pre-processing did not improve algorithmic performance beyond that provided by the push analysis.

A comparison of the computational performance of the 3-opt and the OROPT procedures is also of interest. Recall that the OROPT performs only those reconfigurations of the 3-opt in which the orientation (direction of travel) of the vehicle is preserved. This is accomplished in the OROPT implementation by attempting to insert one, two, or three customers at a time into a position later on the route. While not as comprehensive as a full 3-opt, the execution time required for the procedure is far less. Additionally, the computational experiments of Or [1976] and Baker and Schaffer [1986] suggest that the OROPT's effectiveness in improving the solution is only slightly less than that of the full 3-opt.

The computational results of the experiments performed here confirm our earlier expectations with regard to the OROPT. Overall, the solutions produced from a 2-optimal starting solution by the OROPT procedure were within one percent of the solution produced by the full 3-opt from the same starting point. For the structured problems, NC201 - NC208, five of the eight cases produced identical results. For the randomly generated problems, R201 - R211, two of the

OROPT solutions were actually slightly better than those produced by the 3-opt. Additionally, the execution time required for the most efficient implementation of the OROPT was significantly less than that required for the most efficient 3-opt implementation. For the structured problems, the OROPT was 65 percent more efficient, while for the random problems the OROPT was 83 percent faster.

5. SUMMARY AND CONCLUSIONS

In this paper, we have presented several efficient implementations of branch exchange solution improvement procedures for the vehicle routing and scheduling problem with time window constraints. We have shown, that under certain conditions, the time window constraints may be incorporated without additional increase in computational complexity. Computational tests of the recommended improvements yielded, on average, a 57 percent decrease in required processing time. The efficiency of the computationally burdensome 3-opt procedure was shown to be dramatically improved by the incorporation of pre-processing. Furthermore, the more efficient OROPT procedure was shown to produce improved solutions of comparable quality to those produced by the 3-opt, while requiring significantly less execution time.

The nature of the results produced in this study make them particularly relevant. All of the codes were written in the C programming language and executed on an IBM-PC AT microcomputer. This is the computing environment for much of the development and implementation currently underway for vehicle routing models. The results of this study suggest that in this environment significant improvements in execution time may be obtained for known solution improvement procedures without degradation of solution quality.

One final point should be noted. The tests performed in this study involved problems in which the number of customers per vehicle route was large, generally between 30 and 50 customers per route. It is the authors' experience in smaller problems, 10 to 20 customers per route, that the use of the more efficient OROPT may not be justified and that the full 3-opt procedure should be employed to ensure the desired level of solution quality.

ACKNOWLEDGMENTS

The authors benefited in the early stages of this project from fruitful discussions with Professor William R. Stewart of the School of Business Administration at the College of William and Mary.

The work of Professor Solomon was supported, in part, by the Joseph G. Reisman Research Fund of Northeastern University. The work of Professors Baker and Schaffer was supported, in part, by Ryder Truck Rental, Inc. of Miami, Florida.

REFERENCES

[1] Baker, E., "An Exact Algorithm for the Time Constrained Traveling Salesman Problem," Operations Research 31, 938-945 (1983).

[2] Baker, E. and Schaffer, J., "Computational Experience with Branch Exchange Heuristics for Vehicle Routing Problems with Time Window Constraints," American Journal of Mathematical and Management Sciences 6(3&4), 261-300 (1986).

[3] Bodin, L., Golden, B., Assad, A., and Ball, M., "Routing and Scheduling of Vehicles and Crews, The State of the Art," Computers and Operations Research 10(2), 63-211 (1983).

[4] Christofides, N., Mingozzi, A., and Toth, P., "State Space Relaxation Procedures for the Computation of Bounds to Routing Problems," Networks 20(2), 145-164 (1981).

[5] Cook, T. and Russell, R., "Simulation and Statistical Analysis of Stochastic Vehicle Routing with Timing Constraints," Decision Sciences 9, 673-687 (1978).

[6] Desrosiers, J., Soumis, F., and Desrochers, M., "Routing with Time Windows by Column Generation," Networks 14, 545-565 (1984).

[7] Desrosiers, J., Soumis, F., Desrochers, M., and Sauve, M., "Routing and Scheduling with Time Windows Solved by Network Relaxation and Branch and Bound on Time Variables," Computer Scheduling of Public Transport 2, J. M. Rousseau (ed.) North Holland Press, 451-469 (1985).

[8] Desrosiers, Jr., Soumis, F., and Sauve, "Lagrangian Relaxation Methods for Solving the Minimum Fleet Size Multiple Traveling Salesman Problem with Time Windows," forthcoming in Management Science (1986).

[9] Golden, B. L. and Stewart, W. R., "Empirical Analysis of TSP Heuristics," in The Traveling Salesman Problem, E. Lawler, J. K. Lenstra and A. Rinnooy Kan (eds.), John Wiley and Sons (1985).

[10] Jaw, J. J., Odoni, A. R., Psaraftis, H. N., and Wilson, N. H. M., "A Heuristic Algorithm for the Multi-Vehicle Advance-Request Dial-A-Ride Problem with Time Windows," forthcoming in Transportation Research (1986).

[11] Kolen, A., Rinnooy Kan, A., and Trienekens, H., "Vehicle Routing with Time Windows," Report 8433/0, Erasmus University, Rotterdam (1985).

[12] Lin, S., "Computer Solutions to the Traveling Salesman Problem," Bell System Technical Journal 44, 2245-2269 (1965).

[13] Or, I., "Traveling Salesman-type Combinatorial Problems and Their Relation to the Logistics of Blood Banking," Ph.D. Thesis. Department of Industrial Engineering and Management Sciences, Northwestern University (1976).

[14] Orloff, C., "Route Constrained Fleet Scheduling," Transportation Science 10, 149-168 (1976).

[15] Psaraftis, H. N., "An Exact Algorithm for the Single Vehicle Many-to-Many Dial-A-Ride Problem with Time Windows," Transportation Science 17, 351-357 (1983a).

[16] Psaraftis, H., "K-interchange Procedures for Local Search in a Precedence-constrained Routing Problem," European Journal of Operational Research 13, 391-402 (1983b).

[17] Psaraftis, H., Solomon, M., Magnanti, T., and Kim, T., "Routing and Scheduling on a Shoreline with Release Times," Working Paper MIT-NT-86-1, Massachusetts Institute of Technology (1986).

[18] Russell, R., "An Effective Heuristic for the M-tour Traveling Salesman Problem with Some Side Conditions," Operations Research 25, 517-524 (1977).

[19] Savelsbergh, M. W. P., "Local Search in Routing Problems with Time Windows," Report OS-R8409, Centre for Mathematics and Computer Science, Amsterdam (1984).

[20] Sexton, T. and Bodin, L., "Optimizing Single Vehicle Many-to-Many Operations with Desired Delivery Times: I. Scheduling," Transportation Science 19, 378-410 (1985a).

[21] Sexton, T. and Bodin, L., "Optimizing Single Vehicle Many-to-Many Operations with Desired Delivery Times: II. Routing," Transportation Science 19, 411-435 (1985b).

[22] Solomon, M., "Algorithms for the Vehicle Routing and Scheduling Problem with Time Window Constraints," Operations Research 35, 254-265 (1987).
[23] Solomon, M. M., "On the Worst-Case Performance of Some Heuristics for the Vehicle Routing and Scheduling Problem with Time Window Constraints," Networks 16, 161-174 (1986).
[24] Swersey, A. and Ballard, W., "Scheduling School Buses," Management Science 30(7), 844-853 (1984).
[25] Tovey, C., "Hill Climbing with Multiple Local Optima," SIAM Journal for Discrete Algorithmic Methods 6(3), 384-393 (1985).

Vehicle Routing: Methods and Studies
B.L. Golden and A.A. Assad (Editors)

GENERALIZED ASSIGNMENT METHODS FOR THE DEADLINE VEHICLE ROUTING PROBLEM

Kendall E. Nygard, Department of Computer Science and Operations Research, North Dakota State University, Fargo, North Dakota 58105

Peter Greenberg, Centro de Investigacion y de Estudios Avanzados del IPN, Mexico 14, DF, Mexico

Waverly E. Bolkan, The Boeing Company, Seattle, Washington

Elizabeth J. Swenson, AT&T Bell Labs, Columbus, Ohio

This work concerns vehicle routing problems in which total tour length is to be minimized subject to constraints on vehicle capacities and time deadlines for serving the stops. The methods evaluated are based on the generalized assignment approach to the vehicle routing problem, with two primary extensions: i) use of local optimization procedures to modify tours to produce better time quality characteristics, and ii) iterative cost adjustment techniques which produce clusters of stops which are progressively more likely to produce tours with good time quality measures.

1. INTRODUCTION

The deadline vehicle routing problem (DVRP) has the following characteristics: i) there is a fleet of vehicles, all housed at a single depot, and each has the same capacity and ii) there is a set of stops with known demands and time deadlines (latest allowable arrival times). The goal is to find the shortest possible set of tours for which no deadlines are violated. The same problem with no deadlines is the standard vehicle routing problem (VRP) which has been extensively studied.

The heuristic algorithm for the DVRP which is presented in this paper was applied to two route design problems for newspaper bundle delivery in the upper midwest. In these problems it is critical that bundle delivery be complete by 6:00 AM, so that subsequent door-to-door delivery of the newspapers can be carried out by youths before they leave for school. The time windows are one-sided because early delivery is not an issue in these problems (the bundles are wrapped in plastic for weather protection, and vandalism or theft seldom occurs). The first problem involved delivery of bundles of a non-metropolitan area edition of the newspaper to over 300 relatively small towns dispersed over about 60,000 square miles. All the bundles required for a town are dropped at one location, and the subsequent delivery of the bundles to carriers within the town is handled by a fleet of smaller local trucks under contract. Deadlines at the towns vary because the time required to complete local bundle delivery by 6:00 AM differs among the towns. The second problem involves the metropolitan area edition of the newspaper. In this problem, a fleet of vehicles deliver the bundles from a central depot directly to the carriers.

The generalized assignment approach to the VRP has been used with considerable success (see Fisher and Jaikumar [2] and Nelson [6]). In that approach, stops are first clustered by solving a generalized assignment problem, then routed with a travelling salesman algorithm. In this work, we extend the generalized assignment approach to the DVRP. The basic extensions of the method include: i) successive solution of generalized assignment problems (GAPs) with altered cost coefficients in quest of clusters of stops which can all be visited within time deadlines and ii) use of two types of postprocessors which locally modify tours to better meet time deadlines. Several variations of the extended method were coded and empirically compared. Results indicate that the new approach can generate tours with good time quality characteristics. Section 2 of the paper provides a description of the fundamentals of the generalized assignment approach. Section 3 describes the details of our new method and the variations that were examined. The test problems and empirical work are described in Section 4. Conclusions are presented in Section 5.

2. FUNDAMENTALS OF THE GENERALIZED ASSIGNMENT APPROACH

The generalized assignment approach due to Fisher and Jaikumar [2] is one of the "cluster-first, route-second" heuristic techniques for the VRP. The basic idea is to carry out the following two steps:

Step 1: Clustering. There is a set of vehicles K and a set of stops J. Each vehicle k in set K has an available capacity b_k, and the assignment of a stop j in set J to vehicle k consumes r_{kj} units of this capacity. A cost coefficient c_{kj} is a measure of the desirability of assigning stop j to vehicle k. Given these parameters, the following generalized assignment problem (GAP) is solved to assign stops to vehicles.

$$\text{minimize} \sum_{k\varepsilon K} \sum_{j\varepsilon J} c_{kj}x_{kj}$$

subject to:

$$(1) \quad \sum_{j\varepsilon J} r_{kj}x_{kj} \leq b_k \qquad \text{for all } k\varepsilon K$$

$$(2) \quad \sum_{k\varepsilon K} x_{kj} = 1 \qquad \text{for all } j\varepsilon J$$

$$x_{kj} = 0 \text{ or } 1 \qquad \text{for all } k\varepsilon K,\ j\varepsilon J.$$

The value of the decision variable, x_{kj}, is interpreted as follows:

$$x_{kj} = \begin{cases} 1 \text{ if stop j is assigned to vehicle k,} \\ 0 \text{ otherwise.} \end{cases}$$

Constraint set (2) forces each stop to be assigned to exactly one vehicle. Constraint set (1) limits the assignments by vehicle capacity.

Step 2: Routing. Calculate travelling salesman tours for the stops in each of the clusters from Step 1.

Because tour quality is known only after Step 2, the GAP in Step 1 is solved with objective function cost coefficients that model the influence that individual stops would have on the quality of the travelling salesman tours of the various vehicles. The primary challenge of the approach lies in choosing cost coefficients that translate the linear objective function of the GAP into the highly non-linear travelling salesman tour creation process. In the standard VRP, where tour quality is measured only by total length, the cost coefficient c_{kj} is an estimator of how much of the distance vehicle k travels is attributable to stop j. Since the tour of vehicle k is unknown in Step 1, it is natural to model its tour in some way. A basic way to model a vehicle tour is to identify a single location called a seed, with the vehicle conceptually travelling from the depot to the seed and back. With this model, the extra distance incurred by adding stop j to the tour of vehicle k is given by

$$c_{kj} = \text{DISTANCE (depot, stop j)} + \text{DISTANCE (stop j, seed k)} - \text{DISTANCE (seed k, depot)}.$$

Given this definition of c_{kj}, the clusters produced in Step 1 are fundamentally dependent on the locations of the seed points. In previous research carried out (Bolkan [1], Fisher and Jaikumar [2] and Nelson [6]), the locations for seeds were set by carrying out a radial-arm sweep with a half-ray emanating from the depot, dividing the region into sectors, one for each vehicle. The seeds are placed on the rays which bisect the sector boundaries. The generalized assignment approach consistently produces good tours and has generated the best-known tours for several standard vehicle routing test problems (Fisher and Jaikumar [2] and Nelson [6]).

3. A GENERALIZED ASSIGNMENT APPROACH TO THE DEADLINE ROUTING PROBLEM

In this work, the generalized assignment approach is extended to the DVRP. The complication to the VRP imposed by time constraints in vehicle routing can vary widely in difficulty. Problems in which the stops have easily-met time constraints are little different from the standard VRP, and algorithms which primarily seek to minimize total distance travelled are natural choices. If the stops have difficult-to-meet time constraints, scheduling considerations become dominant, and algorithms which are strongly distance-oriented are likely to generate tours with tardy stops. However, the goals of minimizing total distance and avoiding tardiness are somewhat correlated: short tours are more likely to visit stops promptly than are long tours. These considerations led to an iterative approach, in which we initially ignore deadline information, then progressively consider time as needed to avoid tardy stops.

Our approach is illustrated in the flowchart of Figure 1. Both the clustering and routing steps of the generalized assignment approach to the VRP are modified. At the outset, the GAP clustering ignores the deadlines entirely.

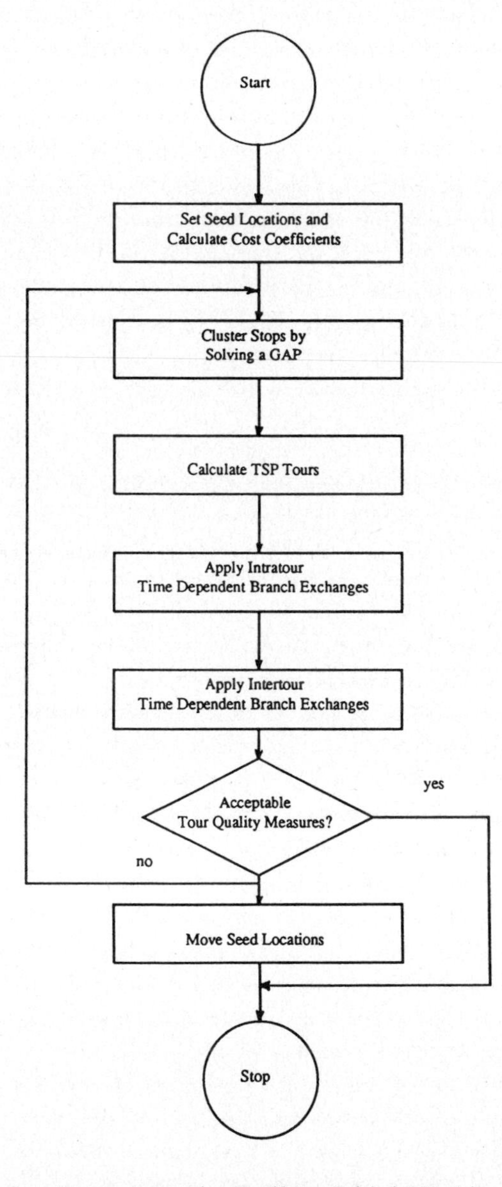

Figure 1. Approach to the Deadline Vehicle Routing Problem

Seeds are set only with geographic information, a GAP is solved and TSP tours are generated. Only the intratour and intertour procedures consider time information. If a tour quality measure is unacceptable, a feedback mechanism is used to move the seeds and produce new cost coefficients for the GAP. The changes are intended to induce clusters for which the resultant tours are more likely to visit the stops on time. The entire process repeats until acceptable tours are found or an iteration count exceeds a pre-specified value.

Each of the procedures which appear in the flowchart is described below, including several variations of intertour branch exchanges and seed movement.

3.1. Setting Seed Locations and Calculating Cost Coefficients

The procedure described in Nelson [6] is used to set the initial seed locations. Figure 2 illustrates the method. The procedure uses an exclusion radius about the depot. Within this radius, stops do not influence seed location. The rationale is that stops close to the depot can easily be served before their deadlines by any of the vehicles, and incur little extra mileage in so doing. The largest radial gap between adjacent stops is calculated. Beginning with a ray from the depot through one of the stops which defines this gap, a radial arm sweep is executed, accumulating the demand of the stops encountered. The goal is to identify a cone for each vehicle. When a specified percentage of the proportionate load of the vehicle is accumulated, a look-ahead process begins. The look-ahead consists of calculating the angle which would be swept if the next stop were included in the cone. If this angle exceeds

(360 degrees)/(number of remaining stops)

then the angle is larger than average and the cone is terminated without including the stop and the next sweep begins. The entire process ends when all stops have been included in a cone. Seeds are placed on the rays which bisect the cones at a distance from the depot which encompasses a specified fraction of the total vehicle load. In this work, the parameters were set as follows: i) exclusion radius of 20 units, with problems in a 100 by 100 unit geographical area, ii) an accumulation of 90% of the proportionate load to trigger the look-ahead process, and iii) seeds placed so that 75% of the vehicle load is closer to the depot than the seed. In empirical testing by Nelson [6], these parameters outperformed others.

3.2. Clustering Stops and Calculating TSP Tours

The algorithm used to solve the generalized assignment problem is essentially the branch and bound procedure of Ross and Soland [9]. Data structures were carefully customized to make the generalized assignment solver as fast as possible. As reported in the computational results section, up to 400 generalized assignment problems with 100 stops and 10 vehicles were solved along with heuristic solutions to 4000 travelling salesman problems in less than 5 minutes

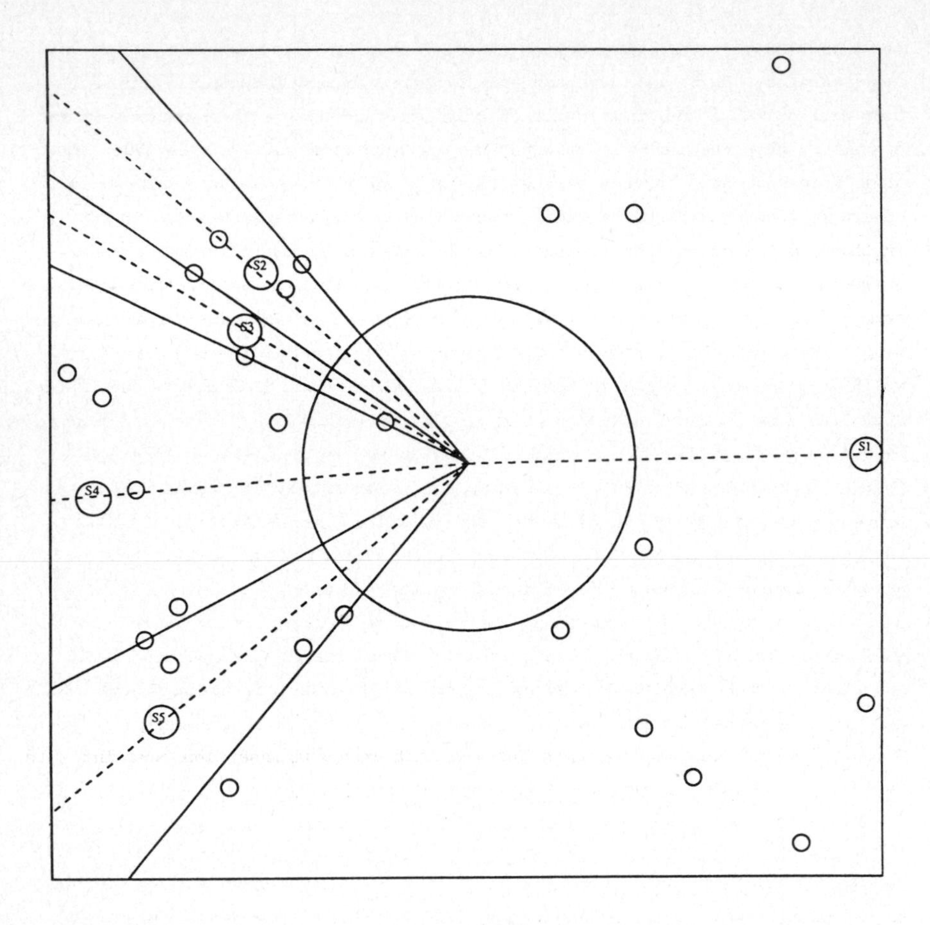

Figure 2. Seed settings in a problem with 5 vehicles, 25 stops, a 100 X 100 area, and 20 unit exclusion radius. Solid lines are the boundaries of the vehicle cones and the seeds S1, S2, S3, S4 and S5 are placed on the dashed lines which bisect the cones.

of CPU time on an IBM 3081 computer. Details of the generalized assignment solver can be found in Nygard, Swenson, Greenberg and Nelson [7].

The travelling salesman tours are generated by a modification of the CCAO heuristic algorithm reported in Golden and Stewart [3]. The CCAO algorithm uses the convex hull of the stops as an initial subtour, builds a full tour using cheapest insertion and greatest angle criteria for selecting and inserting stops, then applies an Or-opt branch exchange tour improvement procedure. Or-opt is a variation of the well known k-opt procedure (Lin and Kernighan [5]), in which the exchanges are restricted to moves of one, two or three adjacent stops from one location in the tour to another. A single branch exchange can drastically modify the time quality measures of a tour. This is because long sequences of stops are left intact by an exchange, but the time at which the vehicle starts these sequences can be greatly changed. A detailed discussion of branch exchanges and time quality of the tours is available in Solomon [10]. In this work, we used a procedure called TOr-opt (Time Or-opt), a modification of Or-opt that rejects any exchange which would increase total tardiness of the tour. Traversing the stops in forward or reverse order does not alter the length of a travelling salesman tour, but does result in different times at which individual stops are visited. The total tardiness calculation in TOr-opt evaluates both forward and reverse traversal of the stops.

3.3. Applying Intertour Time Dependent Branch Exchanges

Two procedures which move stops among tours were developed and tested. A procedure called OFFLOAD seeks time feasibility by transferring stops from routes unable to meet time deadlines to routes which have no tardy stops. For each route with tardy stops OFFLOAD proceeds as follows:

i) Identify the stops whose removal would make the route time feasible.

ii) For each of these stops, calculate the cheapest insertion cost into a time feasible route which does not destroy time feasibility or exceed vehicle capacity.

iii) Transfer the stop with the smallest cheapest insertion cost into its cheapest insertion location on the neighboring route.

iv) Apply TOr-opt to both of the altered routes.

An alternative procedure called SWAP, was also devised and compared with OFFLOAD. In SWAP, pairs of stops from different routes are exchanged if time characteristics of the routes are improved. For each route with tardy stops, SWAP proceeds as follows:

i) Identify the stops whose removal would make the route time feasible.

ii) Create a list of these stops in increasing order of distance to their CLOSEST FEASIBLE NEIGHBOR, the closest stop on a time feasible neighboring route. For each stop, also calculate the distance to the NEXT CLOSEST FEASIBLE NEIGHBOR, the next closest stop on a time feasible neighboring route.

iii) Exchange the stop with its CLOSEST FEASIBLE NEIGHBOR or NEXT CLOSEST FEASIBLE NEIGHBOR, in accordance with whichever exchange would yield the smaller total tardiness after TOr-opt is applied to both routes.

3.4. Moving Seed Locations

In this section three alternative methods of moving seeds are described. Indices n and m refer to stops on a tour carried out by vehicle k. The notation below is used:

$DEADLINE_n$	Latest allowable arrival time at stop n
$MAXDEADLINE_k$	$\max_n\{DEADLINE_n\}$
$TRAVELTIME_{nm}$	Travel time between n and m
$ARRIVALTIME_n$	Time that vehicle k arrives at stop n
$TARDINESS_n$	$\max\{ARRIVALTIME_n - DEADLINE_n, 0\}$
$TOTALTARDINESS_k$	$\sum_n TARDINESS_n$
$SLACK_n$	$\max\{DEADLINE_n - ARRIVALTIME_n, 0\}$
$P_n = (X_n, Y_n)$	Coordinate position of stop n
$S_k = (X_k, Y_k)$	Coordinate position of seed k
$TOURLENGTH_k$	sum of the lengths of all arcs in the tour of vehicle k.

When a cluster of stops produced by solving a GAP results in tours which have tardy stops, the methods discussed here use time-related information to alter the locations of the seed points, seeking clusters for which the tours are on time. Three methods for moving seeds were developed.

The following formula provides a general means of moving seeds in response to a criterion:

$$\Delta S_k = \sum_n Q_{kn} \frac{P_n - S_k}{DISTANCE(P_n, S_k)} .$$

The formula describes a gravity potential model, in which stop n attracts seed k in direct proportion to the factor Q_{kn}, and in inverse proportion to distance. The vector $P_n - S_k$ in the numerator, as illustrated in Figure 3, establishes the direction of the attraction.

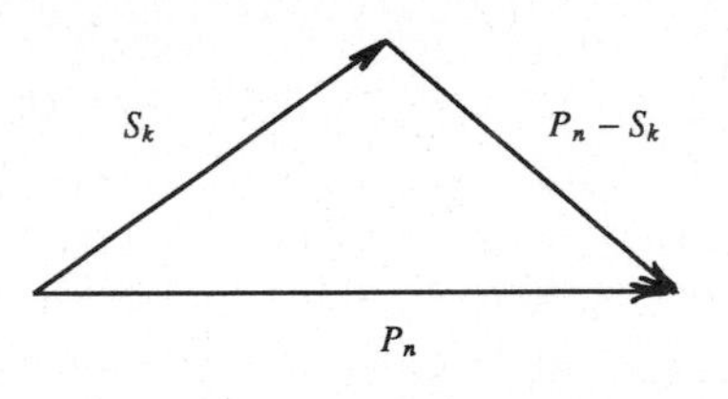

Figure 3. $P_n - S_k$ illustrated.

Defining Q_{kn} in alternative ways provides a family of algorithms for seed movement. In Method 1, Q_{kn} is defined so that the seed of a given tardy tour is drawn toward the stops whose deadlines it is likely to be able to meet. The rationale is that a vehicle is diverted away from stops for which it was tardy in the previous iteration, in the expectation that adjacent vehicles are more likely to be assigned those stops in the new clusters.

For a stop n not on vehicle k's tour in the previous iteration, Method 1 uses the following estimator of the value that $SLACK_{kn}$ would have if vehicle k were to serve stop n.

$$SLACKESTIMATE_{kn} = \max\{DEADLINE_n - TRAVELTIME_{depot,\ seed\ k} - TRAVELTIME_{seed\ k,n}, 0\}.$$

$SLACKESTIMATE_{kn}$ is a calculation of how much time in advance of the stop n deadline vehicle k would arrive if it first travelled directly to seek k, then to stop n. The detailed steps of Method 1 are given in Figure 4.

STEP 1. Carry out STEPS 2 and 3 for each route k which has at least one tardy stop.

STEP 2. For each stop n and seed k, compute the attraction factor Q_{kn} as follows:

$$Q_{kn} = \begin{cases} SLACKESTIMATE_{kn}/MAXDEADLINE & \text{if stop } n \text{ is not on route } k \\ SLACK_{kn}/MAXDEADLINE & \text{if stop } n \text{ is on route } k \end{cases}$$

The division by MAXDEADLINE normalizes the Q_{kn} factors to the interval [0,1].

STEP 3. Move the location of seed k by $(\Delta X_k, \Delta Y_k)$, where

$$\Delta X_k = \sum_n Q_{kn} \frac{(X_n - X_k)}{((X_n - X_k)^2 + (Y_n - Y_k)^2)^{1/2}}$$

$$\Delta Y_k = \sum_n Q_{kn} \frac{(Y_n - Y_k)}{((X_n - X_k)^2 + (Y_n - Y_k)^2)^{1/2}}$$

Figure 4. Steps in Seed Movement Method 1.

Figure 5 illustrates Method 1 for an example with 25 stops, 5 vehicles, a central depot at the origin and a 100 by 100 region. Part A of the figure shows that the initial tour for vehicle 5 is tardy in visiting stops 15, 16, 17 and 19. The seed for vehicle 1 is moved in each of the first 27 iterations. In Part B of the figure (iteration 27), the clusters produced by the GAP solution have assigned stops 12 and 13 to vehicle 2 instead of vehicle 1. Stops 15, 16, 17 and 19 are no longer tardy, but stop 13, now on tour 2, is tardy for the first time. This means that seed 2 will be moved on the next iteration. Part C shows the movement of seed 2 at iteration 28, resulting in stop 13 being reassigned to vehicle 1, and vehicle 3 being assigned 2 stops formerly served by vehicle 2. At iteration 29 (Part D), seed 1 has again moved, stop 13 is again assigned to vehicle 2, and no tardy stops remain. Note that there is considerable variation in total tour length for the various iterations.

In Method 2, seeds are drawn toward difficult stops. The rationale is that the seed is a model of vehicle travel, and that stops with difficult deadlines can possibly be served on time by sending a vehicle directly toward them from the depot. An estimator of the tardiness which would occur if vehicle k served a stop n that is not on the current tour is given below:

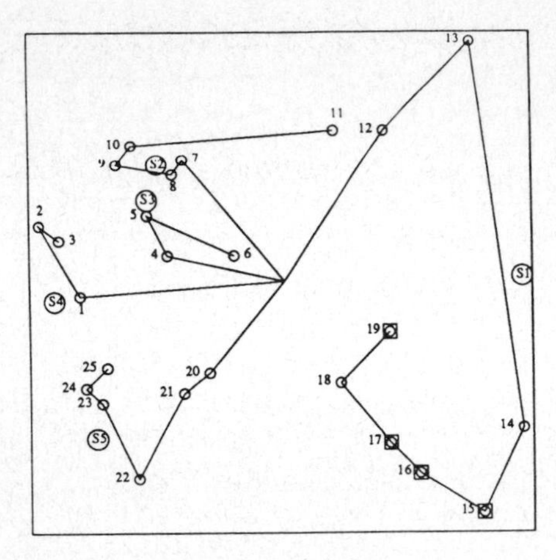

A. Initial Tours. Distance = 654.0 miles

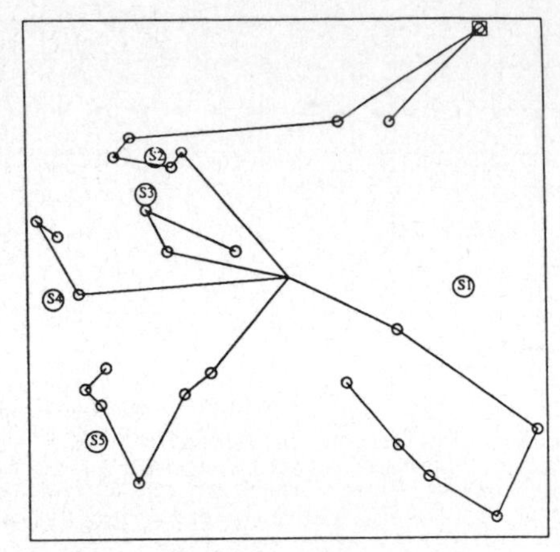

B. Tours after 27 iterations. Distance = 624.5 miles

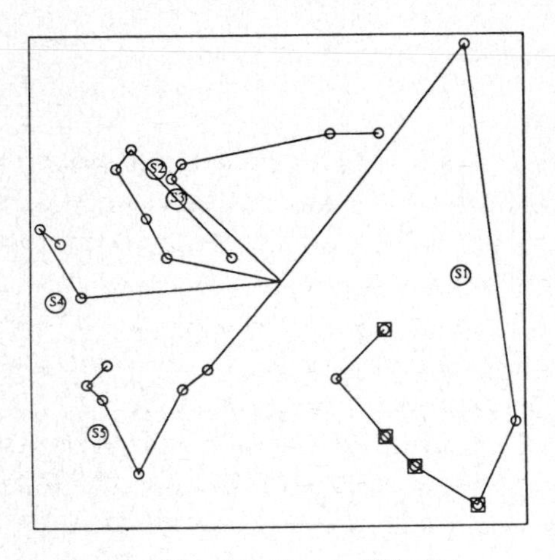

C. Tours after 28 iterations. Distance = 677.9 miles

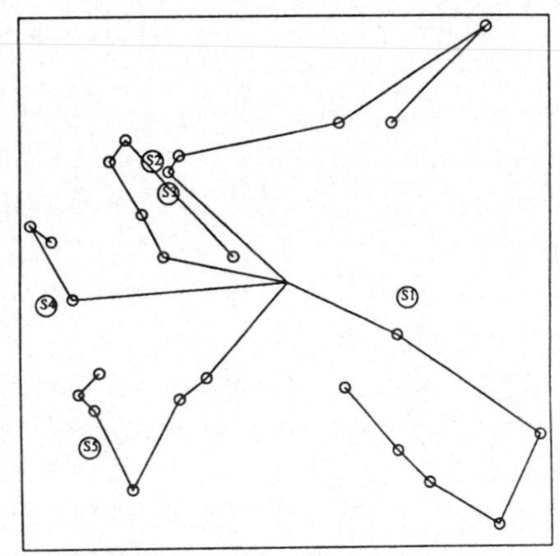

D. Tours after 29 iterations. Distance = 627.7 miles

Figure 5. Example of seed movement in Method 1. Tardy stops are boxed.

$$\text{TARDYESTIMATE}_{kn} = \max\{-\text{SLACKESTIMATE}_{kn}, 0\}$$

$\text{TARDYESTIMATE}_{kn}$ is a calculation of how late vehicle k would arrive at stop n if it first travelled directly to seed k, then to stop n. Method 2 differs from Method 1 only in Step 2, where $\text{SLACKESTIMATE}_{kn}$ and SLACK_{kn} are replaced by $\text{TARDYESTIMATE}_{kn}$ and TARDINESS_{kn}.

Method 3 is a simple rotation of the initial seeds, and is included to determine if forcing a seed dispersion with no time calculations is a competitive approach. We chose rotation increments of 5 degrees for all seeds, calculated clusters with the GAP and found TSP tours. The process stops if a set of tours with no tardy stops is found, or 71 rotations is carried out (return to the starting point).

4. COMPUTATIONAL RESULTS

This section describes the following experimental work: i) a comparison of procedures OFFLOAD and SWAP, and ii) comparisons among Methods 1, 2 and 3 with OFFLOAD in place. The outcome of the experiments shows clear superiority of OFFLOAD over SWAP, and that Method 1 is better with OFFLOAD than without. The comparison of seed movement procedures shows that both Methods 1 and 2 outperform Method 3, and that Method 1 has a slight advantage over Method 2.

4.1. Test Problems

The test problems are fully dense (complete graph) with 100 stops, 10 vehicles and the Euclidean metric. Stop locations were generated randomly in a square, 100 miles on each side. Resolution is one mile, allowing the use of integer coordinates. The depot is located at the center. The demand of each stop is a random integer value between 1 and 100. All vehicles have fixed capacities of 1000. The problem generator code is listed in Nelson [6]. The distributions illustrated in Figure 6 were used to randomly assign deadlines to each of the stops. Deadline distributions 1-4 impose successively tighter time deadline conditions on the same geographical set of stops. Different random number generator seeds were chosen for each deadline distribution, to avoid artificial correlation among the deadlines. Ten different geographical sets of stops were generated, producing a basic set of 40 test problems.

4.2. Comparison of OFFLOAD and SWAP

For each of the 40 basic test problems, OFFLOAD and SWAP were applied after clustering with the GAP and solving the travelling salesman problems. No feedback was invoked, to isolate the effect of OFFLOAD and SWAP. Vehicle speeds were set at 55 miles per hour. This is consistent with non-metropolitan area delivery problems in the upper midwest. Given that the deadline distributions create problems which range from very loose to very tight time constraints, the results would be expected to hold for other speeds as well.

The methods were compared using tour length and 5 measures of time quality. Although both procedures were effective, OFFLOAD clearly outperformed SWAP in all the measures of time quality. For example, OFFLOAD produced time feasible solutions to 17 of the 40 problems, in sharp contrast to 8 of 40 problems for SWAP. Both procedures increased tour length by about 3% above the initial tours that disregarded time deadlines. Both methods dramatically reduce total

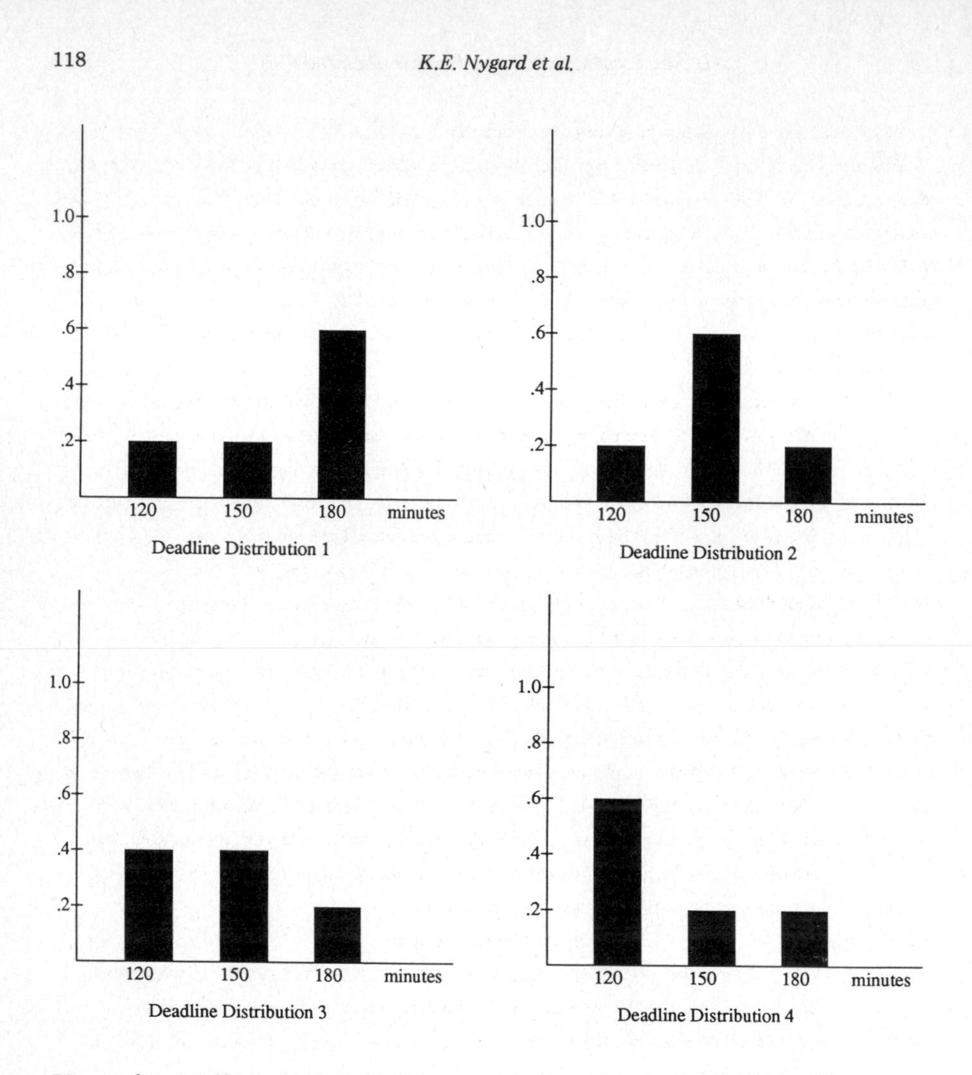

Figure 6. Deadline distributions applied to each of the geographical problems.

tardiness in the looser problems with deadline distributions 1 and 2, and reduce tardiness somewhat even in the tightest problems. OFFLOAD generated the tours with the least total tardiness for 36 of the 40 problems. Similarly, OFFLOAD generated tours with the smallest number of tardy stops for 38 of the 40 problems.

Both procedures require very little CPU time. Over all 40 problems, execution time averaged 0.302 seconds for OFFLOAD and 2.14 seconds for SWAP. We conclude that both methods can detect small changes to tours which improve measures of time quality with little sacrifice in tour length. OFFLOAD was chosen for subsequent experiments.

4.3. Comparisons Among Methods 1, 2, and 3

The same set of test problems used in the comparison of OFFLOAD and SWAP were used to compare Methods 1, 2, and 3. In addition, each of the 40 problems was made increasingly tighter by successively reducing vehicle speeds to 52.5, 47.5, 42.5, and 37.5 miles per hour. Thus, 160 test problems were solved for each of the three methods. Tables 1-4 show the results for each of the four vehicle speeds. Several points of comparison among the alternative methods are discussed below.

1. Methods 1, 2, and 3 successfully produced time feasible solutions for 53, 47, and 22 percent of the 160 problems. In the loosest problems, Methods 1 and 2 are highly likely to find time feasible solutions, and even Method 3 is reasonably successful. Over all vehicle speeds, Method 1 is uniformly more successful in finding time feasible tours than the other methods.
2. Tour length increase may be thought of as a price to pay to achieve tours with better time performance characteristics. Methods 1 and 2 were little different in this respect, with average tour length increases over all problems of 4.1 and 4.4 percent, respectively. Method 3 increased tour lengths by only about one percent.
3. Methods 1 and 2 perform very closely in reduction of total tardiness, averaging 78 and 79 percent reduction respectively over all problems. Again Method 3 is not competitive, achieving an average 51 percent reduction. Even in the tightest problems, these methods can significantly reduce total tardiness.
4. In total number of tardy stops, Methods 1 and 2 are again close, lending credence to the effectiveness of the ideas underlying these methods. Method 1 is uniformly better than the other methods, producing an average decrease in tardy stops of 67 percent, compared with 62 and 38 percent for Methods 2 and 3. As expected, these percentages are somewhat less than the percentages for total tardiness reduction. This is because as problem tightness increases, the methods tend to find small reductions in the tardiness of the individual stops without actually meeting all deadlines.
5. Maximum CPU time allowed was set at 300 seconds for all runs. The tables reveal that required CPU time rapidly increases with problem tightness. For the very tightest problems with speed of 37.5 mph (Table 4), all the Method 1 runs timed out and all but one of the Method 2 runs timed out. In these problems, about 400 iterations through the feedback loop are made in the available 300 seconds. Within each loop, about 40% of the time is spent solving the generalized assignment problem, 40% in the OFFLOAD procedure, and 20% in

solving the travelling salesman problems and adjusting seed locations. In the tightest problems, Method 3 rotates the seeds 5 degrees and loops a total of 71 times, before terminating upon return to the starting seed locations.

To evaluate the value of the OFFLOAD procedure in the overall scheme, Method 1 was run with OFFLOAD suppressed for 10 problems with deadline distribution 2. With OFFLOAD, an average of 5.7 loops were executed, all 10 problems became feasible, and average CPU time was 4.6 seconds. Without OFFLOAD, an average of 44.5 loops were executed, 6 of the 10 problems became feasible within 5 minutes of CPU time, and average CPU time averaged 114.6 seconds. We conclude that OFFLOAD makes an important contribution to the overall method.

Table 1. Comparison of 3 seed movement methods with speed set at 52.5 mph.

	Deadline Distribution	Number of Problems	Initial Solution	Method 1	Method 2	Method 3
Number of problems which attained time feasibility	1	10	0	10	10	10
	2	10	0	10	10	10
	3	10	0	10	10	3
	4	10	0	10	7	2
TOTAL	-	40	0	40	37	25
Average tour length (miles per vehicle)	1	10	130.0	135.2	135.7	134.8
	2	10	130.0	138.5	138.7	134.2
	3	10	130.0	139.8	139.2	130.9
	4	10	130.0	141.1	138.9	130.9
Average total tardiness (minutes)	1	10	50.2	0.0	0.0	0.0
	2	10	67.5	0.0	0.0	0.0
	3	10	138.1	0.0	0.0	52.2
	4	10	175.6	0.0	18.8	77.0
Average number of tardy stops	1	10	2.9	0.0	0.0	0.0
	2	10	4.8	0.0	0.0	0.0
	3	10	7.2	0.0	0.0	3.9
	4	10	9.6	0.0	2.3	5.7
Average CPU seconds, IBM 3081 (execution only)	1	10	-	0.72	1.46	1.21
	2	10	-	5.82	5.92	4.24
	3	10	-	13.35	9.14	32.19
	4	10	-	22.74	87.80	38.87

Table 2. Comparison of 3 seed movement methods with speed set at 47.5 mph.

	Deadline Distribution	Number of Problems	Initial Solution	Method 1	Method 2	Method 3
Number of problems which attained time feasibility	1	10	0	10	10	8
	2	10	0	10	9	1
	3	10	0	8	8	1
	4	10	0	4	2	0
TOTAL	-	40	0	32	29	10
Average tour length (miles per vehicle)	1	10	130.0	137.0	137.4	149.5
	2	10	130.0	140.8	138.0	130.0
	3	10	130.0	138.9	139.6	128.1
	4	10	130.0	135.1	135.0	128.7
Average total tardiness (minutes)	1	10	114.0	0.0	0.0	10.7
	2	10	172.7	0.0	3.1	76.4
	3	10	274.5	13.1	16.6	149.3
	4	10	344.7	72.9	89.0	212.6
Average number of tardy stops	1	10	5.1	0.0	0.0	0.9
	2	10	9.0	0.0	0.4	5.7
	3	10	11.3	1.2	1.0	8.1
	4	10	13.5	5.1	5.8	9.1
Average CPU seconds, IBM 3081 (execution only)	1	10	-	5.80	5.44	21.69
	2	10	-	21.29	34.45	43.61
	3	10	-	83.56	69.80	41.50
	4	10	-	202.36	226.15	43.36

Table 3. Comparison of 3 seed movement methods with speed set at 42.5 mph.

	Deadline Distribution	Number of Problems	Initial Solution	Method 1	Method 2	Method 3
Number of problems which attained time feasibility	1	10	0	8	7	0
	2	10	0	4	1	0
	3	10	0	1	0	0
	4	10	0	0	0	0
TOTAL	-	40	0	13	8	0
Average tour length (miles per vehicle)	1	10	130.0	137.7	137.0	128.2
	2	10	130.0	136.1	133.3	128.4
	3	10	130.0	133.3	134.3	128.3
	4	10	130.0	133.3	132.7	128.6
Average total tardiness (minutes)	1	10	252.3	13.8	33.8	151.8
	2	10	396.5	80.8	74.9	240.5
	3	10	528.0	160.7	167.2	361.7
	4	10	640.4	195.6	259.3	420.3
Average number of tardy stops	1	10	10.6	1.0	2.3	7.9
	2	10	15.0	5.8	6.2	11.7
	3	10	17.8	9.5	11.5	14.6
	4	10	20.1	15.2	13.1	15.6
Average CPU seconds, IBM 3081 (execution only)	1	10	-	78.18	96.66	44.29
	2	10	-	209.32	253.91	42.67
	3	10	-	257.64	281.21(a)	41.78
	4	10	-	281.21(a)	281.21(a)	40.85

(a) Average CPU time of 281.21 seconds indicates all problems in the category exhausted the maximum CPU time allowed.

Table 4. Comparison of 3 seed movement methods with speed set at 37.5 mph.

	Deadline Distribution	Number of Problems	Initial Solution	Method 1	Method 2	Method 3
Number of	1	10	0	1	1	0
problems which	2	10	0	0	0	0
attained time	3	10	0	0	0	0
feasibility	4	10	0	0	0	0
TOTAL	-	40	0	1	1	0
Average tour	1	10	130.0	134.7	133.6	129.0
length	2	10	130.0	132.0	132.3	128.6
(miles)	3	10	130.0	133.0	132.7	128.5
	4	10	130.0	135.6	132.7	128.6
Average total	1	10	568.7	255.8	202.1	401.7
tardiness	2	10	796.4	383.5	333.5	569.4
(minutes)	3	10	874.9	623.0	518.2	587.2
	4	10	1129.3	787.6	655.3	816.1
Average number	1	10	17.2	10.7	10.4	15.2
of tardy stops	2	10	21.8	16.1	15.0	18.9
	3	10	24.0	20.2	21.0	20.7
	4	10	26.7	23.9	23.6	23.7
Average CPU	1	10	-	281.21(a)	254.25	42.38
seconds, IBM 3081	2	10	-	281.21(a)	281.21(a)	42.53
(execution only)	3	10	-	281.21(a)	281.21(a)	38.74
	4	10	-	281.21(a)	281.21(a)	38.41

(a) Average CPU time of 281.21 seconds indicates that all problems in the category exhausted the maximum CPU time allowed.

5. SUMMARY AND CONCLUSIONS

This study shows that a generalized assignment approach can be successfully extended to vehicle routing problems with deadlines. We believe that the OFFLOAD postprocessor and the feedback technique enjoy a symbiotic relationship in the approach. The use of OFFLOAD to carry out local modifications of the tours significantly reduces the number of iterations through the feedback loop, thereby reducing the number of GAP's and travelling salesman problems which are solved before a time feasible tour is produced. On the other hand, the global adjustments to seed locations are typically carried out several times before clusters which are routed without tardy stops are produced.

Time constraints add considerable complexity to already difficult routing problems, and it is not surprising that problems with difficult-to-meet deadlines do not easily succumb to the methods discussed here. We think that the fundamental approach of progressively moving from good tours to modestly altered tours which do measurably better in meeting time constraints fits many practical problems. In addition, the approach would be an ideal basis for a man-machine interactive approach to routing. It would be easy to allow human control and intervention at any iteration in the feedback process.

Much additional work could be done. Many alternative schemes for seed movement could be developed and evaluated. Step sizes could be varied in an attempt to more quickly find good seed locations. A natural extension is to become more selective about choosing which seed-to-stop-assignment costs to alter. For example, it would be possible to not change the seed locations as such, but only alter the costs of stops which are identified by some criterion.

REFERENCES

[1] Bolkan, W. E., The Vehicle Routing Problem with Time Windows, Master's thesis, North Dakota State University, Fargo, North Dakota, May, 1986.

[2] Fisher, M. and Jaikumar, R., "A Generalized Assignment Heuristic for Vehicle Routing," Networks 11, pp. 109-124, 1981.

[3] Golden, B. L. and Stewart, W. R., "Empirical Analysis of Heuristics," in The Travelling Salesman Problem, Lawler, E., Lenstra, J., Rinnooy Kan, A., and Shmoys, D. B., eds., Wiley, pp. 207-249, 1985.

[4] Laporte, G., Desrochers, M., and Norbet, Y., "Two Exact Algorithms for the Distance-Constrained Vehicle Routing Problem," Networks 14, pp. 161-172, 1984.

[5] Lin, S. and Kernighan, B., "An Effective Heuristic Algorithm for the Travelling Salesman Problem," Operations Research 21, pp. 498-516, 1973.

[6] Nelson, M. D., "Implementation Techniques for the Vehicle Routing Problem," Master's thesis, North Dakota State University, Fargo, North Dakota, May, 1983.

[7] Nygard, K. E., Swenson, E., Greenberg, P., and Nelson, M., "Implementation and Comparison of Lagrangian Relaxation Algorithms for the Generalized Assignment Problem," preprint, Department of Computer Science and Operations Research, North Dakota State University, December, 1986.

[8] Or, I., "Travelling Salesman Type Combinatorial Problems and Their Relation to the Logistics of Blood Banking," Ph.D. thesis, Department of Industrial Engineering and Management Science, Northwestern University, 1976.
[9] Ross, G. T. and Soland, R. M., "A Branch and Bound Algorithm for the Generalized Assignment Problem," Mathematical Programming 8, pp. 91-103, 1975.
[10] Solomon, M., "Vehicle Routing and Scheduling with Time Window Constraints: Models and Algorithms," Ph.D. Dissertation, Northeastern University, 1984.
[11] Stewart, W., "New Algorithms for Deterministic and Stochastic Vehicle Routing Problems," DBA dissertation, University of Maryland, 1981.
[12] Swenson, E. J., "The Vehicle Routing Problem with Time Constraints," Master's thesis, North Dakota State University, Fargo, North Dakota, May 1986.

Vehicle Routing: Methods and Studies
B.L. Golden and A.A. Assad (Editors)

VEHICLE ROUTING WITH BACKHAULS: MODELS, ALGORITHMS, AND CASE STUDIES

Daniel O. Casco, General Electric Information Services,
Rockville, MD 20850

Bruce L. Golden, College of Business and Management,
University of Maryland, College Park, MD 20742

Edward A. Wasil, Kogod College of Business Administration,
American University, Washington, D.C. 20016

The purpose of this paper is to examine the vehicle routing problem with backhauls. Very little has been written about this problem in the operations research literature. To begin, we define the problem and explore the ways in which this problem is related to the standard vehicle routing problem and the general pickup and delivery problem. Next, several first-generation algorithmic procedures are surveyed, and key observations are highlighted. A variety of backhaul situations and their associated modeling assumptions are discussed. Finally, a sampling of real-world practice is described.

1. INTRODUCTION

In the basic version of the vehicle routing problem (VRP), a set of delivery customers with known demands is to be serviced by a homogeneous fleet of fixed-capacity vehicles from a single distribution center or depot. The objective is to provide each vehicle with a sequence of deliveries so that all customers are serviced, and the total distance traveled by the fleet is minimized. Typically, vehicles leave the distribution center with a full load or nearly full load and return empty. The basic VRP is an extremely complex combinatorial problem that has been studied extensively in the literature over the last twenty years. The paper by Bodin et al. [2] provides the reader with a comprehensive survey of the basic problem and its many variations.

A variation of the VRP which has received little research attention concerns the routing of vehicles over a set of mixed customers; that is, some customers are delivery points but others are pickup, or backhaul, points. When a vehicle services a backhaul point, an amount of product (bound for the distribution center) is loaded onto the vehicle. As an example, such a customer mix occurs routinely in the grocery industry. Supermarkets are the delivery points and grocery suppliers, such as poultry processors or vegetable and fruit vendors, are the backhaul points.

In recent years, the grocery industry in the USA has recognized the cost-cutting potential of servicing backhaul points on predominantly delivery routes. The industry has saved upwards of $160 million a year since 1982 on its distribution costs by allowing empty vehicles returning from deliveries to pick up large volumes of inbound products. Keep in mind that most vehicles are rear-loaded so that rearranging on-board delivery loads is difficult and deliveries are higher-priority stops than pickups so that practical considerations usually dictate that the number of backhauls be small and that they be serviced near the end of a route.

Given this information as background, the vehicle routing problem with backhauls (VRPB) can be stated as follows: Find a set of vehicle routes that services the delivery and backhaul customers such that vehicle capacity is not violated and the total distance traveled is minimized.

We point out that the VRPB is a special case of the general pickup and delivery problem. In this problem, a vehicle makes a pickup at one location and delivers the product to a second location that is not necessarily the distribution center. This type of operation occurs with messenger services.

In the remainder of this paper, we focus on the VRPB. In the next section, we present a number of important issues that must be addressed when modeling the VRPB. Then, several heuristic procedures that solve the VRPB are discussed. The paper concludes with an examination of real-world practice in dealing with backhauls.

2. MODELING THE BACKHAUL PROBLEM

When thinking about designing a heuristic method to solve the VRPB, it is necessary to first specify the type of backhaul situation under consideration. A number of important questions, such as the following, come to mind.

1. What is the percentage of stops that are backhauls?
2. Must backhauls always come after all deliveries are made?
3. If the answer to the second question is no, then are there specific rules defining when a backhaul is allowed?
4. What is the size of a typical backhaul relative to vehicle capacity?
5. Is the number of trucks (drivers) fixed in advance?
6. Can vehicles run multiple trips per day?
7. Must all backhauls be serviced or can some be either postponed or left for the supplier to handle?
8. Can backhauls be split?
9. Will a vehicle ever service backhauls only?

The answers to these questions will dictate the type of backhaul problem that needs to be addressed. One key point that we make at this time is that a wide variety of backhaul problems is encountered in practice. We return to this point in Section 5 on real-world applications. This observation may partially explain why so little research in this area has been published.

3. SOLVING THE VRPB: ALGORITHMS AND ISSUES

Several algorithms for solving the backhaul problem have been developed and tested computationally. For the most part, these algorithms are modified versions of three procedures that have been used to solve more basic routing problems: Clarke and Wright Savings [3], Cheapest Insertion [7], and Space-filling Curves [1]. The modified algorithms that have been developed recently are summarized in Table 1. For each of these modified algorithms, a different type of backhaul problem is considered. For example, Goetschalckx and Horsley [5] assume that backhauls always come after all deliveries. Table 1 includes information about the type of backhaul problem that each algorithm can solve. It is important to keep this in mind when reading the following discussions of each algorithm, the key developmental issues, and the associated computational results.

A straightforward attempt to handle backhauls using the Clarke and Wright method was first proposed by Deif and Bodin [4]. In the well-known Clarke and Wright procedure for the basic vehicle routing problem, a savings term governs the formation of routes. If delivery stops i and j are joined on the same route, then a savings of

$$S(i,j) = C(1,i) + C(1,j) - C(i,j) \qquad (1)$$

results, where C(i,j) is the cost of linking i and j and 1 is the distribution center. As a first step, Deif and Bodin computed the usual savings in (1) for all customer pairs but imposed the condition that backhauls must occur after deliveries. Therefore, once a backhaul point is placed at the end of a route, no delivery points can be added after the backhaul. In general, this modification does not work well since individual routes may be terminated quite early, resulting in a large number of relatively short routes.

To form longer routes that satisfy the backhauls-only-after-deliveries condition, Deif and Bodin proposed to delay the connection of backhauls to deliveries. This was accomplished by altering the savings definition in (1) to include a term that penalizes (and, hence, delays) the connection of a backhaul and a delivery. As a result, routes grow to be much fuller. The backhaul savings definition is given by

$$BS(i,j) = C(1,i) + C(1,j) - C(i,j) - \pi s, \qquad (2)$$

where

s is an estimate of the maximum savings,

π is penalty multiplier, and

i is a backhaul and j is not, or vice versa.

A computational experiment was performed to test the effectiveness of the savings modification in solving the backhaul problem with various choices of the penalty multiplier. Test problems were generated which included 100, 200, or 300 customers of which 10, 30, or 50 percent were specified as backhauls.

Table 1. Algorithms for Solving the VRPB

Article	Routing Algorithm	Modifications	Constraints	Modeling Assumptions
Deif and Bodin [4]	CW	Delay connection of pickups to deliveries by altering savings definition to include a penalty multiplier.	· Vehicle capacity · Route length	· No deliveries after pickups · 10%, 30%, 50% of all points are backhauls
Golden et al. [6]	CW CI	CI inserts backhauls onto CW all-delivery routes. Insertion costs altered to penalize the number of deliveries after a pickup.	· Vehicle capacity · Route length	· Deliveries after pickups are allowed · Many more deliveries than pickups · About 10% of all points are backhauls
Goetschalckx and Horsley [5]	SFC	Use SFC and GM or KM to generate all-delivery and all-backhaul routes. Routes are then merged and improved using 2-opt and 3-opt procedures.	· Vehicle capacity	· No deliveries after pickups · About 20%, 33%, 50% of all points are backhauls
Yano et al. [9]	SC	Generate initial feasible solution heuristically. Set-covering problem solved with branch-and-bound.	· Vehicle weight and volume · Driving time · Total time on road · Open/close times at destination	· No deliveries after pickups · Maximum of 4 deliveries and 4 pickups per route
Casco et al.	CW CI	CW with reduced vehicle capacity generates delivery routes. Backhauls incur radial distances. Insertion costs modified to penalize delivery load after pickup. Backhaul service rules identify good solutions.	· Vehicle capacity · Route length	· Deliveries after pickups are allowed · About 20% of all points are backhauls

CI Cheapest Insertion
CW Clarke-Wright
GM Greedy Method
KM K-Median Method
SC Set Covering
SFC Spacefilling Curves

Ten values of π, from .05 to .50, were tried. Four values of π (.05, .10, .15, and .20) produced the best (shortest distance) routes.

A different heuristic which solves the backhaul problem using an insertion-based procedure was proposed by Golden et al. [6]. In this procedure, routes are first developed for all of the deliveries, and then backhaul points are considered for insertion between stops on these routes. The cost of inserting a backhaul point k on a route between stops i and j is specified by

$$C(k) = C(i,k) + C(k,j) - C(i,j) + Pd, \tag{3}$$

where

P is a penalty multiplier, and

d is the number of deliveries following k.

The penalty multiplier serves to delay insertion of the backhauls until the end of a route. For each backhaul k, the insertion cost is specified by the minimum C(k) for all links (i,j). We point out that once a backhaul is inserted, the direction of that route is <u>fixed</u> in the i-to-j direction that generated the minimum C(k). Since the decision to insert a backhaul is based on d, the number of delivery stops remaining after the backhaul is serviced, we refer to this heuristic as a stop-based backhaul insertion procedure.

Golden et al. also experimented with a variant of the Deif and Bodin approach in which the requirement that backhauls must be at the end of a route is relaxed. In other words, deliveries can follow backhauls if backhaul savings, BS(i,j), is positive. Golden et al. then evaluated the performance of the Deif and Bodin and the stop-based backhaul insertion procedures on a set of three problems, each consisting of 50 deliveries and 5 backhauls. In both procedures, as the penalty terms are increased (π is increased over the range .05 to .50, P over the range .50 to 15.50), the number of deliveries after backhauls decreases to zero while the total route length increases. The Deif and Bodin procedure produced shorter routes when backhauls were required to be at the end of the route. The stop-based backhaul insertion procedure generally produced shorter routes when a backhaul was allowed to be inserted as soon as vehicle capacity allowed.

The heuristic procedure used by Goetschalckx and Horsley [5] to solve the backhaul problem is based on the notion of spacefilling curves developed by Bartholdi and Platzman [1]. Delivery and backhaul points are first transformed from the plane to a line using the spacefilling curve transformation and then sorted in non-decreasing order according to their positions along the line. That is, the transformation assigns a number between 0 and 1 to each point. Next, each set of points is clustered independently to form all-delivery and all-backhaul routes. (We point out that clustering and routing are essentially accomplished simultaneously--once points are clustered they are then visited in the order in which they are sorted.) Two different methods, one a greedy method and the other an approach based on the solution to the K-median problem, are used to cluster the points. The clustering step using the greedy method proceeds as follows. A new route is formed starting with the point not yet on a route having the smallest assigned number. Larger and larger numbered points are added to this route until the truck is full. A new route is then started and the method is repeated until all points are placed on a route.

To form clusters using the K-median method first requires specifying K, the number of routes to be formed. The line is then divided into K identical intervals and the medians are selected to be the points closest to the mid-points of the intervals. Clustering begins by selecting the point not yet on a route having the smallest assigned number. If truck capacity remains, this point is assigned to route R, where R is the closest median point. If the truck is overloaded, this point is assigned to route R+1. The method is repeated until all points are placed on a route. If all points cannot be placed on a route, K is increased by one and the clustering is repeated. With this method, points which violate vehicle capacity are "pushed" to the next cluster. If there are an insufficient number of clusters, the line is then further divided until enough clusters are formed so that all points may be placed on a route.

After applying the clustering step to obtain all-delivery and all-backhaul routes, each delivery route is merged with a corresponding (nearby) backhaul route to form one route in which backhauls take place after all deliveries have been made. To examine the performance of the two spacefilling curve heuristics, 15 random test problems, ranging in size from 25 to 200 points, with 20% to 50% specified as backhauls, were generated. For each problem, truck capacity was varied over several values to produce a total of 57 test problems. Routes produced by the greedy method and the K-median method for this set of problems were improved using a 2-opt procedure followed by a 3-opt procedure. The authors concluded that both methods generated a set of routes with comparable total lengths, but the greedy method used fewer trucks and utilized their capacities more effectively.

Yano et al. [9] report on the development of a microcomputer-based routing system that incorporates backhaul locations into truck routes for a retail chain consisting of 40 stores with a fleet of 11 vehicles. The route generation procedure is a two-step operation that consists of finding an initial feasible solution using a route-construction heuristic followed by a set-covering routine that uses a branch-and-bound framework to find the optimal solution. They report finding solutions to problems with 20 to 40 pickup and delivery locations in 10 to 30 minutes on a microcomputer. All pickups followed deliveries and the authors point out that because of truck volume and route distance constraints, routes with as many as 4 stops were rare. The small number of stops per route makes an optimal solution procedure feasible in this setting. The authors also discuss a model that incorporates common carrier trucks when pickups and deliveries exceed the capacity of the 11 truck fleet.

4. RECENT EXPERIMENTS IN SOLVING THE VRPB

Recently, the authors have experimented with a <u>load-based backhaul insertion procedure</u> for the VRPB. Our heuristic (coded in Pascal and run on a Macintosh microcomputer) enhances the stop-based backhaul insertion procedure of Golden et al. [6] in four ways.

- An initial feasible solution to the VRPB is generated by using a Clarke-Wright solution for the delivery points and a straight line (backhaul to distribution center) route for each pickup point.

- Initial Clarke-Wright solutions can be generated with route capacities smaller than the actual vehicle capacities in order to increase flexibility in assigning backhauls to routes.
- The decision to insert a backhaul considers the delivery load that remains on the vehicle after the backhaul is serviced.
- Solutions to the VRPB can be filtered through a set of multi-criteria rules that can help to identify high quality solutions.

In the remainder of this section, we focus on the effect of these enhancements on the solutions generated by the load-based procedure for the VRPB.

4.1. Generating an Initial Solution in the Load-Based Insertion Procedure

As in the stop-based procedure, a Clarke-Wright heuristic is used to generate a set of routes for the delivery locations. To incorporate the backhaul locations into the initial solution, each backhaul is joined to the distribution center by a direct (straight line) link. These links are known as radial links, and they incur a radial distance. The rationale for including the radial distances as part of the total distance is straightforward. Instead of servicing the backhaul with a vehicle in the existing fleet, the company could choose to have a common carrier pick up and deliver the load. In the heuristic, the cost of this operation is chosen to be proportional to the straight line distance between the backhaul location and the distribution center. For example, the total cost of linking a backhaul and the distribution center might be 1.5 times the straight line distance. The additional cost above the straight line distance is referred to as the radial surcharge, and it represents the carrier's profit.

To illustrate the radial link concept, consider the set of delivery and backhaul customers shown in Figure 1. Applying the Clarke-Wright procedure to the deliveries and linking the backhauls to the distribution center via the radial links produces the routes shown in Figure 2. The total distance for this set of routes is the sum of the distances of the all-delivery routes produced by Clarke-Wright and the radial distances which link the backhauls to the distribution center.

4.2. Modifying the Backhaul Insertion Cost

Recall that in the stop-based insertion procedure, the decision to insert a backhaul was based on the number of delivery stops remaining after the backhaul. Applying this procedure could result in a large delivery following a backhaul since the specific load at each remaining delivery is not taken into account. We have modified the insertion cost in (3) to include a penalty term based on delivery load after pickup. The cost of inserting a backhaul k on a route between stops i and j that also incorporates the backhaul's radial distance is now given by:

$$C'(k) = C(i,k) + C(k,j) - C(i,j) - (1+R) \times C(1,k) + PD, \qquad (4)$$

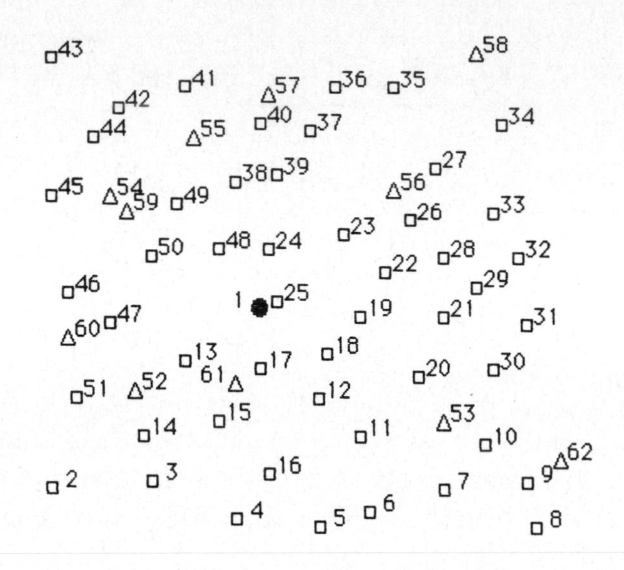

Locations	Number	Symbols
Distribution Center	1	●
Delivery	50	□
Backhaul	11	△

Parameters	Settings
Vehicle Capacity	160
Vehicle Speed (mph)	20
Pickup Load	40
Radial Surcharge	50%

Figure 1. VRPB Example

where

R is the radial surcharge,

C(1,k) is the radial distance between the distribution center and k,

P is a penalty multiplier, and

D is the number of units delivered after servicing k.

This load-based insertion cost will allow more flexibility in servicing backhauls than does the stop-based procedure. The latter procedure treats all delivery stops equally without regard to the delivery load. The new procedure recognizes that not all deliveries are of the same size and that insertions prior to the end of a route might be considered reasonable if the delivery load

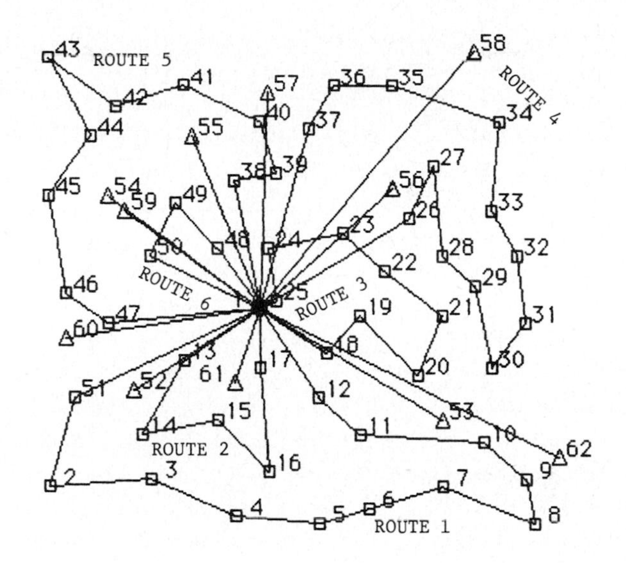

Clarke-Wright distance = 585
Radial distance = 284
Radial surcharge = 142 (.5 x 284)
Total distance = 1011

Figure 2. Initial Solution Using Clarke-Wright and Radial Links

that remains is small enough. With the cost shown in (4), inserting a backhaul k before several small-load delivery stops might now look appealing because D would be small. The driver can make the pickup and is not inconvenienced by having to shuffle large on-board delivery loads to accommodate the backhaul load.

To illustrate this concept, consider the route composed of deliveries 48, 49, and 50 in Figure 2. Suppose we wish to insert backhauls 54 and 59 onto this route. Using the stop-based insertion procedure, these backhauls are inserted between deliveries 49 and 50. This procedure places the backhauls towards the end of the route (1-48-49-54-59-50-1). Given that the demands of deliveries 48, 49, and 50 are 25, 9, and 41 units, respectively, applying the load-based procedure essentially reverses the direction of the route to 1-50-59-... In the load-based route, the total delivery load after pickup is 25 + 9 = 34, as compared to 41 in the stop-based route. This illustrates one of the shortcomings of the stop-based approach. Although the total distance traveled is the same for both routes, the sensitivity of the load-based procedure to delivery stop demand allows greater flexibility in placing backhauls onto routes. In structuring larger routes, this increased flexibility should help produce better quality solutions.

Although our computational analysis is preliminary at this stage and the problems tested have been small, the load-based insertion approach seems to compare favorably with the stop-based insertion approach in terms of total distance.

4.3. Identifying High Quality Solutions

By varying the penalty multiplier in the load-based insertion procedure, many candidate solutions to the VRPB can be generated. Some of the solutions might insert backhauls too early on a route, nearly filling the truck with delivery and backhaul loads, and thus, inconveniencing the driver by forcing him to rearrange the on-board load. Usually, this increases the amount of service time required at each customer location. In most distribution operations, deliveries receive higher priority than pickups. Management insists that the vast majority of units on a route be delivered before pickups are considered. In accordance with these observations, we propose a set of rules to help identify high quality, feasible solutions. These rules are somewhat ad hoc, but, to a large extent, they follow what common sense would dictate.

Rule 1. A vehicle servicing a particular set of customers should deliver sixty percent of its units before accepting any backhauls.

Rule 2. If a delivery follows a backhaul on a proposed route, then accept the backhaul as long as no more than eighty percent of the vehicle is full after doing so.

Rule 3. If no deliveries remain, then all backhauls are eligible to the extent permitted by vehicle capacity.

The first rule ensures that a substantial portion of the units are delivered before backhauls are accepted. The second rule ensures that some vehicle capacity for "load shuffling" remains after accepting a backhaul. Since many vehicles are rear-loaded and rearranging the on-board load to accommodate the pickup can be difficult and time-consuming, the idea is to reserve a portion of the vehicle's capacity (in this case, 20%) to allow for load shuffling. The percent of vehicle capacity that must remain empty can be viewed as a parameter to be set by the dispatcher. The final rule allows the entire capacity of the vehicle to be used for backhauls, once all deliveries have been completed.

To illustrate the application of these rules to candidate solutions, consider the configuration shown in Figure 3. This set of routes, known as Routes Configuration 1 (RC1), was obtained by inserting the backhauls onto the Clarke-Wright routes (shown in Figure 2) using the load-based insertion procedure.

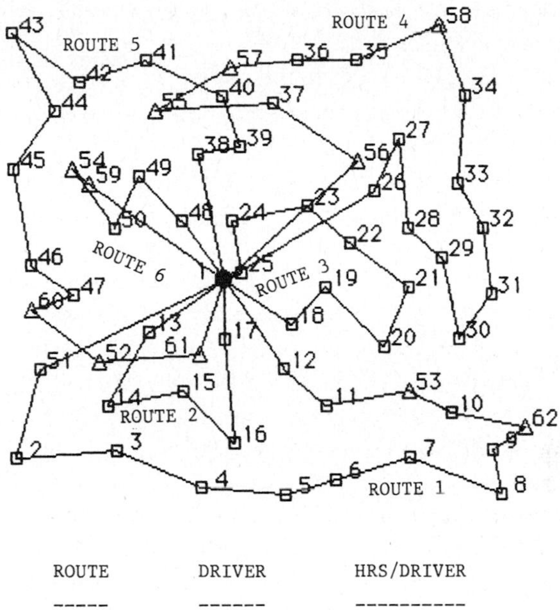

ROUTE	DRIVER	HRS/DRIVER
1	1	7.7
2	2	6.6
3	2	-
4	3	9.6
5	4	6.9
6	5	2.9
		AVG 6.74 HRS/DRIVER

Figure 3. Routes Configuration 1 (RC1)

The total travel distance for these routes is 672, but one route violates Rule 2. On this route, 85% of the vehicle's capacity is used after accepting the backhaul. This solution and several others that we discuss in the next section are summarized in Table 2.

We should also point out that, in Figure 3, the second and third routes are combined (by inspection) to provide the second driver with an average-duration workday. This "packing" of routes can be seen in Figures 4, 5, and 6, as well.

4.4. Making Backhaul Insertions More Flexible

In the initial solution phase, where the Clarke-Wright procedure generates routes that service a large number of delivery locations, it sometimes happens that the routes offer little flexibility for insertion of backhauls. For

Table 2. Summary of Computational Results with Load-Based Insertions

	Configurations			
	RC1	RC2	RC3	RC4
Initial Solution Phase				
Initial Vehicle Capacity	160	120	120	140
Clarke-Wright Distance	585	641	641	640
Radial Distance	284	284	284	284
Radial Surcharge (0.5 × radial distance)	142	142	142	142
Total Distance	1011	1067	1067	1066
Insertion Phase				
Backhaul Penalty	30	25	30	20
Number of Routes	6	8	8	6
Total Distance After Insertion	672	684	686	682
Average Hours per Driver	6.74	6.82	6.88	6.78
Range in Hours per Driver	6.7	1.2	1.3	1.9
Feasibility Phase				
Rule 1 Conformity	Yes	Yes	Yes	Yes
Rule 2 Conformity	No*	Yes	Yes	Yes
Rule 3 Conformity	Yes	Yes	Yes	Yes

*missed by 5%

example, with only a few long distance routes composed of many deliveries, there are very few low-cost places to insert backhauls; the backhauls are usually squeezed onto the end of the route.

To facilitate the insertion of backhauls, we decided to initiate the Clarke-Wright procedure with an artificial vehicle capacity that is smaller than the actual vehicle capacity. For example, vehicles might have an actual capacity of 160 units; however, in order to produce an initial solution that is composed of more routes with fewer deliveries per route, the capacity of the vehicles might be set at 120 units during the Clarke-Wright initial solution phase. More routes with fewer deliveries per route usually result in unused vehicle capacity, and backhauls can often be inserted more freely than in tightly constrained situations. In addition, since there are more routes, the number of backhauls per route decreases which, again, reduces the complexity and interdependence of the insertions.

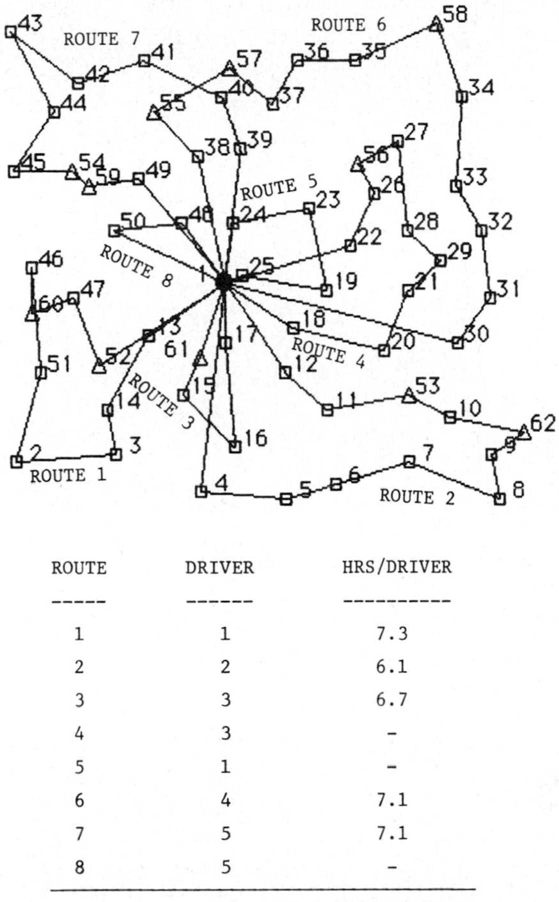

ROUTE	DRIVER	HRS/DRIVER
1	1	7.3
2	2	6.1
3	3	6.7
4	3	-
5	1	-
6	4	7.1
7	5	7.1
8	5	-

AVG 6.82 HRS/DRIVER

Figure 4. Routes Configuration 2 (RC2)

The effect of initiating the Clarke-Wright procedure with a reduced vehicle capacity is illustrated in Figures 4, 5, and 6 and summarized in Table 2. The initial vehicle capacity is chosen to be smaller than the actual capacity of 160. Once backhauls have been considered for insertion, the capacity of the vehicle "returns" to 160.

In Table 2, four route configurations (RC1 through RC4) are presented as candidate solutions to the problem illustrated in Figure 1. Each of these solutions was generated from the load-based insertion procedure which consists of three phases--initial solution phase, insertion phase, and feasibility phase.

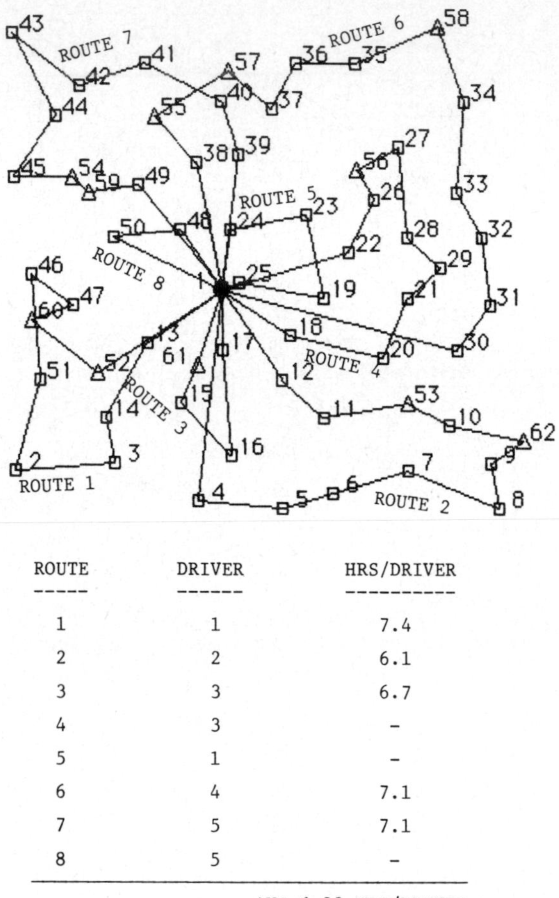

ROUTE	DRIVER	HRS/DRIVER
1	1	7.4
2	2	6.1
3	3	6.7
4	3	-
5	1	-
6	4	7.1
7	5	7.1
8	5	-

AVG 6.88 HRS/DRIVER

Figure 5. Routes Configuration 3 (RC3)

For RC1, the initial solution consists of the Clarke-Wright routes servicing only the delivery locations with a distance of 585 miles, a radial distance to the backhauls of 284 miles, and a radial surcharge of 124 miles. This results in a total distance of 1011 miles. During the insertion phase, the configuration shown in Figure 3 was developed. This configuration consists of six routes and a total distance of 672 miles. In order to service this set of routes, five drivers are expected to be on the road from 2.9 to 9.6 hours. Note that this configuration violates Rule 2 because it allows the usage of truck capacity to exceed 80% after accepting a backhaul.

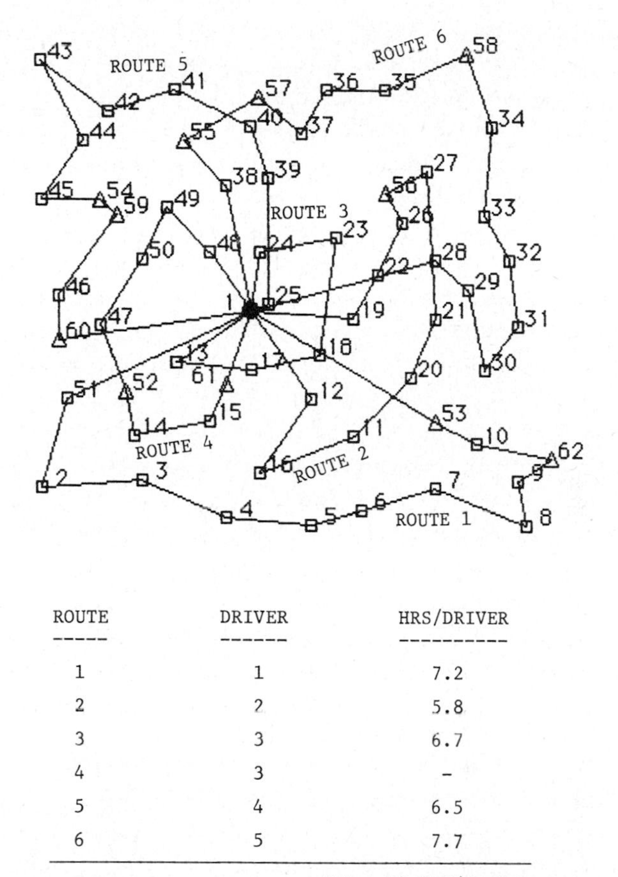

ROUTE	DRIVER	HRS/DRIVER
1	1	7.2
2	2	5.8
3	3	6.7
4	3	-
5	4	6.5
6	5	7.7
		AVG 6.78 HRS/DRIVER

Figure 6. Routes Configuration 4 (RC4)

Configurations RC2 and RC3 were developed using a vehicle capacity of 120 during the initial phase. This resulted in an initial solution with more routes than in RC1 and a total distance of 1067 miles including radial distances and surcharges. During the insertion phase, backhaul penalties of 25 and 30 were used to generate the configurations shown in Figures 4 and 5, respectively. The RC2 configuration consists of eight routes with a total distance of 684 miles. Five drivers are expected to be on the road from 6.1 to 7.3 hours each to serve these routes. Note that this configuration satisfies all of the specified rules. The RC3 configuration has a total distance of 686 miles after insertions, the same number of routes, a similar driver work schedule, and it also satisfies the specified rules.

Configuration RC4 was developed using a vehicle capacity of 140 during the initial phase. This results in an initial solution with a total distance of 1066 miles including radial distances and surcharges. The insertion procedure yielded a configuration (see Figure 6) with six routes and a total distance of 682 miles. Five drivers are expected to be on the road from 5.8 to 7.7 hours each, and all of the rules are fully satisfied.

A comparison of these four configurations leads us to favor the set of routes represented by either RC2 or RC4 for several reasons. First, the two configurations are cost-effective in the sense that the total distance traveled is nearly minimal. Second, the driver loads are balanced. Each driver does about the same amount of work. Third, the rules are fully satisfied.

5. CASE STUDIES

In this section, three rather diverse case studies are presented. Each study describes a real-world distribution environment. (Although these applications describe land-based distribution problems, we point out that backhaul problems are also present in ocean-borne transportation systems. For example, Stott and Douglas [8] encountered backhaul cargo in designing a decision support system for Bethlehem Steel's marine operations.) The studies illustrate the importance and pervasiveness of backhaul and backhaul-like opportunities. They also exhibit a variety of responses to the backhaul problem.

5.1. Case Study 1: Vendors Supply of America

The Mid-Atlantic division of Vendors Supply of America (VSA) makes deliveries of items such as candy bars, juice, soda, coffee, and paper cups to approximately 1000 customers located in Virginia, Maryland, Delaware, New Jersey, and Pennsylvania from its Bridgeport, New Jersey location. About 600 customers are considered "regular" since they require deliveries every week, every other week, or perhaps once every six weeks. The remaining 400 customers are serviced infrequently, some only on a seasonal basis.

VSA currently operates a fleet of twelve tractor/trailer trucks to service its customers. Truck routes are manually constructed with each truck making several delivery stops per day. In addition, whenever possible, the dispatcher tries to incorporate backhaul points onto the routes. At these points, drivers will pick up supplies such as cranberry juice or potato chips destined for the depot.

VSA stocks about 1200 items at the New Jersey depot for delivery to vendors, users of coffee machines, and movie theaters. VSA also distributes items such as candy bars to various organizations for fund raising activities and has recently extended its assortment of delivery items to include fresh bakery pastry. All delivery items (except pastry) are pre-sorted and shrink wrapped into individualized pallets at the depot. This permits quick and easy delivery at the customer locations.

During the delivery portion of the route, fully-loaded trucks carry between 15 and 25 pallets. VSA prefers not to "cube-out" its trucks (i.e., a full load of 44 pallets, double stacked) since this practice can result in damage to both delivery items and trucks. Once deliveries are completed, no more than 18 pallets are accepted for the in-bound trip. The capacity slack is necessary to accommodate empty pallets or a refused load that might remain in the truck from the delivery activities. Once the merchandise is picked up from the supplier, VSA is responsible for safely transporting it back to the depot. Also, with fewer loaded pallets on board, product damage tends to be minimized.

The dispatcher manually constructs truck routes on a daily basis. Typically, a truck visits seven to twelve customer locations in a particular geographic area on a designated day of the week. For example, on a recent Monday, five trucks serviced customers in Norfolk and Richmond, two serviced Roanoke, and five serviced Delaware and Pennsylvania. Most routes are local trips requiring a total of eight to nine driving hours, while some routes (especially those involving pastry deliveries) are overnight trips requiring ten to fifteen driving hours. On the long trips, drivers might leave on Sunday for example, stay overnight, service the delivery stops on Monday, and then return to the depot. The depot tracks trucks throughout the day by requiring the driver to periodically call in. Drivers are provided with a routing sheet that lists the order of the delivery stops and the expected arrival time. Customers know when to expect deliveries and deliveries are usually made within one hour of the expected time.

It is VSA's policy to schedule all backhaul points at the end of the driver's route. (In all cases, drivers have appointed hours at which to pick up the product at the backhaul location.) Since VSA is a service company and its customers have come to expect timely deliveries, it does not want to delay customer deliveries because of problems at a backhaul location (such as lack of a loading dock). In rare instances, if a truck is running late, another truck may be dispatched to the backhaul location.

VSA recognizes that a significant cost-savings can result by incorporating backhaul locations onto delivery routes and has encouraged its suppliers to provide allowances for product pickup. For example, one candy supplier offers an $800 credit, others offer a $1.00 per case allowance or $1.15 per hundred weight when VSA picks up the product. Over a recent 3 month period, VSA estimates that it has saved over $41,000 by picking up products destined for the depot.

5.2. Case Study 2: Ace Fire Extinguisher Service

Ace Fire Extinguisher Service, Inc. services and sells fire extinguishers in the Washington, D.C. metropolitan area. Their customer base includes about 14,000 locations and extends from Baltimore, MD in the north to Manassas, VA in the south. Each customer is visited at least once annually for on-site service, delivery, or pickup of extinguishers. The various on-site activities are listed in Table 3.

Table 3. On-Site Work Classification

- Service
  - inspect and tag extinguishers
  - recharge water-pressure extinguishers
  - tear down and/or recharge dry-chemical units
  - install extinguishers and cabinets
  - service fire alarm systems
- Delivery
  - serviced extinguishers
  - new extinguishers and cabinets
- Pickup
  - extinguishers to be serviced at the depot
  - old unusable extinguishers
  - trade-ins and loaners

Each day servicemen are given a stack of work orders to complete. These work orders fall into the following five categories listed in the order of their importance (priority):

- appointments for service,
- annual service repeat visit,
- VIP calls,
- delivery of new extinguishers, and
- annual service first visit.

This list has been ranked by management according to the priority of the work. The most important calls are the appointments that require the driver to visit a customer location at a specified time. Customers with service contracts are visited (usually without an appointment) so that annual service work (ASW) may be performed. ASW may require two (or more) visits to complete a job--an initial visit to pick up extinguishers and a repeat visit to deliver and/or install serviced units. In order to ensure timely work completion, the repeat ASW visit ranks second on this list. Next on the list are the VIP calls. Customers in this category require service within 48 hours. Delivery of new extinguishers, cabinets, and other activities follow in importance. At the bottom of the list are the first visit annual service calls.

In a typical week, servicemen receive close to 500 work orders and complete approximately 80% of them (about 400 calls are completed and 100 are missed). Table 4 illustrates a mix of calls for a recent week. The first column gives the distribution of the 400 completed calls and the second column gives the distribution of the 100 missed calls among the five work-order categories.

Table 4. Work Order Distribution

	Calls completed	Calls missed
Service appointments	21.9%	8.4%
ASW (repeat visit)	25.7%	7.3%
VIP calls	14.3%	0%
Delivery of equipment	12.3%	5.3%
ASW (first visit)	25.8%	79.0%

The company does not employ a computer-assisted method to route its fleet of ten trucks. Servicemen are responsible for sequencing the customer visits. The clustering of the work orders is based upon customer location, an eight-hour shift, and dispatcher experience. At present, there are ten customer clusters, four encompass the immediate Washington, D.C. area, four are located in suburban Maryland, and the remaining two are in suburban Virginia. We also point out that trucks are seldom strained in terms of their extinguisher-carrying capacity. It is customary for Ace to split up large work orders at the same location over several visits.

Although some of the traditional backhaul characteristics are not present in Ace's distribution environment (such as customers with in-bound loads constrained by vehicle capacity), several features of its problem lend themselves to a backhaul-like model. Recall that in a typical week servicemen are missing about 100 calls--79 ASW first visit calls and 21 higher-priority calls. A detailed examination of the actual daily routes covered during a recent week revealed that servicemen often do not follow management's work order ranking when sequencing their stops. Low-priority stops (such as ASW first visits) are frequently visited early in the route (especially when they are located near customers with appointments). In many cases, however, this precludes more important, higher-priority stops (such as ASW repeat visits) from being visited later in the route.

In order to serve its client base more effectively, management would like low-priority calls (ASW first visit and delivery of new extinguishers) to be visited near the end of a route instead of early-on. We propose to model the low-priority calls as backhaul points--customers that are visited towards the end of a route in order that more important customers can be serviced first. Treating the low-priority calls in this way should help to produce routes that serve customers in a more timely manner by decreasing the number of missed high-priority calls.

5.3. Case Study 3: Nassau Suffolk Express, Inc.

Nassau Suffolk Express is a small trucking company located in Bohemia (Suffolk County), New York. The company delivers business forms, computer paper, and related items from its warehouse to customer locations in and around New York City. In particular, deliveries are made to New Jersey, Connecticut, Westchester County, Nassau County, and New York City on a regular basis. There are also occasional deliveries to Boston and Pennsylvania.

On a typical day, 5 drivers each make about 18 deliveries. Each driver also makes 1 or 2 pickups per week on his way back to the depot in Suffolk County. Additional pickups are made by a sixth (dedicated pickup) vehicle. The driver of this vehicle spends half of his time working in the warehouse. The other half is spent driving primarily between the depot and various stops in New York City; each route consists of 10 to 12 pickups. Reserving vehicles for pickups only is a strategy we have seen employed in other industries as well (such as distribution in the fast food and grocery industries). In other words, Nassau Suffolk Express seeks to handle backhauls effectively by employing two complementary strategies.

5.4. Case Study Overview

In Section 2, nine general questions regarding the characteristics of a particular backhaul situation were posed. In Table 5, we attempt to answer these questions for the three case studies presented in this paper. Our answers are somewhat tentative since we don't always have all the information that we would like. Nonetheless, we have done our best to provide correct responses.

Table 5 is more or less self-explanatory. Although we cannot say that the responses are indicative of the vast majority of backhaul or backhaul-like scenarios, we can observe that definite differences do exist among the three cases.

6. CONCLUSIONS

In this paper, we have focused on the vehicle routing problem with backhauls. Since this important problem has received relatively little attention in the literature, we have summarized the few solution approaches that have been published. In addition, we have included several case studies that underscore the diversity and importance of this routing problem. This paper represents a preliminary examination of this problem. There is much additional algorithmic work that needs to be done.

Table 5. Response to Section 2 Questions

Question Number	Case 1	Case 2	Case 3
1.	approx. 10-30%	approx. 50%	approx. 10-20%
2.	Yes	No	Yes
3.	NA	No†	NA
4.	NK	NA	NK
5.	Yes	Yes	Yes
6.	No	No	Rarely, but possible
7.	No	No	No
8.	No	No	No
9.	No	Rarely, but possible	Yes

NA -- Not Applicable

NK -- Not Known

† This is due to the fact that vehicle capacity is seldom the binding constraint.

REFERENCES

[1] Bartholdi, J. and Platzman, L., "An O(N log N) Planar Travelling Salesman Heuristic Based on Spacefilling Curves," Operations Research Letters, 1, 121-125 (1982).

[2] Bodin, L., Golden, B., Assad, A., and Ball, M., "Routing and Scheduling of Vehicles and Crews: The State of the Art," Computers & Operations Research, 10, 63-211 (1983).

[3] Clarke, G. and Wright, J., "Scheduling of Vehicles from a Central Depot to a Number of Delivery Points," Operations Research, 12, 568-581 (1964).

[4] Deif, I. and Bodin, L., "Extension of the Clarke and Wright Algorithm for Solving the Vehicle Routing Problem with Backhauling," Proceedings of the Babson Conference on Software Uses in Transportation and Logistics Management (A. E. Kidder, editor), Babson Park, MA, 75-96 (1984).

[5] Goetschalckx, M. and Horsley, C., "The Vehicle Routing Problem with Backhauls," Working Paper, Material Handling Research Center, Dept. of Industrial and Systems Engineering, Georgia Institute of Technology (1986).

[6] Golden, B., Baker, E., Alfaro, J., and Schaffer, J., "The Vehicle Routing Problem with Backhauling: Two Approaches," Proceedings of the Twenty-First Annual Meeting of S. E. TIMS (R. D. Hammesfahr, editor), Myrtle Beach, SC, 90-92 (1985).

[7] Rosenkrantz, D., Sterns, R., and Lewis, P., "An Analysis of Several Heuristics for the Traveling Salesman Problem," SIAM Journal on Computing, 6, 563-581 (1977).

[8] Stott, K. and Douglas, B., "A Model-Based Decision Support System for Planning and Scheduling Ocean-Borne Transportation," Interfaces, 11(4), 1-10 (1981).

[9] Yano, C., Chan, T., Richter, L., Cutler, T., Murty, K., and McGettigan, D., "Vehicle Routing at Quality Stores," Interfaces, 17(2), 52-63 (1987).

Vehicle Routing: Methods and Studies
B.L. Golden and A.A. Assad (Editors)

VEHICLE ROUTING WITH SITE DEPENDENCIES

Barindra Nag, School of Business and Economics, Towson State University, Towson, MD 21204

Bruce L. Golden, College of Business and Management, University of Maryland, College Park, MD 20742

Arjang Assad, College of Business and Management, University of Maryland, College Park, MD 20742

Given a vehicle routing problem with a heterogeneous fleet, suppose that each site can be serviced by some, but not necessarily all, vehicle types. Customers with high demands may require large vehicles. Others in congested areas may require small or medium vehicles. In this paper, we analyze several heuristics for solving this variant of the vehicle routing problem where the incompatibility of customer sites with certain vehicle types acts as a complicating constraint.

1. INTRODUCTION

The classical vehicle routing problem asks for a set of delivery routes for vehicles housed at a central depot, which services all the customers and minimizes total distance traveled. The demand at each customer site (node) is assumed to be deterministic and each vehicle has the same known capacity.

Now suppose we complicate the problem in the following way in order to represent some real-world concerns. First, assume we have a heterogeneous fleet, composed of several vehicle types. For example, we might have five small, seven medium-sized, and three large vehicles. No distinction is made between two vehicles of the same type.

Next, suppose that each site can be serviced by some, but not necessarily all, vehicle types. In other words, it is not the case that all vehicle types can service all customers. Some customers with extremely high demands may require large vehicles. Other sites in heavily congested areas may require small or medium-sized vehicles. Table 1 illustrates the form of these restrictions. Note that an entry of 1 indicates compatibility and an entry of 0 indicates otherwise. Thus, customer 3 of Table 1 can be served by either a medium or a large vehicle, but not a small one. Finally, we assume that route duration constraints (e.g., 10 hour workday) are in effect in addition to vehicle capacity constraints.

In this paper, we propose several heuristic procedures for solving this complex real-world problem. We begin with an elementary, first-cut heuristic and proceed to more advanced heuristics afterwards. The elementary approach is sequential in that it deals with one vehicle type at a time, moving from

TABLE 1. Form of Site/Vehicle Type Dependencies

Customer	Demand	Small	Medium	Large
1	30	1	1	1
2	10	1	0	0
3	20	0	1	1
4	40	0	0	1
.	.	.	.	.
.	.	.	.	.
.		.	.	.
Vehicle Capacity		30	40	50
Number of Vehicles		4	3	1

smallest to largest. The procedure has been coded and we report on some computational experiments performed to date. In addition, we discuss three more advanced heuristics that are based on the generalized assignment problem and we test these heuristics on a variety of test problems.

2. OUTLINE OF FIRST-CUT SOLUTION PROCEDURE

In this variant of the vehicle routing problem, there are two key decisions that need to be made. First, vehicles must be assigned to customers. Second, standard vehicle routing problems must be solved for the fleet of vehicles of the same type. Associated with these key decisions are the two algorithmic goals of balanced utilization levels and low routing costs. In our experiments, we have found that the best solutions tend to utilize the various vehicle types uniformly. If smaller vehicles are overutilized, then the larger, more costly and flexible vehicles are in some sense "wasted." If the smaller vehicles are underutilized then it may be difficult to find a feasible solution to the overall problem. Obviously, we are interested in generating compact and cost-effective routes regardless of the vehicle type. These two algorithmic goals will drive our solution approach.

The approach that we adopt is a one-pass procedure that begins with the smallest vehicle type and works up to the largest vehicle type, one type at a time. For type t, a sweep is performed (see Gillett and Miller [3] for details) so that the load is spread evenly amongst all the vehicles of type t. Routes are then generated for the clusters formed by the sweep procedure. Some of the routes may be too full or too long. If this is the case, a number of customers are removed. Customers removed are ones that can be served by larger vehicles. Their inclusion on routes is postponed, therefore, until larger vehicles are considered, unless t is already the largest vehicle type. In Figure 1, a flowchart for the overall approach is provided.

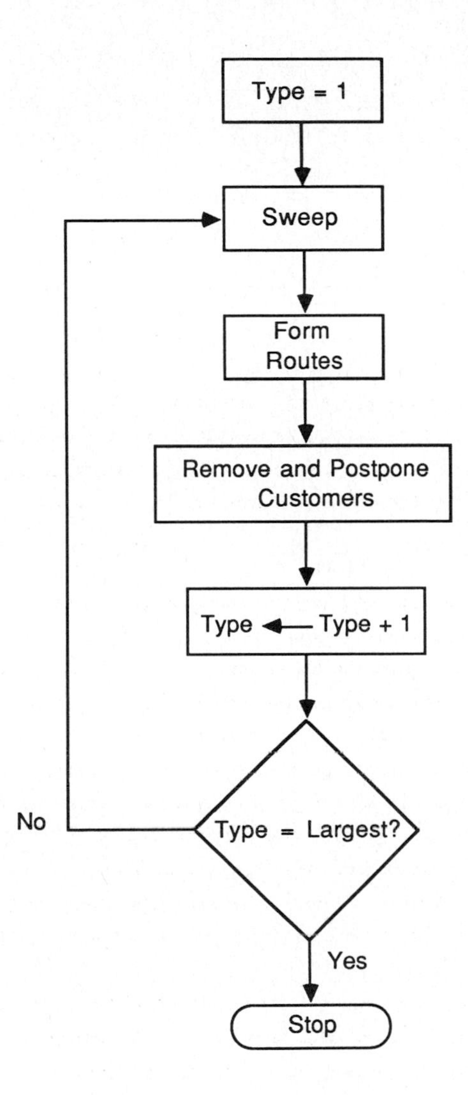

FIGURE 1. Flowchart of the First-Cut Algorithm

3. THE SWEEP PROCEDURE

In applying the sweep procedure, we need to specify an "artificial" capacity based on the following principle: The target load defined for each vehicle should encourage balanced allocation of the demand compatible with a vehicle type to the individual vehicles of that type in the fleet. If we let

D_j = the total demand that can be served by vehicle type j,

and

N_j = the number of type j vehicles, then the

Target Load for Vehicle Type j = D_j/N_j.

With this as background, the artificial capacity is initialized as

Artificial Capacity for Vehicle Type j = 1.1 * Target Load for Vehicle Type j.

The extra 10% in the above formula takes into account the complexity of the packing component in vehicle routing problems.

We remark that it is certainly possible for the artificial capacity to exceed the real vehicle capacity. This is reasonable since a vehicle receives an "excess load" anticipating that the algorithm will shift demands to larger vehicles later. The only other special consideration is as follows. The fixed demands in a cluster are all those customers who are <u>not</u> compatible with any larger vehicle. Clearly, it is not possible to remove and postpone these customers.

The total load of each cluster constructed by the sweep routine must satisfy two conditions:

a) it must not exceed the artificial capacity of the vehicle, and
b) the total of fixed demands in the cluster must be no larger than the real vehicle capacity.

As the sweep routine is applied, two cases need to be identified. In the first case, demands are left over (the sweep runs out of vehicles). If this happens, then the 10% packing buffer needs to be increased slightly. In the second case, the load on the last route is low relative to the load on the first route indicating an imbalance. If this happens, then the 10% packing buffer needs to be decreased slightly. In either case, the sweep routine is rerun. If neither case applies, then the clusters are ready to be converted into routes.

4. THE FORMATION OF ROUTES

Customers within a cluster may be sequenced on a route using any standard traveling salesman heuristic. In our work, we implemented the well-known 2-opt procedure due to Lin [5] and the Or-opt procedure (see Golden and Stewart [4]).

A route is deemed acceptable if the total travel time is less than the upper bound, say TMAX, and the total load is less than an appropriately defined route utilization level. Let TUTIL = Total Free Demand/Total Free Capacity. Free demand refers to customers who are compatible with more than one vehicle type, and free capacity is total capacity minus the sum of demands that are compatible with only the one vehicle type. Then

Route Utilization for Vehicle Type j =
UPAR * TUTIL * Vehicle Type j Capacity.

In this expression, UPAR is a user-specified parameter which is typically set at or near 1.1. This route utilization level for each vehicle type helps to balance loads amongst the various vehicle types.

Suppose we are routing type j vehicles and that j is <u>not</u> the largest vehicle type. If all routes are acceptable, we pass on to the next type. If a route is unacceptable, we pass on to the "Remove and Postpone Customers" routine.

If the current vehicle type is the largest, then we remove customers only if the real vehicle capacity is exceeded. These removed customers are, therefore, not serviced in the associated solution. In our computational experiments, this happened only rarely.

5. THE REMOVE AND POSTPONE CUSTOMERS ROUTINE

For an unacceptable route, we compute the savings associated with removing a customer that can be serviced by the next type also. If customers are numbered in the order in which they appear on the route, then the savings associated with removing i is given by

$$\text{Save}(i) = \text{Dist}(i-1,i) + \text{Dist}(i,i+1) - \text{Dist}(i-1,i+1).$$

The procedure iteratively removes the customer with the largest savings from the route until the route becomes acceptable. The savings need to be updated from one removal to the next. For example, before the removal of customer 6, we have

$$\text{Save}(5) = \text{Dist}(4,5)+\text{Dist}(5,6)-\text{Dist}(4,6), \text{ and}$$
$$\text{Save}(7) = \text{Dist}(6,7)+\text{Dist}(7,8)-\text{Dist}(6,8).$$

After the removal of customer 6,

$$\text{Save}(5) = \text{Dist}(4,5)+\text{Dist}(5,7)-\text{Dist}(4,7)$$
$$\text{Save}(7) = \text{Dist}(5,7)+\text{Dist}(7,8)-\text{Dist}(5,8).$$

We point out that as customers are removed, it may make sense to periodically resequence the remaining customers on the route.

6. PRELIMINARY COMPUTATIONAL RESULTS WITH FIRST-CUT HEURISTIC

In testing this approach, which we refer to as VEHTYPE, we worked with twelve problems. Four of these contain 50 nodes, four contain 75 nodes, and four contain 100 nodes. The problems differ with respect to number of vehicle types and degree of site dependencies. Details are provided in Nag [6]. We refer to the problems as P50A, P50B, P75A, P75B, P100A, P100B, P50AU, P50BU, P75AU, P75BU, P100AU and P100BU; the U (for unrestricted) indicates that the site/vehicle type dependencies have been relaxed. This relaxed set of problems enables us to approximate the cost of enforcing the site/vehicle type constraints.

In Table 2, some typical computational results are provided. In particular, Table 2 focuses on problems P50A and P50AU. Three different starting seed points in the sweep routine are used along with three values for UPAR.

TABLE 2. Typical Computational Results with VEHTYPE

		P50A			P50AU		
UPAR	Seed Pt.	Univac 1100/82 CPU Seconds	Cost	Demands Not Delivered, Percent by Weight	Univac 1100/82 CPU Seconds	Cost	Demands Not Delivered, Percent by Weight
1.08	1	5.30	732+	1,2.3%	3.03	765+	
	2	2.60	788	-	5.29	723+	1,2.3%
	3	3.54	776	1,2.3%	4.79	789+	1,2.2%
1.10	1	2.60	[766+]	-	3.03	[705]	-
	2	2.45	781	-	2.59	745	-
	3	2.47	810+	-	2.62	768	-
1.12	1	2.60	[766+]	-	3.03	[705]	-
	2	2.45	781	-	2.67	745	-
	3	2.49	795	-	2.64	768	-

+All routes are feasible, but some are unacceptable.

The best solution (with acceptable routes) found to each problem is enclosed in a small rectangle. Similar tables are available for the other four problems, but we do not present them here.

From Table 2, it is clear that the procedure is extremely fast, requiring only seconds of running time on a Univac 1100/82. Note that in the unrestricted case for UPAR = 1.1, the cost or distance is 705. This goes up to 766 (less than a 9% increase) in the restricted case. In our experiments, we found that UPAR = 1.1 works consistently well.

Obviously, solution cost and running time are two key measures of algorithmic performance. Another measure of performance is balance of load amongst vehicle types. In P50A, the utilization levels for the three vehicle types are 87%, 87%, and 92%. Similar results apply for the other restricted problems.

In another experiment, we wanted to evaluate the route effectiveness of the algorithm. A convenient post-processor idea comes to mind. Take the solution generated by VEHTYPE as the starting point. This solution allocates customers to vehicle types. If the subset of customers assigned to vehicles of the same type by VEHTYPE is fixed, the problem separates into three standard vehicle routing problems--one for each vehicle type. For all nodes assigned to a small vehicle type, apply the Clarke-Wright algorithm [1] or the sweep algorithm. Next, do the same for medium and large vehicles. We asked the following question: Is it possible to significantly improve upon routes found by VEHTYPE while fixing the subset of customers served by each vehicle type? From Table 3, we see that in only one out of twelve cases, does the Clarke-Wright post-processor find better routes while using the same number of vehicles. The sweep post-processor finds better solutions in six out of twelve cases. The best solution found to each problem is enclosed in a small rectangle.

Overall, VEHTYPE is quick (always less than 10 seconds on a Univac 1100/82) and it seems to perform reasonably well as a first-cut heuristic.

TABLE 3. Comparison of Routing Costs

Problem	VEHTYPE Solution	Clarke-Wright Solution	Sweep Solution
P50A	766.1	785.6	**745.7**
P50AU	**704.6**	741.2+	708.2
P50B	**686.0**	719.8	694.9
P50BU	**659.0**	707.8	659.5
P75A	1219.1	**1159.7**	1294.7
P75AU	**1158.5**	1085.4+	1163.2
P75B	1033.4	1052.3	**1021.1**
P75BU	978.5	997.3	**971.5**
P100A	1368.7	1320.7−	**1260.9**
P100AU	1231.4	1245.4−	**1231.0**
P100B	1292.0	1253.0+	**1237.3**
P100BU	**1120.4**	1110.6+	1120.9

+Clarke-Wright procedure uses more vehicles than available.

−Clarke-Wright procedure uses fewer vehicles than available.

7. IMPROVED HEURISTICS

In this section, we try to improve upon the first-cut heuristic. In particular, we seek a heuristic that is more accurate than the first-cut heuristic and still reasonably fast. With this in mind, we take advantage of the generalized assignment heuristic due to Fisher and Jaikumar [2].

At this point, we briefly review the generalized assignment heuristic. To begin, we assume that the fleet consists of m vehicles and that the fleet may be heterogeneous. The generalized assignment heuristic clusters the customers around seed points.

If m seed points are selected (one for each vehicle) and we denote the jth seed point by $u(j)$, the capacity of the associated vehicle by b_j, and we let

$$z_{ij} = \begin{cases} 1 & \text{if customer } i \text{ is assigned to the vehicle with seed } u(j) \\ 0 & \text{otherwise,} \end{cases}$$

then we have the following generalized assignment problem:

$$\text{Min} \quad \sum_{i=1}^{n} \sum_{j=1}^{m} d_{ij} z_{ij} \tag{1}$$

$$\text{s.t.} \quad \sum_{i=1}^{n} d_i z_{ij} \leq b_j \qquad j = 1,\ldots,m \tag{2}$$

$$\sum_{j=1}^{m} z_{ij} = 1 \qquad i = 1,\ldots,n \tag{3}$$

$$z_{ij} \in \{0,1\} \qquad i = 1,\ldots,n;\ j = 1,\ldots,m\ . \tag{4}$$

The cost coefficients d_{ij} reflect the cost of inserting customer i onto the route which extends from the depot to the single stop u(j) and back. Specifically, if customer i is compatible with the vehicle assigned to the jth seed, we set $d_{ij} = \text{Dist}(0,i) + \text{Dist}(i,u(j)) - \text{Dist}(0,u(j))$ where node 0 is the central depot. In (1), an estimate of the actual travel costs is minimized. Constraints (2) ensure that the total demand assigned to a particular vehicle may not exceed its capacity. Equations (3) guarantee that each customer is assigned to exactly one vehicle.

The solution to this generalized assignment problem determines the assignment of the customers to the different vehicles. To determine the delivery sequence a TSP is solved for each vehicle. If there are a small number of customers per vehicle, the TSP's can be solved optimally. For larger problems a good heuristic might be more appropriate.

The performance of the generalized assignment heuristic depends heavily on seed point selection. In general, seed points are selected so as to spread out the routes over the relevant geographic area. Several techniques for seed selection are discussed in Fisher and Jaikumar [2]. We apply a somewhat specialized seed selection strategy that seems to perform well in practice. For a homogeneous fleet it works as follows.

Consider a set of available vehicles and a demand set of specified customer sites as input. The customer farthest from the depot serves as the first seed point. Subsequent seed points are chosen iteratively by selecting the customer farthest from the set formed by the depot and all previously selected seed points. The selection continues until the number of seed points reaches the number of available vehicles. In this manner, a widely distributed set of seed points among the customers results.

We have coded and tested three heuristics that take advantage of the generalized assignment procedure described above--GAP1, GAP2, and GAP3. We now describe each of these heuristics. In each of these cases, the generalized assignment problem is solved using the branch and bound approach developed by

Ross and Soland [7] and the traveling salesman problems are solved using 2-opt and Or-opt heuristics.

GAP1 works as follows. Initially, the vehicle routing problem with site dependencies is solved using VEHTYPE. The assignment of customers to vehicle types recommended by VEHTYPE is taken as input for GAP1. One vehicle type at a time, GAP1 applies the generalized assignment approach to the derived homogeneous fleet problem. The seed selection also proceeds by type taking the customers assigned by VEHTYPE to that type as input.

GAP2 also builds upon the VEHTYPE solution, but uses this solution only to select seed points for each type of vehicle. More specifically, the assignment of customers to types from VEHTYPE is used to determine seed points for each vehicle type. Then, one large generalized assignment problem is solved over the entire fleet of vehicles. In this generalized assignment problem, we let $d_{ij} = \infty$ if customer i is not compatible with the vehicle type associated with seed point j. After assignments are decided upon, we sequence stops on each route using the traveling salesman heuristic.

GAP3 works directly from the compatibility table (e.g., see Table 1). Starting with the largest vehicle type, seed points are selected. This is done by taking the available vehicles to be all the large ones and the customer set to be all the customers compatible with a large vehicle in the seed selection rule given before. Next, we eliminate those customers selected as seeds and repeat this procedure for the next largest vehicle type. One large generalized assignment problem is then solved and routes are sequenced via a traveling salesman heuristic.

Several other variants of the generalized assignment approach were attempted, but the three strategies already described turned out to be most revealing. Note that GAP1 and GAP2 require an initial run of VEHTYPE whereas GAP3 does not.

8. COMPUTATIONAL RESULTS FOR IMPROVED HEURISTICS

Computational results for GAP1, GAP2, and GAP3 are presented in Table 4, along with the VEHTYPE solutions. In comparing GAP1 with VEHTYPE, we observe that GAP1 does not always provide improved results. In particular, GAP1 solutions are from 1.0 to 9.5 percent longer than VEHTYPE solutions in the first six problems. In general, the last six problems are less tightly constrained than the first six. As a result, GAP1 has more freedom in assigning individual vehicles of each type and the procedure seems to outperform VEHTYPE consistently.

GAP2 and GAP3 are the procedures where the careful (global) assignment of vehicles to customers is expected to pay off. By "global," we refer to the fact that the assignment of vehicles to customers takes place simultaneously and not sequentially by type, as in GAP1. GAP2 and GAP3 each outperform VEHTYPE in 10 of 12 cases.

GAP2 outperforms GAP3 in 7 out of 12 cases. However, given the rather small number of problems tested, it would be premature to declare GAP2 the better of the two procedures. They both seem to perform well on a consistent basis. The solutions, displayed in Nag [6] for all twelve problems, are typically hard to improve upon by inspection. Running times for the three procedures are remarkably fast--always under 1 second on a Univac 1100/90.

TABLE 4. Route Lengths and Percent above VEHTYPE Solution

Problem	VEHTYPE Length	GAP1 Length (%)	GAP2 Length (%)	GAP3 Length (%)	Best Procedure
P50A	766.1	773.8 (+1.0)	767.8 (+0.2)	746.4 (-2.6)	GAP3
P50AU	704.6	751.4 (+5.6)	720.5 (+2.3)	723.8 (+2.7)	VEHTYPE
P50B	686.0	736.4 (+9.5)	624.3 (-9.0)	701.5 (+2.3)	GAP2
P50BU	659.0	702.4 (+6.6)	639.3 (-3.0)	645.7 (-2.0)	GAP2
P75A	1219.1	1311.9 (+7.6)	1169.6 (-4.1)	1199.7 (-1.6)	GAP2
P75AU	1158.5	1185.5 (+2.3)	1081.8 (-6.6)	1126.6 (-2.8)	GAP2
P75B	1033.4	988.0 (-4.3)	974.8 (-5.7)	944.4 (-8.6)	GAP3
P75BU	978.5	960.6 (-1.8)	923.4 (-5.6)	907.7 (-7.2)	GAP3
P100A	1368.7	1350.5 (-1.3)	1203.4 (-12.0)	1264.4 (-7.6)	GAP2
P100AU	1231.4	1268.9 (-3.0)	1100.4 (-11.0)	1174.6 (-4.6)	GAP2
P100B	1292.0	1175.2 (-9.0)	1209.1 (-6.4)	1178.8 (-8.7)	GAP1
P100BU	1120.4	1120.9 (+0.1)	1092.8 (-2.4)	1047.9 (-6.4)	GAP3

9. CONCLUSIONS

This paper demonstrates that solution procedures for the classical vehicle routing problem can be adapted to handle an important real-world variant where site-vehicle dependencies exist. VEHTYPE, a first-cut procedure that combines savings and sweep ideas, is presented along with more powerful generalized assignment procedures. The latter procedures are very fast computationally and seem to generate relatively low-cost, visually appealing routes.

There is much more work to be done in this problem area. The heuristics tested in this paper are very fast. More advanced heuristics that might take more time but produce even better routes need to be investigated. In addition, the quality of the heuristic solutions needs to be assessed in a more comprehensive manner.

REFERENCES

[1] Clarke, C. and Wright, J., "Scheduling of Vehicles from a Central Depot to a Number of Delivery Points," Operations Research, 12 (4), 568-581 (1964).

[2] Fisher, M. and Jaikumar, R., "A Generalized Assignment Heuristic for Vehicle Routing," Networks, 11 (2), 109-124 (1981).

[3] Gillett, B. and Miller, L., "A Heuristic Algorithm for the Vehicle Dispatch Problem," Operations Research, 22, 340-349 (1974).

[4] Golden, B. and Stewart, W., "Empirical Analysis of Heuristics," in The Traveling Salesman Problem (E. Lawler, J. K. Lenstra, A. Rinnooy Kan, D. Shmoys, eds.), John Wiley & Sons, 207-249 (1985).

[5] Lin, S., "Computer Solutions of the TSP," Bell Systems Technical Journal, 44, 2245-2269 (1965).

[6] Nag, B., "Vehicle Routing in the Presence of Site/Vehicle Dependency Constraints," Ph.D. Dissertation, College of Business and Management, University of Maryland (1986).

[7] Ross, G. T. and Soland, R., "A Branch and Bound Algorithm for the Generalized Assignment Problem," Mathematical Programming, 8 (1), 91-103 (1975).

MODELS FOR
COMPLEX ROUTING ENVIRONMENTS

Vehicle Routing: Methods and Studies
B.L. Golden and A.A. Assad (Editors)

LOCATION-ROUTING PROBLEMS

Gilbert LAPORTE

Centre de recherche sur les transports, Université de Montréal
C.P. 6128, Succursale "A", Montréal (Québec) CANADA H3C 3J7*

Location-routing problems involve locating a number of facilities among candidate sites and establishing delivery routes to a set of users in such a way that the total system cost is minimized. This paper reviews this class of problems that is characterized by the interaction between location and routing decisions. The main examples and applications found in the literature are described. Mathematical models as well as heuristic and exact algorithms for their solution are then presented.

1. INTRODUCTION

1.1. Location-Routing Problems

There exist several practical situations where the need arises to locate one or several facilities and to construct associated (delivery) routes including several users or customers. In such contexts, location and routing are intertwined decisions which must be modelled and often optimized simultaneously. These **location-routing problems** (LRPs) belong to the general class of network optimization problems (Golden et al., 1981) and to the family of arc-node problems (Schrage, 1981) since their solution can be described by a selection of arcs and nodes in a network. Research on LRPs is relatively new and the literature on it is much less developed than that associated with either pure location or pure routing problems. This can be explained partly by the fact that the need to combine the two subjects has been recognized only fairly recently (Webb, 1968; Christofides and Eilon, 1969) and partly because location-routing problems are in general NP-hard. It is therefore not surprising that most algorithms for LRPs are approximate; only recently have exact procedures been proposed and most of these can only solve very special classes of LRPs.

One possible approach to tackling location-routing problems is to treat them as pure location problems in which the (marginal) cost of making a delivery to a user from a given facility is approximated by a linear function of the radial distance between the user and the facility. Webb (1968)

*This work was supported by the Canadian Natural Sciences and Engineering Research Council (Grant A4747) and by the Office de Coopération France-Québec (Project 20020686FQ). Thanks are due to Arjang Assad, Pierre Dejax, Yves Nobert and two anonymous referees for their valuable comments.

pointed out that such an approach is in general inadequate. Christofides and Eilon (1969) examined the same question by considering the case of n customers randomly and uniformly distributed in a square of side a. These are served by m vehicles based at the same facility located within the square and it is further assumed that the m routes do not intersect. Letting z^* denote the optimal total length of the m routes, and z_R, the sum of the radial distances between the customers and the facility, they showed that z^* can be approximated by the following formula:

$$z^* \doteq Amz_R/n + B\sqrt{a}\sqrt{z_R} \quad (1)$$

where A and B are two constants related to the position of the facility within the square. This result was derived from the asymptotic formula of Bearwood et al. (1959) for the expected optimal length of a traveling salesman tour in the plane.

When the first term of (1) dominates the second, i.e. when

$$z_R >> a\left(\frac{Bn}{Am}\right)^2 \quad (2)$$

it is valid to approximate z^* by a multiple of z_R. Further developments based on this analysis can be found in Eilon et al. (1971).

When condition (2) is not satisfied, the need to develop techniques specific to LRPs remains. This need may persist even when (2) holds since

i) customers are not always uniformly distributed in the plane;
ii) the value of m may not be known a priori;
iii) routes may intersect;
iv) various side constraints which were not taken into account in the development of (1) sometimes exist; and
v) the facility is sometimes located outside the customer area.

1.2. A General Framework

If we consider users or customers on one hand, and facilities to be located on the other hand, we can theoretically arrive at the definition of four different types of problems depending on whether users or facilities are assumed to belong to discrete or continuous sets. However, like most authors in the fields of location and routing, we limit ourselves to the discrete versions. We also temporarily assume that users and facilities belong to disjoint sets, but some examples where this assumption does not hold will be presented.

Several distribution systems can be represented by a layer diagram such as the one depicted by figure 1. In this example, there are three layers which have been identified as **primary facilities, secondary facilities** and **users.** Frequently, the primary facilities will represent factories, secondary facilities will correspond to depots or warehouses and users will be customers. Primary facilities and also users are usually situated at known and fixed locations. On the other hand, the location of secondary facilities will frequently not be determined a priori : the number or location of these facilities, together with the associated distribution routes constitute decision variables. There can of course be only two layers [as in the transportation problem, the classical location-allocation problem (Cooper, 1963), the p-median problem (Hakimi, 1964), the vehicle routing problem (VRP) (Laporte and Nobert, 1987)] or more than three layers (Mercer et al., 1978; Dejax, 1986).

Several problems can be defined according to the distribution mode M_t used by vehicles based at a facility located at layer t. Usually, these vehicles will make trips to layer t + 1, although some authors (e.g. Watson-Gandy and Dohrn (1973)) consider the possibility of making direct trips from layer 1 to layer 3. We consider two distribution modes :

M_t = R : all trips from layer t must be return trips (i.e. trips to and from a single user or facility);

M_t = T : trips from layer t may be tours (i.e. round-trips through several users or facilities) or return trips.

The distribution modes used for the whole system will be represented by the expression $\lambda/M_1/\ldots/M_{\lambda-1}$, where λ is the number of layers. Thus, in a three-layer system, we will have the four following possibilities :

2/R/R : occurs mostly when shipments of a generally bulky material (e.g. lumber, cement) have to be made in full loads between successive layers.

2/R/T : here, large shipments arriving at the secondary facilities are broken up and dispatched in smaller loads to customers. This situation is encountered in the food industry, for example.

2/T/R : in this situation, trips are often made from the users who bring goods to the secondary facilities (Labbé and Laporte, 1986; Nambiar et al., 1981). These goods are then collected in round- trips and brought to primary facilities.

2/T/T : this case is frequently encountered in the newspaper industry. Here, the primary facilities are printing plants; newspapers are dispatched daily to secondary facilities (transfer points) and then again to retail outlets (see Jacobsen and Madsen, 1980).

In location-routing problems,

i) locational decisions must be made for at least one layer (otherwise, the problem reduces to a pure routing problem);

ii) tours (M_t = T) must be allowed at least once (otherwise the problem becomes a pure location problem).

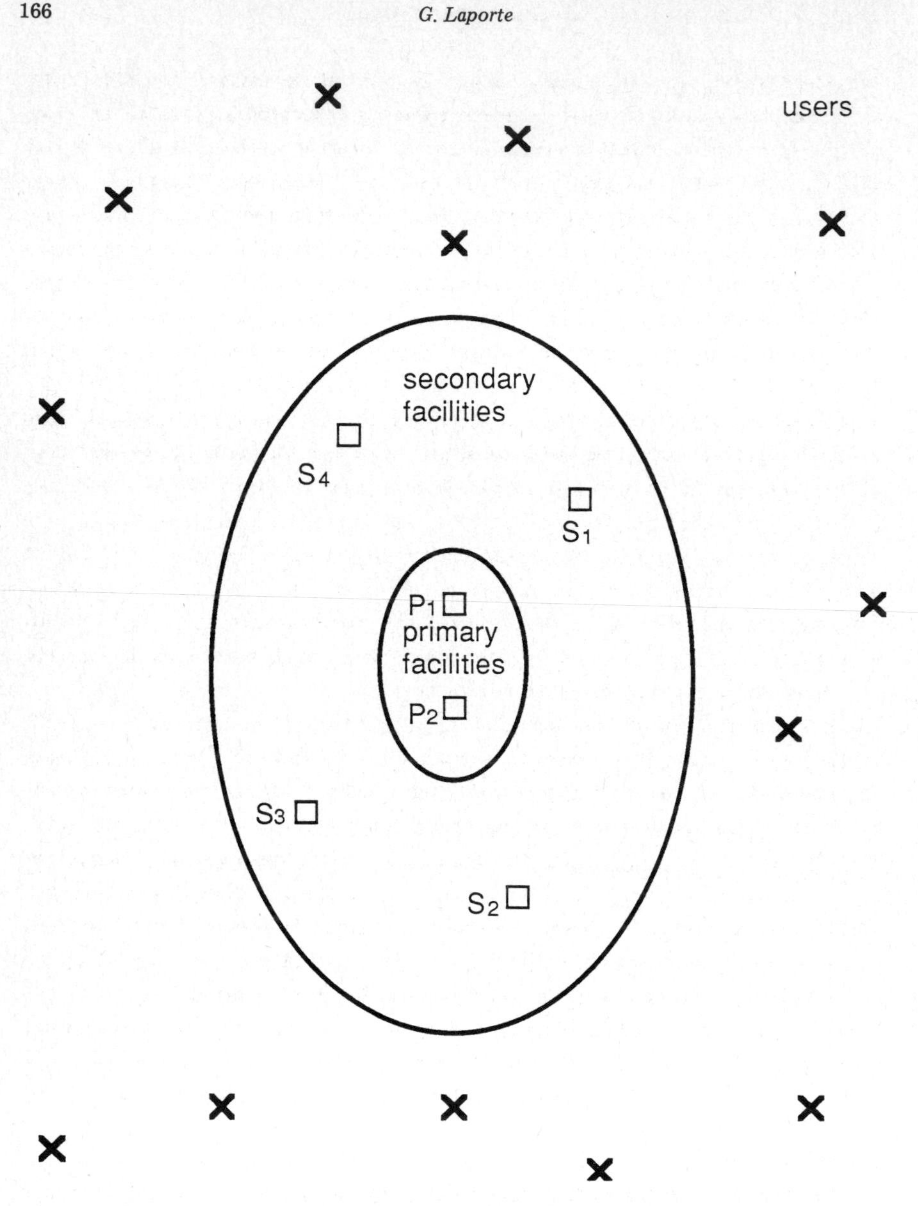

FIGURE 1
A three-layer distribution system

Therefore, problems of the form λ/R/.../R will not be covered in this study. All papers surveyed may however be viewed as an extension of this case.

The most common objective in LRPs is to minimize a linear combination of (i) routing costs, (ii) vehicle fixed costs and (iii) depot operating costs. These have to be scaled down so that they relate to the same time horizon. However, some authors note the danger of locating distribution centers without considering the effect on sales revenues of their proximity to customers, since customers located far from distribution centers tend to be less profitable (see for example Watson-Gandy and Dohrn, 1973; Mercer, 1970; Mercer et al. 1978). Finally, problems where costs are shared between users and the organization providing service yield multicriteria objectives (see the case 2/T/R above). Several approaches are available for such problems; one possible strategy is to combine the various criteria into a single objective (Labbé and Laporte, 1986).

If we now turn to the various constraints involved in location- routing, a wide variety of problems can arise. We are dealing here with a class of problems rather than with a single problem. For the sake of clarity and of simplicity, these and the associated notations will be presented in the course of the text, as the need arises.

1.3. Organization of the Paper

The remainder of this paper is organized as follows. In the next section, we present some applications and examples of LRPs. Section 3 contains a description of the main heuristics developed for LRPs. In section 4, we review briefly the various algorithms used for the exact solution of VRPs since these have inspired most research on the development of exact algorithms for LRPs. Then, in sections 5, 6 and 7 a classification and review of the various integer linear programming (ILP) formulations and algorithms for LRPs are presented: (i) section 5 : three-index vehicle flow formulations; (ii) section 6 : two-index vehicle flow formulations and algorithms for symmetrical problems; (iii) section 7 : two-index vehicle flow formulations and algorithms for asymmetrical problems. Conclusions are presented in section 8.

2. APPLICATIONS

The general framework description of section 1.2 was left sufficiently general so that it would account for most problems involving at some stage simultaneous locating and routing decisions. We review in this section two major classes of LRP applications : (i) generic problems involving location and routing and (ii) actual implementations. Most problems of the first class are of the type 2/T.

Burness and White (1976) and Drezner et al. (1984a, 1984b) describe the problem of selecting a central depot location among n points so that the expected length of a tour through the depot and a subset of randomly selected points is minimized. Applications mentioned by Burness and White include the location of a storeroom in an industrial plant which delivers orders to departments located throughout the plant area, the location of a district sales office when the salesman services a number of customers and the location of a truck depot which provides pick-up and delivery service for a number of firms.

Chan and Francis (1976) analyze the problem of optimally locating a depot in a tree in order to minimize the maximum distance travelled by n vehicles making two deliveries each. Berman and Simchi Levi (1986) consider the case where vehicles make n deliveries each. Chan and Hearn (1977) solve the Chan and Francis problem in the plane. Ghosh et al. (1981) show that the optimal location is sensitive to the choice of distance norm. Ichimori and Nishida (1985) and Drezner (1985) provide exact $O(n \log n)$ algorithms for the Chan and Hearn problem. Finally, Kolen (1985) studies the problem of locating several depots in a tree, under the same assumptions, in order to minimize the maximum round-trip cost. He then studies the problem of determining the minimum number of depots required in order to ensure that the length of a round-trip does not exceed a given bound. Both problems are solved in polynomial time.

Laporte et al. (1987a) studied the problem of locating a central depot and of establishing several delivery routes for vehicles of given capacity to customers with a stochastic demand. Two variants of the problem are considered : (i) the probability that the demand of a planned route exceeds the vehicle capacity is bounded by a preset value, (ii) penalties associated with "route failures" are computed; the planned routes must be such that their expected penalty does not exceed a preset treshold. This problem is an extension of pure stochastic routing problems treated by Stewart and Golden (1983) and Dror and Trudeau (1986).

The generalized traveling salesman (GTSP) (Laporte and Nobert, 1983) can be stated as follows : given a depot and several sets of nodes (which may intersect), determine the shortest Hamiltonian circuit through the depot and at least one node from every set. This problem may be viewed as an 2/T LRP with a single node at the first layer and sets of potential locations at the second layer : at least one location must be chosen within each set. This problem has several applications, e.g. the optimal location of post boxes in an urban area.

Current et al. (1986) analyzed hierarchical network design problems which constitute an important class of 3/T/R LRPs. The problems consist of locating, in a network, a primary path with predetermined start and end nodes or a closed path, as well as secondary paths linking users to the primary path. Such problems occur in the design of expressway and feeder road networks, for example.

Case studies describing various LRP implementations have been reported by a number of authors. These cover several fields of government and economic activities. Some of the most interesting cases are summarized in table 1.

TABLE 1
LRP Applications

TYPE	APPLICATION	LAYERS (with number of sites; * indicates locational decision)				REFERENCES
		1	2	3	4	
2/T	Optimal location of blood banks in the Chicago area	blood banks* (3)	hospitals (17)			Or and Pierskalla (1979)
2/T	NATO aircraft operating locations	military airports* (5)	military bases (84)			McLain et al. (1984)
3/R/T	Distribution of consumer goods	factories	depots*	customers		Watson-Gandy & Dohrn (1973) TFD (1977) Von Bednar & Strohmeier (1979)
3/T/R	Rubber collection in Malaysia	rubber processing factories* (8)	collection stations (50)	smallholders (3750)		Nambiar et al. (1981)
3/T/T	Newspaper delivery in Denmark	printing plants (21)	transfer points* (37)	retailers (4500)		Jacobsen and Madsen (1980)
4/R/R/T	Distribution of consumer goods	factories	warehouses*	depots*	customers	Mercer et al. (1978, chap. 2 and 3)

3. HEURISTIC ALGORITHMS

LRPs are often described as a combination of three distinct components : (i) facility **location;** (ii) **allocation** of users to facilities; and (iii) vehicle **routing** (see for example Madsen (1983), McLain et al. (1984), Perl and Daskin (1985) and Salhi and Rand (1986)). These three subproblems are closely interrelated and cannot be optimized separately without running the risk of arriving at a suboptimal solution. Comprehensive mathematical programming formulations which incorporate simultaneously all aspects of the problem will, in general, contain too many variables and constraints to be easily solved using an exact algorithm. Examples of these formulations and exact algorithms are presented in sections 5, 6 and 7. But most practical problems call for a suboptimal algorithm.

Not surprisingly, heuristic algorithms for LRPs exploit the decomposition of the problem into its three components. A typical algorithm will consist of a sequence of routines designed to solve well-known problems such as location-allocation, the p-median problem, clustering, and the VRP. Basically, most algorithms described are of two types : location-allocation-routing (LAR), and allocation-routing-location (ARL), with various combinations of steps being performed sequentially or simultaneously, with possible omission or repetition of some steps, and with the eventual inclusion of feedback loops.

LAR methods are easily described : facilities are located, users are then allocated to facilities and routes are finally defined. Watson- Gandy and Dohrn (1973) execute these three steps sequentially while other authors combine the location and allocation steps (see, for example, Von Bednar and Strohmeier (1979), Or and Pierskalla (1979), the Tree-Tour and the ALA-SAV heuristics of Jacobsen and Madsen (1980), McLain et al. (1984)).

In ARL algorithms, the allocation and routing steps are often simultaneous : sets of routes are first constructed, assuming all facilities "open"; locations are then selected by dropping various facilities from the system and by updating the location and routing decisions. As noted by Feldman et al. (1966), ARL algorithms will be preferred in tightly constrained problems where it is difficult to construct feasible routes if only a limited number of facilities are open. The SAV-DROP heuristic of Jacobsen and Madsen (1980) and the algorithm of Srikar and Srivastava (1984) constitute examples of ARL algorithms.

Some other heuristics do not easily fit in one of the above categories. Nambiar et al. (1984) do not have an allocation phase. Perl and Daskin (1984) decompose the original problem into three subproblems which they solve sequentially : (i) allocation-routing, (ii) location-allocation and (iii) allocation-routing. Labbé and Laporte (1986) also use a decomposition approach which repeatedly solves a GTSP with some facility weights and allocates users to the closest open facility, thus changing the weights. The procedure ends when a stable solution has been obtained.

4. EXACT ALGORITHMS FOR THE VEHICLE ROUTING PROBLEM

Relatively few exact algorithms have been developed for LRPs and most have been published over the last seven or eight years. These are usually based on ILP formulations for the VRP, to which a location dimension has been added. For this reason, we find it useful to briefly review the most frequently used exact algorithms for the VRP in this section. We also mention some VRP algorithms with no counterparts in location-routing at the present time, hoping that their inclusion will inspire researchers to develop new types of exact algorithms for LRPs.

4.1. Classification

According to a recent survey (Laporte and Nobert, 1987), exact algorithms for the VRP fall into three main categories : (i) direct tree search algorithms, (ii) dynamic programming algorithms, (iii) ILP algorithms. The latter category can be subdivided, as in Magnanti (1981), into (iiia) set partitioning algorithms; (iiib) vehicle flow algorithms; (iiic) commodity flow algorithms.

4.2. Direct Search Algorithms

Direct tree search algorithms consist of sequentially building vehicle routes by means of a branch and bound tree. Two basic strategies are used within this category : (i) branching on arcs (a branch corresponds to the inclusion or the exclusion of an arc) - an early example of this type of branching can be found in Christofides and Eilon (1969); (ii) branching on routes (here, a branch corresponds to a feasible vehicle route) - the Christofides (1976) algorithm is of this type. Christofides et al. (1981a) provide improvements on these early tree search algorithms: these are based on the derivation of sharp lower bounds on the optimum (computed from k-degree center trees and q-routes), in conjunction with Lagrangean relaxation.

4.3. Dynamic Programming

Dynamic programming (DP) has been applied to the solution of VRPs by a variety of authors. Eilon et al. (1971) provide a general formulation. Consider a VRP with a fixed number $\bar{m}$ of vehicles, a set $N=\{1,...n\}$ of points (users or cities) and a depot located at city 1. Let $c(S)$ denote the cost of the optimal single route through the depot and all the users of a subset S of $N-\{1\}$. We wish to minimize

$$z = \sum_{j=1}^{\bar{m}} c(S_j) \qquad (3)$$

over all feasible partitions $\{S_1,...,S_{\bar{m}}\}$ of $N-\{1\}$. Let $f_k(U)$ be the minimum cost achievable using k vehicles and delivering to a subset U of $N-\{1\}$. Then, the minimum can be determined through the following recursion formula :

$$f_k(U) = \begin{cases} c(U) & (k=1) \\ \min\limits_{U^* \subseteq U \subseteq N-\{1\}} [f_{k-1}(U-U^*) + c(U^*)] & (k>1) \end{cases} \quad (4)$$

The solution cost is equal to $f_{\bar{m}}(N-\{1\})$ and the optimal partition corresponds to the optimizing subsets U^* in (4).

Unfortunately, this formulation is generally of little use due to the very large number of states it involves. However, the performance of DP algorithms can be considerably improved by reducing the number of states through state-space relaxation techniques (Christofides et al., 1981b) or by the elimination of dominated and infeasible states (Desrosiers et al., 1986).

4.4. Set Partitioning Formulations

ILP algorithms constitute by far the most widely used solution tool for VRPs. Among these, set partitioning formulations cover a wide range of problems. Balinski and Quandt (1964) provide one of the earliest examples of this approach. Consider all feasible routes j and let a_{ij} be a 0-1 coefficient taking the value 1 if and only if city i appears on route j. Let c_j^* be the optimal cost of route j and x_j be a 0-1 variable equal to 1 if and only if route j is used in the optimal solution. Then the problem is to

(SP) $$\text{Minimize} \sum_j c_j^* x_j$$

subject to

$$\sum_j a_{ij} x_j = 1 \qquad (i \in N-\{1\}) \qquad (5)$$

$$x_j = 0,1 \qquad (\text{all } j) \qquad (6)$$

However, such formulations suffer the same drawback as the general DP formulation : the number of their variables, even for relatively small problems, is in most cases astronomical. Moreover, the determination of the c_j^*'s usually requires the optimization of a difficult problem such as the TSP. These difficulties can be overcome by using column generation techniques and linear lower bounds on the c_j^*'s. Successful applications of one or both of these techniques can be found in Bell et al. (1983) and in Desrosiers et al. (1984).

4.5. Flow Formulations

Commodity flow and vehicle flow models constitute the other two major types of ILP formulations. Commodity flow models explicitly consider the quantity of goods travelling on every route and passing through every facility. These models generally include upper bounds on the flow of goods which can travel through any part of the system, as well as volume dependent processing and transportation costs. Classical examples of such models are the capacitated plant location problem (Kuehn and Hamburger, 1962), the transportation problem and the transshipment problem. Vehicle flow models, on the other hand, deal with the optimal circulation of vehicles and users in the system and do not include costs and contraints directly related to the actual flow of goods. The TSP and the p-median problem fall into this category. Some problems, such as the capacitated VRP, are hybrid cases. They have been modelled as vehicle flow problems by some authors (Laporte et al., 1985) and as commodity flow problems by others (Gavish and Graves, 1982).

It is often convenient to classify VRP formulations according to the number of indices of the flow variables. Common cases are :

i) $x_{ij} = \begin{cases} 1 \text{ if a vehicle travels on arc } (i,j) \\ 0 \text{ otherwise} \end{cases}$

ii) $x_{ijk} = \begin{cases} 1 \text{ if vehicle } k \text{ travels on arc } (i,j) \\ 0 \text{ otherwise} \end{cases}$

iii) w_{ij} = quantity of goods travelling on arc (i,j)

iv) $w_{ij\ell}$ = quantity of goods travelling on arc (i,j) and destined for city ℓ.

v) $z_{ik} = \begin{cases} 1 \text{ if city } i \text{ is served by vehicle } k \\ 0 \text{ otherwise .} \end{cases}$

Moreover, depending whether the distance (or travel time) matrix of the problem under study is symmetrical or not, (i,j) will represent an undirected arc (edge) or a directed arc. In symmetrical problems, it is usually unnecessary to define variables for $i > j$.

There are certain advantages and disadvantages in using two-index or three-index variables in VRP or for that matter, LRP formulations. Two-index formulations are concise and involve a relatively small number of variables (particularly in the symmetrical case). However, they cannot take into account different vehicle costs and characteristics and can therefore only be applied under the assumption that the vehicle fleet is homogeneous. Successful implementations of such formulations (using x_{ij} variables) are reported by Fleischmann (1982) and Laporte et al. (1985), for example. Three-index formulations are more versatile but more costly. Classical examples of such formulations are the vehicle flow models proposed by Golden et al. (1977) (using x_{ijk} variables) and by Fisher and Jaikumar (1978) (using x_{ijk} and z_{ik} variables). However, to the author's knowledge, three-index formulations have never led to the successful implementation of an **exact** algorithm for the VRP. In the next three sections, we examine some three-index and two-index flow formulations for LRPs. In the latter case, successful exact algorithms are also described.

5. THREE-INDEX VEHICLE FLOW FORMULATIONS

Three-index vehicle flow formulations appear to be the most widely used in the location-routing literature. Since they are very flexible, they facilitate the inclusion of a variety of constraints and the definition of very general objective functions. In general, their size and structure are such that they cannot be solved to optimality by using ILP or network optimization codes. They do however provide valuable insights into the problem structure and reveal mathematical properties which can be exploited in the design of heuristic algorithms (as in Perl and Daskin (1985), for example).

Four papers (Golden et al., 1977; Or and Pierskalla, 1979; Srikar and Srivastava, 1983; Perl and Daskin, 1985) propose such formulations. The first three papers consider a 2/T problem, with depots to be located at the first layer and users at known locations at the second layer. The Perl and Daskin article treats the "classical" 3/R/T problem, with known supply sources at the first layer, and the next two layers as in the 2/T model. These formulations all rest upon the same principle and differ mainly in the number of cost components and restrictions they incorporate. If we remove from these formulations elements which are specific to the given problem, the similarities become even more striking. By disregarding the secondary local features contained in the formulations, and by making minor adjustments to the Perl and Daskin model, we arrive at a fairly comprehensive formulation which includes the preceding four models as special cases.

5.1. Objective, Constraints and Notation

The classical 3/R/T LRP consists of simultaneously

i) selecting facility sites at the first layer;

ii) determining the composition of the vehicle fleet based at each facility;

iii) constructing optimal delivery routes from the supply sources to the facilities located at the second layer and from these facilities to users;

The solution must be such that

iv) the total system cost is minimized;

v) all user requirements are satisfied without exceeding vehicle capacities;

vi) the number of vehicles used does not exceed a given bound;

vii) route lengths are within a given maximum distance;

viii) route durations do not go beyond a preset time;

ix) vehicle routes pass through only one facility;

x) facility throughput capacities are respected.

In order to formally express these constraints and objective, the following notation is introduced. Define

Sets

L : the set of all supply sources (first layer)

I : the set of all potential facility sites (second layer)

J : the set of all users (third layer)

(Note : I, J and L are pairwise disjoint.)

K : the set of all vehicle routes.

Parameters

$\bar{m} = |K|$: the maximum allowed number of vehicle routes

$C = (c_{ij})$ $(i,j\varepsilon I \cup J)$: a distance matrix (undefined if $i,j\varepsilon I$)

$T = (t_{ijk})$ $(i,j\varepsilon I \cup J;\ k\varepsilon K)$: a three-dimensional travel time array for all routes and for all pairs of facilities and users

g_i : the fixed cost of establishing facility $i\varepsilon I$

v_i : the variable cost per throughput unit at facility $i\varepsilon I$

V_i : the maximum throughput at facility $i\varepsilon I$

$s_{\ell i}$: the unit transportation cost from $\ell\varepsilon L$ to $i\varepsilon I$

d_j : the requirement of user $j\varepsilon J$

τ_{kj} : the time required by the vehicle used on route k to unload at user $j\varepsilon J$

D_k : the capacity of the vehicle used on route k

E_k : the maximum allowable length (in distance units) of route k

T_k : the maximum allowed duration of route k

p_k : the cost per distance unit of delivery vehicle on route k

q_{ij} : a fixed cost incurred for delivering from facility $i\varepsilon I$ to user $j\varepsilon J$.

Variables

$x_{ijk} = \begin{cases} 1 & \text{if } i \text{ immediately precedes } j \text{ on route } k\ (i,j\varepsilon I \cup J;\ k\varepsilon K) \\ 0 & \text{otherwise} \end{cases}$

$y_i = \begin{cases} 1 & \text{if a facility is located at site } i\varepsilon I \\ 0 & \text{otherwise} \end{cases}$

$z_{ij} = \begin{cases} 1 & \text{if user } j\varepsilon J \text{ is served from facility } i\varepsilon I \\ 0 & \text{otherwise} \end{cases}$

$w_{\ell i}$ = quantity of goods shipped from supply source $\ell\varepsilon L$ to facility $i\varepsilon I$.

5.2. Formulation

The problem formulation is as follows :

$$\text{(LRP1)} \quad \text{Minimize} \quad \sum_{i\varepsilon I} g_i y_i + \sum_{\ell\varepsilon L} \sum_{i\varepsilon I} s_{\ell i} w_{\ell i} + \sum_{i\varepsilon I} \sum_{j\varepsilon J} (v_i d_j + q_{ij}) z_{ij} + \sum_{k\varepsilon K} \sum_{i\varepsilon I\cup J} \sum_{j\varepsilon I\cup J} p_k c_{ij} x_{ijk}$$

subject to

$$\sum_{k\varepsilon K} \sum_{i\varepsilon I\cup J} x_{ijk} = 1 \qquad (j\varepsilon J) \qquad (7)$$

$$\sum_{j\varepsilon J} d_j \sum_{i\varepsilon I\cup J} x_{ijk} \leq D_k \qquad (k\varepsilon K) \qquad (8)$$

$$\sum_{i\varepsilon IUJ} \sum_{j\varepsilon IUJ} c_{ij}x_{ijk} \leqslant E_k \qquad (k\varepsilon K) \qquad (9)$$

$$\sum_{j\varepsilon J} \tau_{kj} \sum_{i\varepsilon IUJ} x_{ijk} + \sum_{i\varepsilon IUJ} \sum_{j\varepsilon IUJ} t_{ijk}x_{ijk} \leqslant T_k \qquad (k\varepsilon K) \qquad (10)$$

$$\sum_{k\varepsilon K} \sum_{i\varepsilon S} \sum_{j\varepsilon (IUJ)-S} x_{ijk} \geqslant 1 \quad (2\leqslant |S| \leqslant |IUJ|\ ;\ S\subseteq IUJ\ ;\ S\cap J\neq\emptyset) \qquad (11)$$

$$\sum_{j\varepsilon IUJ} x_{ijk} - \sum_{j\varepsilon IUJ} x_{jik} = 0 \qquad (k\varepsilon K,\ i\varepsilon IUJ) \qquad (12)$$

$$\sum_{i\varepsilon I} \sum_{j\varepsilon J} x_{ijk} \leqslant 1 \qquad (k\varepsilon K) \qquad (13)$$

$$\sum_{\ell\varepsilon L} w_{\ell i} - \sum_{j\varepsilon J} d_j z_{ij} = 0 \qquad (i\varepsilon I) \qquad (14)$$

$$\sum_{\ell\varepsilon L} w_{\ell i} \leqslant V_i y_i \qquad (i\varepsilon I) \qquad (15)$$

$$-z_{ij} + \sum_{u\varepsilon IUJ} (x_{iuk}+x_{ujk}) \leqslant 1 \qquad (i\varepsilon I,\ j\varepsilon J,\ k\varepsilon K) \qquad (16)$$

$$x_{ijk} = 0,1 \qquad (i\varepsilon I,\ j\varepsilon J,\ k\varepsilon K) \qquad (17)$$

$$y_i = 0,1 \qquad (i\varepsilon I) \qquad (18)$$

$$z_{ij} = 0,1 \qquad (i\varepsilon I,\ j\varepsilon J) \qquad (19)$$

$$w_{\ell i} \geqslant 0 \qquad (\ell\varepsilon L,\ i\varepsilon I) \qquad (20)$$

In this formulation, the objective function is the sum of facility fixed costs, first level delivery costs, variable warehousing costs and delivery costs, respectively. Constraints (7) ensure that every user belongs to one and only one route. Constraints (8), (9) and (10) guarantee that vehicle capacities, maximum route lengths and maximum route durations, respectively, are respected. Constraints (11) are connectivity constraints : they ensure that every user is on a route connected to the set of facilities. Constraints (12) are flow conservation equations : any point of IUJ must be entered and left by the same vehicle. Constraints (13) stipulate that a vehicle can depart only once from a facility. This prevents cases where a vehicle leaves the same facility for two different users and also passes through more than one facility. Constraints (14) ensure that the flow entering a facility is equal to the flow out of that facility. Constraints (15) limit the flow through a facility to the capacity of that facility. Constraints (16) specify that z_{ij} must be equal to 1 if facility i and user j belong to the same route k. Finally, constraints (17) to (20) impose bounds and integrality conditions on the variables.

5.3. Special Cases

As mentioned earlier, the formulation just presented contains (with minor simplifications) the main models described in the literature. The Perl and Daskin formulation is very close to (LRP1) : it suffices to remove constraints (10) and to set the q_{ij}'s equal to zero. The other three papers considered (Golden et al., 1977; Or and Pierskalla, 1979; Srikar and Srivastava, 1983) describe two-layer problems. Therefore, in all three cases, we can set $L = \emptyset$, eliminate the corresponding variables ($w_{\ell i}$) and parameters ($s_{\ell i}$) as well as constraints (14), (15) and (20). Apart from the fact that these three papers possess slightly different objective functions, (all special cases of that used in (LRP1)), they contain the following differences :

i) Golden et al. do not consider maximum distance restrictions (9), pairing variables z_{ij}, nor constraints (16).

ii) Or and Pierskalla do not include time limits (10) and omit constraints (13) (thus allowing routes through several facilities and in turn, because of constraints (16), possibly assigning a user to more than one facility).

iii) Srikar and Srivastava use, instead of constraints (11), the subtour elimination constraints of Miller et al. (1960); they do not define variables y_{ij}, nor constraints (16). Finally, they specify the number of facilities which are to be opened.

6. TWO-INDEX VEHICLE FLOW FORMULATIONS AND ALGORITHMS FOR SYMMETRICAL PROBLEMS

Exact algorithms for symmetrical traveling salesman problems (TSPs) and VRPs (i.e. problems having a symmetrical distance matrix) are traditionally based on ILP models and algorithms, while network flow approaches are more often used for asymmetrical problems (see, for example, Lawler et al. (1985); Laporte and Nobert (1987)). For the symmetric case, the models are usually derived from the classical TSP formulation first proposed by Dantzig et al. (1954). Not surprisingly, models for symmetrical LRPs also belong to the same class.

One common feature to all models presented in this section is that, despite their large size, they are sometimes relatively easy to solve by means of constraint relaxation algorithms. Such algorithms initially operate on a relaxed problem possessing an easily-solved structure (usually close to that of a 2-matching problem). The relaxed constraints are then gradually introduced as violations are detected. If the total number of constraints which have to be generated remains relatively low, optimality can be reached within reasonable computing effort.

In this section, we present three classes of LRP formulations and algorithms based on the work of the author. We show how they relate to some well-known TSP and VRP models. For the sake of brevity, we restrict ourselves to two-layer problems, omitting other multi-layer problems and various classes of LRPs such as the GTSP (an exact algorithm for symmetrical GTPSs is described in Laporte and Nobert (1983)). We then summarize computational results for all LRP algorithms presented in this section.

6.1. A Classical TSP Formulation and Some Extensions

The first ILP formulation for the TSP is due to Dantzig et al. (1954). Define :

$N = \{1,\ldots,n\}$ a set of nodes (points, customers or cities)

$\bar{S} = N-S$

$C = (c_{ij})$ $(i,j \in N)$, a symmetrical distance matrix

x_{ij} : a 0-1 variable equal to 1 if and only if edge (i,j) is chosen in the optimal solution. x_{ij} need only be defined if $i<j$. Here and elsewhere, x_{ij} must be interpreted as x_{ji} whenever $i>j$.

The problem is then to

(TSP) Minimize $\sum_{i,j \in N} c_{ij} x_{ij}$

subject to

$$\sum_{i<k} x_{ik} + \sum_{j>k} x_{kj} = 2 \qquad (k \in N) \qquad (21)$$

$$\sum_{i,j \in S} x_{ij} \leq |S|-1 \qquad (S \subset N;\ 2 \leq |S| \leq n-2) \qquad (22)$$

$$x_{ij} = 0,1 \qquad (i,j \in N) \qquad (23)$$

In this formulation, constraints (21) are **degree constraints** : they specify the degree of each node. Constraints (22) are **subtour elimination constraints** : they prevent the formation of subtours over proper subsets of N. Constraints (23) are **integrality constraints.** It is well known (Dantzig et al., 1954) that, if (21) holds, constraints (22) are equivalent to the following **connectivity constraints** :

$$\sum_{\substack{i \in S, j \in \bar{S} \\ \text{or } i \in \bar{S}, j \in S}} x_{ij} \geq 2 \qquad (S \subset N;\ 2 \leq |S| \leq n-2) \qquad (24)$$

When, as in (TSP) for example, the degree of every node of S is equal to 2, this equivalence can be extended to :

$$\sum_{i,j \in S} x_{ij} \leq |S| - r \quad \Leftrightarrow \quad \sum_{\substack{i \in S, j \in \bar{S} \\ \text{or } i \in \bar{S}, j \in S}} x_{ij} \geq 2r \quad (S \subset N;\ r \in \mathbb{R}) \qquad (25)$$

Depending on r, (25) specializes to various subtour elimination constraints used in some VRP formulations (see, for example, Laporte et al. (1985)). Now define $\Delta(S)$, the **degree of S** :

$$\Delta(S) = \sum_{i \in S} \left(\sum_{k>i} x_{ik} + \sum_{k<i} x_{ki} \right) \qquad (26)$$

It can be shown (Laporte, 1986) that when $\Delta(S) \neq 2|S|$, (25) no longer holds. However the following equivalence is still valid :

$$\sum_{i,j \in S} x_{ij} \leq \frac{1}{2} \Delta(S) - r \quad \Leftrightarrow \quad \sum_{\substack{i \in S, j \in \bar{S} \\ \text{or } i \in \bar{S}, j \in S}} x_{ij} \geq 2r \quad (S \subset N; 2 \leq |S| \leq n-2; r \in \mathbb{R}) \qquad (27)$$

We will now use this result to formulate a class of LRPs.

6.2. The Single Depot LRP

In this very simple form of the LRP, we consider a set N of users and a subset I N of potential sites for a single facility, in order to minimize the total routing cost for a fleet of exactly m vehicles based at that facility. (We deviate here from the standard LRP framework presented in section 1.2 since now, the facility set and the user set are not disjoint).

Define a binary variable y_i equal to 1 if and only if the facility is located at node i. The remaining notation is as in (TSP). The problem is then to

(LRP2) Minimize $\sum_{i,j \in N} c_{ij} x_{ij}$

subject to

$$\sum_{i \in I} y_i = 1 \qquad (28)$$

$$\sum_{i<k} x_{ik} + \sum_{j>k} x_{kj} = 2 + 2(m-1) y_i \quad (k \in I) \qquad (29)$$

$$\sum_{i<k} x_{ik} + \sum_{j>k} x_{kj} = 2 \quad (k \in N - I) \qquad (30)$$

$$\sum_{i,j \in S} x_{ij} \leq \frac{1}{2} \Delta(S) - 1 \quad (S \subset N;\ 2 \leq |S| \leq n-2) \qquad (31)$$

$$y_i = 0,1 \quad (i \in I) \qquad (32)$$

$$x_{ij} = \begin{cases} 0, 1, 2 & (i \text{ or } j \in I) \\ 0,1 & (i,j \in N-I) \end{cases} \quad (33)$$

Most constraints of this model are self-explanatory. Constraints (31) are derived as follows : any feasible solution must correspond to a connected graph, therefore the following constraint must hold :

$$\sum_{\substack{i \in S, j \in \bar{S} \\ \text{or } i \in \bar{S}, j \in S}} x_{ij} \geq 2 \qquad (S \subset N;\ 2 \leq |S| \leq n-2) \quad (34)$$

By equivalence (27), for r=1, (34) can be rewritten as (31). Note that it would not have been valid to use (22) instead of (31), since $\Delta(S) > 2$ whenever $\sum_{i \in S} y_i = 1$.

Laporte (1986) shows that (LRP2) is equivalent to the model presented in Laporte and Nobert (1981) for the same problem, and how it can be extended to the case where m is a variable bounded above by $\bar{m}$.

In Laporte and Nobert (1981), (LRP2) is solved by means of a constraint relaxation method inspired by the branch and bound algorithm for the TSP devised by Miliotis (1976). (LRP2) is solved by initially relaxing the subtour elimination constraints (31) and the integrality conditions on the variables (but by retaining their upper bounds). Integrality is gradually regained by branch and bound; at an integer solution, a check for violated subtour elimination constraints is made, the constraint corresponding to the subtour involving the least number of nodes is introduced and the problem is reoptimized.

6.3. A Family of Multi-Depot LRPs

Laporte et al. (1983) studied a variant of (LRP2) in which at most $\bar{P}$ facilities, instead of only one, have to be located. The triangle inequality is assumed to be satisfied in I, i.e.

$$c_{ki} + c_{ij} \geq c_{kj} \qquad (i \in I) \quad (35)$$

Under this condition, it can be shown that there will be only one vehicle per opened facility in the optimal solution (i.e. the m-TSP solution is then dominated by a TSP tour). As in (LRP1), g_i was defined as the cost of opening a facility at site i. The problem can be formulated as (LRP2) with m=1, $\bar{P}$ for the right-hand side of (28), and a new objective :

$$\text{(LRP3)} \quad \text{Minimize} \quad \sum_{i,j \in N} c_{ij} x_{ij} + \sum_{i \in I} g_i y_i .$$

Two variants of (LRP3) were considered :

i) $g_i > 0$ for all $i \in I$ and $\bar{P} = |I|$;

ii) $g_i = 0$ for all $i \in I$ and $\bar{P} < |I|$;

where g_i and $\bar{P}$ are input parameters.

Problems were solved by adapting Miliotis' (1978) REVERSE algorithm for the TSP that checks for violated subtour elimination constraints even at fractional solutions. The procedure may be summarized in three steps :

Step 1. Solve (LRP3) by relaxing constraints (31) and the integrality conditions on the variables.

Step 2. Identify violated subtour elimination constraints (even if the solution at hand is not necessarily integer) and introduce one such constraint for each illegal component (i.e. a set of nodes disjoint from I and connected by edges (i,j) for which $x_{ij} > 0$). Reoptimize until no illegal component remains.

Step 3. Reach integrality by gradually introducing Gomory cuts. Stop if a feasible solution has been reached. Otherwise, proceed to step 2.

A simple variant of (LRP3) occurs when $g_i = 0$ for all $i \in I$ and $\bar{P} = |I|$. Then the problem (LRP4) consists of "covering" the nodes with a set of Hamiltonian cycles, each containing at least one node belonging to I. (LRP4) is in fact a relaxation of (TSP) in which some subtours are allowed. Its formulation is that of (TSP) with the exception that constraints (22) are replaced by

$$\sum_{i,j \in S} x_{ij} \leq |S| - 1 \qquad (S \subset N;\ S \cap I = \emptyset;\ 2 \leq |S| \leq n-2) \qquad (36)$$

It should be noted that (LRP4) is not, properly speaking, an LRP. The facilities having no fixed costs, their precise locations on the various Hamiltonian cycles contained in the solution are immaterial, as long as they correspond to elements of I. In a recent paper, Branco and Coelho (1986) proposed a set partitioning formulation as well as four heuristics for (LRP4). The quality of these heuristics is difficult to assess since no comparisons with the global optima are made.

Finally, Laporte et al. (1983) also studied two LRPs belonging to the same family. In the first of these, condition (35) is relaxed, and it is assumed that if a facility is opened at site i, then exactly m_i vehicles will be based at that facility. The problem appears to be considerably more difficult than (LRP3). In the second case, it is assumed that the costs do not satisfy the triangle inequality and that multiple visits to a node are allowed. The problem is relatively easy to solve; it constitutes in fact a relaxation of the problem which consists of determining the shortest complete cycle in a graph.

6.4. A Multi-Depot Capacitated LRP

We now revert to our original framework (section 1.2) and consider an 2/T problem in which I, the set of potential facilities, is disjoint from J, the set of users. Laporte et al. (1986b) have studied such a problem belonging to this class and involving (i) simultaneous location and routing, (ii) user demands and capacity constraints, (iii) fixed costs in facilities and vehicles, (iv) bounds on the number of opened facilities, and (v) bounds on the number of vehicles per facility. One interesting feature of the paper is the description of a procedure for generating cuts in order to eliminate solutions in which some vehicles start their journey at a given facility and terminate it at a different facility. These cuts known as **chain barring constraints** are also used in other types of multi-depot VRPs.

Let I, J, d_j, g_i, y_i be defined as in (LRP1), section 5.1. Here and in the remainder of this paper, $C=(c_{ij})$ is a matrix of travel **costs.** We also use the following notation :

D : the capacity of a vehicle

P : the number of facilities in the optimal solution. P lies between two prespecified bounds $\underline{P} \geqslant 1$ and $\bar{P} \leqslant |I|$.

$\lceil x \rceil$: the smallest integer greater than or equal to x if $x > 0$; 1 otherwise.

m_i : the number of vehicles based at facility i; m_i lies between two bounds: $1 \leqslant \underline{m}_i \leqslant m_i \leqslant \bar{m}_i$.

x_{ij} : the number of times (0, 1 or 2) edge (i,j) is used in the optimal solution; x_{ij} is undefined if $i \geqslant j$, if $i,j \varepsilon I$ or if $d_i + d_j > D$.

The problem involves selecting facility sites (when $\underline{P} < |I|$), determining how many vehicles to base at each opened facility, and establishing vehicle routes to achieve minimal total cost subject to the following constraints :

i) each route starts and ends at the same facility;

ii) all of the service requirement of a user is met by only one vehicle (the same user may be visited more than once if this saves distance, but only one of the vehicle visits is used to meet the user's requirement);

iii) the sum of all requirements satisfied by any vehicle does not exceed D;

iv) the numbers of facilities and of vehicles per facility lie within their respective bounds.

The problem can be formulated as follows :

(LRP5) Minimize $\sum_{i,j \varepsilon I \cup J} c_{ij} x_{ij} + \sum_{i \varepsilon I} (g_i y_i + f_i m_i)$

subject to

$$\sum_{i<k} x_{ik} + \sum_{k>j} x_{kj} = 2 \qquad (k\varepsilon J) \qquad (37)$$

$$\sum_{i<k} x_{ik} + \sum_{j>k} x_{kj} = 2m_k \qquad (k\varepsilon I) \qquad (38)$$

$$\sum_{i,j\varepsilon S} x_{ij} \leqslant |S| - \left\lceil \sum_{k\varepsilon S} d_k/D \right\rceil \qquad (S\subseteq J; |S| \geqslant 3) \qquad (39)$$

$$x_{i_1 i_2} + 3x_{i_2 i_3} + x_{i_3 i_4} \leqslant 4 \qquad (i_1, i_4 \varepsilon I;\ i_2, i_3 \varepsilon J) \qquad (40)$$

$$x_{i_1 i_2} + x_{i_{h-1} i_h} + 2 \sum_{i,j\varepsilon\{i_2,\ldots i_{h-1}\}} x_{ij} \leqslant 2h-5 \qquad (41)$$

$$(h \geqslant 5;\ i_1,\ i_h \varepsilon I;\ i_2,\ldots,i_{h-1} \varepsilon J)$$

$$y_i \leqslant m_i \leqslant My_i \qquad (i\varepsilon I) \qquad (42)$$

$$\underline{m}_i \leqslant m_i \leqslant \bar{m}_i \qquad (i\varepsilon I) \qquad (43)$$

$$\underline{P} \leqslant \sum_{i\varepsilon I} y_i \leqslant \bar{P} \qquad (44)$$

$$y_i = 0,1 \qquad (i\varepsilon I) \qquad (45)$$

$$x_{ij} = \begin{cases} 0,1 & (i,j\varepsilon J) \\ 0,1,2 & (i \text{ or } j\varepsilon I) \end{cases} \qquad (46)$$

The objective function and most constraints of this formulation are self-explanatory. We provide details on constraints (39)-(41). Constraints (39) are imposed to eliminate either subtours which are disconnected from I or vehicle routes whose sum of requirements exceeds D. They are derived as follows : any subset S of J must be entered and left by at least $r = \left\lceil \sum_{k\varepsilon S} d_k/D \right\rceil$ vehicles, hence the following constraint is valid :

$$\sum_{\substack{i\varepsilon S, j\varepsilon \bar{S} \\ \text{or } i\varepsilon \bar{S}, j\varepsilon S}} x_{ij} \geqslant 2r \qquad (S\subseteq J;\ |S| \geqslant 3) \qquad (47)$$

When (37) is satisfied, (47) is equivalent to (39) (see (25)). Constraints (40) and (41) are chain barring constraints : they ensure that each route starts and ends at the same facility. They prohibit solutions containing chains of edges between two facilities i_1, i_h, and h-2 nodes $i_2,\ldots i_{h-1}$ belonging to J. The constraints are derived by considering three cases :

$h \leq 3$, $h = 4$ and $h \geq 5$. The first case is of no interest since it cannot occur in an optimal solution. For $h = 4$ and $h \geq 5$, the required constraint is derived as follows. Define

$$X = x_{i_1 i_2}, \quad Y = x_{i_{h-1} i_h} \quad \text{and} \quad Z = \sum_{i,j \in \{i_2, \ldots, i_{h-1}\}} x_{ij} \tag{48}$$

and let $\bar{X}$, $\bar{Y}$, $\bar{Z}$ be the values taken by X, Y and Z respectively. In a feasible solution to (LRP5), $\bar{X}, \bar{Y} \in \{0,1,2\}$ and $\bar{Z}$ is an integer in $[0, u(\bar{X},\bar{Y})]$ where $u(\bar{X},\bar{Y})$ is an upper bound on $\bar{Z}$ whose value depends on h, $\bar{X}$ and $\bar{Y}$. Details of the computation of $u(\bar{X},\bar{Y})$ are provided in Laporte et al. (1986b). We seek the strongest constraint of the form

$$aX + bY + Z \leq d \tag{49}$$

which satisfies

$$d - a\bar{X} - b\bar{Y} \geq u(\bar{X},\bar{Y}) \qquad (\bar{X},\bar{Y} \in \{0,1,2\}) \tag{50}$$

and

$$a, b \in \mathbb{R}; \quad d \geq 0 \tag{51}$$

while cutting off any infeasible solution for which $\bar{X} = \bar{Y} = 1$ and $\bar{Z} = h-3$. The strongest cut is obtained by minimizing (d-a-b), the value taken by the left-hand side of (50) when $\bar{X} = \bar{Y} = 1$, subject to (50) and (51).

It is worth noting that the problem of illegal chains does not occur in three-index formulations such as (LRP1) since constraints (13) prevent the same vehicle from passing through two facilities. The difficulty in two-index formulations arises from the fact that vehicles are not identified by the variables.

The algorithm proposed for the solution of (LRP5) also belongs to the class of constraint relaxation methods. Here, the constraints which are initially relaxed are (i) the subtour elimination constraints; (ii) the chain barring constraints, and (iii) the integrality constraints. In addition, checks are carried out in the course of the branch and bound process to fix at zero certain variables which could not enter the solution without violating the vehicle capacity constraints. Branching on a fractional variable takes place only when no other violated constraint can be identified.

6.5. Computational Results

We have summarized in table 2 the computational results related to the problems and exact algorithms described earlier in this section. Most of the information contained in this table is straightforward and requires no explanation. In the column "Type of ILP algorithm", BB means that integrality was reached by branch and bound, and CUTS means that Gomory cutting planes were used. Using Miliotis' terminology, STRAIGHT refers to an algorithm in

TABLE 2
Computational results for various types of LRPs

PROBLEM IDENTIFICATION (REFERENCE)	I : DEPOT SET J : USER SET	DEPOT FIXED COSTS	MAXIMUM NUMBER OF DEPOTS IN SOLUTION	TRIANGLE INEQUALITY ASSUMED	TYPE OF ILP ALGORITHM	SIZE	CPU TIME IN SECONDS (CYBER 173)
LRP2 : single depot LRP (Laporte and Nobert, 1981)	$I \subseteq J$	no	1	in I	BB, STRAIGHT	$\|I\| = \|J\| = 50$	671
				no	BB, STRAIGHT	$\|I\| = \|J\| = 50$	434
LRP3 : multi-depot LRP (Laporte et al., 1983)	$I \subseteq J$	yes	2-7	in I	CUTS, REVERSE	$\|I\| = 2, \|J\| = 40$	19-31
						$\|I\| = 5, \|J\| = 40$	18-63
						$\|I\| = 7, \|J\| = 40$	17-60
		no	3-8	in I	CUTS, REVERSE	$\|I\| = 5, \|J\| = 40$	26
						$\|I\| = 7, \|J\| = 40$	19-24
						$\|I\| = 15, \|J\| = 40$	26
LRP4 : multi-depot LRP (Laporte et al., 1983)	$I \subseteq J$	no	3-15	in I	CUTS, REVERSE	$\|I\| = 3, \|J\| = 50$	28-33
						$\|I\| = 5, \|J\| = 50$	32-41
						$\|I\| = 10, \|J\| = 50$	35-45
						$\|I\| = 15, \|J\| = 50$	23-36
LRP5 : capacitated multi-depot LRP (Laporte et al., 1986b)	$I \cap J = \emptyset$	no	3-5	no	BB, REVERSE	$\|I\| = 8, \|J\| = 20$	43-298
		yes	3-5	no	BB, REVERSE	$\|I\| = 8, \|J\| = 20$	39-217

which subtour elimination constraints are introduced only at an integer solution whereas in REVERSE algorithms, violated subtour elimination and chain barring constraints are generated as soon as they can be identified, even at fractional solutions. In most cases, the column "Size" gives the maximum problem size which could be consistently reached within 300 CPU seconds on the CYBER 173. One exception is LRP4 : since this problem is a TSP relaxation, much larger sizes could have been attained within the prescribed time. Problems LRP2, LRP3, and LRP4 were solved on the Université de Montréal CYBER 173; LRP5 was solved on a CYBER 855 but the reported times are CYBER 173 equivalent.

These results indicate that using a constraint relaxation approach, relatively large problems can be solved to optimality. With the exception of LRP4, most LRPs are more difficult than the corresponding problems without any locational dimension. For example, LRP2 with $|I|=1$ is a multiple TSP (similar in difficulty to a single TSP) whereas LRP5 with $|I|=1$ is a standard capacitated VRP which can often be solved for up to 50 customers (Laporte and Nobert, 1987).

7. TWO-INDEX VEHICLE FLOW FORMULATIONS AND ALGORITHMS FOR ASYMMETRICAL PROBLEMS

We present in this section a general methodology for solving asymmetrical LRPs. The fundamental idea consists of transforming the original LRP into a problem containing an easy-to-solve relaxation; constraints not included in the relaxation can be dealt with by a branch and bound procedure. More precisely, we transform the problems into TSPs with specified nodes, under a variety of side constraints. This approach was recently used with success by Laporte et al. (1987b). The proposed methodology will be illustrated on a three-layer problem. By adjusting two parameters, the four scenarios 3/R/R, 3/R/T, 3/T/R, 3/T/T can be covered. The model can easily be adapted to other situations.

7.1. Problem Definition

As previously, let L be the set of primary facilities, I, the set of secondary facilities and J, the set of users. These three sets are pairwise disjoint. Assume all elements of J and L have to be included in the solution whereas only a subset of I has to be used; the elements of I represent potential locations. The cost of establishing a facility at site $i \in I$ is equal to g_i. A maximum of $\bar{M}_\ell$ vehicles can be based at primary facility $\ell \in L$; similarly, a maximum of $\bar{m}_i$ vehicles can be based at secondary facility $i \in I$. These vehicles have fixed costs of F and f respectively. Various constraints derived from vehicle capacities, maximum route length restrictions, etc. may be imposed on the routes emanating from any layer. Let $C = (c_{ij})$ $(i \neq j)$ be the travel cost matrix defined on N x N, where $N = L \cup I \cup J$. In (LRP6), one seeks a minimum cost solution that selects secondary facility sites, determines the size of the vehicle fleet at all opened facilities, and establishes vehicle routes subject to the restrictions below :

i) each route starts and ends at the same facility;

ii) all side constraints on the routes are satisfied;

iii) the number of vehicles based at any facility does not exceed its upper bound.

We represent this problem by a directed graph $G = (N,A,C)$ with node set N, arc set $A = \{(i,j);i,j \in N, i \neq j\}$, and costs C.

7.2. Graph Extension

Several authors (see, for example, Lenstra and Rinnooy Kan (1975)) have suggested graph transformations to solve a variety of TSP extensions. These transformations are such that every feasible solution to the original problem corresponds to a Hamiltonian circuit on a subgraph of the new graph. Any feasible solution on the transformed graph can be interpreted as a solution on the original graph. The desired transformation for (LRP6) can be outlined as follows. Full mathematical details are provided in Laporte (1987).

Graph G is transformed into $G' = (N',A',C')$. The new set of nodes N' is the union of three pairwise disjoint sets of nodes :

i) a set L' consisting of the union of $|L|$ sets of nodes corresponding to the primary facilities; the ℓth set contains $\bar{M}_\ell + 1$ nodes;

ii) a set I' consisting of the union of $|I|$ sets of nodes associated with the secondary facilities; the ith set contains $\bar{m}_i + 1$ nodes;

iii) a set J of user nodes.

It is convenient to further partition N' into $\{N_1', N_2'\}$. All elements of N_1' are designed as **specified nodes** : $N_1' = L' \cup J \cup I^*$ where I^* is the union of all sets of nodes corresponding to secondary facilities which must appear in the optimal solution; N_2' is a set of **unspecified nodes.**

The new set of arcs A' can be defined as the union of the following sets :

A_1 : arcs from primary to secondary facilities;

A_2 : arcs from secondary to primary facilities;

A_3 : arcs from secondary facilities to users;

A_4 : arcs from users to secondary facilities;

A_5 : intra-primary facility arcs;

A_6 : inter-primary facility arcs;

A_7 : intra-secondary facility arcs;

A_8 : inter-secondary facility arcs;

A_9 : loops associated with the first node of every secondary facility which does not necessarily appear in the solution;

A_{10}: loops associated with other nodes of N_2';
A_{11}: inter-user arcs.

The cost matrix $C' = (c'_{uv})$ associated with A' is then defined by the following formulas :

$$c'_{uv} = \begin{cases} F + c_{uv} & (u,v)\varepsilon A_1 \\ f + c_{uv} & (u,v)\varepsilon A_3 \\ -g_r & (u,v)\varepsilon A_9,\text{ for the } r^{th} \text{ secondary facility} \\ \theta_1 c_{uv} & (u,v)\varepsilon A_8 \\ \theta_2 c_{uv} & (u,v)\varepsilon A_{11} \\ c_{uv} & (u,v)\varepsilon A_2 \cup A_4 \\ 0 & (u,v)\varepsilon A_5 \cup A_6 \cup A_7 \cup A_{10} \\ \infty & \text{otherwise} \end{cases} \quad (52)$$

In (52), $\theta_t (t=1,2)$ corresponds to the distribution mode used between layer t and layer $t+1$:

$$\theta_t = \begin{cases} 1 & \text{if } M_t = T \\ \infty & \text{if } M_t = R \end{cases} \quad (53)$$

7.3. Interpretation

Before proceeding to the formulation of the problem, it seems useful to illustrate a feasible solution on G' (figure 2) and its translation on G (figure 3). In this example, there are two primary facilities (with a maximum of two vehicles each) and four secondary facilities (with a maximum of three vehicles each). These facilities are represented by rectangles containing as many dots as the maximum number of vehicles plus one. A black dot indicates that the corresponding facility must necessarily appear in the optimal solution. The solution on G' corresponds to a Hamiltonian circuit on a subset of N'. Primary vehicle routes are represented by dashed lines whereas secondary vehicles tours correspond to full lines. The dotted arrows between primary facilities have no particular meaning. Loops shown on secondary facility S_3 mean that this facility is not used. Arrows emanating from a node within a facility and pointing away from the facility correspond to used vehicles; arrows having their two ends within the same facility corrpespond to unused vehicles. Note that

i) vehicle tours start and end at the same facility;
ii) vehicle nodes are only used if the associated facility is opened.

7.4. The Model

In order to formulate the problem, first define 0-1 variables x_{ij} :

i) if $i \neq j$ (i.e. $(i,j) \in A'-A_9-A_{10}$), $x_{ij}=1$ if arc (i,j) is used in the optimal solution and $x_{ij}=0$ otherwise;

ii) if $i=j$ (i.e. $(i,j) \in A_9 \cup A_{10}$), $x_{ij}=0$ if node i of N_2' is used in the optimal solution and $x_{ii}=1$ otherwise.

In what follows, variables x_{ij} which are undefined must be interpreted as 0. The formulation is then :

(LRP6) Minimize $\sum_{i,j \in N'} c'_{ij} x_{ij}$

subject to

$$\sum_{i \in N'} x_{ik} = 1 \qquad (k \in N') \qquad (54)$$

$$\sum_{j \in N'} x_{kj} = 1 \qquad (k \in N') \qquad (55)$$

$$\sum_{i,j \in S} x_{ij} \leq |S|-1 \qquad (S \in N, |S| \geq 2, \text{ i or } j \in N_1') \qquad (56)$$

If a route (defined on G) starts at a given facility, it must end at the same facility. (57)

Routes (defined on G) must satisfy all side conditions. (58)

$$x_{uu} \leq x_{vv} \qquad (59)$$

(u is the first node of a secondary facility and v is another node of the same facility)

$$x_{ij} = 0,1 \qquad (i,j \in N') \qquad (60)$$

In this formulation, constraints (54) and (55) specify that all nodes which appear in the solution must be entered and left once; if the variable x_{ii} associated with a node of N_2' is equal to 1, the node is unvisited and the sums of incoming and outgoing arcs are then equal to zero. Constraints (56) are subtour elimination constraints; they need not be defined for sets of S which are subsets of N_2'. Constraints (57), (58) and (60) are self-explanatory. Constraints (59) forbid solutions in which vehicles based at unopened facilities are used. These correspond to constraints (42) in (LRP5).

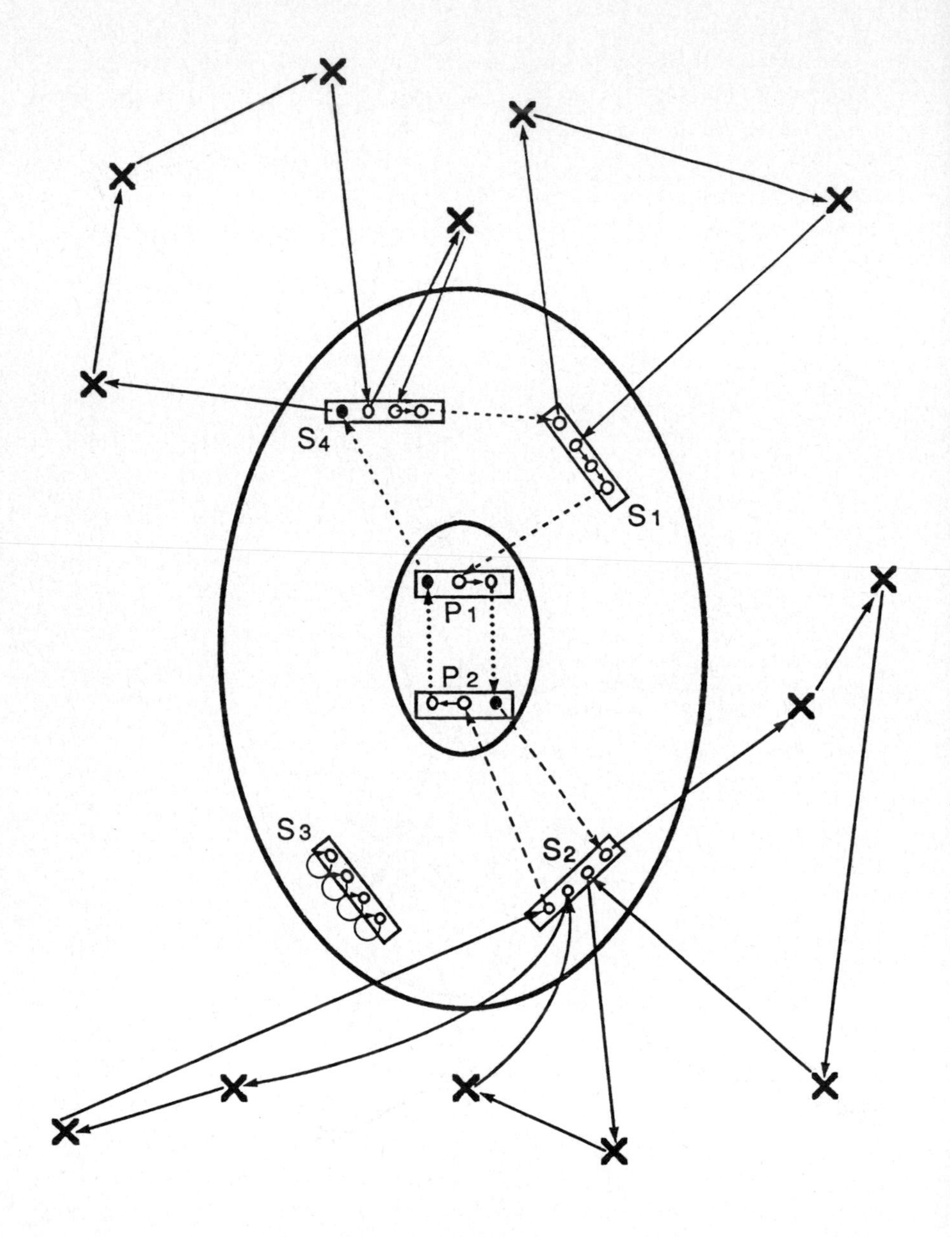

FIGURE 2
Feasible solution on G'

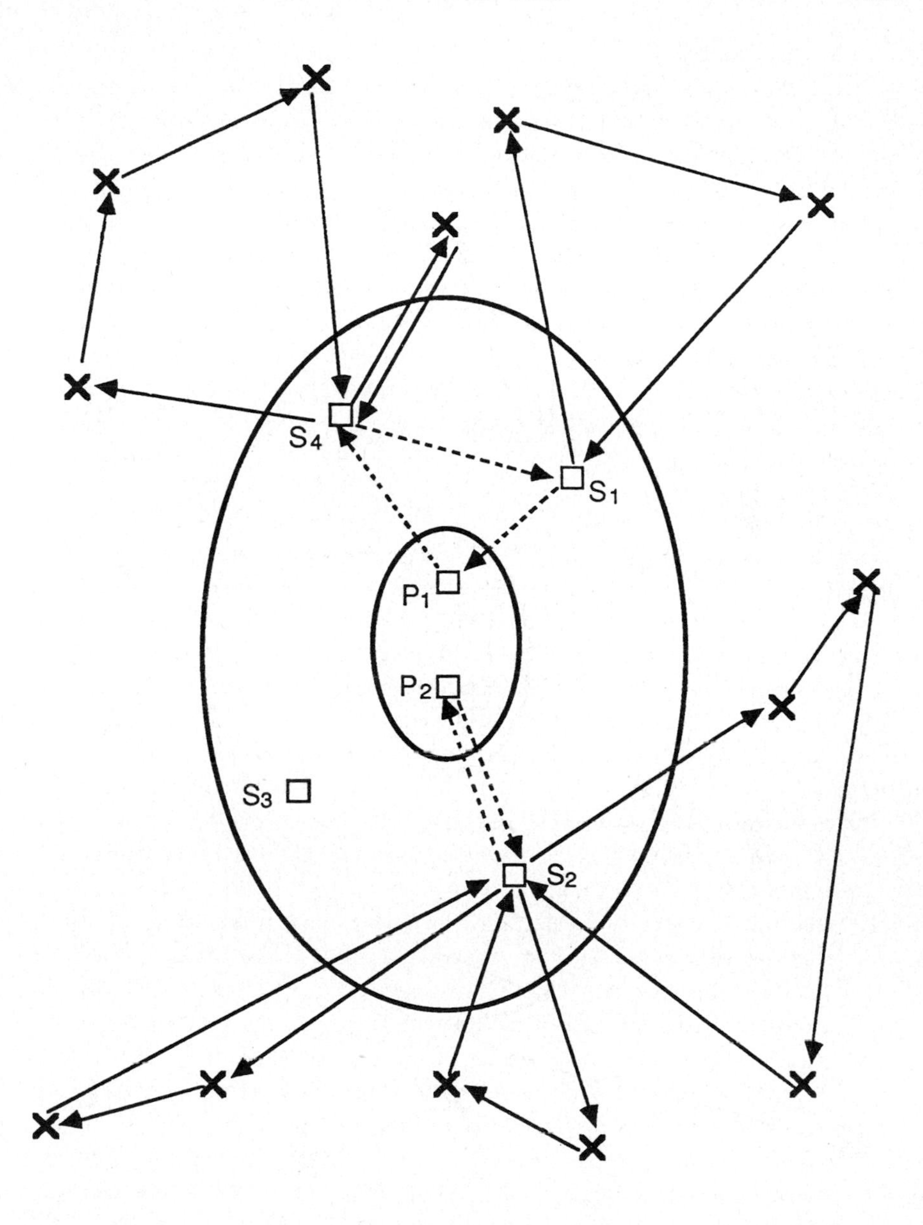

FIGURE 3
Feasible solution on G

7.5. Algorithm

The problem obtained by relaxing constraints (56) to (59) is an assignment problem (AP) for which efficient algorithms exist. A sensible approach is to solve (LRP6) by means of a branch and bound tree : at the first node, the AP relaxation of LRP6 is solved. Then, and at every subsequent node of the search tree, a check for violated relaxed constraints is made. If at least one such constraint can be identified, a number of subproblems are created according to the rule proposed by Carpaneto and Toth (1980). The development of the search tree follows the usual branch and bound rules. Such an approach was used successfully by Laporte et al. (1986a, 1987b) for the capacitated VRP and for a class of asymmetrical LRPs. In the latter case, problems containing up to 80 nodes (before transformation of G into G') were solved to optimality.

8. CONCLUSIONS

This survey has shown that LRPs are encountered in a variety of practical situations and that their study constitutes a fast growing research area. Most developments have occurred over the last few years and can be primarily attributed to algorithmic advances in the field of vehicle routing. By and large, research on LRPs is still relatively fragmented, exhibiting several gaps that further research must fill.

At the end of this study, we can identify a number of promising research areas :

i) **Development and systematic analysis of heuristics** for basic types of LRPs (following the work of Perl and Daskin (1985) and Branco and Coelho (1986), for example).

ii) Location and routing constitute, in several practical contexts, decisions which must be made at different levels. Very often, locational decisions involve large capital expenditures (such as in the optimal siting of factories, warehouses, or depots), whereas routing is operational in nature and may have to be revised frequently. In such cases, LRPs may be solved in two relatively independent phases : facilities can be located first by approximating the routing costs (such as in the work of Daganzo (1984), for example) and optimal routing can be determined in a second phase. In this spirit, the study of **hierarchical LRPs** along the lines of Daganzo and Newell (1986) or of Marchetti Spaccamela et al. (1982) would constitute a welcome contribution to the field.

iii) In several contexts, however, facilities are inexpensive or free but their locations have a major impact on routing costs. This is the case, for example, of several LRPs encountered in the mail collection and delivery context and in school bus routing where transfer points have to be located. In such contexts, combined location-routing models are more meaningful and it then makes sense to develop models and algorithms which simultaneously take into account two major components of the problem. More effort should therefore be spent on the **development of new types of exact algorithms.** So far, most exact algorithms for LRPs have been derived from two-index vehicle flow formulations, coupled with constraint relaxation. The problem size which can be tackled using this type of approach appears however to be limited to 80 nodes. Moreover, the efficiency of constraint relaxation methods diminishes in the case of problems which are highly constrained (and this appears to be the case of several practical situations). Thus the usefulness of this approach appears to be somewhat limited. Theoretical gains may be made by developing a better understanding of the polyhedral structure of LRPs (similarly to Padberg and Rinaldi (1987) for the TSP and to Laporte and Nobert (1984) for the VRP), but given the complex constraint structure of LRPs, this may be unrealistic and the computational payoff is not obvious. A more promising approach is to use Lagrangean relaxation techniques or dynamic programming (using state-space relaxation, see Christofides et al. (1981b)). Both of these techniques have shown great success for the VRP have not yet been used for LRPs. In our opinion, they could have a great potential.

iv) From a practical point of view, several interesting developments which have occured in recent years in the field of vehicle routing (see Golden and Assad (1986) for a recent discussion) can be applied to LRPs. Here, the use of **visual interactive models** naturally comes to mind. A computer system (HOTOCARS) now being developed at the Centre de recherche sur les transports, Université de Montréal, is of this type. Users can select seed locations for facilities and choose through parameters a suitable heuristic algorithm for the routing phase. Solutions are shown on screen and can be modified directly by users. For details on the algorithmic content of HOTOCARS, see Potvin (1987).

BIBLIOGRAPHY

BALINSKI, M. and QUANDT, R. (1964), "On an integer program for a delivery problem", Operations Research 12, 300-304.

BEARWOOD, J., HALTON, J.H. and HAMMERSLEY, J.M. (1959), "The Shortest Path through many Points", Proceedings of the Cambridge Philosophical Society 55, 299-327.

BELL, W.J., DALBERTO, L.M., FISHER, M.L., JAIKUMAR, R., KEDIA, P., MACK, R.G., and PRUTZMAN, P.J. (1983), "Improving the distribution of industrial gases with an on-line computerized routing and scheduling optimizer", Interfaces 13, 4-23.

BERMAN, O. and SIMCHI LEVI, D. (1986), "Minisum location of a traveling salesman", Networks 16, 239-254.

BRANCO, I.M. and COELHO, J.D. (1986), "The Hamiltonean P-median Problem", presented at the Euro VIII Conference, Lisbon.

BURNESS, R.C. and WHITE, J.A. (1976), "The traveling salesman location problem", Transportation Science 10, 348-360.

CARPANETO, G., TOTH, P. (1980), "Some new branching and bounding criteria for the asymmetric travelling salesman problem", Management Science 26, 736-743.

CHAN, A.W. and FRANCIS, R.L. (1976), "A round-trip location problem on a tree graph", Transportation Science 10, 35-51.

CHAN, A.W. and HEARN, D.W. (1977), "A rectilinear distance minimax round-trip location problem", Transportation Science 11, 107-123.

CHRISTOFIDES, N. (1976), "The vehicle routing problem", R.A.I.R.O. (recherche opérationnelle) 10, 55-70.

CHRISTOFIDES, N. and EILON, S. (1969), "Expected distances in distribution problems", Operations Research Quarterly 20, 437-443.

CHRISTOFIDES, N., MINGOZZI, A. and TOTH, P. (1981a), "Exact algorithms for the vehicle routing problem based on spanning tree shortest path relaxations", Mathematical Programming 20, 255-282.

CHRISTOFIDES, N., MINGOZZI, A. and TOTH, P. (1981b), "State-space relaxation procedures for the computation of bounds to routing problems", Networks 11, 145-164.

COOPER, L. (1963), "Location-allocation problems", Operations Research 11, 331-343.

CURRENT, J.R., REVELLE, C.S. and COHON, J.L. (1986), "The hierarchical network design problem", European Journal of Operational Research 27, 57-66.

DAGANZO, C. (1984), "The distance traveled to visit N points with a maximum of C stops per vehicle: an analytical model and an application", Transportation Science 18, 331-350.

DAGANZO, C. and NEWELL, G.F. (1986), "Configuration of physical distribution networks", Networks 16, 113-132.

DANTZIG, G.B., FULKERSON, D.R. and JOHNSON, S.M. (1954), "Solution of a large scale traveling salesman problem", Operations Research 2, 393- 410.

DEJAX, P. (1986), "Methodology for Facility Location and Distribution Systems Planning", Technical Report LEIS 86-04, Ecole Centrale de Paris, Chatenay-Malabry, France.

DESROSIERS, J., DUMAS, Y. and SOUMIS, F. (1986), "A dynamic programming solution of the large scale single vehicle dial-a-ride problem with time windows", American Journal of Mathematical and Management Sciences 6, 301-326.

DESROSIERS, J., SOUMIS, F. and DESROCHERS, M. (1984), "Routing with time windows by column generation", Networks 14, 545-565.

DREZNER, Z. (1985), "O(NlogN) algorithm for the rectilinear round-trip location problem", Transportation Science 19, 91-93.

DREZNER, Z., STEINER, G. and WESOLOWSKY, G.O. (1984a), "The Euclidean Tour Location Problem", presented at the Third International Symposium on Locational Decisions, Boston.

DREZNER, Z., STEINER, G. and WESOLOWSKY, G.O. (1984b), "Facility Location with Rectilinear Tour Distances", presented at the TIMS/ORSA Conference, Boston.

DROR, M. and TRUDEAU, P. (1986), "Stochastic vehicle routing with modified savings algorithm", European Journal of Operational Research 23, 228-235.

EILON, S., WATSON-GANDY, C.D.T. and CHRISTOFIDES, N. (1971), "Distribution Management : Mathematical Modelling and Practical Analysis", Griffin, London.

FELDMAN, E., LEHRER, F.A. and RAY, T.L. (1966), "Warehouse location under continuous economies of scale", Management Science 12, 670-684.

FISHER, M. and JAIKUMAR, R. (1978), "A Decomposition Algorithm for Large-Scale Vehicle Routing", Working paper 78-11-05, Department of Decision Sciences, University of Pennsylvania.

FLEISCHMANN, B. (1982), "Linear Programming Approaches to Travelling Salesman and Vehicle Scheduling Problems", presented at the XI International Symposium on Mathematical Programming, Bonn.

GAVISH, B. and GRAVES, S.C. (1982), "Scheduling and Routing on Transportation and Distribution Systems : Formulations and New Relaxations", Working paper, Graduate School of Management, University of Rochester, NY.

GHOSH, J.K., SINHA, S.B. and ACHARYA, D. (1981), "A Generalized Reduced Gradient Based Approach to Round-Trip Location Problem", in N.K. Jaiswal, ed., Scientific Management of Transport Systems, North-Holland Publishing, Amsterdam.

GOLDEN, B.L. and ASSAD, A. (1986), "Perspectives on vehicle routing: exciting new developments", Operations Research 34, 803-810.

GOLDEN, B.L., BALL, M. and BODIN, L.D. (1981), "Current and future research directions in network optimization", Computers & Operations Research 8, 71-81.

GOLDEN, B.L., MAGNANTI, T.L. and NGUYEN, H.Q. (1977), "Implementing vehicle routing algorithms", Networks 7, 113-148.

HAKIMI, S.L. (1964), "Optimum locations of switching centers and the absolute centers and medians of a graph", Operations Research 12, 450- 459.

ICHIMORI, T. and NISHIDA, T. (1985), "Note on a rectilinear round-trip location problem", Transportation Science 19, 84-91.

JACOBSEN, S.K. and MADSEN, O.B.G. (1980), "A comparative study of heuristics for a two-level routing-location problem", European Journal of Operational Research 5, 378-387.

KOLEN, A. (1985), "The round-trip p-center and covering problem on a tree", Transportation Science 19, 222-234.

KUEHN, A.A. and HAMBURGER, M.J. (1963), "A heuristic program for locating warehouses", Management Science 9, 643-666.

LABBE, M. and LAPORTE, G. (1986), "Maximizing user convenience and postal service efficiency in post box location", Belgian Journal of Operations Research, Statistics and Computer Science 26, 21-35.

LAPORTE, G. (1986), "Generalized subtour elimination constraints and connectivity constraints", Journal of the Operational Research Society 37, 509-514.

LAPORTE, G., (1987), "Location-Routing Problems", Cahier du G.E.R.A.D. G-87-05, École des Hautes Études Commerciales de Montréal.

LAPORTE, G. LOUVEAUX, F., and MERCURE, H. (1987a), "Models and Exact Solutions for a Class of Stochastic Location-Routing Problems", Cahier du G.E.R.A.D. G-87-14, École des Hautes Etudes Commerciales de Montréal.

LAPORTE, G., MERCURE, H. and NOBERT, Y. (1986a), "An exact algorithm for the asymmetrical capacitated vehicle routing problem", Networks 16, 33-46.

LAPORTE, G. and NOBERT, Y. (1981), "An exact algorithm for minimizing routing and operating costs in depot location", European Journal of Operational Research 6, 224-226.

LAPORTE, G. and NOBERT, Y. (1983), "Generalized travelling problem through n sets of nodes : an integer programming approach", INFOR 21, 61-75.

LAPORTE, G. and NOBERT, Y. (1984), "Comb Inequalities for the Vehicle Routing Problem", Methods of Operations Research 51, 271-276.

LAPORTE, G. and NOBERT, Y. (1987), "Exact Algorithms for the Vehicle Routing Problem", in S. Martello et al., eds., Surveys in Combinatorial Optimization, North-Holland Publishing, Amsterdam.

LAPORTE, G., NOBERT, Y. and ARPIN, D. (1986b), "An exact algorithm for solving a capacitated location-routing problem", Annals of Operations Research 6, 293-310.

LAPORTE, G., NOBERT, Y. and DESROCHERS, M. (1985), "Optimal routing under capacity and distance restrictions", Operations Research 33, 1050-1073.

LAPORTE, G., NOBERT, Y. and PELLETIER, P. (1983), "Hamiltonian location problems", European Journal of Operational Research 12, 82-89.
LAPORTE, G., NOBERT, Y., and TAILLEFER, S. (1987b), "Solving a Family of Multi-Depot Vehicle Routing and Location-Routing Problems", Cahier du G.E.R.A.D. G-87-10, École des Hautes Études Commerciales de Montréal.
LAWLER, E.L., LENSTRA, J.K., RINNOOY KAN, A.H.G. and SCHMOYS, D.B. (1985), "The Traveling Salesman Problem. A Guided Tour of Combinatorial Optimization", Wiley, Chichester, U.K.
LENSTRA, J.K. and RINNOOY KAN, A.H.G. (1975), "Some simple applications of the travelling salesman problem", Operational Research Quarterly 26, 717-734.
MADSEN, O.B.G. (1983), "Methods for solving combined two level location-routing problems of realistic dimensions", European Journal of Operational Research 12, 295-301.
MAGNANTI, T.L. (1981), "Combinatorial optimization and vehicle fleet planning : perspectives and prospects", Networks 11, 179-214.
MARCHETTI SPACCAMELA, A., RINNOOY KAN, A.H.G. and STOUGIE, L. (1982), "Hierarchical Vehicle Routing Problems", Report 8236/0, Erasmus University, Rotterdam.
McLAIN, D.R., DURCHHOLZ, M.L. and WILBORN, W.B. (1984), "U.S.A.F. EDSA Routing and Operating Location Selection Study", Report XPSR84-3, Operations Research Division, Directorate of Studies and Analysis, Scott Air Base, IL.
MERCER, A. (1970), "Strategic planning of physical distribution systems", International Journal of Physical Distribution 1, 20-25.
MERCER, A., CANTLEY, M.F. and RAND, G.K. (1978), "Operational Distribution Research", Taylor and Francis Ltd, London.
MILIOTIS, P. (1976), "Integer programming approaches to the travelling salesman problem", Mathematical Programming 10, 367-378.
MILIOTIS, P. (1978), "Using cutting planes to solve the symmetric travelling salesman problem", Mathematical Programming 15, 177-188.
MILLER, C.E., TUCKER, A.W. and ZEMLIN, R.A. (1960), "Integer programming formulation of traveling salesman problems", Journal of the ACM 7, 326-329.
MOLE, R. (1979), "A survey of local delivery vehicle routing methodology", Journal of the Operational Research Society 30, 245-252.
NAMBIAR, J.M., GELDERS, L.F. and VAN WASSENHOVE L.N. (1981), "A large scale location-allocation problem in the rubber industry", European Journal of Operational Research 6, 183-189.
OR, I. and PIERSKALLA, W.P. (1979), "A transportation location-allocation model for the regional blood banking", AIIE Transactions 11, 86-95.
PADBERG, M. and RINALDI, G. (1987), "An LP-based Algorithm for the Resolution of Large-Scale Traveling Salesman Problems", Research Report, New York University.
PERL, J. and DASKIN, M.S. (1985), "A warehouse location-routing problem", Transportation Research B 19B, 381-396.
POTVIN, J.-Y. (1987), "Un système informatique pour le développement et l'expérimentation d'algorithmes de génération de tournées", Ph.D. thesis, département d'Informatique et de Recherche opérationnelle, Université de Montréal.
SALHI, S. and RAND, G.K. (1986), "The Interdependance of Depot Location and Vehicle Routing", presented at the ORSA/TIMS Conference, Miami Beach, FL.
SCHRAGE, L. (1981), "Formulation and structure of more complex / realistic routing and scheduling problems", Networks 11, 229-232.
SRIKAR, B. and SRIVASTAVA, R. (1983), "Solution Methodology for the Location Routing Problem", presented at the ORSA/TIMS Conference, Orlando, FL.
STEWART Jr., W.R. and GOLDEN, B.L. (1983), "Stochastic vehicle routing : a comprehensive approach", European Journal of Operational Research 14, 371-385.

TFD (Transportforskningsdelegationen) (1977), "Distribution Planning Using Mathematical Methods" (in Swedish), Report 1077:1, Contract Research Group for Applied Mathematics, Royal Institute of Technology, Sweden.

VON BEDNAR, L. and STROHMEIER, E. (1979), "Lagerstandortoptimierung und Fuhrparkeinsatzplanung in der Konsumgüter-Industrie", Zeitschrift für Operations Research 23, B89-B104.

WATSON-GANDY, C.D.T. and DOHRN, P.J. (1973), "Depot location with van salesman - a practical approach", Omega 1, 321-329.

WEBB, M.H.J. (1968), "Cost functions in the location of depot for multiple-delivery journeys", Operational Research Quarterly 19, 311-320.

Vehicle Routing: Methods and Studies
B.L. Golden and A.A. Assad (Editors)

ALLOCATION/ROUTING: MODELS AND ALGORITHMS

Michael O. Ball
College of Business and Management
University of Maryland at College Park

Allocation/routing problems involve determining a set of routes for a fleet of vehicles over a multiple day time horizon. Thus, these problems can be viewed as containing two components, one that allocates deliveries to days of the week and a second that forms routes over each day of the week. Allocation/routing problems arise in such diverse settings as garbage collection, industrial gas distribution and beer and soft drink distribution. In this paper, we describe application environments in which these problems arise, present formulations for several versions of the problem and review solution procedures.

1. INTRODUCTION

Vehicle routing and scheduling problems arise in contexts where goods must be picked up and/or delivered for a geographically dispersed set of customers. Solution algorithms for them typically must address both spatial considerations related to customer geography and temporal considerations arising from constraints on the timing of deliveries. The standard definition of a routing problem usually explicitly or implicitly is restricted to a very limited time horizon, such as a day. The allocation/routing problem, which we address in this paper, involves explicit consideration of a longer time horizon. The typical time horizon is several (5 to 10) days. During this time horizon, it may be necessary to deliver to particular customers more than once. Inventory considerations may influence the choice of the day of delivery. It is most natural to think of the problem as having two components: an allocation component, which allocates pickups and/or deliveries to days and a routing component, which determines vehicle routes for each day.

In the typical application environment where a standard (non-allocation) vehicle routing problem arises, the customer specifies the day on which the pickup or delivery is to be made. Consequently, at the start of each day a demand set would be known which had an absolute requirement for pickup and/or delivery on that day. In the typical application environment for an allocation/routing problem, such as fuel oil delivery, there is flexibility with respect to the day of delivery. The fuel oil customer does not particularly care on which day the delivery is made as long as the customer does not run out of fuel. The manager of a distribution system, or alternatively an optimization algorithm, can use this flexibility to design more efficient routes by, for example, assigning deliveries to customers in the same geographic area on the same day of the week.

In this paper we describe models and algorithms for addressing allocation/routing problems. We do not mean to give a comprehensive survey, but rather, to present a set of diverse approaches which we feel covers the different application settings in which these problems arise. In the Section 2 of the paper, we describe three real-world allocation/routing problems. In each case, the allocation component of the problem was of such significance that computerized methods that addressed only the single day routing problem were felt to be inadequate. Section 3 presents several formulations of allocation/routing problems, leading to a description of algorithms in Section 4. The final section gives a set of areas where further research is required.

Our intent in this paper is to cover algorithms that produce specific vehicle routes given the customer characteristics for the current planning horizon/day. We should mention that there is another body of work on this problem that has produced analytical formulas, route structure information and inventory policy statements. We feel this work is quite exciting and innovative but do not attempt to review it here. Some relevant references are: Anily [1], Burns, Hall, Blumenfeld and Daganzo [7] and Haimovich and Rinnooy Kan [16].

2. APPLICATION ENVIRONMENTS

We now describe three different application settings in which allocation/routing problems arise. These are particularly appropriate for a number of reasons. First, they represent a diverse set of problem features. Secondly, in all cases, there has been substantial study of these problems, which has resulted in the use of computer based systems for their solution.

2.1. Garbage Collection (reference: Beltrami and Bodin [5])

The garbage collection problem is one that is "close to home" and well understood by most people. It also represents an early example of routing problem to which computerized procedures were applied (Beltrami and Bodin [5]). There is a high degree of flexibility related to the day on which pickups are made. The principal factor that determines the frequency of pickups is insuring that the amount of garbage that is generated at a customer location between pickups does not become too large. For a number of institutional/management reasons a periodic solution to the allocation problem is almost always required. That is, certain days of the week are designated as pickup days for each customer and this schedule is adhered to each week. One advantage of such a solution is that the customer knows, in advance, the day of delivery.

2.2. Fuel Oil and Industrial Gas Distribution (references: Assad et al [2,3], Bell et al [4], Dror and Ball [11], Dror et al [10])

Fuel oil and industrial gas distribution systems are driven by the requirement that a supply of some product be maintained in a storage tank at each customer location. The distributor maintains a fleet of vehicles which are dispatched from a central depot daily. Generally there are few restrictions placed on the precise day on which customer deliveries are made. The distributor's primary concern is that customer stockouts be eliminated or kept as low as possible. A stockout decreases customer satisfaction and, in the long term, will reduce the number of customers. In addition, stockouts are usually handled by immediate costly emergency deliveries.

In two well studied examples (Assad et al. [2,3], Bell et al [4]), the distributor did not have precise knowledge of customer product levels, nor did customers generally request deliveries. Rather, the distributor had the responsibility for estimating customer product levels and for planning deliveries appropriately. Using inventory theory terminology, these were "push" systems rather than "pull" systems.

Due to the variability in demand over time as well as the variability in demand from one customer to another a periodic solution is usually not desired. Ideally, one would like to plan deliveries over a very long time horizon, e.g. one year. This is impractical for several reasons, including the unreliability of data estimates over such a long horizon and the fact that the

Mondays, Wednesdays and Fridays, etc. Assuming that territories have been fixed prior to the assignment of customer delivery patterns, the assignment of patterns to customers and the formation of routes can be performed on a territory by territory basis. Thus, the problem reduces to a number of single vehicle/multiple day problems.

3. MODELS

We now describe formulations for three classes of allocation/routing problems. The three formulations are:

(DFD): Dynamic Formulation with Fixed Delivery Size
(DVD): Dynamic Formulation with Variable Delivery Size
(P): Periodic Formulation

(DFD) and (DVD) apply to variants of the fuel oil and industrial gas distribution applications described in the previous section. In this context dynamic refers to a routing environment in which a periodic solution is not used. Rather, the route structure changes from planning horizon to planning horizon. Dynamic formulations are appropriate in contexts where customer demand characteristics experience significant fluctuation and/or where the spacing between deliveries to a particular customer is relatively large, e.g. more than one month. Typically a problem is solved over a short time horizon, e.g. one week. However, the mechanism used within the problem to choose which customers are serviced during that time horizon must reflect long term considerations.

(DFD) was originally developed to model the problem setting analyzed by Assad et al [2,3], Dror and Ball [11] and Dror et al [10] (propane distribution), and (DVD) was originally developed to model the setting analyzed by Bell et al [4] (industrial gas distribution). The propane distribution application environment and the industrial gas distribution application environment differed in the following two ways:

i.) The customer tank sizes in the propane distribution problem were much smaller than the vehicle capacities. This resulted in routes with a large number of customers, e.g. as many as 60, whereas in the industrial gas distribution case vehicle capacities were close to customer tank sizes which resulted in short routes (1,2 or 3 customers).

ii.) For the majority of customers, the customer usage rate was small relative to tank capacity in the propane distribution case, resultant optimization problem would be of impractical size. Typically, a problem is solved for a planning period of modest size, e.g. one week. In a given planning horizon, the distributor is faced with the following tradeoff for each customer. Making deliveries earlier in the period tends to decrease the possibility of customer stockouts, whereas later deliveries tend to increase the possibility. On the other hand, if in the long term deliveries are made to customers sooner than necessary, then more deliveries will be made, thus increasing long term distribution costs. An effective solution approach should be able to consider this tradeoff as well as the efficiencies to be gained by assigning geographically contiguous deliveries to the same delivery day.

2.3. "Driver Sell" Beer and Soft Drink Distribution (reference: Golden and Wasil [15])

In "driver sell" distribution systems, delivery orders are not known a-priori, rather the driver visits a set of customers and then sells as much product as possible to each customer. All customers visited are regular customers who normally can be expected to buy some product -- the question is how much. For route planning purposes, an estimated demand can usually be determined for each customer, although when the routes are actually implemented a certain amount of customer demand variation should be allowed for.

The distribution area is usually divided into geographically contiguous regions. Each region is assigned to a particular driver and is then considered that driver's "territory" so that all customers and potential customers within the territory can only be visited by the assigned driver. Due to the special characteristics that a driver territory must have, the construction of territories is usually performed as an independent first step. It is likely that a high degree of user interaction would be necessary in order to generate a politically acceptable set of territories.

The customer locations are typically retail outlets such as liquor stores and supermarkets or establishments that allow on premises consumption, such as taverns and restaurants. Thus, product is used up on a fairly continuous basis. As in the garbage collection case, a periodic solution is usually sought. For example, with a week-long planning horizon, certain customers might be visited daily, others Mondays and Thursdays, others

which resulted in infrequent visits to most customers, e.g. once every 45 days. Visits to customers were much more frequent in the industrial gas distribution case.
These differences led to the following differences in modeling assumptions:
1.) In (DFD) it is assumed that each customer receives at most one delivery during the planning horizon,
2.) In (DFD) customer delivery size is a constant that depends only on the day of delivery, the time of the last visit and customer usage rate,
3.) In (DVD) it is assumed that routes contain a very small number of customers, e.g. 1,2, or 3.
Note that assumption 1.) is valid as long as the length of the planning horizon is somewhat shorter than the time required for a customer to use all of the fuel in the customer's tank. Assumption 2.) is valid assuming that whenever a delivery is made then the customer's tank is filled up. These two conditions were consistent with management policy and are quite reasonable given i.) and ii.).

(P) models the garbage collection and driver-sell application environments. It generates a periodic solution which would be used on a regular basis.

Prior to describing each formulation, we define notation common to all formulations.

M = the set of customer (demand) locations,
w = the number of vehicles,
q_k = the capacity of vehicle k ,
Q_i = the set of vehicles that can service customer i ,
n = the number of days in the planning horizon .

The planning horizon is the period of time over which the optimization problem is to be solved.

All models require the eventual generation of individual vehicle routes. Each route should minimize the cost of a single vehicle delivering to the set of demand locations assigned to that vehicle. This cost function is given by,

$TSP_k(S)$ = the fixed costs required for vehicle k to make deliveries to all customers in set S (note: in this context fixed costs refer to those delivery costs that do not depend on delivery size).

Of course it is a bit presumptuous to call $TSP_k(S)$ a cost function since an optimization problem must be solved to evaluate it. In particular, the evaluation of $TSP_k(S)$ requires the solution of a traveling salesman problem over the locations in set S. We also note that this may be a constrained traveling salesman problem in that constraints may be imposed on the timing of deliveries.

We also find it convenient to represent compactly the cost of a vehicle routing solution. Before doing so we define,

d_{ih} = customer's i delivery size on day h.

This delivery size will be described more precisely for each specific formulation. The variables of the vehicle formulation are,

$$y_{ik} = \begin{cases} 1 \text{ if vehicle } k \text{ visits customer } i, \\ 0 \text{ otherwise}. \end{cases}$$

S_k = the set of customers visited by vehicle k.

We now define the vehicle routing problem over customer set S on day h by,

$$VRP(S,h) = \text{Min} \sum_{k=1}^{w} TSP_k(S_k)$$

$$\text{s.t.} \quad \sum_{k \in Q_i} y_{ik} = 1 \quad \text{for all } i \in S, \tag{1}$$

$$\sum_{i \in S} d_{ih} \, y_{ik} \leq q_k \quad \text{for } k=1,2,\ldots,w, \tag{2}$$

$$S_k = \{i \in S : y_{ik} = 1\} \quad \text{for } k=1,2,\ldots,w, \tag{3}$$

$$y_{ik} \in \{0,1\} \quad \text{for all } i,k. \tag{4}$$

The constraints insure that each customer receives one delivery and that vehicle capacity is respected.

The principal characteristic that differentiates the three formulations is the manner in which the customer delivery sizes are determined. Before describing each formulation we discuss the manner in which this particular problem feature is modeled.

3.1. A Dynamic Formulation with Fixed Delivery Sizes

This dynamic formulation was developed for the inventory/routing application by Dror et al [10] and Dror and Ball [11]. Based on assumption 2.), the customer delivery size,

d_{ih}, is a constant that depends on the customer and day of delivery. In general, there will be a set of customers to whom deliveries must be made during the planning horizon. These are the customers who would consume all of their available product before the end of the planning horizon. Furthermore, it may be necessary to deliver to these customers before the end of the planning horizon. Thus, we define,

M^* = the set of customers that must receive a delivery during the planning horizon,

n_i = the last day in the planning horizon on which a delivery can be made to a customer $i \in M^*$.

It is entirely possible that we may wish to deliver to customers not in M^*. For example, if customer $i \notin M^*$ was a neighbor of customer $j \in M^*$ and customer i's tank was 1/4 full, then it would probably make economic sense to deliver to i. For similar reasons, it could easily make economic sense to deliver to a customer $i \in M^*$ prior to the day n_i. Thus, we define a second customer subset,

M' = those customers not in M^* for whom deliveries should be considered,

and a set of penalties and rewards,

c_{ih} = an estimate of the increase in long-term costs associated with making a delivery to customer i on day h for $i \in M^*$,

g_{ih} = an estimate of the savings in long-term costs associated with making a delivery to customer i on day h for $i \in M'$.

The purpose of the c_{ih}'s is to encourage deliveries later in the planning horizon to save long term costs. Thus, it is reasonable to assume the properties: $c_{ih} \geq 0$, $c_{in_i} = 0$ and $c_{ih} > c_{ih+1}$. The purpose of the g_{ih}'s is to provide an economic incentive to deliver to customers in M'. The incentive is driven by future delivery savings, so it is reasonable to assume the properties: $g_{ih} \geq 0$ and $g_{ih} < g_{ih+1}$. Dror and Ball [11] give procedures for deriving c and g values. The values are based on a fair allocation of the cost of a typical route used to deliver to a customer i as well as the expected amount of product remaining in customer i's tank on day h.

The variables of the model are:

$$y_{ih} = \begin{cases} 1 \text{ if customer } i \text{ receives a delivery on day } h, \\ 0 \text{ otherwise} . \end{cases}$$

S_h = the set of customers serviced on day h.

We may now formulate the dynamic allocation/routing problem with fixed delivery sizes as,

$$\text{(DFD):} \quad \text{Min} \sum_{h=1}^{n} \left\{ VRP(S_h, h) + \sum_{i \in M^*} c_{ih} \, y_{ih} - \sum_{i \in M'} g_{ih} \, y_{ih} \right\}$$

$$\text{s.t.} \quad \sum_{h=1}^{n_i} y_{ih} = 1 \qquad \text{for } i \in M^* , \qquad (5)$$

$$\sum_{h=1}^{n} y_{ih} \leq 1 \qquad \text{for } i \in M' , \qquad (6)$$

$$S_h = \{ i \in M : y_{ih} = 1 \} \text{ for } h=1,2,\ldots n, \qquad (7)$$

$$y_{ih} \in \{0,1\} \quad \text{for all } i,k,h . \qquad (8)$$

Constraint (5) ensures that all customers who require deliveries will receive their delivery before day n_i. Constraint (6) insures that no more than one delivery will be made to customers that do not require a delivery during the planning horizon. Constraint (7) defines the delivery set for day h. The objective function includes three components. The first represents the routing costs. The second component encourages later deliveries so as to reduce long term delivery costs and the third provides an incentive for delivering to customers in M', which is also based on long term considerations.

3.2. A Dynamic Formulation with Variable Delivery Sizes

In general, one might expect (DVD) to be much more complicated than (DFD) since assumptions 1.) and 2.) must be relaxed. However, Bell et al [4] used 3.) to devise a very different formulation that leads to a much different solution approach. They partitioned the customer set into three subsets,

M' = those customers that should be considered for at most one delivery during the planning horizon,

M'' = those customers that can receive any number of deliveries during the planning horizon,

M^* = those customers that must receive exactly one delivery during the planning horizon.

As in the previous model we also define,

n_i = the last day in the planning horizon on which a delivery can be made to a customer $i \in M^*$.

Demand requirements were modeled by defining lower and upper bounds on the cumulative amount delivered over the first h days of the planning horizon for h=1,2,...,n, i.e.,

d'_{ih} = the minimum amount that must be delivered to customer i on days 1 through h,

d''_{ih} = the maximum amount that can be delivered to customer i on days 1 through h.

Assumption 3.) was used in that (DVD) is a set partitioning based formulation. Set partitioning based formulations require the enumeration of all, or a large number of, the possible routes. If routes are very long then the number of possible routes is very large. On the other hand, if routes are short, as they are in this case, then the number of possible routes, m, is relatively small. We denote the set of possible routes by,

$\{S_1, S_2, ..., S_m\}$ = the customer sets associated with the set of vehicle routes enumerated.

The additional problem data are:

g_i = the <u>per unit</u> value of deliveries to customer i,

L_i = the set of routes containing customer i.

The variables are,

$$y_{khr} = \begin{cases} 1 & \text{if vehicle drives route r on day h,} \\ 0 & \text{otherwise ,} \end{cases}$$

x_{ikhr} = the amount delivered by vehicle k to customer i on day h using route r .

$$\text{(DVD):} \quad \text{Max} \sum_{k=1}^{w} \sum_{h=1}^{n} \sum_{r=1}^{m} \{ \sum_{i \in S_r} g_i \, x_{ikhr} - TSP_k(S_r) \, y_{khr} \}$$

$$\text{s.t.} \quad d'_{ih} \leq \sum_{k \in Q_i} \sum_{j=1}^{h} \sum_{r \in L_i} x_{ikjr} \leq d''_{ih} \quad \text{for } i \in M'' \text{ , and } h=1,2,\ldots,n \tag{9}$$

$$x_{ikhr} \geq d'_{in} \, y_{khr} \quad \text{for } i \in M^* \text{ and all k,h and r ,} \tag{10}$$

$$x_{ikhr} \leq d''_{ih} \quad \text{for } i \in M^* \cup M' \text{ and all k,h,r,} \tag{11}$$

$$\sum_{k \in Q_i} \sum_{h=1}^{n_i} \sum_{r \in L_i} y_{khr} = 1 \quad \text{for } i \in M^* \tag{12}$$

$$\sum_{k \in Q_i} \sum_{h=1}^{n} \sum_{r \in L_i} y_{khr} \leq 1 \quad \text{for } i \in M' \tag{13}$$

$$\sum_{i \in S_r} x_{ikhr} \leq q_k \, y_{khr} \quad \text{for all k,h,r} \tag{14}$$

$$y_{khr} \in \{0,1\} \text{ for all i,h,r ,} \tag{15}$$

$$x_{ikhr} \geq 0 \text{ for all i,k,h,r.} \tag{16}$$

The objective function contains a value component that provides a motivation for delivering product. Note that the value function used for (DFD) changed with the day of delivery whereas the one used here does not. Constraint (9) provides cumulative upper and lower bounds on the amount delivered for customers in **M"**. Constraint (10) insures that the single delivery made to customers in M^* satisfies the minimum time horizon requirement and constraint (11) insures that the single delivery to customers in $M^* \cup M'$ is below the maximum allowed up to the day of delivery.

Constraints (12) and (13) ensure that the number of deliveries made to customers in M^* and M' are as specified in their definitions. Constraint (14) insures that vehicle capacity is respected and that the x variables are positive only when the corresponding y variable is one. The viability of this formulation depends very much on the requirement that the number of possible routes (m) is small, i.e. depends on assumption 3.).

We note that this formulation is actually a simplification of the one presented by Bell et al [4]. In particular, the h index in their formulation corresponds to vehicle start times rather than days. The set of feasible start times is defined in such a way that routes of varying length are allowed. For example, h indices might represent start times 6:00 AM, 7:00 AM, 8:00 AM, 10:00 AM, 12:00 PM, etc. Routes 2 hours, 4 hours and 8 hours in length might be generated. This approach requires additional constraints to ensure that a vehicle is not assigned to "overlapping" work, e.g. if a 4 hour route starting at 6:00 AM is assigned to a vehicle then that vehicle will not be available for any other assignments until 10:00 AM. Defining the h index in this way allows for the explicit modeling of multiple routes per day per vehicle. Among other advantages this provides a very convenient way of handling time window constraints. While this is certainly a very important aspect of their formulation, we did not include it to simplify our presentation and to maintain some consistency among formulations.

3.3. A Periodic Formulation

We now describe a formulation appropriate for periodic routing problems. This formulation can be viewed as a variant of the one presented by Christofides and Beasley [8]. It also uses ideas implemented in the Roadnet Technologies routing system (see Golden and Wasil [15]). We start with the notion of a pattern, which specifies a particular set of delivery days within the planning horizon, to be used for certain types of customers. A total demand for the time horizon is specified for each customer. Together with the pattern, we also specify the fraction of total demand that is delivered each day. After associating a customer with a pattern, the fractional demand distribution specified with the pattern will determine the amount of demand handled each day for that customer. In general, a large number of potential patterns will be specified together with feasible customer/pattern pairs. An example of the pattern concept is illustrated in Figure 1.

PATTERNS

PAT #	M	T	W	TH	F	S
A	.4		.3		.3	
B		.4		.25		.35
C		.5			.5	
D	.5			.5		
E		1				
F			1			

DEMANDS

CUST #	WEEKLY DEMAND
1	100
2	120
3	200
4	150
5	40
6	30
7	80

FEASIBLE DEMAND DISTRIBUTIONS

CUST #	PAT #	M	T	W	TH	F	S
1	A	40		30		30	
1	B		40		25		35
1	C		50			50	
2	A	48		36		36	
2	B		48		30		42
3	A	80		60		60	
3	B		80		50		70
4	A	60		45		45	
5	E		40				
5	F			40			
6	E		30				
6	F			30			
7	C		40			40	
7	D	40			40		

FIGURE 1
Example of Demand Patterns

We now describe notation necessary to represent patterns

formally.

P = set of patterns ,

f_{ph} = fraction of total customer demand allocated to day h by pattern p ,

$$e_{ph} = \begin{cases} 1 \text{ if } f_{ph} > 0 , \\ 0 \text{ otherwise,} \end{cases}$$

d'_i = total demand of customer i for planning horizon ,

g_{ip} = value of customer i/pattern p combination .

It is interesting to note that Christofides and Beasley [8] did not include a customer/pattern value (g_{ip}) in their model. However, the Roadnet Technologies practical experience (see Golden and Wasil [15]) indicated that while several patterns might be feasible to a particular customer, there was a definite preference amongst them. In particular, the driver-salesmen indicated that certain patterns had sales advantages over others. One advantage was increased customer satisfaction in situations where customers had preferences relative to the day of delivery. Another related to increased sales (demand) over the long term. The typical situation involved cases where different salesmen competed for product space. If more space were empty when a delivery was made then the salesman could deliver (and sell) more product over the long term.

The variables of the model are:

$$z_{ip} = \begin{cases} 1 \text{ if pattern p is assigned to customer i ,} \\ 0 \text{ otherwise .} \end{cases}$$

$$y_{ikh} = \begin{cases} 1 \text{ if customer i is assigned to vehicle k on day h ,} \\ 0 \text{ otherwise .} \end{cases}$$

x_{ih} = the amount delivered to customer i on day h.

S_h = the set of customers serviced on day h .

We may now formulate the period routing problem as,

$$(P): \quad \text{Min} \sum_{h=1}^{n} VRP(S_h, h) - \sum_{i \in M} \sum_{p \in P} g_{ip} \; z_{ip}$$

$$\text{s.t.} \quad \sum_{p \in P} z_{ip} = 1 \qquad \text{for all } i \in M \; , \qquad (17)$$

$$d_{ih} = \sum_{p \in P} (f_{ph} \, d'_i) \, z_{ip} \qquad \text{for } h=1,2,\ldots,n \text{ and } i \in M \; , \qquad (18)$$

$$y_{ih} = \sum_{p \in P} e_{ph} \, z_{hp} \qquad \text{for } h=1,2,\ldots,n \text{ and } i \in M \; , \qquad (19)$$

$$S_h = \{ \, i \in M : y_{ih} = 1 \, \} \qquad \text{for } h=1,2,\ldots,n \; , \qquad (20)$$

$$y_{ih}, \, z_{ip} \in \{0,1\} \qquad \text{for all } i,h \text{ and } p \; , \qquad (21)$$

$$d_{ih} \geq 0 \qquad \text{for all } i \text{ and } h \; . \qquad (22)$$

Constraint (17) insures that each customer will be assigned one pattern. Constraint (18) defines the daily demands implied by the pattern allocation. Note that the definition of demands by this constraint is not explicitly needed in this formulation. However, demands are implicitly needed in determining the feasibility of the sets S_h. More importantly, when we relax this formulation in the next section constraint (18) will become very important. Constraint (19) insures that if a delivery is assigned to day h then a vehicle will visit the customer on that day. Constraint (20) defines the customer set that will be serviced on each day.

4. ALGORITHMS

Models (DFD) and (P) contained assignment variables. That is, variables that assigned customers/demand to vehicles, days or patterns. This was done intentionally to emphasize the generalized assignment or packing aspects of the problems. Such a point of view naturally suggests two step algorithms, where the first step forms clusters of customers and the second step forms routes over the clusters. Approaches which start by forming vehicle clusters are known as "cluster first, route second"

(Bodin et al [6]). Fisher and Jaikumar [13] use the term "generalized assignment" approach for a cluster first approach that solves the clustering problem as a generalized assignment problem. In this section, we first describe approaches that take this point of view, then discuss approaches that use a substantially different philosophy and end with a discussion of improvement routines that can be used as a final step after any other approach.

4.1. Cluster First/Generalized Assignment Approaches

We start by reviewing the generalized assignment heuristic for the vehicle routing problem due to Fisher and Jaikumar (1981). That heuristic proceeds in two steps. The first step assigns customers to vehicles and the second solves a traveling salesman problem for each vehicle. The first step requires a surrogate objective function which includes a cost, b_{ik}, for assigning each customer i to each vehicle k. The surrogate objective suggested is obtained by assigning a "seed" customer to each vehicle and then determining the cost of assigning customer i to route k as the incremental cost of a route that includes the seed customer from route k and customer i. A generalized assignment algorithm is applied to the problem:

$$\text{Min} \quad \sum_{i \in M} \sum_{k=1}^{w} b_{ik} \, x_{ik}$$

$$\text{s.t.} \quad (1), (2), (4) .$$

A traveling salesman algorithm is then applied to the set of customers assigned to each vehicle.

We now consider (DFD). Dror et al [10] suggest two cluster first approaches, one that assigns customers to days and then solves a vehicle routing problem for each day and one that assigns customers to vehicle/day pairs and then solves a traveling salesman problem for each vehicle/day pair. For the customer to day assignment approach they use the following formulation:

$$\text{Min} \quad \sum_{h=1}^{n} \{ \sum_{i \in M^*} c_{ik} \; y_{ik} \; - \sum_{i \in M'} g_{ik} \; y_{ik} \}$$

s.t. (5), (6), (8)

$$\sum_{i \in M \times \cup M'} d_{ih} \; y_{ih} \leq q^* \qquad \text{for } h=1,2,\ldots,n \; ,$$

where q^* is an artificial "daily" capacity set equal to an estimate of the total load that the fleet of vehicles could handle in a day. Dror et al applied a linear programming based approach to solve this generalized assignment problem approximately. A vehicle routing heuristic was then applied to determine vehicle routes for each day. A final improvement routine exchanged customers between routes and between days. Note that Dror et al did not provide any surrogate objective function component to represent geographic considerations related to which customers were assigned to which day in the generalized assignment formulation. This was felt to be unnecessary since the exchange procedure executed at the end took such considerations into account. Dror et al did not test an approach based on a vehicle/day assignment for two reasons: i.) there did not appear to be a natural surrogate objective function with which to represent the cost of assigning customers to vehicle/day pairs, ii.) it was felt that the associated generalized assignment problem would be considerably harder to solve.

Christofides and Beasley [8] propose a p-median approach for (P) which can be interpreted as a variant on the generalized assignment approach. They employ a customer/day surrogate assignment cost which is very similar in spirit to the Fisher-Jaikumar seed concept. In particular, based on the result due to Christofides and Eilon [9] that expected total vehicle routing costs increase as the total customer radial distances from a center increase, Christofides and Beasley propose associating a center with each day and then using the distance to the center as a surrogate for the cost of assigning a customer to a day. This would tend to have the effect of clustering Monday's routes in one geographic section, Tuesday's in another, etc. If a center were assigned in advance to each day then the following problem would result,

$$\text{Min} \quad \sum_{i \in M} \sum_{h=1}^{n} b_{ih}\, y_{ih} \;-\; \sum_{i \in M} \sum_{p \in P} g_{ip}\, z_{ip}$$

$$\text{s.t.} \quad (17),\ (18),\ (19),\ (21),\ (22)$$

$$\sum_{i \in M} d_{ih} \le q^{*} \qquad \text{for } h=1,2,\ldots,n\ , \qquad (23)$$

where q^* is as defined previously and b_{ih} is the cost of assigning customer i to the center associated with day h. The formulation we have just derived differs from Christofides and Beasley's in the following way. First, their demands do not depend on the day of the planning horizon so equation (18) is not necessary. Second, they do not specify the center assigned to a day in advance. Rather, for each day, a set of possible centers is given. The choice of a center for each day is allowed to vary as part of the optimization process. Finally, their model does not include the g_{ip} component in the objective function.

Christofides and Beasley find a feasible solution to their problem by first ordering the customers by decreasing demand. Then, they assign each customer in order to the pattern that increases the surrogate objective least. This greedy algorithm is then followed by an interchange algorithm. Once the above problem is solved a vehicle routing problem remains for each day.

Russell and Igo [17] employ a similar approach. In their problem setting, the delivery day for a significant number of deliveries was specified in advance. As a result a set of deliveries could be associated with each day prior to the execution of a clustering/generalized assignment algorithm. Russell and Igo used this set of deliveries as attraction points in assigning a cost for assigning a delivery to a particular day of the week. This eliminated the need for seeds.

4.2. Other Approaches

We now mention two other approaches which were developed for the specific problems we have described. The approach of Assad et al [2] and Golden et al [14] was devised to address the same application modeled in Dror and Ball [11] and Dror et al [10]. However, it did not employ the specific formulation described in the previous section. In particular, the sets M^* and M' were not defined. A value similar to g_{ih} was defined for each customer. Rather than specifying that certain customers must be serviced on a particular day, they gave a very high value to those customers that absolutely required service on a given day. When this system was actually implemented the dispatcher had control over which customers were put into this category. The g_{ih} value associated with other customers increased as the expected customer tank level approached zero. To determine the set of customers to be serviced on a particular day a variant of the prize collecting traveling salesman problem was solved. This problem involves forming a traveling salesman tour through a subset of the customers. The objective is to maximize the value (based on the g_{ih}'s) of the customers serviced minus the routing costs. Constraints are placed on the total route time and on the total customer demand. The customers placed on this tour were the ones chosen for service on a given day. The right hand sides of the route time and total customer demand constraints were set based on experiments so as to yield a set of customers that could be feasibly serviced on a single day. Once this problem was solved a vehicle routing problem was solved to produce the routes actually used on the day.

Bell et al [4] used a Lagrangian Relaxation approach to solve (DVD). In particular, they dualize constraints (9), (12) and (13). These are the constraints that force the deliveries to customers to be within certain bounds. When they are deleted no deliveries are required. The Lagrangian Relaxation approach associates multipliers with these constraints that essentially provide economic incentives for delivering to customers. The resultant relaxation can be solved very easily. We should note the actual relaxation solved by Bell et al is more complicated then the one based on the formulation we described. As we mentioned in the previous section their formulation involved the possibility of overlapping time intervals. Bell et al devised a dynamic programming algorithm to solve their relaxation. The solution to the relaxation provides a lower bound on the optimal solution but not a feasible solution. The complete procedure used by Bell et al employed a heuristic that was based on the results of the relaxation as well as branch and bound. The authors noted that the heuristic and the relaxation were quite accurate so that branch and bound played a minor part in determining solutions.

4.3. Improvement Algorithms

A number of authors, including Christofides and Beasley [8] Dror et al [10] and Russell and Igo [17] propose improvement algorithms. These procedures start with a feasible solution to the problem and then execute solution improving exchanges. These employ simple extensions of standard exchange algorithms so we will not give the details here. We should note however, that in some cases, such procedures significantly improved the quality of the solution produced by the procedure that constructed the initial feasible solution.

5. RESEARCH TOPICS

We feel the work to date has laid a firm foundation for the allocation/routing problem. Certainly, it is clear that it has led to many practical successes. On the other hand, we feel that it is an area that deserves increased research attention. We now outline some important areas for investigation.

i. Models for Delivery of Home Heating Fuel: (DFD) was derived to model a delivery setting that included some home heating fuel customers. However, the seasonal fluctuations in their demand characteristics were handled implicitly by adjusting parameters from one time horizon to another. It is well known that demand for heating fuel is correlated with degree days. Thus, a model that explicitly incorporated the degree day concept would be quite useful.

ii. Explicit Consideration of Multiple Routes per Day: All the formulations given in this paper assume that each vehicle covers one route per day, i.e. the total demand handled by a vehicle in one day is no more than the vehicle's capacity. We noted that the more complicated formulation given by Bell et al [4] was able to handle this aspect of the problem very nicely. However, the other models could only handle multiple routes per day through artificial daily capacities. For example, a vehicle with a capacity of 100 units might be assigned a daily capacity of 300 units anticipating that the vehicle would be able to handle three routes in a single day. While such an approach allowed the practical use of many models we feel their accuracy would be enhanced if a more explicit consideration of multiple routes per day were devised.

Finally, we note that all of the algorithms described were approximate in nature. With the exception of the algorithm of Bell et al [4], none provided a bound the deviation from optimality. It certainly seems that additional research into optimization algorithms for all of these problems would be quite worthwhile.

REFERENCES

[1] Anily, S., "Integrating Inventory Control and Transportation Planning", PhD Dissertation, Columbia, University (1986).

[2] Assad, A., Golden, B., Dahl, R. and Dror, M., "Design of an Inventory/Routing System for a large Propane Distribution Firm", in C. Gooding, ed., Proceedings of the 1982 Southeast TIMS Conference (1982), Myrtle Beach, 315-320.

[3] Assad, A., Golden, B., Dahl, R. and Dror, M., "Evaluating the Effectiveness of an Integrated System for Fuel Delivery", in J. Eatman, ed., Proceedings of the 1983 Southeast TIMS Conference (1983), Myrtle Beach, 153-160.

[4] Bell, W., Dalberto, L., Fisher, M., Greenfield, A., Jaikumar, R., Kedia, P., Mack, R. and Prutzman, P., "Improving the Distribution of Industrial Gases with an On-Line Computerized Routing and Scheduling Optimizer", Interfaces 13 (1983), 4-23.

[5] Beltrami, E. and Bodin, L., "Networks and Vehicle Routing for Municipal Waste Collection" Networks 4 (1974), 65-94.

[6] Bodin, L., Golden, B., Assad, A. and Ball, M., "Routing and Scheduling of Vehicles and Crews", Computers and Operations Research 10 (1983), 62-212.

[7] Burns, L., Hall, R., Blumenfeld, D. and Daganzo, C., "Distribution Strategies that Minimize Transportation and Inventory Costs", Operations Research 33 (1985) 469-490.

[8] Christofides, N. and Beasley, J., "The Period Routing Problem", Networks 14 (1984), 237-256.

[9] Christofides, N. and Eilon, S., "Expected Distances in Distribution Problems", Operational Research 20 (1969), 437-443.

[10] Dror, M., Ball, M. and Golden, B., "A Computational Comparison of Algorithms for the Inventory Routing Problem", Annals of Operations Research 4 (1986), 3-23.

[11] Dror, M. and Ball, M., "Inventory/Routing: Reduction from an Annual to a Short-Period Problem", Naval Research Logistics Quarterly 34 (1987), 891-905.

[12] A. Federgruen and Zipkin, P., "A combined Vehicle Routing and Inventory Allocation Problem", Operations Research 32 (1984), 1019-1032.

[13] Fisher, M. and Jaikumar, R., "A Generalized Assignment Heuristic for the Vehicle Routing Problem", Networks 11 (1981) 109-124.

[14] Golden, B., Assad, A. and Dahl, R., "Analysis of a Large-Scale Vehicle Routing Problem with an Inventory Component", Large Scale Systems 7 (1984), 181-190.

[15] Golden, B. and Wasil, E., "Computerized Vehicle Routing in the Soft Drink Industry", Operations Research 35 (1987) 6-17.

[16] Haimovich, M. and Rinnooy Kan, A., "Bounds and Heuristics for Capacitated Routing Problems", Mathematics of Operations Research 10 (1985), 527-542.

[17] Russell, R. and Igo, W., "An Assignment Routing Problem", Networks 9 (1970), 1-17.

Vehicle Routing: Methods and Studies
B.L. Golden and A.A. Assad (Editors)

DYNAMIC VEHICLE ROUTING PROBLEMS*

Harilaos N. PSARAFTIS

Massachusetts Institute of Technology, Room 5-211, Cambridge, MA.02139, U.S.A.

The purpose of this paper is to put dynamic vehicle routing into perspective within the broader area of vehicle routing, as well as provide a flavor of recent progress in this area. We identify the important issues that delineate the dynamic case vis-a-vis the static one, comment on methodological issues, review generic design features that a dynamic vehicle routing procedure should possess, discuss the adaptation of static approaches to a dynamic setting, and describe an algorithm for the dynamic routing of cargo ships in an emergency situation. We conclude by recommending directions for further research in this area.

1. INTRODUCTION

By "dynamic vehicle routing" one traditionally means the dispatching of vehicles to satisfy multiple demands for service that evolve in a real-time ("dynamic") fashion. The vehicles may be taxicabs, trucks, ships, aircraft, etc. The service provided may consist of dropping off a passenger to the airport, picking up and/or delivering small parcels, delivering gases to industrial customers, shipping troops and materiel in case of a mobilization situation, or, in general, satisfying a wide variety of other transportation or distribution requirements in a broad spectrum of settings.

For all the explosive growth in the vehicle routing literature over the past several years (see Bodin et al. (1983), and, more recently, Golden and Assad (1987)), in a strict sense (see definition in Section 2) very little has been published on dynamic variants of vehicle routing problems. Of the 62 references cited in Golden and Assad (1987), only three include phrases such as "dynamic", "real-time", or "on line" in their titles. This state of affairs is to be contrasted with the real-world picture, in which a significant proportion of applications are dynamic rather than static. Among publications that have explicitly addressed a dynamic vehicle routing situation we may mention several papers or reports in the paratransit (or "demand responsive" transportation) area, such as Wilson et al. (1971, 1976, 1977) and Psaraftis (1980), a paper by Brown and Graves (1981) on the real-time dispatching of

*Research supported in part through Contract No. N00016-83-K-0220 of the Office of Naval Research and through a UPS Foundation grant to the MIT Center for Transportation Studies.

petroleum tank trucks, the award-winning work of Bell et al. (1983) on the bulk delivery of industrial gases, a report by Powell (1985) on the dynamic allocation of trucks under uncertain demand, and a report by Psaraftis et al. (1985) on the problem of cargo ship routing in a mobilization situation. This situation reflects the rather scant methodological base in dynamic vehicle routing as compared to static; indeed, most real-time implementations of vehicle routing problems are straightforward adaptations of static approaches. By contrast, the state-of-the-art in other areas that can be conceivably considered "close relatives" to dynamic vehicle routing, such as the dynamic dispatching of mobile servers (e.g., ambulances, fire engines, even tugboats - see Larson and Odoni (1980), Minkoff (1985), etc.), or the dynamic routing in communications networks (see Bertsekas and Gallager (1987), among others), is relatively rich in specialized methodologies explicitly developed for these problems.

The purpose of this paper is to put dynamic vehicle routing into perspective within the broader area of vehicle routing, as well as provide a flavor of recent progress in this area. It is not the intention of the paper to be encyclopaedic. Rather, the scope of the paper is to identify the important issues that delineate the dynamic case vis-a-vis the static one, comment on methodological issues, and describe one specific context and algorithm in the dynamic vehicle routing area.

In Section 2 of this paper we explore the relationship between static and dynamic routing, by identifying factors that make these two problems drastically different, and by commenting on the methodological implications of these differences. In Section 3 we review generic design features that a dynamic vehicle routing procedure should possess, discuss the transferrability of static approaches to a dynamic setting, and end by describing an algorithm for the dynamic routing of cargo ships. Finally, Section 4 recommends directions for further research in this area by introducing and briefly discussing the Dynamic Traveling Salesman Problem.

2. DIFFERENCES BETWEEN STATIC AND DYNAMIC VEHICLE ROUTING

In this section we explore the relationship between static and dynamic vehicle routing problems by focusing on those elements that make dynamic routing different from static, and hence generally necessitate specialized solution procedures for dynamic vehicle routing problems.

To make our discussion more clear, we define a vehicle routing problem as "static" if the assumed inputs to this problem do not change, either during the execution of the algorithm that solves it, or during the eventual execution of the route. By contrast, in a "dynamic" vehicle routing problem, inputs may (and, generally, will) change (or be updated) during the execution of the algorithm and the eventual execution of the route. Actually, algorithm execution and route execution are processes that evolve concurrently in a dynamic situation, in contrast to a static situation in which the former process clearly precedes (and has no overlap with) the latter.

Dynamic vehicle routing differs from static in several ways, some of them obvious, some less obvious. The main differences are listed below. For dynamic vehicle routing:

(1) Time dimension is essential;
(2) Problem may be open-ended;
(3) Future information may be imprecise or unknown;
(4) Near-term events are more important;
(5) Information update mechanisms are essential;
(6) Resequencing and reassignment decisions may be warranted;
(7) Faster computation times are necessary;
(8) Indefinite deferment mechanisms are essential;
(9) Objective function may be different;
(10) Time constraints may be different;
(11) Flexibility to vary vehicle fleet size is lower;
(12) Queueing considerations may become important.

We now discuss each of these points and their implications in some detail.

(1) Time dimension is essential

In static vehicle routing, the time dimension may or may not be an important factor in the problem. If it is, the problem is usually termed a routing and scheduling problem. However, not all static situations have a scheduling component. Actually, most classical generic routing problems such as the Traveling Salesman Problem (single or multiple TSP), and the Vehicle Routing Problem (VRP) do not have a scheduling component. In all of these problems, times are assumed proportional to distances traveled, and therefore do not have to be considered explicitly and separately in the formulation and solution of the problem.

By contrast, in every dynamic vehicle routing situation, whether it be time-constrained or not, the time dimension is essential. At a minimum, we need to know the spatial location of all vehicles within our fleet at any given point in time during their schedule, and particularly when new customer or cargo requests or other information are made known. A fortiori, and in more common situations, we need to keep track of how vehicle schedules and scheduling options dynamically evolve in time.

(2) Problem may be open-ended

In contrast to a static situation, in which the duration of the routing process is more or less bounded or known in advance, the duration of such process in a dynamic situation may neither be bounded, nor known. In fact, a typical dynamic vehicle routing scenario is that of an open-ended process, going on for an indefinite period of time. An implication of this is that whereas in a static problem one usually considers tours (vehicles return to their depot), in a dynamic problem one considers (open) paths. Other implications regard the types of objective functions that are relevant in a dynamic routing problem (see also (9) below).

(3) Future information may be imprecise or unknown

In a static case there may be no "past", "present" or "future", particularly if the problem has no scheduling component. But even if it has, information about all problem inputs is assumed to be of the same quality, irrespective of where within the schedule this input happens to be (beginning, middle, or end). This is not the case in a dynamic problem, in which information on any input is usually precise for events that happen in real time, but more tentative for events that may occur in the future. As in any real life situation, the future is almost never known with certainty in a dynamic vehicle routing problem. Probabilistic information about the future may be available (e.g., we may know the probability that a certain customer will request service on a particular day), but in many cases even that type of information may not exist (a taxicab company waiting for customers is a typical example).

(4) Near-term events are more important

An implication of the previous point is that in terms of making decisions in a dynamic vehicle routing situation, near-term events are more important than longer-term ones. This is not the case in a static setting, where because of uniformity of information quality and lack of input updates all events (whether in the beginning, in the middle, or at the end of a vehicle's route) carry the same "weight". In dynamic routing, it would be unwise to immediately commit vehicle resources (i.e., decide to assign a vehicle, or make routing decisions) to requirements that will have to be satisfied way into the future, because other intermediate events may make such decisions suboptimal, and because such future information may change anyway. Focusing more on near-term events (of course without adopting a totally myopic policy) is therefore an essential aspect of a dynamic vehicle routing problem.

(5) Information update mechanisms are essential

Virtually all inputs to a dynamic routing problem are subject to revision at any moment during the execution of the route. For instance, a vehicle may break down. A customer requesting service may change the time he or she wishes to be picked up. "No-show" situations may occur. Due to unpredictable weather conditions, a ship may not be able to arrive at a certain port as scheduled. And so on. It is therefore imperative that information update mechanisms be an integral part of the algorithm structure and input/output interface in a dynamic situation. Data structures and database management techniques that help efficiently revise problem inputs as well as efficiently figure out the consequences of such revisions (see also (6) below) are central to a dynamic routing scheme. By contrast, in a static scenario, the scope of such mechanisms is either nonexistent, or, at best, tangential to the core of the problem (e.g., perform sensitivity analysis, play "what if" games, etc.).

(6) Resequencing and reassignment decisions may be warranted

In a dynamic vehicle routing situation, the appearance of a new input may render decisions already made before that input's appearance suboptimal (with respect to a certain objective). This fact concerns both sequencing decisions (decide sequence of stops to serve a given set of points) and assignment decisions (allocate vehicles to demand points). Thus, the appearance of a new input (such as a new customer request) may necessitate either the resequencing of the stops of one (or more) vehicle(s), or the reassignment of those vehicles to demands requesting service (or both).

Figures 1 and 2 help further illustrate this point. Euclidean space is assumed in both cases. In Figure 1, a dial-a-ride vehicle starts from point A to service customers 1 and 2 (a pickup point is denoted by a "+" and a delivery point by a "-"). The objective is to minimize the total distance traveled by the vehicle until the last customer is delivered. Figure 1(a) shows the optimal route. If now customer 3 requests service - while the vehicle is still at A (Fig. 1(b)), the new optimal route is shown in Figure 1(c). Notice that under the presence of customer 3, it is no longer optimal to adhere to the same sequence of pick ups and deliveries deemed optimal for customers 1 and 2 alone. Put another way, if we were to keep the same sequence and were simply to find the best insertion of customer 3 into the previously optimal route, we would arrive at a suboptimal solution, shown in Figure 1(d).

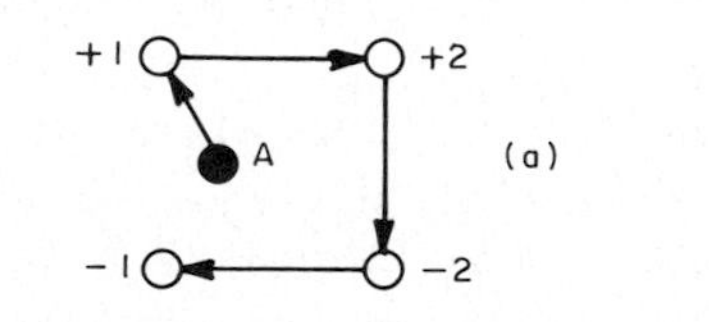

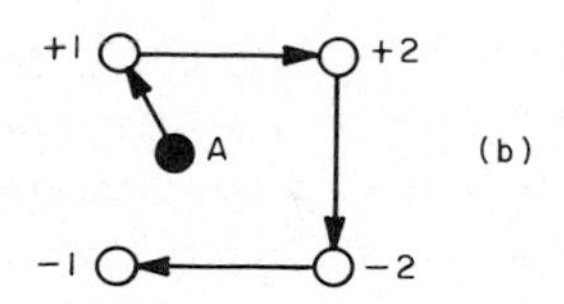

FIGURE 1 (a) & (b)

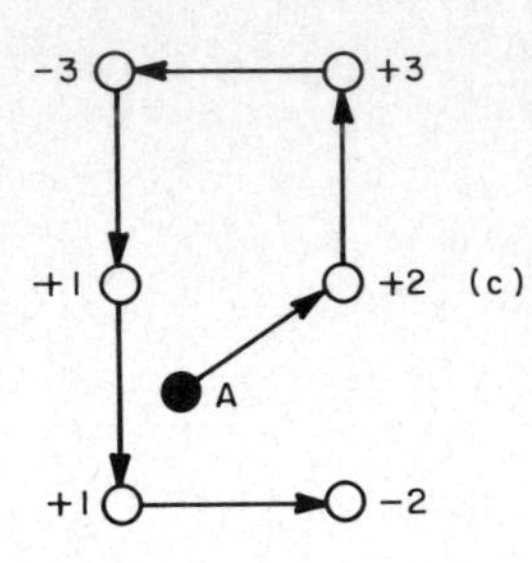

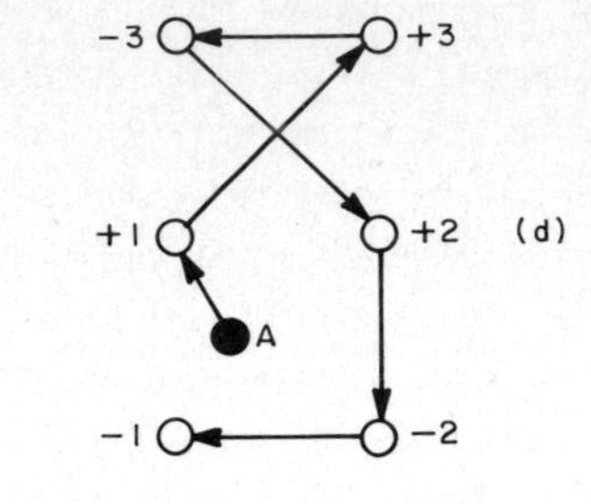

FIGURE 1 (c) & (d)

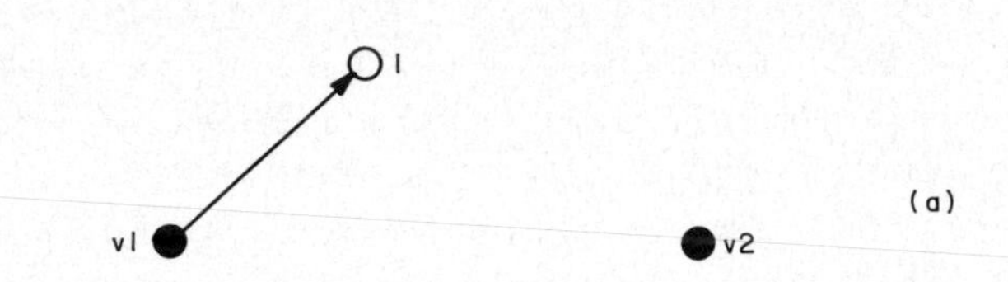

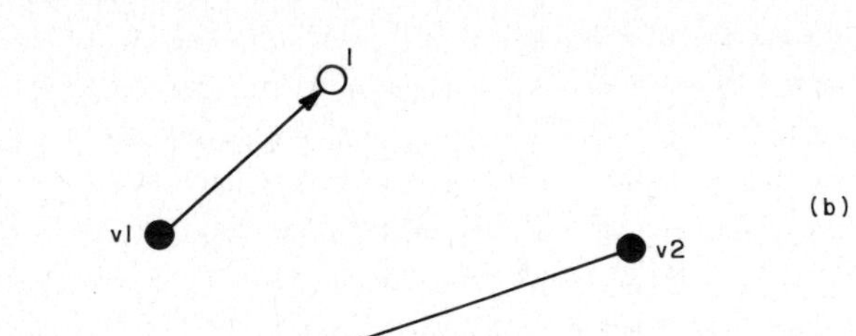

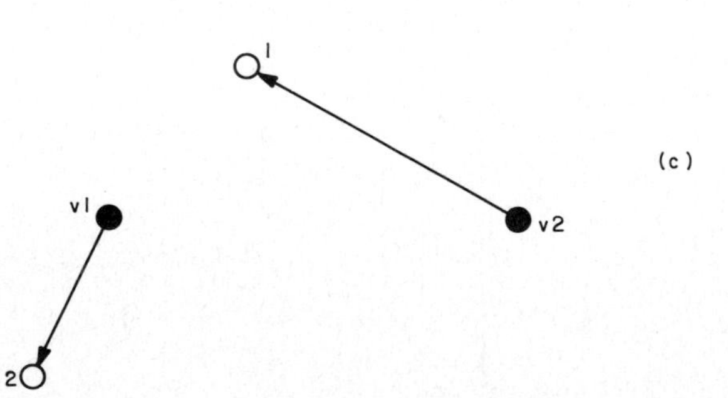

FIGURE 2

Figure 2 illustrates our point in a reassignment situation. Figure 2(a) shows the optimal allocation of two vehicles, v1 and v2 to satisfy the demand originating at point 1 (imagine that the vehicles are fire engines and point 1 is a fire). "Optimal" here means minimizing the maximum distance traveled by a vehicle. Under this assignment, v1 is dispatched to 1 and v2 idles. If now there is a second demand at point 2, the previous allocation is clearly suboptimal, for if we were to adhere to it we would have to have v2 travel a very long distance to go to point 2 (see Figure 2(b)). The optimal allocation in this case is the one shown in Figure 2(c). (Note that exactly the same observations are true if the objective is to minimize the total distance traveled by all vehicles).

A similar argument can be made whenever an input disappears (for instance a request is cancelled, a customer is a no-show, etc). As before, the deletion of an input will generally warrant a resequencing or a reassignment consideration.

(7) Faster computation times are necessary

The need to reoptimize routes and/or vehicle assignments on a continual basis in real-time necessitates computation times faster than those necessary in a static situation. In a static routing setting one may indeed afford the luxury of waiting for a few hours in order to get the output of the code solving the problem at hand. In such a setting, the problem may be solved exactly, and the code run in batch mode, perhaps overnight. This is not the case in a dynamic routing situation, in which the dispatcher wishes to know as soon as possible (i.e., in a matter of minutes, not hours) what the solution to a particular problem is in the presence of new information. The dispatcher may also want to run (again, in real-time) a few "what if" scenarios before deciding on the final action to take. The usual implication of this "running-time" constraint is that rerouting and reassignment decisions tend (by necessity) to be made on a heuristic and "local" fashion. Fast heuristics such as insertion, k-interchange and other improvement routines lend themselves to such a scheme (see also Section 3).

(8) Indefinite deferment mechanisms are essential

By indefinite deferment we mean the eventuality that the service of a particular demand be postponed indefinitely because of that demand's unfavorable geographical characteristics relative to other demands. An example of indefinite deferment for the single-vehicle case is depicted in Figure 3. As long as there are no time or priority constraints, and as long as there are unserviced requests near the current location of the vehicle, customer 1 (located far away from the central area) will always be scheduled to be serviced last (objective is to minimize total distance traveled until last customer is serviced).

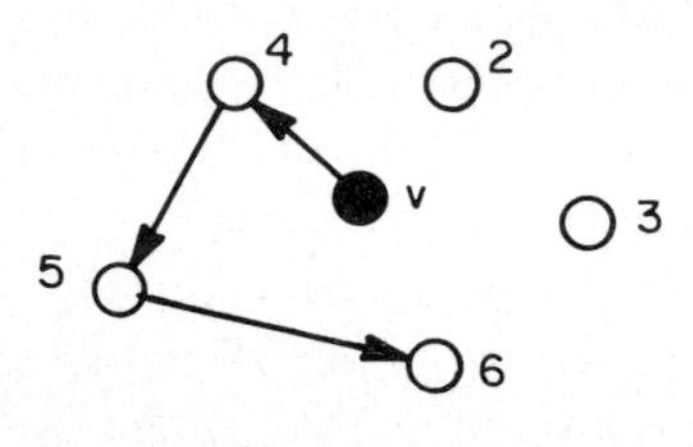

FIGURE 3

There are a variety of ways to alleviate this problem. Time constraints (in the form of time windows for instance) can generally force the vehicle to service a particular demand point on time, irrespective of that point's geographical location. A nonlinear objective function that penalizes excessive wait may also achieve the same goal (see also (9) below). Finally, priority constraints such as for instance limiting the number of positions each demand point can be shifted (up or down) away from its First-Come, First-Served (FCFS) position, can do the job as well (see Psaraftis (1980)).

(9) *Objective function may be different*

In a strict sense, the traditional "static" objectives of minimizing total distance traveled, or maximum distance traveled, or the overall duration of the schedule, may be meaningless in a dynamic setting. After all, if the process is open-ended, the overall duration of the schedule will be unbounded too. Measures of performance that have more meaning in a dynamic situation are more "throughput" or "productivity" - related. For instance, in a share-a-cab system we may wish to maximize the long-term average number of serviced customers per vehicle hour. Unfortunately, such an objective function does not always lend itself to algorithmic implementation, and one typically ends up replacing it with a set of surrogate objectives, sometimes identical, or closely related to the traditional static objectives, sometimes more complicated. Such criteria are typically applied to *parts* of the overall problem (decomposition by time or by space).

Optimizing only over *known* inputs might be a reasonable way to proceed if *no* information about future inputs is available. However, if *some* information about future inputs is available, it would make sense if such information is explicitly considered by the objective function. Since such information is usually vague, algorithms typically devise surrogate criteria (sometimes nonlinear) which attempt to "predict" future system workload (see also Wilson and Weissberg (1976), Jaw et al. (1986)).

Nonlinear objective functions are also used to induce the algorithm to avoid certain undesirable phenomena such as indefinite deferment in the absence of "hard" time constraints (see Tharakan and Psaraftis (1980)). Queueing considerations may also warrant a nonlinear objective treatment (see Psaraftis et al. (1985) and (12) below).

(10) Time constraints may be different

The main difference between static and dynamic vehicle routing as far as time constraints are concerned is that in dynamic routing inputs such as earliest pickup (or delivery) times or (especially) latest pickup (or delivery) times tend to be softer than in a static situation. A time constraint is "soft" (as opposed to "hard") not only if it can be violated at a penalty - and this can happen in a static problem too - but also if it is subject to update and eventual revision. In the eyes of the dispatcher, a customer-imposed latest delivery time is essentially a soft constraint (even if in the eyes of the customer it is a hard one). This is so because denying service to that customer if that constraint cannot be met is usually a less viable alternative. If a "hard" deadline makes a routing problem infeasible, it is far better to renegotiate that deadline so as to make it feasible, than to declare infeasibility and quit. However, a deadline that is relaxed is, by definition, a "soft" constraint. Of course, some of these constraints may indeed be hard (or harder than others). However, in dynamic routing we would expect at least some of these constraints to be soft.

(11) Flexibility to vary vehicle fleet size is lower

In theory, another alternative to denying service to a customer if a time constraint cannot be met is to add an additional vehicle to serve that customer, at a cost. However, this proposition may not necessarily be viable in dynamic vehicle routing, because it may not be possible to have access to backup vehicle resources in real time. In a static situation, the time gap between execution of the algorithm and execution of the route is usually long enough to allow such determination to be made. Such a high flexibility does not generally exist in dynamic routing. Implications: if vehicle resources are scarce, some customers will receive lower quality of service (their due dates will be shifted, etc.) This may also result in queueing phenomena, which we discuss next.

(12) Queueing considerations may become important

A dynamic vehicle routing system may at times become saturated (or congested). This will happen if the rate of customer demand exceeds a certain threshold, beyond which the system simply cannot handle all of the requests without creating excessive delays. In this case, any algorithm which tries to make assignment and routing decisions according to classical static surrogate criteria is bound to produce meaningless results. Unfortunately, whereas queueing theory and vehicle routing are two especially rich disciplines, very little is known about their interface. Under the current state of the art, including queueing considerations in vehicle routing is limited to empirical modeling (see also the MORSS algorithm described in Section 3 and the discussion of the Dynamic Traveling Salesman Problem introduced in Section 4).

It is clear that many of the above points are interrelated. For instance, (4) is an implication of (3), (6) and (7) are generally in conflict with each other, (9) and (10) may be used to take care of (8), (10) is due, in part, to (11), and (9) is true in part because of (12). Of course, specific dynamic environments vary, in general, with respect to each of these 12 points.

3. DYNAMIC VEHICLE ROUTING: SOLUTION METHODS

In this section we first review some generic design features that a dynamic vehicle routing procedure should possess in order to be useful in practice. We then discuss the adaptability of static approaches to a dynamic setting. We conclude by presenting the MORSS procedure developed by the author and his colleagues for the dynamic routing of cargo ships in a mobilization situation. Some of the concepts below draw from Psaraftis et al. (1985).

Design Features

In a generic sense, a dynamic vehicle routing procedure should possess, by design, the following features:

(1) It is obvious that such a procedure should be interactive. One should always have the "human in the loop" and enable him/her to override the computer at will. Various options should be designed, ranging from a completely "manual" approach where all major allocation (and possibly routing) decisions are made by the human operator, to more sophisticated modes where the computer deals with more difficult problems (e.g., routing) but still allows user discretion for "key" decisions. A fully automated mode, in which incoming data (e.g., new customer requests) are directly fed into the computer (say, by telephone, or by entering data in "checkpoint stations" in the area of service) can also be considered, so long as the human has the capability to override the machine in case an "unpredictable" situation occurs.

(2) The procedure should have a "restart" capability, that is, should be able to efficiently update routes and schedules at any time within the execution of a plan, without compromising "key decisions" already made. For

instance, if a certain cargo is en route from its origin to its destination at the time of the update, it would be nonsensical to have the procedure recommend that the cargo be reassigned on a different vehicle. On the other hand, if such a determination is made before that cargo is picked up, the reassignment recommendation is potentially implementable. New demands should be able to be inserted quickly into existing schedules. This means that the consequences of such an insertion (or of all potential insertions) should be able to be accounted for quickly. Efficient list processing techniques should be implemented for fast database manipulations.

(3) The procedure should be hierarchically designed, that is, allow the user to start the decision making process with "first-cut" gross feasibility analyses (possibly in several levels of aggregation) and only then proceed with detailed scheduling. Such a feature is considered important because a "quick and dirty" feasibility analysis may establish that, say, some due dates are infeasible, and hence allow the user to inquire for adjustments before further decisions are made.

(4) Finally (and perhaps obviously) we consider it important that the procedure be user-friendly. This is much more important in a dynamic setting than it is in a static one. In particular graphics aids are significant features that can enhance the efficiency of the man-machine interaction.

It is clear from the above list that in a dynamic vehicle routing procedure, information management and user interface issues become probably even more important than the theoretical performance and efficiency of the routing/scheduling (or, decision) module of the procedure. This module may still be considered to be the "core" of the overall procedure. However, in light of the discussion thus far, it is important to conclude that an efficient dynamic routing procedure implies much more than just an efficient core module. Actually, the core module itself must be designed in such a way that the above features can be easily implemented. This may have profound implications on the methodology used in the "core" module.

Adaptations of Static Approaches

Can a static routing approach, after suitable modifications, be efficiently used in a dynamic setting? The answer of course depends on the specific approach and setting.

A successful real-time implementation of a static approach has been reported in Bell et al. (1983), for the routing and scheduling a fleet of vehicles delivering a bulk product stored at a central depot. The routing "core" of the procedure is the static algorithm of Fisher et al. (1982), which is based on a mixed integer programming formulation of the problem and a solution using Lagrangian relaxation and a multiplier adjustment method. The core algorithm first heuristically generates a menu of possible vehicle routes taking into account the geographical location of customers and the amounts of demands and truckloads. A set packing problem is then formulated so as to select from this menu of routes the subset that would actually be driven, specifying the time each route should start, the vehicle to be used, and the amount to be delivered to each customer. A typical scenario is for this overall system to be run once a day, so as to determine the schedules for the next two to five days. A separate "schedule change" module takes care of updates in input data that may occur in real-time.

In general, and in all fairness to a specific static vehicle routing algorithm, it would be unreasonable to expect that the procedure, as it stands, can handle any dynamic situation. In many cases, the algorithm would have to undergo a significant degree of redesign, most of it heuristic, to be tailored to the nature of the dynamic scenario.

There are generally two ways to adapt a static algorithm to a dynamic case. The first is to rerun the procedure virtually from scratch each time a (significant) revision of the input occurs (say, a new customer appears, or another one cancels his request, or a vehicle breaks down, etc.). This would involve generating a new set of routes at each input update, while guaranteeing that decisions already made (e.g., allocation of cargoes already enroute) are not compromised. Both steps would involve "freezing" many of the variables of the problem to values determined at previous iterations. Running a static algorithm in such a way could present several nontrivial challenges, one of which would be how to cope with the excessive computational burden of rerunning the algorithm over and over again while results are needed in real-time.

An _alternative_ and more commonly used adaptation would be to handle dynamic input updates via a series of "local" operations, applied via the execution of an insertion heuristic (possibly followed by an interchange heuristic), _after_ the static core algorithm is run. This would involve running the static algorithm just to initialize the process (say, once every day), and rely on "local" operations for all subsequent input updates. This author believes that this approach would work reasonably well if both the time horizon of the initial input is relatively long _and_ subsequent input updates are infrequent (that is, if the overall problem is closer to static than dynamic). If the time horizon of the initial input is short or if subsequent input updates are numerous, the overall schedule would be less influenced by its initial solution and more by the subsequent local improvements. Such a scenario would drastically reduce the role of the static core in a dynamic situation, and shift the emphasis to the efficiency of the local operation method.

Local operations provide a reasonable way to handle dynamic input updates, their principal advantage being execution speed. The fastest local operation method is an <u>insertion</u> approach, in which a new request is inserted within the current schedule, without perturbing the sequence of visits already planned. An insertion approach can also work <u>in reverse</u>, that is, whenever a request is deleted from the list for some reason (e.g., a cancellation). However, as mentioned in the previous section, the main drawback of the insertion method is that it cannot take care of the need of possible resequencing or reassignment operations. Among the various insertion methods we refer to the work of Wilson et al. (1971, 1976, 1977) that was specifically developed for the dynamic version of the dial-a-ride problem, and its more sophisticated variant for the advance request case with time windows (Jaw et al. 1986). Both algorithms keep track of how much (up or down the schedule) the service time of a customer can be shifted so that feasibility is maintained. The second reference can be readily implemented in a "mixed-demand" scenario.

<u>Interchange</u> methods can be used after the insertion to further improve the set of routes and schedules. Such methods are based on the concept of "k-interchange" made widely known by Lin (1965) and by Lin and Kernighan (1973) for the TSP (plus, their extensions for the multiple vehicle case). Resequencing and reassignment can be effectively performed by such methods, some variants of which are particularly powerful (e.g., for the TSP the $k = 3$ case is much more powerful than the $k = 2$ case, and the $k = 4$ case is only marginally better than the $k = 3$ case). Unfortunately, a drawback of such methods is that they tend to become computationally expensive as k increases. Also, computational effort has to be spent to check whether the improved route maintains feasibility. Sophisticated adaptations of the interchange concept that do not result in a substantial CPU time increase (in an order-of-magnitude sense) to check feasibility are due to Psaraftis (1983) for the dial-a-ride problem and Savelsbergh (1985) for the time-window TSP. Also, we note the approach of Or(1976) that looks only at a subset of possible interchanges and ignores those that are unlikely to result in an improvement. All of the above methods can be adapted in a dynamic situation.

<u>The MORSS Algorithm</u>

The remainder of this section is devoted to giving an overview of MORSS, a dynamic vehicle routing algorithm developed by this author, Jim Orlin, and their colleagues to assist schedulers of the U.S. Military Sealift Command (MSC) to route cargo ships in an emergency situation. Complete details on the MSC problem and MORSS (which stands for MIT Ocean Routing and Scheduling System) can be found in Psaraftis et al. (1985).

The MSC is the agency responsible for providing sealift capability for the Department of Defense. To do this, it provides peacetime logistical sealift support of U.S. military forces worldwide, it develops contingency plans for the expansion of the peacetime sealift cargo fleet in case a military emergency occurs, and it has the operational control of this expanded fleet in mobilization situations.

Under conditions of military emergency, the objective of the MSC is to allocate cargo ships under its control (which can be as many as 1,000 in serious situations) to cargoes (whose number can be several thousands) so as to ensure that all cargoes, dry and liquid, arrive at their destinations as planned. Constraints that have to be satisfied include time windows for the cargoes, ship capacity, and cargo/ship/port compatibility. In addition, the scheduler has to allocate ships to cargoes so that three criteria are satisfied: First, cargoes should not be delivered (too) late. Second, ship utilization should be high. And third, port congestion should be avoided. The problem is dynamic in nature, as in a mobilization situation anything can change in real time. After extensive discussion with MSC personnel, an abstract model was developed for this problem. A simplified characterization of the input is the following (see Psaraftis et al. (1985) for more details).

For each cargo:

POE : Port of embarkation (origin)
POD : Port of debarkation (destination)
EPT : Earliest pickup time at POE (hard constraint)
EDT : Earliest delivery time at POD (hard constraint)
LDT : Latest delivery time at POD (soft constraint)
WEIGHT: Weight
VOL : Volume
SQFT : Deck surface area

For each ship:

LOC : Initial geographical location at time zero
W : Weight capacity
V : Volume capacity
S : Deck area capacity
SPEED: Speed
LOAD : Cargo loading/unloading rate
D : Draft

For each port:

DRAFT: Draft
THRU : Throughput characteristics of berths and terminals
DIST : Distances to all other ports and ship initial locations.

MORSS is based on the "rolling horizon" principle. In Figure 4, t_k is the "current time", that is, the time at the kth iteration of the procedure. At t_k, MORSS considers only those known cargoes whose EPT's are between t_k

and t_k+L, where L, the length of the rolling horizon, is another user input (say, L = 2 weeks). It then makes a tentative assignment of those cargoes to eligible ships (more on how assignments are made shortly). However, only cargoes within the "front end" of L are considered for permanent assignment. Those are those whose EPT's fall between t_k and t_k + aL, where a is another user input between 0 and 1. Thus, if L = 2 weeks and a = 0.5, MORSS will "look" at two weeks of cargo data into the future, but will commit to assign cargoes only within the first week. In such a fashion, MORSS places less (but, still, some) emphasis on the less reliable information on future cargo movements, since such information is more likely to change as time goes on.

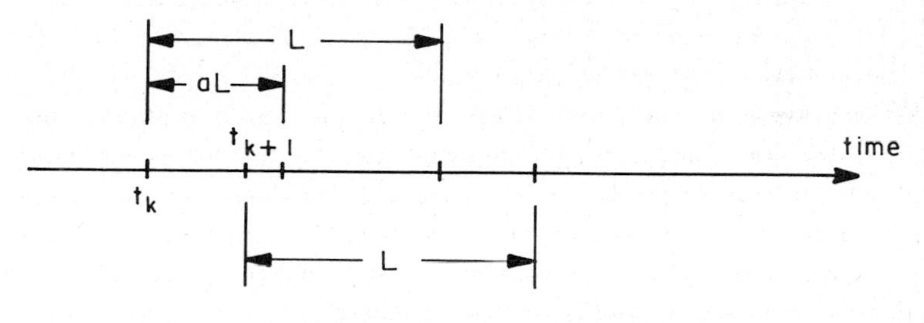

FIGURE 4

Iteration k+1 will move the "current time" to t_{k+1} (see Figure 4), which is equal to the time a significant input update has to be made (say, new cargo movement requirements are made known), or to the lowest EPT of all yet unassigned cargoes, whichever of the two is the earliest.

More formally, MORSS works as follows:

STEP 0 : Initialize locations of available ships.
Initialize "master list" of unassigned cargoes.
Select length L of individual time horizons.
Select fraction a ($0 < a < 1$).
Set k = 1, t_1 = 0.

STEP 1 : Set up next horizon (t_k, t_k+L).
Form list of cargoes eligible for assignment (all cargoes in master list whose EPT's are between t_k and t_k+L).

STEP 2 : Calculate assignment utilities for all eligible cargo/ship pairs (see Note (1) below).

STEP 3 : Form and optimize a transportation network using assignment utilities as arc costs. Resulting assignment forms the "tentative assignment" for (t_k, t_k+L) (see Note (2) below).

STEP 4 : Return to master list of unassigned cargoes (a) all unassigned cargoes by Step 3, (b) all tentatively assigned cargoes whose EPT's are between $t_k + aL$ and $t_k + L$ and (c) all tentatively assigned cargoes which "interact unfavorably" with one another or with cargoes assigned at previous iterations (see Note (3) below).
Make all other cargo/ship assignments in $(t_k, t_k + aL)$ "permanent". Remove permanently assigned cargoes from master list of unassigned cargoes.

STEP 5 : "Roll" time horizon. Set t_{k+1} = min (lowest EPT of cargoes in master list of unassigned cargoes, time "significant" input update occurs). Update ship locations at t_{k+1}.
Set $k = k+1$ and go to Step 1.

The above sequence of steps refers only to the "core" of MORSS, that is, does not include descriptions of preprocessing and postprocessing modules. Preprocessing modules conduct gross feasibility analyses, and postprocessing modules perform local improvements to the schedules produced up to Step 5. Also in Steps 3 and 4 the human scheduler can override the procedure and "force" it to make assignments of his/her choice.

The following notes provide additional details on the algorithm:

Note (1): The "utility" u_{ij} of assigning ship i to cargo j (Step 2) is defined by:

$$u_{ij} = u_{ij}(1) + u_{ij}(2) + u_{ij}(3) + u_{ij}(4).$$

(a) $u_{ij}(1)$ measures the assignment's effect on the delivery time of cargo j and by ship i. Dropping subscripts i and j and defining t as the tardiness of cargo j (= arrival time - LDT if > 0, zero otherwise), and V_{min}, V_{max}, t_0 and b as user-specified (and, generally, cargo-dependent) parameters, a "reasonable" functional expression for $u_{ij}(1)$ is

$$u_{ij}(1) = V_{min} + (V_{max} - V_{min})e^{-2(t/t_0)^b} \tag{1}$$

The motivation of this functional form, which is typically bell-shaped with the utility very close to V_{max} if t is small and then dropping to a level very close to V_{min} if t increases ($u_{ij}(1) = V_{min} + 0.135(V_{max} - V_{min})$

for $t = t_0$, independent of b), -is the following. If the cargo is delivered early or on time (t = 0), its utility is maximum (V_{max}). If it is only a few days late (say, 1 or 2 days after its LDT), we assume that its value is close to V_{max}, but a bit lower. If it is more than a few days late, we assume its utility decreases rapidly with tardiness, until it reaches a "bottom" value (V_{min}) which is practically independent of t, if t is large ("if it's delivered after two weeks, it might as well be delivered after a month"). We have arrived at this form after discussions with MSC personnel, however we have also experimented with other, less CPU-intensive formulas that essentially exhibit the same features (see Psaraftis et al. (1985) for more details).

(b) $u_{ij}(2)$ measures the assignment's effect on the delivery times of all other cargoes already assigned to ship i. It does this by calculating the net change (if any) of the utilities of all such cargoes. These are computed according to the same formula suggested above (1).

(c) $u_{ij}(3)$ measures the assignments effect on the "efficiency" of use of ship i. Such an efficiency has two dimensions. First, we would like the ship to sail as full as possible. Second, we would like to keep some slack in the ship's schedule, so that additional cargoes can be carried by the ship in future iterations. Again dropping subscripts i and j and defining as C the ship's capacity, as R the residual capacity of the ship after the cargo has been picked up (C and R are expressed in weight, volume or area units depending on which of the three capacities is binding), and as F the slack in the ship's schedule averaged over all future stops, a reasonable formula for $u_{ij}(3)$ is:

$$u_{ij}(3) = V_s e^{-2(R/C)^c(1-fF/L)^d} \tag{2}$$

where V_s, c,d and f are user-specified (and, generally, ship-dependent) parameters ($c, d > 0$, $0 < f < 1$).

The motivation for this function (which is two-dimensionally bell-shaped) is that the utility reaches its maximum value (V_s) if R = 0 (ship is already full, so ship utilization is maximum), independent of the slack in the ship's schedule. It drops to lower values (for F = constant) if R increases. At the same time, the utility is a non-decreasing function of F, for R = constant, for, everything else being equal, one would prefer more flexibility in a ship's schedule than less. See Psaraftis et al. (1985) for more details.

(d) Finally, $u_{ij}(4)$ measures the assignment's effect on the system's port resources, as manifested by the increase in port queueing and congestion caused by ship i's visit to the POE and POD of cargo j. Calculating queueing delays in this problem is very complicated. Instead, we chose to use the following empirical formula:

$$u_{ij}(4) = \begin{cases} V_p & \text{if cargo j's POE/POD coincide with ports that have to be visited anyway,} \\ V_p e^{-2(mN/P)^n} & \text{otherwise.} \end{cases} \quad (3)$$

In (3), N is the anticipated number of visits in cargo j's port during the current time horizon (by all ships), P is the throughput capacity of the port (expressed in equivalent number of visits), and V_p, m, and l are non-negative user-specified parameters. According to (3), $u_{ij}(4)$ drops from its maximum value V_p to zero if cargo j's port is not in the previous schedule and if N is high (see Psaraftis et al. (1985) for more details).

It is obvious that finding an effective set of values for the numerous user-specified parameters used in (1), (2) and (3) is in itself an extremely difficult (and probably data-dependent) task. We have carried out such a calibration to some extent, many times assigning "reasonable" but essentially arbitrary values to these parameters. More calibration would be carried out in a possible implementation phase (see also at the end of this section).

Note (2): The problem that determines the maximum utility "tentative assignment" (Step 3) is formulated as follows:

$$\begin{aligned} \text{Maximize} \quad & \sum_i \sum_j u_{ij} x_{ij} \\ \text{s.t.} \quad & \sum_i x_{ij} \leq 1 \quad \text{for all j} \\ & \sum_j x_{ij} \leq K \quad \text{for all i} \\ & x_{ij} \geq 0 \quad \text{for all i and j} \end{aligned} \quad (4)$$

where x_{ij} = 1 if ship i is assigned to cargo j and zero otherwise. In (4), K is a user-specified integer (usually no more than 2 or 3). This formulation forbids more than K cargoes to be simultaneously assigned to the same ship (per iteration). This "artificial" constraint has been imposed so that one can justify adding all utilities in the objective function and limit the chance of "bad" cargo interactions. Note also that (4) does not explicitly incorporate ship capacity constraints. These constraints come into play in Step 4 of the procedure (moving from tentative to permanent assignment - see also Note (3)).

Note (3): It is clear that the "true" utility of a cargo-ship assignment is directly dependent upon the assignment of other cargoes to the same ship. This nonlinearity is patently neglected in our formulation (4) and may cause significant errors in case multiple cargoes are selected to be assigned to the same ship (put in another way, if each of cargoes 1 and 2 alone is a good assignment for ship 3, it does not necessarily mean that both of them together are). Setting a low value for K limits (but cannot eliminate) the chance of unfavorable cargo interactions. However, should such interactions occur, some of the cargo assignments will have to be cancelled, and the corresponding cargoes will have to return to the pool of unassigned cargoes (see also example below).

Also when a permanent assignment is made, it may happen that the cargo is too large for the available residual ship capacity. In this case the cargo is split. As much as possible of it goes on the ship, and the remaining amount is returned to the pool of unassigned cargoes for the next iteration.

We now illustrate the MORSS approach by a rather rudimentary example. Figure 5 shows two ships (S1 and S2) and four cargoes (1 to 4, pluses are origins, minuses are destinations) in a Euclidean geographical area. Sailing times (days) are also shown in the figure. Assume that cargoes 1 to 3 are known at time t=0, but that cargo 4 appears only at time t=12. EPT's, EDT's and LDT's are given as follows:

Cargo	EPT	EDT	LDT
1	0	0	12
2	1	1	21
3	0	0	22
4	12	12	32

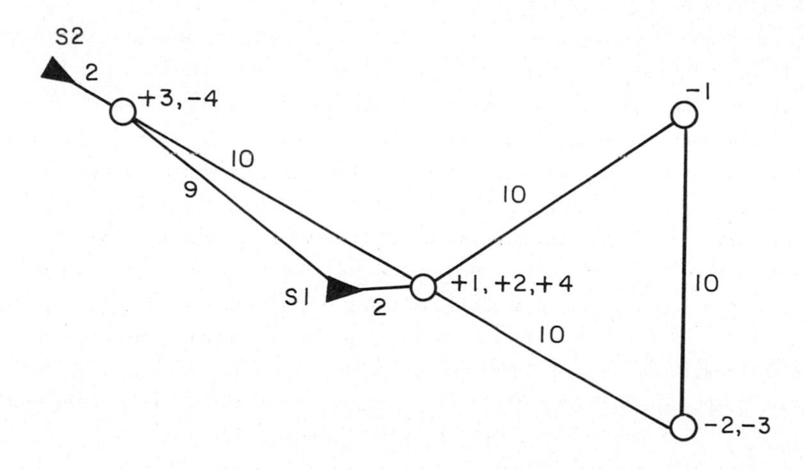

FIGURE 5

For the sake of the example, assume that no capacity constraints exist, and that the only relevant components of the utility are the delay-related ones, that is, the ones given by formula (1) shown earlier. Also assume the following values for MORSS parameters: L = 5, a = 0.5, V_{min} = 0, V_{max} = 1, t_0 = 5, b = 2, and K = 2.

Let us now see how MORSS behaves. The first time horizon (k = 1) is the interval (0, 5), which includes cargoes 1 to 3 (cargo 4 is not known at that time). MORSS then calculates utilities for all eligible cargo/ship pairs. It is straightforward to check that these are as follows:

Ship 1

Cargo 1: Tardiness = 0, Utility = 1

Cargo 2: Tardiness = 0, Utility = 1

Cargo 3: Tardiness = 7, Utility = 0.02

Ship 2

Cargo 1: Tardiness = 10, Utility = 0.00

Cargo 2: Tardiness = 1, Utility = 0.923

Cargo 3: Tardiness = 0, Utility = 1.

Notice the relatively small decrease in utility if the tardiness is small (cargo 2 by ship 2), and the precipitous utility drop if the tardiness increases considerably (cargo 3 by ship 1 and cargo 1 by ship 2).

The maximization of total utility via transportation problem (4) results in the following tentative assignment:

Ship 1: Gets cargoes 1 and 2 (utility = 2),

Ship 2: Gets cargo 3 (utility = 1).

Notice however that the simultaneous assignment of both cargoes 1 and 2 on ship 1 actually results in a lower total utility than the sum of individual utilities, because cargoes 1 and 2 interact unfavorably with one another (see Figure 5). Thus, MORSS will have to return either cargo 1, or cargo 2, to the pool of unassigned cargoes.

Ties can be broken arbitarily, or by using some secondary criteria. One of such criteria is to cancel the assignments of cargoes with higher EPT values. If this is so, the assignment of cargo 2 is cancelled, and MORSS will complete iteration 1 by permanently assigning cargo 1 to ship 1 and cargo 3 to ship 2. Cargo 2 is still unassigned.

Iteration 2 is for horizon interval (1, 6). MORSS here assigns cargo 2 to ship 2, because this is clearly the maximum utility assignment, given the previous assignments. The resulting schedules are established as follows:

Ship 1: Pick up cargo 1 on day 2, deliver it on day 12,

Ship 2: Pick up cargo 3 on day 2, pick up cargo 2 on day 12, deliver these cargoes on day 22.

The third iteration of MORSS will occur at time t = 12, when cargo 4 appears. At that time, ship 1 is just delivering cargo 1, and ship 2 is just picking up cargo 2. Despite the fact that the appearance of cargo 4 coincides exactly (location- and time-wise) with ship 2, it turns out that it is better in terms of utility to assign cargo 4 to ship 1, and let ship 2 proceed with the deliveries of cargoes 2 and 3. The rest of the schedules are thus established as follows:

Ship 1: Pick up cargo 4 on day 22, deliver it on day 32,

Ship 2: Pick up cargo 2 on day 12, deliver cargoes 2 and 3 on day 22.

It can be seen from the above that making simultaneous cargo-to-ship assignments is advantageous from a computational viewpoint, (vis-a-vis a one-by-one sequential assignment procedure), but that care should be taken to avoid unfavorable cargo interactions. Again, since K is low, such bad interactions are expected to occur relatively rarely.

There has been more refinement, testing, calibration, and computational experience with MORSS. The procedure has been coded in Pascal, and developed on both an IBM mainframe system (CMS) and on an Apollo workstation. Details can be found in Psaraftis et al. (1985) and in forthcoming publications (in preparation). The MSC has been very satisfied with the structure and generic features of the procedure, and is planning to proceed with an implementation phase in the foreseeable future. This implementation phase would link MORSS with "real" databases on ships, cargoes and ports, and would typically use MORSS for simulation and training purposes, so as to be prepared for the (undesirable) case in which a real mobilization situation occurs. Of course, many details on the actual operation of MORSS (such as, for instance, the time lag between execution of the algorithm and implementation of scheduling decisions, or how frequently the solution will be updated, or the degree of aggregation of the solution, or what kind of computer system would be used, etc.), will be determined in the implementation phase.

4. DIRECTIONS FOR FURTHER RESEARCH

As mentioned earlier, the state-of-the-art in dynamic vehicle routing methodologies is nowhere near that of the equivalent static case. This paper has attempted to put dynamic vehicle routing into perspective within the broader area of vehicle routing, as well as identify methodological and algorithmic design issues that are likely to be important in the dynamic case.

Methodologically, some effort should be spent to first develop a taxonomy of dynamic vehicle routing problems that parallels the traditional static classification structure. On this score, this author believes that one should really start from scratch, for there isn't that much there to begin with. For instance, the classical (static) TSP is considered to be the "archetypal" (static) vehicle routing problem, in the sense that most other vehicle routing problems are its extensions and generalizations. What is known about the equivalent "dynamic" TSP? To this author's knowledge, the dynamic TSP (DTSP) is a problem that has not even been explicitly defined, let alone investigated or solved. We shall give one definition of the DTSP below, so as to hopefully stimulate the development of such a dynamic vehicle routing taxonomy, and the subsequent buildup of a methodological base in this area.

The Dynamic Traveling Salesman Problem (DTSP)

Let G be a complete graph of n nodes. Demands for service are independently generated at each node of G according to a Poisson process of parameter λ. These demands are to be serviced by a salesman who takes a (known) time of t_{ij} to travel from node i to node j of G, and spends a (known) time of t_0 servicing each demand (on location). If at time zero the salesman is at node 1, what should his "optimal" routing policy be? "Optimal" here may be with respect to a number of objectives (as will be further clarified below).

With the possible exception of the Probabilistic Traveling Salesman Problem (see also below), we are aware of no work by others on problems similar to the DTSP, as defined above. The work of this author is no exception, and he claims no special expertise on this problem. However, a very cursory investigation can reveal a number of interesting issues:

(a) As in dynamic routing in communications networks (see chapter 5 of Bertsekas and Gallager (1987)), there are two main classes of performance measures that are affected by routing decisions in this problem: (i) throughput measures and (ii) delay measures. According to (i), we may want to maximize the average expected number of demands serviced per unit time, that is, the limit, as T goes to infinity, of the ratio of the expected number of demands serviced within T, divided by T. According to (ii), we may want to minimize the average, over all demands, expected time from the appearance of a demand until its service is completed.

(b) Each of the two measures defined above is relevant or irrelevant, in the following sense. If the demand rate λ is "relatively low", then the vehicle will be able to keep up with the demand, and the throughput will be equal to $n\lambda$, irrespective of the routing policy. However, the average expected delay will definitely depend on the routing policy. Figure 6 further clarifies this point (Euclidean travel times are assumed and shown on the network links-in hours). Assume that t_0 = 1 hour and that λ = 0.01 demands/hour. Then it is clear that both the policy "service the (probably sole) demand as soon as it appears and then wait" <u>and</u> the policy "service demands as-you-go, by performing tour 1-2-3-4-1 ad infinitum" achieve a throughput of 0.04 demand services/hour (other routing policies can exhibit the same throughput as well). However, in terms of average expected delay, the above two policies differ. This delay is approximately equal to (1/4) x (1 + 2 + 2.4 + 2) = 1.85 hours for the first policy (there are 4 equal-chance possibilities for the location of the next demand, relative to the current location of the vehicle, and the service time is one hour in all cases). For the second policy, the delay is approximately equal to 3 hours (2 hours average waiting time until vehicle comes to the demand point plus one hour service time). In the above calculations the probability of more than one active demand was ignored.

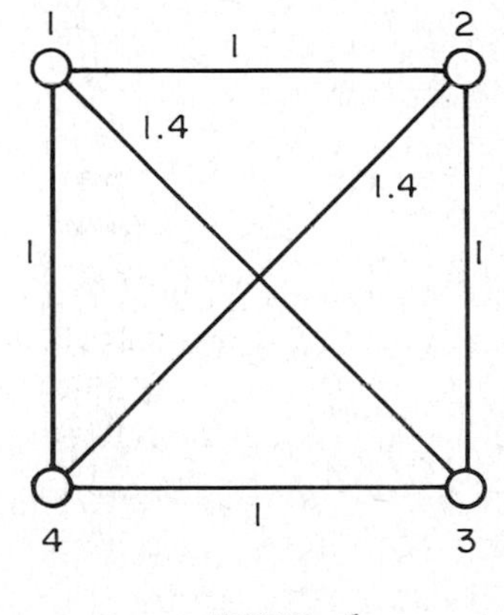

FIGURE 6

(c) Things become more complicated if λ takes on higher values. In this case, not only will the delay depend on the routing policy (as before), but the vehicle may not be able to cope with the demand, that is, the throughput may be forced to be <u>lower</u> than $n\lambda$. In this case, the queue of demands will become unstable, and then the option will be for customers to either suffer an infinite delay, or be denied service (rejected from the system). It is important to realize that whereas for very high values of λ no routing policy will be able to cope with the demand, for intermediate values of λ it is precisely the routing policy which will determine whether the system can handle the demand (and result in realistic delay values), or cannot handle it (and result in infinite delays). For instance, with respect to the previous example (Figure 6), and if λ is high, the policy "perform tour 1-2-3-4-1 ad infinitum, servicing only one demand at a time" achieves a throughput of 4/8 = 0.5 demand services/hour. However, if this policy is applied via tour 1-3-4-2-1, the throughput drops to 4/8.8 = 0.45 demand service/hour. Notice that both throughputs are independent of λ and much lower than the theoretical maximum value of 4λ, if λ is high. In both cases, delays are expected to be infinite (if demands are not rejected). What happens however, if λ is in the vicinity of 0.1 demands/hour? It seems that the former policy is superior to the latter, because of the higher throughput. Does this mean that the optimal policy if λ is very high is always to perform the optimal TSP tour?

(d) The other side of the coin is the following. If λ is low, it may make sense for the vehicle to move to a strategically located node (such as the graph's median, for instance), in anticipation of the next demand. To our knowledge, this issue has been addressed in the context of facility location in a congested network (see Berman et al. (1985) and Chiu et al. (1985)). It would seem that if λ is extremely low and if the objective is to minimize average expected delay, the DTSP resembles the 1-median problem. However, and in contrast to the locational problems examined in the above two references, in the DTSP the server is not restricted to return to a prespecified node after the service of a given demand. Further research is necessary to explore the relationship between these two classes of problems (especially for intermediate values of λ).

Several other variants of the DTSP can be considered. For instance, the graph can be incomplete, symmetric, or Euclidean. Morever, each node can have its own Poisson parameter λ_i. Actually, the demand generation process does not have to be Poisson.

As stated above, the closest problem to the DTSP that comes to mind is the Probabilistic Traveling Salesman Problem (PTSP) (the reader is referred to Jaillet (1985) and to the paper by Jaillet and Odoni in this volume). In the PTSP, a demand at each node occurs (with probability p), or does not occcur (with probability 1-p) during a given day. The PTSP calls for the construction of a route R through all of the nodes so that the expected travel time of the <u>actual</u> route that will be traveled is minimized. The convention here is that the actual route will be based on R so that nodes that have no demand on a particular day will simply be skipped.

Note that for all the resemblance between DTSP and PTSP, the PTSP is really a <u>static</u> problem, for the determination of the optimal R has to be made <u>before</u> actual dispatching of the salesman, even though the actual route to be traveled depends on which (if any) of the demands is actually "active" that day. By contrast, in the DTSP the salesman's decisions have to be based on the current, and, generally, on the probable future states of the system as well.

How can this basic problem be solved? Under what circumstances is a myopic policy (optimize over known demands only) optimal? What happens if service time is zero? Does it make sense to let demands accumulate before the vehicle departs from a node? And so on. Answers to these (and similar) questions would increase our knowledge about dynamic vehicle routing problems, and might motivate further work on more realistic (and more difficult) variants.

ACKNOWLEDGEMENTS

The author is indebted to the Editors and referees for their comments, and to Amedeo Odoni for pointing out the relationship between the Dynamic Traveling Salesman Problem and some locational problems on congested networks.

REFERENCES

Bell, W., L.M. Dalberto, M.L. Fisher, A.J. Greenfield, R. Jaikumar, P. Kedia, R.G. Macj, and P.J. Prutzman, 1983. Improving the Distribution of Industrial Gases with an On-Line Computerized Routing and Scheduling Optimizer. Inferfaces 13, 4-23.

Berman, O., R.C. Larson, and S. Chiu, 1985. Optimal Server Location on a Network Operating as an M/G/1 Queue. Operations Research 33, 746-771.

Bertsekas, D.P., and R. Gallager, 1987. Data Networks, Prentice-Hall.

Bodin, L.D., B.L. Golden, A. Assad and M.O. Ball, 1983. Routing and Scheduling of Vehicles and Crews: The State of the Art. Computers and Operations Research 10, 62-212.

Brown, G., and G. Graves, 1981. Real-Time Dispatch of Petroleum Tank Trucks, Management Science 27, 19-31.

Chiu, S., O. Berman, and R.C. Larson, 1985. Stochastic Queue Median Problem on a Tree Network. Management Science 31, 764-772.

Fisher, M.L., A.J. Greenfield, R. Jaikumar, and P. Kedia, 1982. Real-Time Scheduling of a Bulk-Delivery Fleet: Practical Application of Lagrangian Relaxation. Report 82-10-11, Decision Sciences Dept., University of Pennsylvania.

Golden, B.L., and A. Assad, 1987. Perspectives on Vehicle Routing: Exciting New Developments. Operations Research 34, 803-810.

Jaillet, P., 1985. The Probabilistic Traveling Salesman Problem. PhD thesis, Department of Civil Engineering, MIT.

Jaw, J.J., A.R. Odoni, H.N. Psaraftis, and N.H.M. Wilson, 1986. A Heuristic Algorithm for the Multi-Vehicle Advance-Request Dial-A-Ride Problem with Time Windows. Transportation Research 20B, 243-257.

Larson, R.C., and A.R. Odoni, 1980. Urban Operations Research, Prentice-Hall.

Lin, S., 1965. Computer Solutions to the Traveling Salesman Problem. Bell System Technical Journal 44, 2245-2269.

Lin, S., and B.W. Kernighan, 1973. An Effective Heuristic Algorithm for the Traveling Salesman Problem. Operations Research 21, 498-516.

Minkoff, A.S., 1985. Real-Time Dispatching of Delivery Vehicles. PhD Thesis, Operations Research Center, MIT.

Or, I., 1976. Traveling-Salesman-type Combinatorial Problems and Their Relation to the Logistics of Blood Banking. PhD Thesis, Dept. of Industrial Engineering and Management Sciences, Northwestern University.

Powell, W., 1985. An Operational Planning Model for the Dynamic Vehicle Allocation Problem with Uncertain Demands. Department of Civil Engineering Working Paper EES-85-15, Princeton University.

Psaraftis, H.N., 1980. A Dynamic Programming Solution to the Single-Vehicle Man-to-Many Immediate Request Dial-A-Ride Problem. Transportation Science 14, 130-154.

Psaraftis, H.N. 1983. k-Interchange Procedures for Local Search in a Precedence-Constrained Routing Problem. European Journal of Operational Research 13, 391-402.

Psaraftis, H.N., J.B. Orlin, D. Bienstock, and P.M. Thompson, 1985. Analysis and Solution Algorithms of Sealift Routing and Scheduling Problems: Final Report. Working Paper No. 1700-85, Sloan School of Management, MIT.

Savelsbergh, M.W.P., 1985. Local Search in Routing Problems with Time Windows. Annals of Operations Research 4, 285-305.

Tharakan, G.G., and H.N. Psaraftis, 1981. An Exact Algorithm for the Exponential Disutility Dial-A-Ride Problem. Working Paper, MIT.

Wilson, N.H.M., J.M. Sussman, H.K. Wang, and B.T. Higonett, 1971. Scheduling Algorithms for Dial-A-Ride systems. Urban Systems Laboratory Report USL TR-70-13, MIT.

Wilson, N.H.M., and H. Weissberg, 1976. Advanced Dial-A-Ride Algorithms Research Project: Final Report. Report R76-20, Dept. of Civil Engineering, MIT.

Wilson, N.H.M., and N.H. Colvin, 1977. Computer Control of the Rochester Dial-A-Ride System. Report R77-31, Dept. of Civil Engineering, MIT.

Vehicle Routing: Methods and Studies
B.L. Golden and A.A. Assad (Editors)

A COMPARATIVE REVIEW OF ALTERNATIVE ALGORITHMS FOR THE DYNAMIC VEHICLE ALLOCATION PROBLEM †

Warren B. POWELL

Princeton University, Department of Civil Engineering and Operations Research, Princeton, New Jersey 08544

The dynamic vehicle allocation problem involves managing a generally large fleet of vehicles over time to maximize total profits. The problem is reviewed in the context of truckload trucking with special attention given to dispatching and repositioning trucks in anticipation of forecasted future demands. Four different methodological approaches are reviewed: deterministic transshipment networks, stochastic/nonlinear networks, Markov decision processes and stochastic programming. The methods are contrasted in terms of their formulation of the objective function and decision variables, the degree to which actual practices can be represented, and computational requirements. The paper provides an example of how a particular problem can be approached from significantly different perspectives.

1. INTRODUCTION

The dynamic vehicle allocation problem arises in industries where a fleet of vehicles must be managed over time responding to known or forecasted demands for capacity. Motor carriers, railroads, container shipping lines, and auto or truck rental companies are immediate examples of this problem. Different industries, however, exhibit unique characteristics which lend themselves to different modelling approximations. For this reason, the discussion here uses truckload trucking as the industry context, although the basic concepts and algorithms are more general. Truckload trucking is also one of the simplest of these modes and as a result poses the least overhead in terms of industry minutiae.

Briefly, and somewhat simplistically, the problem faced by truckload motor carriers can be described as follows. A shipper will call a carrier with a load going from city A to city B. The carrier must deadhead a truck to the shipper where the trailer is loaded and then run to city B where the delivery is made. The carrier must then decide what to do with the truck once it arrives in B. At any point in time, a truck must either be assigned to a load, repositioned empty to a city in anticipation of loads to be called in later, or simply held at its current location. It is important to realize that there is no consolidation function, as typically arises in vehicle routing problems. This property significantly simplifies the problem and focuses attention on the fleet allocation side of the problem.

As with auto rental companies, truckload trucking is characterized by a high level of competition in the marketplace compounded by a high degree of uncertainty regarding future demands. Typically, 60 percent of the loads called in are for pickup the same day, implying

† This research was supported by the National Science Foundation under a Presidential Young Investigator Award under Grant ECE-8408044.

that at the beginning of the day the carrier will know only 40 percent of the loads that will be carried that day. At the same time, if a carrier does not have a truck close by when a load is called in, that load will generally be given to a competitor. Since the largest company captures only 1 percent of the market, shippers regularly work with several carriers. For this reason, a carrier also has some freedom to turn down freight and as a result must learn to be selective about the freight it carries. This process, known as load acceptance, is an important part of the fleet management problem.

The fleet dispatching process at most truckload carriers works as follows. At any given point in time, there will be a set of unassigned drivers and unassigned loads. Dispatchers will typically have access to a computer terminal that might show, for example, all the loads within 80 miles of a given driver. The dispatcher must then assign the driver to one of these loads, taking care to match equipment types. This decision is often made while the driver is waiting on the telephone, and it is not unusual for the driver to refuse to pull a load, preferring another load moving in a desired direction (the driver may be trying to get home). At other times, the dispatcher may simply ask the driver to call back after 30 minutes if no load is available. If it is unlikely that a load will become available in his area, the dispatcher may tell the driver to move empty to a better location and call back when he arrives.

The process of assigning a driver to a load generally focuses on minimizing empty miles, which is often interpreted to mean that a driver should be assigned to the nearest load for which the driver has the correct equipment. Only one carrier, to this author's knowledge, actually solves a transportation problem to assign drivers to loads optimally. Separate from the dispatching process is the planning function which determines how to reposition empty trucks. This is a largely heuristic process where experienced planners will move trucks from traditional surplus regions to traditional deficit regions. The difficulty is in forecasting the number of loads yet to be booked out of a region which must be balanced against the profitability of loads out of each region. For the most part, dispatchers and planners respond to activities like trucks and loads, and have considerably more difficulty comparing the profitability of different actions.

The last important step in fleet management is load acceptance and load solicitation. Unlike classical vehicle routing problems, where the objective is to carry all the freight at least cost, the highly competitive truckload market allows carriers to refuse freight occasionally which does not appear profitable. In addition, the sales force may attempt to solicit freight to help fill empty backhaul movements. Almost all carriers approach load acceptance and load solicitation by trying to attract all the freight they can and accepting everything they have the capacity to handle. However, most carriers have a good sense that a load paying $1.30 per mile in one direction may be good while another load paying $0.90 in a backhaul lane may be even better. Just the same, under the philosophy that a bird in the hand is worth two in the bush, the usual rule is to accept loads until they run out of capacity. Load solicitation, on the other hand, is a process that varies widely among carriers. It is likely that most carriers evaluate their sales force in terms of total sales (or, in some cases, total loads) which gives little incentive to attract freight in profitable lanes either on a long term or a short term basis. One carrier solves a network transshipment model each night. The network model heuristically turns forecasts of freight into fictitious loads (this process is described below) which might be accepted or rejected. The fictitious loads which are "accepted" by the model become recommendations of places where the sales force should solicit freight.

Fortunately from a planning perspective, most carriers actively collect and maintain all the data that is necessary to feed an optimization model. When a shipper calls in a load, the carrier immediately keys in the exact origin and destination of the load, its pickup time and date, its delivery time, the equipment type required and other special handling costs. Generally, the carrier already has in the computer a previously agreed on tariff for the load (which may vary between shippers and between traffic lanes). At the same time, each driver, after dropping off a load, calls in giving his exact location and status. This way, it is possible to calculate accurately the deadhead miles from each driver to each load.

Separate from real-time data on each driver and each load, it is relatively straightforward to develop detailed forecasts of future activities from extensive records of past activities. Generally each loaded and empty move is recorded on the computer. This is necessary since all drivers are paid by the mile, and there are usually different rates paid for loaded and empty moves. Thus, records of activities must be kept in order to pay the driver.

The dynamic vehicle allocation problem (DVA) has been studied by a number of authors. An excellent review of this work is given by Dejax and Crainic [9]. The earliest efforts in this problem were applied to the repositioning of empty rail freight cars (Misra, [18], Ouimet, [19] and Baker, [1]). Misra used a simple static transportation formulation, even though White and Bomberault [30] had already formulated the dynamic vehicle allocation problem in the now classical time-space framework. Uncertainty in demand forecasts has been incorporated by posing the problem as a spatially separated inventory problem (Philip and Sussman, [20]). Other contributions are reviewed during the presentation.

The primary goal of this research is to review and contrast alternative modelling and solution approaches with particular attention given to the handling of forecasting uncertainties. This review documents a transition from a purely network based approach to stochastic optimization techniques using Markov decision processes and other techniques. The presentation is organized as follows. Section 2 describes deterministic formulations, covering both single stage and dynamic models. The deterministic assumption applies to the modelling of forecasting uncertainties surrounding loads that will have to be carried in the future. Section 3 describes two approaches for extending the basic network formulation to handle forecasting uncertainties, including one formulation where flows on links are handled explicitly as random variables. Section 4 formulates the same problem as a Markov decision process with a very large state space, serving primarily as an approach for developing a better understanding of the problem structure. Section 5 combines insights from Markov decision processes and the classical network formulations to form a hybrid model that reduces to a linear network. Finally, Section 6 provides an overview of alternative models and discusses implementation issues.

2. DETERMINISTIC MODELS

Deterministic models have played an important historical role in the dynamic vehicle allocation problem. At the same time, while highly simplistic in the assumptions these models impose, their inherent simplicity suggests that they will continue to serve as valuable tools in practice. For the purposes of the discussion here deterministic models are divided between single stage models, which simply assign trucks to loads with little or no forecasting, and dynamic models which explicitly track trucks and loads over a given planning horizon.

It is common when modelling flows over a continuum as large as the United States to divide the country into 60 to 100 discrete regions. Let $\mathbf{R} = \{1, ..., R\}$ denote the set of regions. Then define:

$x_{i,j}(t)$ = flow of trucks moving loaded from region i to region j, departing from i in period t,

$y_{i,j}(t)$ = flow of trucks moving empty from region i to region j,

$t_{i,j}$ = travel time in integer time units to travel from i to j (for simplicity, travel times for moving loaded and empty are assumed to be equal),

$r_{i,j}$ = average contribution (revenue minus direct operating cost) for pulling a load from i to j,

$c_{i,j}$ = cost of moving empty from i to j,

$F_{i,j}(t)$ = random variable denoting the number of loads that will be called in from i to j to be picked up at time t,

$f_{ij}(t)$ = $E[F_{ij}(t)]$,

= expected number of forecasted loads from i to j departing at time t,

$L_{i,j}(t)$ = actual number of loads known at time t=0 to be available moving from i to j at time t,

$T_i(t)$ = number of drivers becoming available for the first time in region i at time t,

P = length of the planning horizon.

Using this notation it is possible to describe single stage and dynamic deterministic network models.

2.1 Single stage deterministic models

The simplest single stage model is a transportation problem assigning available drivers to available loads, as depicted in Figure 1. In this figure,

$D_i(t)$ = total outbound loads from region i at time t,

$$= \sum_{j \in R} L_{ij}(t) ,$$

E_r = extra trucks needed to satisfy demand,

$$= \max \{ \sum_{i \in R} [D_i(1) - T_i(1)] , 0 \} ,$$

E_D = extra demand needed to absorb excess trucks,

$$= \max \{ \sum_{i \in R} [T_i(1) - D_i(1)] , 0 \} .$$

This model does nothing more than assign available trucks to available loads and is unable to plan for the future. Any trucks that are assigned to the dummy demand node are simply held in the region. It is possible, of course, to simply augment the demands by adding the number of known loads with forecasted loads, but such a model runs the risk of refusing a known load in one region over a forecasted load in another region. This problem can be alleviated by providing very simple forecasting capabilities. Let

$p_i(t)$ = the average return of a truck in region i on day t until the end of the planning horizon.

The factors $\{p_i(t)\}$ are termed *salvage values*, and a simple approach for calculating them based on historical data is given in the appendix.

Using these salvage values, Figure 2 presents an alternative single stage dispatch model. Here, the demand nodes for known loads have been augmented by separate nodes for forecasted loads. A truck repositioned empty from i to j will arrive on day t_{ij}, at which point it will receive expected profits of $p_j(t_{ij})$. It is necessary to bound the number of vehicles repositioned to region j arriving on day t_{ij} . Let

$$f_{i\cdot}(t) = \sum_{j \in R} f_{ij}(t) \qquad (1)$$

be the total forecasted demand out of i. Since $f_{i\cdot}(t)$ will usually be noninteger, we let $\hat{f}_{i\cdot}(t)$ be the upper bound and set

$$f_{i\cdot}(t) = \left\lfloor f_{i\cdot}(t) + \gamma \right\rfloor \tag{2}$$

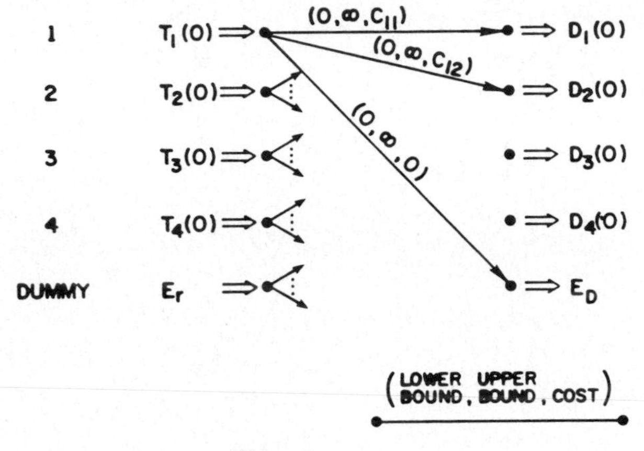

Figure 1
Single stage transportation formulation with overflow nodes

where $\lfloor x \rfloor$ is the largest integer less than or equal to x and $0 \le \gamma < 1$ is a specified parameter. $\gamma = .5$ produces conventional rounding while $\gamma = .8$ tends to round more forecasts up.

The real value of simple assignment models is their ability to handle high levels of detail. Consider the network in Figure 1 where each supply node is a single driver and each demand node is a single load. In this case a link (i,j) represents an assignment of driver i to load j. The cost coefficient c_{ij} would then represent the empty cost from the precise location of driver i to the precise location of load j, thus avoiding the need to aggregate the country into regions. In addition, if truck i represents an equipment type that is incompatible with load j, this link would not be generated. The biggest advantage of this model, then, is its ability to incorporate a high level of detail about a driver's location and characteristics. This more detailed formulation can also be augmented by the forecast nodes given in Figure 2, allowing the model to recommend empty repositioning moves. The only, and major, weakness of the model is that its relatively simplistic representation of the downstream effects limits its ability to reposition empties between regions accurately. Despite this weakness, however, it is quite possible that the ability of the model to choose the best driver for a load may measurably outperform a dispatcher's performance in the same task, overcoming the model's other weaknesses.

Another feature of the transportation formulation is its small size and the speed with which it can be solved. Assume every driver and load were represented explicitly. Typical problem sizes range from 100 to 1000 drivers being assigned to a comparable number of loads. All

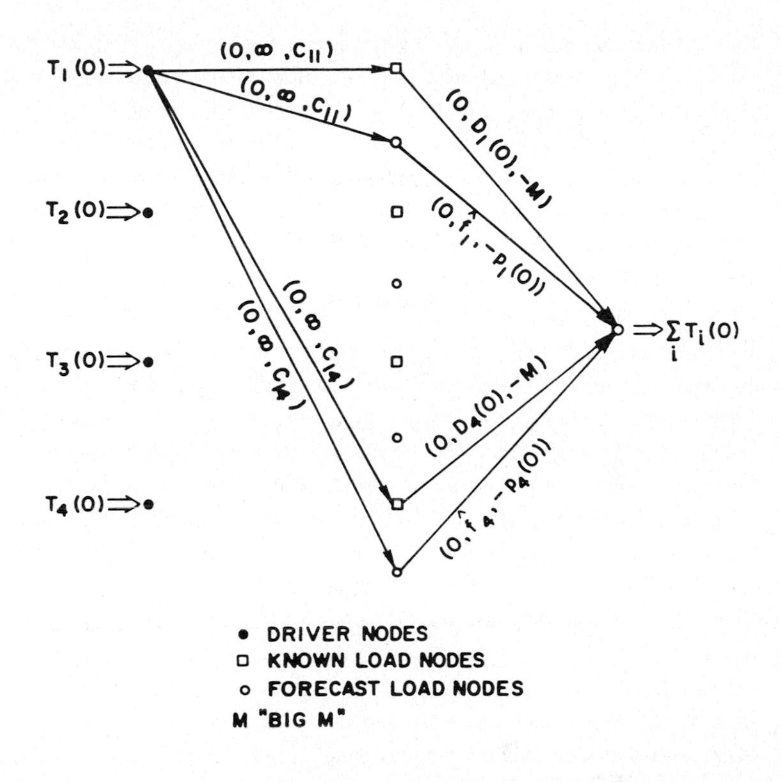

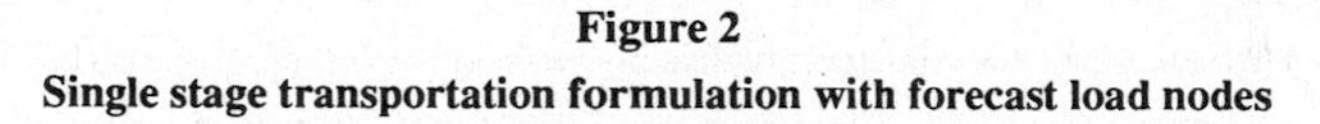

Figure 2
Single stage transportation formulation with forecast load nodes

these models use pruning rules which limit assignments of drivers to the nearest 5 to 10 loads, giving network sizes with 500 to 10,000 links. On a large mainframe, such problems can be optimized from scratch in a few seconds. Furthermore, these codes can be reoptimized from a previous basis following a change in the data usually in one tenth the time.

2.2 Dynamic deterministic models

The biggest limitation of static models is the simple way in which future activities are forecasted. Ideally, the simple linear salvage values used to summarize the value of an additional truck in each region should be replaced with a nonlinear, nonseparable function of the vector of supplies in each region. A straightforward solution to this problem, which is now considered the classical approach to the dynamic vehicle allocation problem, is to form a time-space diagram where each node represents a region at a particular point in time. Between any pair of nodes are three types of links: known loads, where the cost coefficient is minus the direct contribution for that load; empty movements, with a cost equal to the region to region empty repositioning costs; and forecasted loads, with a cost coefficient equal to minus the historical average direct contribution. Typically, loaded movements between the same regions on the same day are modelled as separate links which then would each have an upper bound of one. This allows known loads to be modelled with more detailed information on costs, revenues and travel times. Empty movement links of course have no upper bound.

The data requirements for a dynamic model are relatively straightforward. First is the real-time information on loads and drivers. It is necessary to know when and where each driver will first become available (it is not possible to take equipment types into consideration without sacrificing the pure network structure). Next we need to know the pickup and delivery place and time, and the contribution (revenue minus direct operating cost), of each pending load in the system. Empty movements can be generated by calculating the average empty distance among empty moves actually made between a given pair of regions. It is also possible to calculate distances using the coordinates of zip codes within a region. Normally the list of possible empty moves is restricted to moves within a given radius.

The last and most difficult input to any dynamic model are the forecasts of future loads, represented by the $f_{ij}(t)$. It is beyond the scope of this discussion to discuss this step in detail, but the essence is that the $f_{ij}(t)$ are derived from time series forecasting models. These models might be built from six months or more of past loads, taking into account seasonal and other trends as well as recent activities.

Two important issues arise in the use of deterministic, dynamic network models. First, a method must be developed for representing forecasted loads as links in the network. Second, it is necessary to choose the length of the planning horizon and to develop a procedure for truncating the network. A difficulty arises again with setting the upper bounds on the forecasted loads. Ninety percent of the demand forecasts will normally fall between 0 and 1, with many below .5. Three approaches may be used:

(1) "Integerize" the upper bounds using heuristic rounding rules. The upper bound might be given by $\hat{f}_{ij}(t) = \lfloor f_{ij}(t) + \gamma \rfloor$ as we did in Equation 2. γ should be chosen so that $\sum_{j \in R} \hat{f}_{ij}(t) \approx \sum_{j \in R} f_{ij}(t)$.

(2) Randomly sample forecasted loads by using the mean, $f_{ij}(t)$, to fit a distribution (such as the Poisson) and then sample from this distribution. Each randomly sampled load would then be represented as a link with an upper bound of one.

(3) Use fractional upper bounds. Typically this is handled by changing the units from trucks to tenths or hundredths of trucks.

Since the last approach produces fractional solutions, it will be necessary to "integerize" the final solution. The second approach suffers from the randomness introduced by the sampling process. The first approach suffers from biases introduced by the rounding process, resulting in significantly higher or lower forecasts.

Separate from the issue of dealing with fractional demand forecasts is the serious problem of truncating the planning horizon. Researchers actively applying this approach to problems in rail and trucking have, in private conversations, reported serious distortions when the truncation is not handled properly even when reasonably long planning horizons are used. At some point we are forced to ignore the future impact of a decision, and this often has an amazing ability to ripple back to the beginning of the planning horizon. The simplest example is a network with regions where trucks generally cannot move out loaded (such as Montana). Normally the price to carry a load into such a region is quite high to cover the cost of the empty backhaul. Near the end of the planning horizon, the network model will tend to push trucks into Montana to get the high price since the cost of moving out empty is being ignored.

The deterministic, dynamic network model is simply a special example of an infinite stage linear program. For notational simplicity assume that each time period t, $t = 0, 1, 2, \ldots$, is one day and that all travel times between regions are exactly one day (travel times other than one day are easy to handle but complicate the presentation without contributing any insights). In addition, we will aggregate over a single link all loaded flows from region i on day t to region j, where in practice we would generate different links for each known load and all forecasted loads. To discuss the issues associated with truncation define the following:

$U_{ij}(t)$ = total number of loads expected to be available from i to j leaving on day t,

$= L_{ij}(t) + f_{ij}(t)$,

$S_i(t)$ = supply of trucks at i on day t,

$= \sum_{k \in R} [x_{ki}(t-1) + y_{ki}(t-1)] + T_i(t)$,

$\mathbf{S}(t) = \{S_1(t), \ldots, S_R(t)\}$,

α = a "discount" factor where α^t is the value of a dollar today spent in period t.

Taking advantage of the assumption of one day travel times, we have that $\sum_{i \in R} S_i(t) = \sum_{i \in R} T_i(0) =$ the fleet size, and $T_i(t) = 0$ for $t \geq 1$. This implies that the state of the system at time t is given completely by the vector $\mathbf{S}(t)$. The problem is one of maximizing total profits over time, where we adopt the approach of maximizing total discounted profits with a daily "discount" factor α. The problem then is to solve the following optimization problem:

$$\max_{\{x,y\}} \sum_{t=0}^{\infty} \sum_{i \in R} \sum_{j \in R} [r_{ij}\, x_{ij}(t) - c_{ij}\, y_{ij}(t)]\, \alpha^t, \tag{3}$$

subject to:

$$\sum_{j \in R} [x_{ij}(0) + y_{ij}(0)] = T_i(0) \quad \forall i \tag{3a}$$

$$\sum_{j \in R} [x_{ji}(t-1) + y_{ji}(t-1)] = \sum_{k \in R} [x_{ik}(t) + y_{ik}(t)] \quad t = 1, 2, ..., \quad \forall i \tag{3b}$$

$$x_{ij}(t) \leq U_{ij}(t) \quad \forall i, j, t, \tag{3c}$$

$$x_{ij}(t), y_{ij}(t) \geq 0 \quad \forall i, j, t. \tag{3d}$$

In practice we have to maximize profits over a given planning horizon. Equation 3 can always be rewritten using standard dynamic programming concepts as follows:

$$\max_{\{x,y\}} \sum_{t=0}^{P} \sum_{i \in R} \sum_{j \in R} [r_{ij} x_{ij}(t) - c_{ij} y_{ij}(t)] \alpha^t + \Psi_{P+1}(S(P+1)) \tag{4}$$

where

$$\Psi_{P+1}(S(P+1)) = \max_{\{x,y\}} \sum_{t=P+1}^{\infty} \sum_{i \in R} \sum_{j \in R} [r_{ij} x_{ij}(t) - c_{ij} y_{ij}(t)] \alpha^t \tag{5}$$

subject to:

$$\sum_{j \in R} [x_{ij}(P+1) + y_{ij}(P+1)] = S_i(P+1) \tag{6a}$$

$$\sum_{j \in R} [x_{ji}(t-1) + y_{ji}(t-1)] = \sum_{k \in R} [x_{ik}(t) + y_{ik}(t)] \quad t = 1, 2, ..., \quad \forall i \tag{6b}$$

$$x_{ij}(t) \leq U_{ij}(t) \quad \forall i, j, t, \tag{6c}$$

$$x_{ij}(t), y_{ij}(t) \geq 0 \quad \forall i, j, t. \tag{6d}$$

The problem with (4) is that the function $\Psi_{P+1}(S(P+1))$ is a very complex, nonlinear, nonseparable function ($\Psi_{P+1}(\cdot)$ must also incorporate the effects of flow conservation constraints and upper bounds). The challenge is to replace $\Psi_{P+1}(S(P+1))$ with an easier to estimate function $\hat{\Psi}_{P+1}(S(P+1))$. This problem has been studied in a network context by Hughes and Powell [15] based on more general work by Grinold [10, 11, 12] and Grinold and Hopkins [13]. Three methods are reviewed briefly here, the first two resulting in simple pure network formulations and the last one resulting in a generalized network.

The naive approach

This approximation consists simply of

$$\hat{\Psi}_{P+1}\ (\mathbf{S}(P+1)) \ = \ 0\,. \tag{7}$$

End effects are just ignored, generally producing serious distortions in the optimal flow pattern.

The salvage method

This method puts a price (the salvage value) on a truck left in a region at the end of the planning horizon. As before, let $p_i(t)$ be the value of an additional vehicle in region i on day t, whose calculation is described in the appendix. Then

$$\hat{\Psi}_{P+1}\ (\mathbf{S}(P+1)) \ = \ \sum_{i \in R} S_i(P+1)\, p_i(P+1)\,. \tag{8}$$

This linear approximation lends itself easily to a pure network formulation. Both the naive and salvage approximations involve links from each node in time period $P+1$ to a supersink. The naive approach puts a cost of zero while the salvage method puts a cost of $p_i(P+1)$ from the node for region i. The salvage values $p_i(P+1)$ capture the expected contribution from period $P+1$ out to the end of a second planning horizon P_S. Trucks are assumed to follow historical trajectories in the periods $P+1$, $P+2$, ..., P_S (this is explained more thoroughly in the appendix).

A way to improve the accuracy of $\hat{\Psi}_{P+1}(\cdot)$ in both the naive and salvage methods is to add upper bounds on the flows that terminate in each region. That is

$$S_i(P+1) \le \hat{u}_i(P+1) \tag{9}$$

This constraint helps to mitigate somewhat the nonlinear properties of $\hat{\Psi}_{P+1}(\cdot)$. A common choice of $\hat{u}_i$ is:

$$\hat{u}_i\ (P+1) \ = \ \left\lfloor \sum_{j \in R} f_{ij}\ (P+1) + \gamma \right\rfloor, \tag{10}$$

where again γ is our rounding factor.

Deterministic transshipment networks using the salvage method to mitigate end effects is the most widely used approach for solving dynamic vehicle allocation problems (see, for example, Chih [7] and Shan [25]). At this time, relatively little attention has been given to developing formal methods for calculating salvage values or evaluating their performance. The attraction is that the problems can be efficiently solved using standard network codes. A standard problem will use 60-80 regions with a planning horizon of 10 days, resulting in a network with 20,000 to 60,000 links which can be optimized in less than a minute on a large mainframe.

The dual equilibrium method

Assume, again for simplicity, that while there may be information about loads available for pickup today, nothing is known about tomorrow and, in addition, the forecasts of loads available for days $t = 1, 2, \ldots$, are the same. Thus we may let:

$$U(0) = \text{vector of upper bounds for today}$$

$$U(1) = \text{vector of upper bounds for tomorrow}$$

$$= U(t), \quad t \geq 2 .$$

In Grinold's terminology, day 0 is the transient stage and day 1 is the stationary stage which repeats itself indefinitely. Both the transient and stationary stages may in principle consist of one or more days. The transient stage might reasonably consist of the first 14 days while the first stationary stage might start on day 15 and consist of one week which repeats itself indefinitely.

The dual equilibrium method approximates the flows in the stationary stages as all being equal (that is, in a kind of stationary equilibrium). Using this notion, we may aggregate the flows starting in time period 1 using:

$$\hat{x}_{ij}(\alpha) = (1-\alpha) \sum_{t=1}^{\infty} \alpha^{t-1} x_{ij}(t) . \tag{11}$$

If in fact $x_{ij}(1) = x_{ij}(2) = \ldots = x_{ij}(t)$, then $\hat{x}_{ij}(\alpha) = x_{ij}(1)$. Next we aggregate the other network constraints (3b), (3c) and (3d) by multiplying both sides by $(1-\alpha)\alpha^{t-1}$ and summing over all t, giving:

$$\sum_{j \in R} [(1-\alpha) x_{ij}(0) + \alpha \hat{x}_{ij}(\alpha) + (1-\alpha) y_{ij}(0) + \alpha \hat{y}_{ij}(\alpha)] = \sum_{k \in R} [\hat{x}_{ki}(\alpha) + \hat{y}_{ik}(\alpha)] \tag{12}$$

$$\hat{x}_{ij}(\alpha) \leq u_{ij} \tag{13}$$

$$\hat{x}_{ij}(\alpha)\,,\hat{y}_{ij}(\alpha) \;\geq\; 0 \tag{14}$$

where $u_{ij} = u_{ij}(1) = u_{ij}(2) = \ldots = u_{ij}(t)$.

Finally, the objective function becomes, after a few manipulations:

$$\min_{\{x,y\}} \sum_{i \in R} \sum_{j \in R} [\, r_{ij}\,(x_{ij}(0) + \frac{\alpha}{1-\alpha}\,\hat{x}_{ij}(\alpha)) \;-\; c_{ij}(y_{ij}(0) + \frac{\alpha}{1-\alpha}\,\hat{y}_{ij}(\alpha))\,] \tag{15}$$

subject to (12), (13) and (14).

The key to this approach is the realization that Equation 12 represents the constraints for a generalized network. The graph for a two region network is depicted in Figure 3. Flows from the transient stage are first factored down by $1-\alpha$ when entering the stationary stage. Since each stage consists of only one time period, the links in the stationary stage loop back on themselves, factored down by α to represent the discounting from one stage to the next. In addition, the cost coefficients in the transient stage are factored by $\alpha/(1-\alpha)$.

The intuition behind Grinold's dual equilibrium approach is a little difficult to follow in this problem context. Hughes and Powell present the *generalized summation* method which is virtually equivalent to the dual equilibrium method. This method aggregates the flows using $\hat{x}(\alpha) = \sum_{t=1}^{\infty} \alpha^t\, x(t)$. In it, all flows making a transition from one period to the next are discounted by α. The arc coefficients are unaffected. The flows on the stationary arcs can now be interpreted as the total discounted *flows* and hence the upper bounds must be factored up by $\alpha/(1-\alpha)$. The resulting graph, shown in Figure 4, is much easier to understand intuitively since the arcs moving from one stage to the next always carry the discount factor.

Hughes and Powell report on a set of experiments on randomly generated test networks using discount factors α of .3 and .6. A "brute force" approach was used where a large network was generated covering N stages where N was chosen so that $\alpha^N \leq .05$ (resulting in networks with $N = 3$ and 8 stages, respectively) with 2 or 7 time periods within each stage. The solution to this brute force approach was then compared on the basis of optimal flows in the first time period alone as well as the flows throughout the transient stage. The results demonstrated that the dual equilibrium and generalized summation methods significantly outperformed the naive and salvage value methods.

What is most important about this line of research are the following observations:

(i) The assumption of a deterministic future actually complicates the problem if end effects are handled carefully, and

(ii)forecasting uncertainties must be handled in a highly heuristic fashion through the use of deflated "discount factors."

The Hughes and Powell experiments are limited since they do not actually simulate decision making under uncertainty over a planning horizon. Rather, they report only on a side-by-side

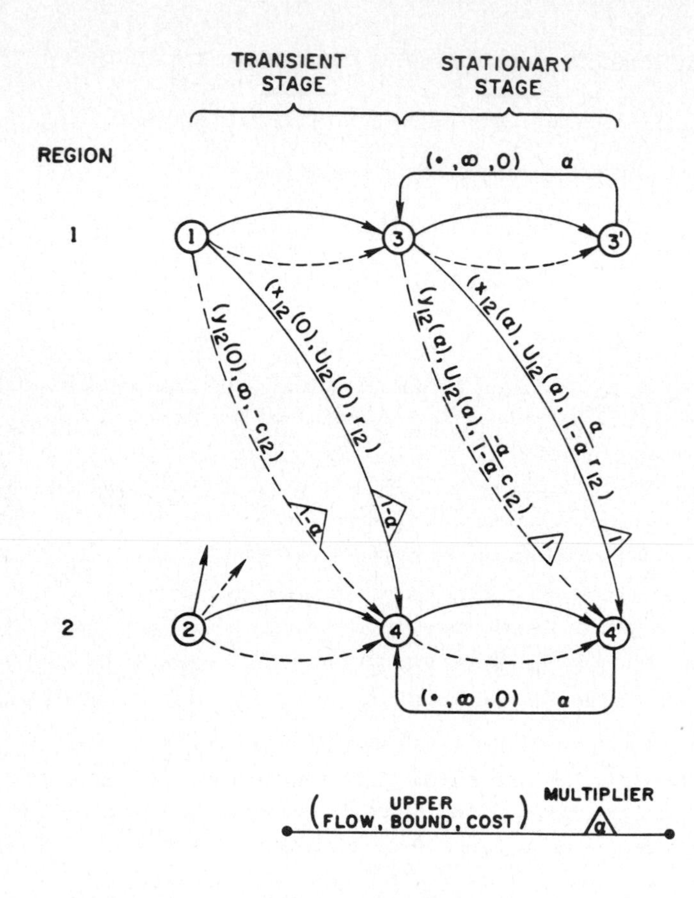

Figure 3

Generalized network for dual equilibrium method, one day per stage

comparison of flows from a single run of a network model. Second, there are several variations to the salvage value method in terms of how the salvage values are computed and the choice of upper bounds on the final link.

A question that often arises in the development of dynamic models is whether an accurate model of future activities is really needed given that the model will be solved repeatedly on a rolling horizon basis. This question must be addressed from two perspectives. First, from the perspective of dispatching trucks, the question is an empirical one that has yet to be carefully and rigorously investigated. Cape [6] compared a simple transportation formulation to a stochastic programming heuristic described below, showing a significant improvement for the

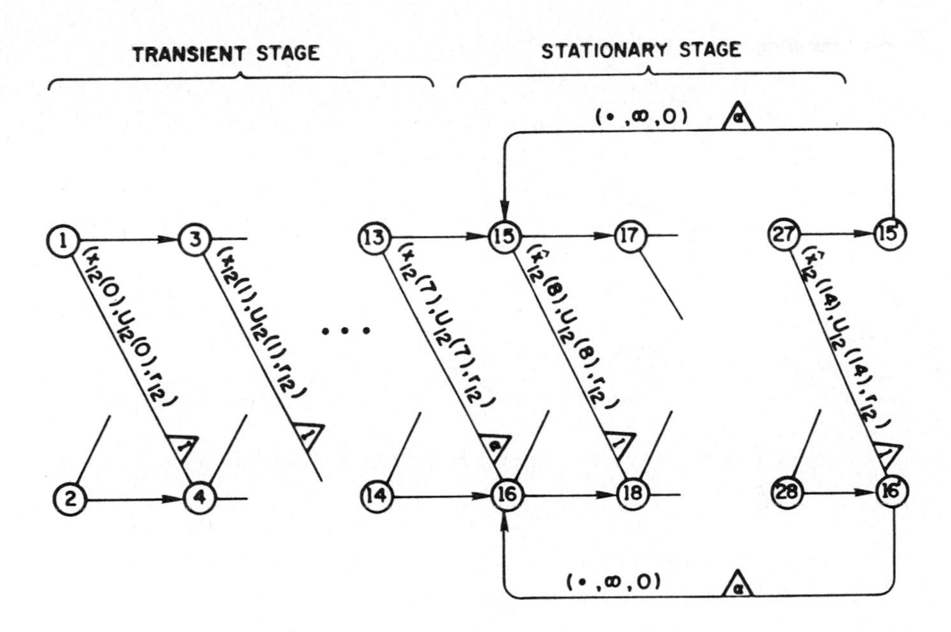

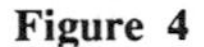

Figure 4

Generalized network for generalized summation method, seven days per stage

latter model. The author is continuing to investigate this issue more thoroughly. As a general rule, good decisions today require good forecasts of future activities, even if we are not going to carry out all of the decisions we optimize for in the future.

The second perspective which is generally ignored is that of pricing and load evaluation. An important but largely neglected aspect of optimization models is their ability to provide estimates of the marginal value of an activity. In the context of the DVA, we might ask what is the value of a load booked from region i, day 1 to region j, day 3. This requires knowing the marginal value of a truck in region j, day 3, which in turn requires that we make our best forecast of what we will be doing on day 3. Our research has shown that dispatching trucks is much easier than pricing their activities, and places much higher demands on the quality of the optimization in the future.

The remainder of this paper reviews different approaches to handling uncertainty in demand forecasts. The next section considers direct modifications to the network formulation, after which Section 4 turns to more classical stochastic optimization based approaches.

3. STOCHASTIC NETWORK FORMULATIONS

The literature on stochastic formulations of the DVA is quite thin. Two approaches have been attempted in the literature, the first resulting in a nonlinear network model with deterministic flows and the second producing a network-based nonlinear math program which represents flows as random variables. In order to build a common framework, the first approach is termed the stochastic DVA with *simple recourse* while the second is the stochastic DVA with *null recourse*, for reasons that are described below.

3.1 The stochastic DVA with simple recourse

Define:

$z_{ij}(t)$ = total flow of trucks assigned to move (loaded or empty) from i to j departing at time t,

$X_{ij}(t)$ = random variable denoting the number of loaded trucks moving from i to j at time t,

$Y_{ij}(t)$ = random variable denoting the number of empty trucks moving from i to j at time t.

From the definition of $z_{ij}(t), X_{ij}(t)$ and $Y_{ij}(t)$, it is clear that

$$X_{ij}(t) = \min [\, z_{ij}(t), F_{ij}(t) \,] \tag{16}$$

and

$$Y_{ij}(t) = z_{ij}(t) - X_{ij}(t) \tag{17}$$

where $F_{ij}(t)$, as before, is a random variable denoting the number of loads that will be available from i to j at time t. The probability distribution of $F_{ij}(t)$ is assumed known. We wish to find an optimal allocation of trucks $\{z\}$ to maximize total expected profits over a given planning horizon:

$$\max_{\{z\}} \sum_{t=0}^{P} \sum_{i \in R} \sum_{j \in R} [\, r_{ij} \, E[X_{ij}(t)] - c_{ij} \, E[Y_{ij}(t)] \,] \tag{18}$$

subject to flow conservation on the flows $\{z\}$ and where $E[X_{ij}(t)]$ and $E[Y_{ij}(t)]$ are derived using (16) and (17). Note that we must still choose a planning horizon P and devise a strategy to manage end effects. It is possible to rewrite (18) by defining a profit function $g_{ijt}(z_{ij}(t))$ for each link using:

$$g_{ijt}(z_{ij}(t)) = E\,[\,r_{ij}\,X_{ij}(t) - c_{ij}\,Y_{ij}(t)\,] \tag{19}$$

which produces the following objective function:

$$\max_{\{z\}} \sum_{t=0}^{P} \sum_{i\in\mathbf{R}} \sum_{j\in\mathbf{R}} g_{ijt}(z_{ij}(t)) \tag{20}$$

subject to flow conservation and nonnegativity constraints on the flows $\{z\}$. Equation (20) is a concave, separable nonlinear network problem which is easily solved using the Frank-Wolfe algorithm which produces a sequence of linear transshipment problems.

An equivalent formulation of the same problem, developed in Powell <u>et al</u>. [21] starts by defining the following:

$S_i(t)$ = total flow through region i at time t,

$\theta_{ij}(t)$ = fraction of the total supply in i at time t that is to be sent to j.

Clearly

$$z_{ij}(t) = \theta_{ij}(t)\cdot S_i(t) \qquad \forall\, i,j,t\,. \tag{21}$$

The flow conservation constraints are now written

$$S_i(0) = T_i(0) \qquad \forall\, i \tag{22}$$

$$S_i(t) = \sum_{k\in\mathbf{R}} \theta_{ki}(t-1)\cdot S_k(t-1) \qquad \forall\, i, t = 1,2,\ldots \tag{23}$$

$$\sum_{j\in\mathbf{R}} \theta_{ij}(t) = 1 \qquad \forall\, i, t = 0,1,2,\ldots \tag{24}$$

$$\theta_{ij}(t) \geq 0 \qquad \forall\, i,j,t\,. \tag{25}$$

The real decision variables are the flow allocation fractions $\{\theta\}$, implying that we may explicitly incorporate constraints (22) and (23) into supply functions $S_i(t,\theta)$ where the vector θ reflects the fact that the supply of vehicles at i at time t may depend on all flow allocation decisions made prior to t. Now the optimization problem may be written as:

$$\max_{\{\theta\}} G(\theta) = \sum_{t=0}^{P} \sum_{i \in R} \sum_{j \in R} g_{ij} (\theta_{ij}(t) \cdot S_i(t, \theta)) \tag{26}$$

subject to (24) and (25). Equation (26) is a concave, *nonseparable* objective function with a very simple constraint set. Taking advantage of the acyclic structure of the network, simple recursions can be developed for taking derivatives using equations of the form:

$$\frac{\partial G}{\partial \theta_{ij}(t)} = \frac{\partial g_{ij}(\theta_{ij}(t) \cdot S_i(t,\theta))}{\partial \theta_{ij}(t)} + \frac{\partial G}{\partial S_j(t+1)} \cdot \frac{\partial S_j(t+1)}{\partial \theta_{ij}(t)} \tag{27}$$

and

$$\frac{\partial G}{\partial S_i(t)} = \sum_{j \in R} \frac{\partial g_{ij}(\theta_{ij}(t) \cdot S_i(t,\theta))}{\partial S_i(t)} + \frac{\partial G}{\partial S_j(t+1)} \frac{\partial S_j(t+1)}{\partial S_i(t)} . \tag{28}$$

Equations (27) and (28) are applied recursively going backwards in time. Using this approach, the Frank-Wolfe algorithm can be applied to (26) yielding a sequence of trivial subproblems. Powell et al. [21] showed that solving (26) using the decision variables $\{\theta\}$ was approximately five times faster than solving (20) using a more classical formulation based on link flows. The extra computational effort required to find the derivatives using (27) and (28) was more than offset by the speed with which the linearized subproblems could be solved. Particularly interesting is that the second formulation (26) was faster even in terms of iterations (that is, ignoring the time required to solve each iteration), requiring 7 iterations to reach the same objective function value the first approach required 18 iterations to reach.

Separate from the pure algorithmic issues, the use of the vector $\{\theta\}$ as decision variables provides a cleaner statement of the modelling assumptions. Specifically, we are asked to decide, in advance (that is, before we can see realizations of the demands $D_{ij}(t)$) the amount of flow that will be moved from i to j at time t. Then, when the demands are known, we are to move as many loads as we can. Excess vehicles move empty and excess loads are lost. The inability to reallocate excess trucks on one link to handle excess demand on another link defines a strategy with simple recourse and represents a very strong assumption. In the case of truckload trucking, motor carriers are highly responsive to real-time conditions. It is useful to consider, however, that while this model assumes a vehicle will move empty even if there is no load available, the deterministic model further assumes that the load will be available (that is, that the forecast will come true). The next model represents an attempt to extend the model to provide for a more flexible response.

3.2 The stochastic DVA with null recourse

The model with null recourse is defined as the strategy where if a vehicle cannot be moved loaded from i to j then it will simply be held until time period $t+1$, a kind of null response for the vehicle. Under this assumption, it is necessary to control the flow of loaded and empty vehicles separately. Define

$\alpha_{ij}(t)$ = fraction of the supply of vehicles at region i at time t which is allocated to be moved loaded from i to j,

$\beta_{ij}(t)$ = fraction of the supply of vehicles at region i, time t which is to be moved empty from i to j.

In this case, the flows of loaded and empty vehicles are given by

$$X_{ij}(t) = \min [\, \alpha_{ij}(t) \cdot S_i(t), F_{ij}(t) \,] \tag{29}$$

$$Y_{ij}(t) = \beta_{ij}(t) \cdot S_i(t) \qquad j \neq i\,. \tag{30}$$

The fractions $\alpha_{ij}(t)$ and $\beta_{ij}(t)$ must satisfy

$$\sum_{j \in R} [\, \alpha_{ij}(t) + \beta_{ij}(t) \,] = 1 \tag{31}$$

where generally $\alpha_{ii}(t) = 0$. Let $Y_{ii}(t)$ be the flow of vehicles held at i until $t+1$. This is given by the number of vehicles that were intentionally held at i plus the overflow from links where demand fell short of the allocated capacity. Thus

$$Y_{ii}(t) = \beta_{ii}(t)\, S_i(t) + \sum_{\substack{j \in R \\ j \neq i}} [\, \alpha_{ij}(t) \cdot S_i(t) - X_{ij}(t) \,]\,. \tag{32}$$

This simple recourse strategy implies that $X_{ij}(t)$ is a random variable (as is $X_{ij}(t) + Y_{ij}(t)$, unlike in the previous model), which in turn means that the supplies $S_i(t)$ are random. Equation 32 is the heart of the difference between the simple recourse model and the null recourse model. Under the simple recourse model, a truck might be "allocated" to move over a traffic lane with 30 percent probability of having a load, meaning a 70 percent probability of moving empty. There is no chance a carrier could operate profitably under such a strategy. The null recourse policy assumes that a truck allocated to move loaded will simply be held in a region if no load arises, thus avoiding the cost of moving empty. This is more accurate than the simple recourse approach but does not model the reality of possibly using a truck on one of multiple outbound traffic lanes, thereby increasing the probability it will move loaded somewhere.

The solution approach follows very closely that used to solve (26). This model was explored in depth by Powell [22]. This work was based in part on earlier work by Jordan [16] and Jordan and Turnquist [17] who considered the empty car distribution problem for railroads. While their model did not consider loaded movements, they did explicitly model stochastic supplies and the concept behind the allocation vectors $\{\alpha\}$ and $\{\beta\}$, while not explicitly stated, is implicit in their formulation.

Both the simple and null recourse models produce nonlinear programs that can be solved fairly efficiently even for large networks. Execution times are roughly 3 to 10 times that for comparable deterministic models (the execution time depends largely on how close to optimality the algorithm is run). Of course the models produce fractional solutions. The more serious problem is that even the null recourse formulation is not a realistic model of actual carrier behavior. Furthermore, while it can be optimized fairly efficiently, the null recourse model is already alarmingly complex due to the need to handle stochastic supplies of trucks. In terms of the mathematical formulation, there does not appear to be much room left for further relaxing the assumptions and still having a workable mathematical model. For this reason, the next section attempts to develop additional insights by formulating the problem as a classical Markov decision process.

4. THE DVA AS A MARKOV DECISION PROCESS

Having effectively run into a dead end with classical network formulations of the stochastic DVA, an alternative approach is to reformulate the problem as a Markov decision process (MDP). Here the simplifying assumptions, particularly that all travel times are one time period, significantly ease the presentation. Assume we have a fleet of K vehicles that may be distributed over R regions, and let $S_i(t)$ be the number of vehicles in region i at time t. If all the travel times between regions are one period, then $\mathbf{S}(t) = \{S_1(t), ..., S_R(t)\}$ completely defines the state of the system at time t. The problem is now to optimize the transitions from one transition to the next to maximize the average reward per time period. That is, if $R(t)$ is the profits earned in the transition from t to $t+1$, then we wish maximize $\lim_{s \to \infty} \sum_{t=0}^{s} E[\, R(t)\,] / s$. An appreciation of the structure of the problem is most easily developed for the case where the fleet $K = 1$ vehicle. After this, the approach is extended to the problem with $K > 1$. The purpose of this exercise is primarily to explore the structure of the decision variables under uncertainty. At the end of this section, we show how the flow splitting approach described in Section 3 can be viewed as a restricted form of the MDP structure.

4.1 The one vehicle MDP

With one vehicle, the state vector $S(t)$ can also be viewed equivalently as a vector giving the current location of the vehicle. Alternatively we may define a new state variable $\overline{S}(t) = i$ if the vehicle is in region i at time t. Now a transition from state i to state j is equivalent to a vehicle moving from region i to region j, earning a random "reward" of R_{ij} where

$$R_{ij} = \begin{cases} r_{ij} & \text{if the movement is made loaded} \\ -c_{ij} & \text{if the movement is made empty.} \end{cases}$$

The problem now is to find the structure of an optimal policy that will determine what a vehicle will do when faced with a set of realizations on the available loads. On a given day, when all the loads are known, the vehicle must accept one of these loads or reject all of them and move empty instead (holding in the region until the next time period is a special case of an empty move). Define:

Λ_{ij} = the decision to move loaded from i to j *if a load is available*, and

E_{ij} = the decision to move empty from i to j.

For a vehicle in region i, the set of all possible actions is given by $\{\Lambda_{ij}, E_{ij}, j \subset R = 1, ..., R\}$. Let Δ_i be the set of all possible permutations of this set and let $\delta_i \in \Delta_i$. Thus, for a three region problem, we might have

$$\delta_1 = [\Lambda_{13}, \Lambda_{12}, E_{11}, \Lambda_{11}, E_{12}, E_{13}]. \tag{33}$$

δ_1 is a particular policy that says the vehicle (in region 1) should move loaded from 1 to 3, if there is a load available, otherwise it should move loaded from 1 to 2 and, failing this, it should move empty from 1 to 1. Since the vehicle can always do the empty option, this is the last option that need be considered.

For a given policy δ_i, there is a vector of rewards $\gamma_i = [\gamma_{i1}, \gamma_{i2}, ..., \gamma_{iN}]$, where $N = 2R$, that corresponds to the reward received if a particular option is used. Thus, for the example in (33) we would have

$$\gamma_1 = [r_{13}, r_{12}, -c_{11}, r_{11}, -c_{12}, -c_{13}]. \tag{34}$$

Next, there are probabilities associated with each option. Assume the options in δ_i are ordered from $n = 1, ..., 2R$, and let

$d_{in}(\delta_i(t))$ = probability the vehicle in i is dispatched on the nth option given a policy $\delta(t)$.

These probabilities can be easily worked out if we know the probability that a loaded option is not available (that is, the probability the number of loads from i to j is zero). Again, for the example in (33) we might have

$$\mathbf{d}_i(\delta_i(t)) = [\,.26\ \ .33\ \ .41\ \ 0\ \ 0\ \ 0\,] \tag{35}$$

where the first empty option receives all the remaining probabilities.

If $R_i(t)$ is the reward earned by a vehicle in region i at time t, then

$$E\,[\,R_i(t)\,] = \sum_{n=1}^{2R} \gamma_{in}(\delta_i(t))\, d_{in}(\delta_i(t)). \tag{36}$$

Assume we are maximizing expected profits $\bar{R}(P)$ over a planning horizon of length P. Our problem is to find a *strategy* $\delta = [\delta(0), \delta(1), ..., \delta(P)]$ that solves

$$\max_{\{\delta\}} \sum_{t=0}^{P} \pi(0) \cdot \mathbf{P}^t \cdot E\,[\,\mathbf{R}(t)\,] \tag{37}$$

where $\pi(0)$ is the initial state vector, $\mathbf{P}$ is the matrix of transition probabilities and $\mathbf{R}(t)$ is the column vector of rewards for being in each state (the convention is used that probabilities are row vectors and rewards are column vectors). The matrix $\mathbf{P}$ is easily derived from the dispatch probabilities.

The MDP in (37) is normally solved through a standard dynamic programming recursion. Let $W_i(t)$ be the optimal expected reward from time t until the end of the planning horizon given the process is in state i.

$$W_i(t) = \max_{\{\delta_i\}} E\,[\,R_i(t,\delta(t))\,] + \sum_{j\in R} p_{ij}(\delta(t)) \cdot W_j(t+1). \tag{38}$$

The recursion in (38) is easily solved. Let $q_i(n)$ denote the *destination* that is implied by the nth option in the vector δ_i, and define

$$w_{in}(t) = \gamma_{in}(\delta_i(t)) + W_{q_i(n)}(t+1). \tag{39}$$

thus $w_{in}(t)$ is the reward if option n is used, taking us to region $q_i(n)$. We can now rewrite (38) to be

$$W_i(t) = \max_{\{\delta\}} \sum_{n=1}^{2R} w_{in}(t) \cdot d_{in}(\delta(t)). \tag{40}$$

It is not hard to see that (40) is solved optimally by choosing a vector $\delta_i(t)$ which satisfies

$$w_{i1}(t) \geq w_{i2}(t) \geq \ldots \geq w_{iN}(t) \tag{41}$$

where as before $N = 2R$. Thus we need simply to rank the options in terms of their direct contribution plus optimal future earnings, thereby maximizing the probability the vehicle will be dispatched on the most lucrative options.

This section describes an optimal algorithm for the finite planning horizon problem for a single vehicle. Under certain conditions, as P becomes large the optimal policy in the first period becomes independent of P (see, for example, Heyman and Sobel [14, pp. 125-138] and Bean and Smith, [2]). Since the state space is so small, the algorithm is easily implementable. Most importantly, the discussion provides insights into the structure of optimal policies that are used in the design of a heuristic in Section 5. Next we consider briefly the problems in extending the classical MDP framework to the K vehicle problem.

4.2 The K vehicle MDP

With K vehicles the state of the system must now be given by

$$S(t) = \{S_1(t), \ldots, S_R(t)\} \tag{42}$$

where $S_i(t)$, as before, is the number of vehicles in region i at time t.

This immediately introduces a problem in terms of the size of the state space. Let S be the set of all possible states. It can be shown that for a problem with K vehicles and R regions that

$$|S| = \binom{K+R-1}{R-1}. \tag{43}$$

For a small fleet we might have $K = 100$ and $R = 50$ giving $|S| = 6.7 \times 10^{39}$, while with a full sized fleet with $K = 2000$ and $R = 100$ we get, using Stirling's approximation $|S| \approx 10^{83}$. Even a toy problem with $K = 20$ vehicles and $R = 10$ gives $|S| = 10^7$. Problems smaller than this are not even interesting for testing purposes.

Aside from the size of the state space, the determination of strategies, rewards and transition probabilities, the problem becomes significantly more complex when $K > 1$. It is useful, nonetheless, to at least formulate the structure of a policy variable in this context. With one vehicle, $\delta_i(t)$ represents a ranking of options to be considered by a vehicle in region i at time t. With K vehicles, we need a policy for each vehicle in each region. Consider a system in *state* $s = \{S_1(t), S_2(t), ..., S_R(t)\}$ and a particular region i. We would then have a policy $\delta_{sik}(t)$ which is identical to (33) but would apply to the kth vehicle in region i when the system is in state s, where $0 < k \le S_i(t)$. Now a vehicle moving from i to j no longer implies a transition from state i to j, and as a result the value of sending a truck from i to j, which earlier we could represent as $w_{in}(t)$ (Equation 39 above), is no longer so easily derived. This in turn takes away our optimal policy of ranking options so that (41) is satisfied.

If the MDP could be solved, the outputs would be identical to those from a network model in terms of instructions for the first time period. Given the deterministic supplies of trucks and deterministic opportunities, each dispatch probability vector will be of the form $d_{ikn}(1) = [1, 0, 0, ...]$ since there will always be a "best" option (given the policies for the other vehicles) for each truck, and this option will be available with probability 1. Thus there is a specific instruction for each vehicle in the first period and a much richer set of strategies in the future (these strategies, however, are not directly implemented).

The K vehicle problem quickly gives us a very large state space, a significantly larger space of possible policies and finally eliminates the special structure we used to determine an optimal policy. Further research may yet reveal structure that will yield a workable algorithm, but at this point this line of investigation does not look promising. The exercise does, however, yield insights into the structure of the problem. The strategy vectors $\delta_{sik}(t)$ for each vehicle k in region i with the system in state s is a relatively general mechanism for controlling the flow of vehicles between regions. A restriction of this formulation would use a policy vector of the form $\delta_i(t)$ as given by (33). In this case, we are forcing all vehicles to be dispatched under the same policy independent of the state or the number of vehicles in the system. Note that such a restriction implies that all empty vehicles move in the same direction. Let $d_{ikn}(\delta_i(t))$ be the probability the kth vehicle in region i is dispatched on the nth option. Let $\eta_{in}(x)$ be the fraction of vehicles dispatched on the nth option given that there are $x = S_i(t)$ vehicles in the region, where

$$\eta_{in}(x) = \frac{1}{x} \sum_{k=1}^{x} d_{ikn}(\delta_i(t)) . \tag{44}$$

Clearly $\eta_{in}(x)$ is a nonlinear function of x. Contrast this with the flow splitting variables $\{\theta\}$ or $\{\alpha, \beta\}$ used earlier where the fraction of flow moved over a particular option is controlled directly as decision variables. The comparison is illustrated in Table 1. Assume we are using a fixed policy vector $\delta_i(t) = \{\Lambda_{i3}, \Lambda_{i2}, \Lambda_{i5}, \Lambda_{i1}, E_{ii}\}$ for all vehicles (we could also use a different vector $\delta_{ik}(t)$ for each vehicle k). Assume that the first four dispatch probability vectors d_{i1}, d_{i2}, d_{i3} and d_{i4} corresponding to $\delta_i(t)$ are as shown in Table 1. Using (44) the vectors $\eta_i(x)$ are also calculated for $x = 1, 2, 3$ and 4. Thus the fraction of vehicles dispatched loaded from region i to region 5 is 0.10 if there is $x = 1$ truck in region i. This changes to 0.11 if $x = 2$, 0.16

if $x = 3$ and 0.22 if $x = 4$. As one might expect, the fraction of trucks moving loaded to region 5 varies as a function of the number of trucks in the region. We may alter these probabilities by manipulating the policy vector $\delta_i(t)$ (or the individual policy vector $\delta_{ik}(t)$ for $k = 1,...,4$). Under the simple recourse strategy, we may change θ_{i5}, thereby directly changing the *total* number of trucks moving from i to 5. Furthermore, this fraction is statistically independent of $S_i(t)$, the supply of trucks in i at time t. The fraction of trucks actually moving loaded is a nonlinear function of $S_i(t)$ which must be worked out from (16). This function, however, will be quite different from $\eta_i(x)$, particularly when one compares the fraction of trucks moving loaded somewhere as a function of x. Under the null recourse strategy, we may use d_{i5} to directly control the fraction of trucks allocated to move loaded from i to 5. This fraction is also independent of $S_i(t)$, although the fraction moving loaded will again be a (strictly decreasing) function of $S_i(t)$. The biggest difference between the simple and null recourse formulations and the MDP formulation is that the latter will yield a much higher probability that a truck will move outbound loaded to some destination.

Table 1
Illustrative dispatch probabilities for a fixed policy vector $\delta_i(t)$ and flow fractions $\eta_i(x)$

Option	Policy $\delta_i(t)$	d_{i1}	d_{i2}	d_{i3}	d_{i4}	$\eta_i(1)$	$\eta_i(2)$	$\eta_i(3)$	$\eta_i(4)$
1	A_{i3}	0.5	0.35	0.15	0.05	0.50	0.43	0.33	0.26
2	A_{i2}	0.4	0.50	0.45	0.30	0.40	0.45	0.45	0.41
3	A_{i5}	0.1	0.12	0.25	0.40	0.10	0.11	0.16	0.22
4	A_{i1}	0.0	0.0	0.10	0.15	0.0	0.0	0.03	0.06
5	E_{ii}	0.0	0.0	0.05	0.10	0.0	0.0	0.02	0.04

It is important in the design of an efficient algorithm that the inherent network structure of the problem be recognized. The enumeration of all possible states required by the MDP approach loses this structure, but it does at least provide a rigorous framework for formulating the problem. Most important is the structure of the policy variables used to handle decision making under uncertainty, although the problem now is to determine how to optimize over the set of possible strategies. In the next section we consider an approach that combines the basic ideas used in the dynamic network formulation in Section 2 with the recourse structure developed in this section.

5. STOCHASTIC PROGRAMMING FORMULATION OF THE DVA

A somewhat different literature has evolved under the general heading of stochastic programming with recourse. Although this literature is addressing the same types of problems as those solved using the MDP framework, the language and orientation is somewhat different. In this section, we present the stochastic DVA as a stochastic programming problem with recourse. We then show how the two time period problem (referred to in this literature as a two stage problem) can be solved optimally as a pure network, and from this we present a simple approximation for the n-stage problem which can also be solved as a pure network.

5.1 Background

Stochastic programming has enjoyed a fairly rich literature since the initial work by Dantzig [8], with important recent contributions by Birge and Wets [4], Birge and Wallace [5], and Wets [27, 28, 29], among others. Most recently Wallace [26] deals directly with stochastic programming problems arising in networks, which forms the basis for the discussion here.

As before, let $\mathbf{x}(t)$ and $\mathbf{y}(t)$ be vectors of loaded and empty flows for time period t, with coefficient vectors $\mathbf{r}$ and $\mathbf{c}$, and let $\mathbf{F}(t)$ be the vector of random demands between regions. We also let $\mathbf{L}(t)$ be the vector of the number of known loads moving between regions. For the discussion here let $\mathbf{L}(t) = 0, t \geq 1$, meaning that we only know the loads in the first time period. Our problem can now be stated as:

$$\max \ \mathbf{r}^T \mathbf{x}(0) - \mathbf{c}^T \mathbf{y}(0) + Q(\overline{\mathbf{x}}, \overline{\mathbf{y}}) \tag{45}$$

$$\sum_{j \in R} [x_{ij}(0) + y_{ij}(0)] = T_i(0) \qquad \forall i \tag{45a}$$

$$\sum_{j \in R} [x_{ji}(t-1) + y_{ji}(t-1)] = \sum_{k \in R} [x_{ik}(t) + y_{ik}(t)] \quad t = 1, 2, \ldots, \qquad \forall i \tag{45b}$$

$$x_{ij}(t) \leq U_{ij}(t) \qquad \forall i, j, t, \tag{45c}$$

$$x_{ij}(t), y_{ij}(t) \geq 0 \qquad \forall i, j, t. \tag{45d}$$

$Q(\overline{\mathbf{x}}, \overline{\mathbf{y}})$ gives total expected profits from time periods $t = 1, \ldots, P$ given the decisions $\overline{\mathbf{x}} = \mathbf{x}(0)$ and $\overline{\mathbf{y}} = \mathbf{y}(0)$ made in time period 0. $Q(\overline{\mathbf{x}}, \overline{\mathbf{y}})$ is the recourse problem and can be stated as follows. Let $\mathbf{F} = \{\mathbf{F}(1), \mathbf{F}(2), \ldots, \mathbf{F}(P)\}$ be the random vector of all loads in the future and let $Q(\overline{\mathbf{x}}, \overline{\mathbf{y}} \mid \mathbf{F})$ be the conditional expected profits given $\mathbf{F}$. Then $Q(\overline{\mathbf{x}}, \overline{\mathbf{y}} \mid \mathbf{F})$ is given by

$$Q\,(\bar{\mathbf{x}}, \bar{\mathbf{y}} \mid \mathbf{F}) \;=\; \max \sum_{t=1}^{P} [\, \mathbf{r}^T \mathbf{x}(t) \;-\; \mathbf{c}^T \mathbf{y}(t) \,] \tag{46}$$

subject to (45b), (45c) and (45d), where $U_{ij}(t) = F_{ij}(t)$ in 45c. Taking expectations gives:

$$Q\,(\bar{\mathbf{x}}, \bar{\mathbf{y}}) \;=\; E_{\mathbf{F}}\,\{Q\,(\bar{\mathbf{x}}, \bar{\mathbf{y}} \mid \mathbf{F})\}\,. \tag{47}$$

The difficulty with this approach is that the sample space for $\mathbf{F}$ is extremely large and each outcome involves solving the network transshipment problem in (46). For a two stage problem Wallace proposes replacing $Q\,(\bar{\mathbf{x}}, \bar{\mathbf{y}})$ with a series of cuts which would each successively produce an improved bound.

A standard technique for solving stochastic programming problems is to use a simple recourse strategy. To state this in more classical terms, let $z_{ij}(t) = x_{ij}(t) + y_{ij}(t)$ be the total flow from i to j as is done in Section 3. Now replace constraint (45c) with

$$z_{ij}(t) \;+\; z_{ij}^{+}(t) \;-\; z_{ij}^{-}(t) \;=\; F_{ij}(t) \tag{48a}$$

$$z_{ij}(t),\, z_{ij}^{+}(t),\, z_{ij}^{-}(t) \;\geq\; 0 \tag{48b}$$

where $z_{ij}^{+}(t)$ is the underage (the extent to which flow falls below demand) and $z_{ij}^{-}(t)$ is the overage, representing empty vehicles. Thus $x_{ij}(t) = z_{ij}(t) - z_{ij}^{-}(t)$ and $y_{ij}(t) = z_{ij}^{-}(t)$. In our problem there is no penalty for underage and the overage cost is the cost of moving empties. The notion of simple recourse is that we decide on the entire vector $\mathbf{z}(t), t = 0, 1, ..., P$, and then use the recourse variables $\mathbf{z}^{+}$ and $\mathbf{z}^{-}$ to respond to uncertainties. Defining the decision variables in terms of $\{z\}$, the recourse function becomes

$$Q\,(\mathbf{z} \mid \mathbf{F}) \;=\; \max_{\{\mathbf{z}^{-}(t), t \geq 1\}} \sum_{t=1}^{P} \mathbf{r}^T\,[\,\mathbf{z}(t) - \mathbf{z}^{-}(t)\,] \;-\; \mathbf{c}^T\,\mathbf{z}^{-}(t) \tag{49}$$

subject to (48a) and (48b). Since the vector $\{\mathbf{z}^{-}\}$ is not constrained by network constraints, we clearly wish to minimize $\mathbf{z}^{-}$ implying that (49) is solved by

$$z_{ij}^{-}(t \mid z_{ij}(t), F_{ij}(t)) \;=\; \max\,[\, z_{ij}(t) \;-\; F_{ij}(t), 0\,]\,. \tag{50}$$

Let

$$g_{ij}(z_{ij}(t)) \;=\; E\,[\, r_{ij}(z_{ij}(t) - z_{ij}^{-}(t)) \;-\; c_{ij}\, z_{ij}^{-}(t)\,] \tag{51}$$

where $z_{i\bar{j}}(t)$ is now given by (50). The recourse function is then

$$Q(\mathbf{z}) = \sum_{t=1}^{P} \sum_{i \in R} \sum_{j \in R} g_{ij}(z_{ij}(t)). \tag{52}$$

The function $g_{ij}(z_{ij}(t))$ is identical to that given in (19). Combining (52) and (45) gives the same nonlinear network formulation as in (20), although the functions in the first period are linear. In other words, stochastic programming with simple recourse in the context of the DVA is equivalent to deciding in advance how many vehicles to send from i to j at time t, and then sending that number anyway when the demand is known, incurring lost revenues or additional costs as required.

To gain an appreciation of how poor simple recourse performs as a model, consider the following situation typical of a large motor carrier. Assume from a given region i that the forecasted number of loads to each of 50 destinations is described by a Poisson distribution with a mean of .1 loads. If there is a single vehicle in region i and the decision is made to move it to region j, (that is, $z_{ij} = 1$), then the probability the vehicle will move empty is $Prob[F_{ij} = 0] = .1^0 e^{-.1} / 0! = .905$. In reality, the carrier might adapt a strategy to accept the best available load. Let $F_i = \sum_{j \in R} F_{ij}$ be the total outbound demand. In the latter strategy the probability that the vehicle will have to move empty (or hold) is $Prob[F_i = 0] = 5^0 e^{-5} / 5! = .000056$. This strategy can be represented using a multiple stage formulation where the decision maker is allowed to respond to the opportunities available to him.

In the next section we show how the two stage DVA can be solved optimally as a pure network.

5.2 The two stage DVA

Remembering that each stage of the DVA is just a transportation problem, it is easy to see that dispatching vehicles out of region i is independent of dispatching out of any other region. Note that this is purely a consequence of the two stage model since in an n stage model the downstream effects of dispatching vehicles implies that dispatching decisions out of region i must be coordinated with those out of all other regions. The two stage problem can be represented using (45) to (47) with $P = 1$.

Begin with the recourse function $Q(\bar{\mathbf{x}}, \bar{\mathbf{y}})$. The solution of (46) can be represented using the same approach used to formulate the problem as an MDP. For trucks in region i, the optimal policy is given by

$$\delta_i(1) = [\Delta_{ik_1}, \Delta_{ik_2}, \ldots, \Delta_{ik_R}, E_{ii}, \ldots] \tag{53}$$

where $r_{ik_1} \geq r_{ik_2} \geq \ldots \geq r_{ik_R}$, and where we assume that the cost of holding the vehicle in the region, c_{ii} , is less than the cost of moving it empty to any other region. The policy vector in (53) states that we will use a vehicle on the available load with the highest revenue and, if none are available, the vehicle will be held in that region. Since we are only considering a two stage problem, the policy in (53) is clearly optimal for all vehicles in region i, whereas in the P-stage problem we would require the much more complex formulation outlined in Section 4.2 for the N vehicle MDP.

Given the simple structure of the second stage problem, we now need to find $Q(\overline{\mathbf{x}}, \overline{\mathbf{y}})$. Let

s_i = supply of vehicles in region i at the beginning of the second stage,

$= \sum_{k \in R} [\, x_{ki}(0) + y_{ki}(0)\,]$,

$Q_i(s_i)$ = expected profits from vehicles dispatched out of region i given a supply of s_i .

Clearly

$$Q(\overline{\mathbf{x}}, \overline{\mathbf{y}}) = \sum_{i \in R} Q_i(s_i) . \tag{54}$$

Thus we have to find $Q_i(s_i)$. Recall that $d_{ikn}(\delta_i(t))$ is the probability the kth vehicle in region i is dispatched on the nth option, and that $\gamma_{in}(\delta_i(t))$ is the value of the nth option. Define

$v_{ik}(\delta_i(t))$ = value of the kth vehicle in region i.

Then

$$v_{ik}(\delta_i(t)) = \sum_{n=1}^{2R} d_{ikn}(\delta_i(t)) \cdot \gamma_{in}(\delta_i(t)) \tag{55}$$

and

$$Q_i(s_i) = \sum_{k=1}^{s_i} v_{ik}(\delta_i(t)) . \tag{56}$$

All that is left is determining the dispatch probabilities.

Let

$$\hat{F}_{in} = \sum_{l=1}^{n} F_{ik_l} \tag{57}$$

where $\hat{F}_{in}$ is the cumulative total number of loads in the *best* n options, $1 \leq n \leq R$. The event that the kth vehicle is dispatched on the nth option is equivalent to the joint event that $\hat{F}_{i,n-1} < k$ *and* $\hat{F}_{in} \geq k$. Thus

$$d_{ikn}(\delta_i(t)) = Prob\ [\hat{F}_{i,n-1} < k, \hat{F}_{i,n} \geq k\ ]$$

$$= Prob\ [\hat{F}_{i,n-1} < k\ ]\ -\ Prob\ [\hat{F}_{i,n} < k\ ]$$

where the second equality follows from the identity $P(A \cap B) = P(A) - P(\overline{B})$ when $\overline{B} \subset A$. Thus the dispatch probabilities reduce to the difference between two cumulative distributions. If a vehicle is not dispatched on one of the first n loaded options, then it is always moved on the first empty option which, following (53), means being "held" in region i until the next time period.

Having determined the recourse function, we can now consider the two stage optimization problem, given by (45). The recourse function $Q(\overline{\mathbf{x}}, \overline{\mathbf{y}})$ is separable in the variables s_i and from (56) we see that the function is piecewise linear. Thus (45) can be solved exactly as a pure network as indicated in Figure 5. The two stages are easily discerned in the network. The first stage, reflecting "known" loads and empty opportunities, forms a transportation problem and the second stage made up of "stochastic links" represents the value of each additional vehicle in a region.

It is useful to contrast the stochastic programming formulation with that based on Markov decision processes. Within the research literature, it is common to use one of the two approaches but apparently less common to compare the two directly. Stochastic programming uses traditional decision variables representing flows on vehicles. It also requires at least implicitly enumerating all possible outcomes of the random vector $\mathbf{F}$ and solving a network problem for each possible outcome F. The computational challenge has been finding the optimal first period flows without actually enumerating all the outcomes for the second period. The addition of a third stage appears to make the problem completely intractable since in principle every outcome of $\mathbf{F}$ in the second period still requires enumerating all possible outcomes of $\mathbf{F}$ in the third stage.

The Markov decision process formulation uses policies as decision variables where a single policy describes what decisions must be made for *all* possible outcomes of $\mathbf{F}$ in the second stage. Of course, a policy vector $\delta(t)$ is much more complex than a set of link flows $x(t)$ and $y(t)$. On the other hand, rather than enumerating all possible outcomes of $\mathbf{F}$, the MDP framework requires enumerating all possible states $\mathbf{S}(t)$. Since multiple outcomes of $\mathbf{F}$ can produce the same state $\mathbf{S}(t)$, the number of states is smaller (in fact, significantly so) than the number of outcomes of $\mathbf{F}$.

5.3 An approximation for the multistage DVP

This last section looks to combine ideas developed for the multistage *deterministic* DVA with the approach just presented for the two stage stochastic DVA presented in the previous

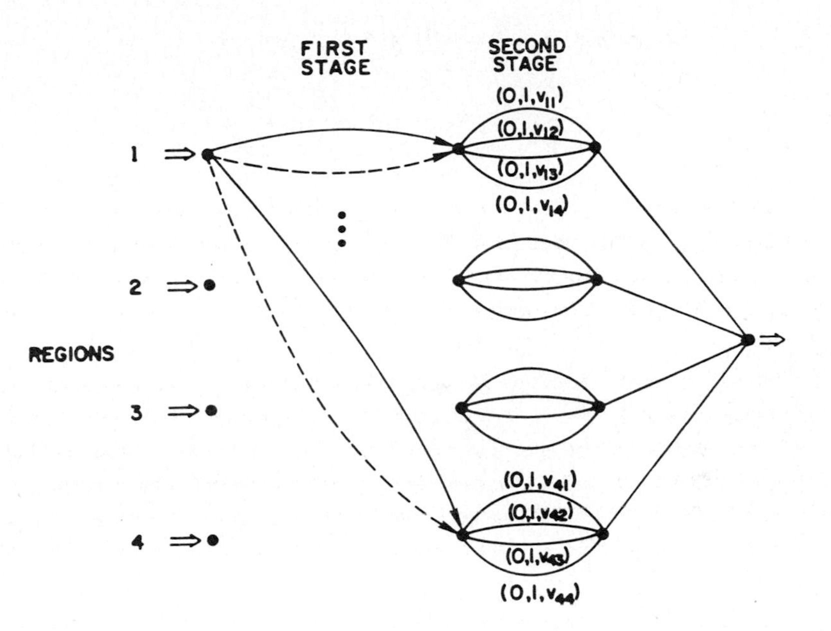

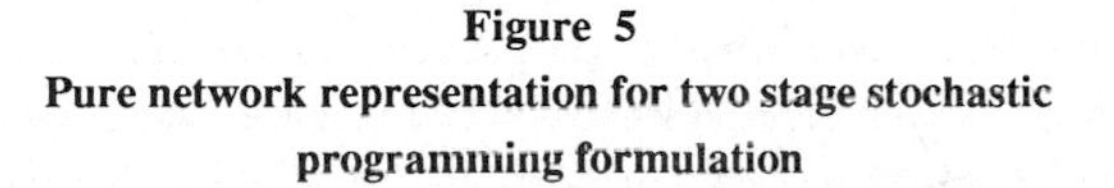

Figure 5
Pure network representation for two stage stochastic programming formulation

section. The ideas in this section were developed in Powell [23] and are the foundation of a vehicle dispatching system implemented at North American Van Lines (Powell *et al.* [24]). The development here, however, is considerably different and serves to synthesize ideas from dynamic networks, MDP's and stochastic programming.

The two stage problem involves two sets of decision variables: the policy vector $\delta(1) = \{\delta_1(1), \delta_2(1), \ldots, \delta_R(1)\}$ for the second stage, and flow variables $\mathbf{x}(0)$ and $\mathbf{y}(0)$ for the first stage. For the multistage problem, let $\mathbf{S}(t) = \{S_1(t), S_2(t), \ldots, S_R(t)\}$ be the state of the system given decisions $\{\mathbf{x}(0), \mathbf{y}(0)\}$ in the first stage and policies $\delta(1), \delta(2), \ldots, \delta(t-1)$ up until time t. Let $\Psi(t \mid \mathbf{S}(t))$ be the optimum expected profits from time t until the end of the planning horizon given a system starting in time t. Then

$$\Psi(t \mid S(t)) = \max_{\delta(t)} \{ \sum_{i \in R} \sum_{k=1}^{S_i(t)} \sum_{n=1}^{2R} d_{ikn}(\delta(t)) \, \gamma_{in}(\delta(t))$$

$$+ E_{S(t+1)} [\Psi(t+1 \mid S(t+1)) \mid S(t), \delta(t)] \} . \tag{61}$$

The expectation reflects the fact that $S(t+1)$ is a random variable, dependent on $S(t)$ and the policy $\delta(t)$. As we did with the deterministic DVA, we would like to replace $\Psi(t+1 \mid S(t+1))$ with a simpler approximation. As before, we will use

$$\hat{\Psi}(t+1 \mid S(t+1)) = \sum_{i \in R} p_i(t+1) \, S_i(t+1) . \tag{62}$$

where, as before, $p_i(t+1)$ is the salvage value giving the expected value of a vehicle in region i at time $t+1$ until the end of the planning horizon. Taking expectations of (62) gives

$$E [\hat{\Psi}(t+1 \mid S(t+1)) \mid S(t), \delta(t)] = \sum_{i \in R} p_i(t+1) \; E [S_i(t+1)] \tag{63}$$

where

$$E [S_i(t+1) \mid S(t)] = \sum_{i \in R} \sum_{k=1}^{S_i(t)} \sum_{n=1}^{2R} d_{ikn}(\delta(t)) . \tag{64}$$

Combining (61)-(64) yields

$$\Psi(t \mid S(t)) = \max_{\delta(t)} \left\{ \sum_{i \in R} \sum_{k=1}^{S_i(t)} \sum_{n=1}^{2R} d_{ikn}(\delta(t)) \; [\gamma_{in}(\delta(t)) + p_{q_i(n)}(t+1)] \right\} . \tag{65}$$

(65) is now very similar to (40) with $w_{in}(t) = \gamma_{in}(\delta(t)) + p_{q_i(n)}(t+1)$ and can be solved by choosing $\delta_i(t)$ for each region i so that (41) is satisfied. One important difference, however, is that we no longer can guarantee that all loaded moves will be ranked above all empty moves (also, the best empty move may not be to hold in a region until the next day). It is certainly intuitively more reasonable that some empty moves would be ranked above some loaded moves, but in our logic the highest ranked empty option within the vector $\delta_i(t)$ is the lowest option to receive any probability. This behavior is a direct consequence of the use of the linear approximation implicit in (62). We would like some vehicles to move empty to a given destination but we would like the model to recognize that there are declining marginal returns for each additional empty vehicle sent.

This problem can be mitigated heuristically. Let $U_{ij}(\delta(t))$ be a random variable denoting the maximum number of vehicles that we wish to allow to be used for the nth option. The choice of $U_{in}(\delta(t))$ must satisfy

$$U_{ij} \leq \begin{cases} F_{ij}(t) & \text{if option } n \text{ represents moving loaded region } j \\ \infty & \text{if option } n \text{ represents moving empty region } j\,. \end{cases}$$

If n is a loaded option, then it is natural (though not necessarily optimal) to use $U_{ij} = F_{ij}(t)$. If n is an empty option, one possible approximation is to use $U_{ij} = E_{ij}(t)$ where $E_{ij}(t)$ is a random variable denoting the *historical* number of empties that have been moved from i to j. This approach has been implemented and works quite well in practice though it imposes additional data requirements that can be hard to explain, as well as creating problems in certain situations.

Using the random variables $\{U_{in}(\delta(t))\}$, it is straightforward to work out the dispatch probabilities $\{d_{ikn}(\delta(t))\}$. Let the indices, $k_1, k_2, \ldots, k_{2R}$ rank the options W_{in} as in (41), and let

$$\hat{U}_{in} = \sum_{l=1}^{n} U_{ik_l}, \tag{66}$$

similar to (57). Then the same arguments leading to (60) gives

$$d_{ikn}(\delta(t)) = Prob\,[\,\hat{U}_{i,n-1} < k\,] - Prob\,[\,\hat{U}_{i,n} < k\,]. \tag{67}$$

The problem can now be solved in a manner similar to the two stage DVA. Let

$$v_{ik}(\delta_i(t)) = \text{the value of the } k^{th} \text{ vehicle in region } i \text{ under policy } \delta_i(t)$$

$$= \sum_{n=1}^{2R} d_{ikn}(\delta(t)) \cdot w_{in}(t)\,. \tag{68}$$

The recourse function $Q(\overline{x},\overline{y})$ can now be approximated as a separable, piecewise linear function. Let $\hat{Q}(S(t))$ denote this approximation. Then

$$\hat{Q}(S(t)) = \sum_{i\in R} \hat{Q}_i(S_i(t)) \tag{69}$$

where

$$\hat{Q}_i(S_i(t)) = \sum_{k=1}^{S_i(t)} v_{ik}(\delta_i(t))\,. \tag{70}$$

This approximation is only useful when the state vector $S(t)$ is known with certainty. The latest time period for which this is true is period $t = 1$. The complete optimization problem from the first stage onward is given by

$$\max \mathbf{r}^T \mathbf{x}(0) - \mathbf{c}^T \mathbf{y}(0) + \hat{\mathbf{Q}}(\mathbf{S}(1)) . \tag{71}$$

This can be solved using the same pure network shown in Figure 5. The flows on links in the first stage are given by $\{\mathbf{x}(0), \mathbf{y}(0)\}$ while the flows on links starting in the second stage into the supersink are given by $\mathbf{S}(1)$. The difference is that the coefficients on these links capture the approximate value of an additional vehicle starting at time $t = 1$ out to $t = P$.

A particularly important property of this formulation is that the salvage values $p_i(t)$ satisfy

$$p_i(t) \rightarrow \eta_i + \eta \cdot (P - t) \quad \text{as } P \rightarrow \infty \tag{72}$$

where η_i is a region specific factor (reflecting the relative value of a vehicle in one region over the other), and η is a system parameter giving the limiting expectation of the daily contribution of a vehicle. This result is a well known property of Markov reward processes (see, for example, Bhat [3]). Thus, for P sufficiently large, a further increase in P will increase the costs on all arcs leading into the supersink by a constant. Experiments reported in Powell [23] show that for a particular truckload motor carrier that (72) becomes quite accurate for $P > 10$ days. A simple recursion for calculating the values $\{p_i(s)\}$ is described in the appendix. The idea is that future activities will approximately follow recent history. Also, errors in the salvage values will be partially mitigated by the fact that the value of the kth truck in region i, $v_{ik}(\delta(t))$, is formed by a weighted average of salvage values, reducing the effect of an error in any one value.

6. SUMMARY AND CONCLUSIONS

The principal goal of this paper has been to expose a range of modelling frameworks and approximations. It was not possible to describe every variation, and attention was focussed on models which recognized the dynamic nature of the problem and uncertainties in forecasting. The large majority of the vehicle routing literature assumes a deterministic problem, and stochastic routing problems incorporate uncertainty in customer demands but not in the tours.

Table 2 provides an overview of the different modelling approaches and their characteristics in terms of problem size, ability to handle details about specific drivers and loads, and the resulting optimization problem and solution algorithm. Several of the models (1, 3, 4, 6, 7, 9 and 10) are single commodity network flow problems which are unable to handle details about

TABLE 2

Comparison of alternative modelling formulations

FORMULATION	PROBLEM SIZE[1]	LEVEL OF DETAIL	FORECASTING	OBJECTIVE FUNCTION	MODEL OUTPUT	SOLUTION ALGORITHM
1: TRANSPORTATION	1,600 LINKS 122 NODES	SINGLE COMMODITY	NONE, OR SIMPLE SALVAGE VALUES	LINEAR NETWORK	INTEGER LINK FLOWS	NETWORK SIMPLEX
2: ASSIGNMENT	7,500 LINKS 660 NODES	DRIVERS AND LOADS MODELLED EXPLICITLY	NONE, OR SIMPLE SALVAGE VALUES	LINEAR NETWORK	INTEGER DRIVER ASSIGNMENTS	NETWORK SIMPLEX
3: TRANSSHIPMENT INTEGER FORECASTS	15,000 LINKS 1,200 NODES	SINGLE COMMODITY	HEURISTIC GENERATION OF FORECASTER	LINEAR NETWORK	INTEGER LINK FLOWS	NETWORK SIMPLEX
4: TRANSSHIPMENT FRACTIONAL FORECASTS	30,000 LINKS 1,200 NODES	SINGLE COMMODITY	DETERMINISTIC FORECASTS	LINEAR NETWORK	FRACTIONAL LINK FLOWS	NETWORK SIMPLEX
5: ASSIGNMENT/ TRANSSHIPMENT[2]	22,500 LINKS 1,800 NODES	DRIVERS/LOADS MODELLED EXPLICITLY IN FIRST TIME PERIOD	HEURISTIC GENERATION OF FORECASTED LOADS	LINEAR NETWORK	INTEGER DRIVER ASSIGNMENTS AND LINK FLOWS	NETWORK SIMPLEX
6: STOCHASTIC NETWORK- SIMPLE RECOURSE	30,000 LINKS 1,200 NODES	SINGLE COMMODITY	STOCHASTIC DEMANDS	NONLINEAR, SEPARABLE NETWORK	FRACTIONAL LINK FLOWS	FRANK-WOLFE
7: STOCHASTIC NETWORK NULL RECOURSE	30,000 LINKS 1,200 NODES	SINGLE COMMODITY	STOCHASTIC DEMANDS	NONLINEAR, NONSEPARABLE NETWORK	FRACTIONAL LINK FLOWS	FRANK-WOLFE
8: MARKOV DECISION	10^{85} STATES[3]	SINGLE COMMODITY	STOCHASTIC DEMANDS	NONLINEAR	INTEGER FLOWS/ DISPATCH POLICIES	DYNAMIC PROGRAMMING
9: STOCHASTIC PROGRAMMING	10,000 LINKS 500 NODES	SINGLE COMMODITY	STOCHASTIC DEMANDS- HISTORICAL ACTIVITIES IN THE FUTURE	LINEAR NETWORK	INTEGER LINK FLOWS	NETWORK SIMPLEX
10: ASSIGNMENT/STOCHASTIC PROGRAMMING (N-STAGE HEURISTIC)	17,500 LINKS 1,100 NODES	DRIVERS/LOADS MODELLED EXPLICITLY IN FIRST TIME PERIOD	STOCHASTIC DEMANDS- HISTORICAL ACTIVITIES IN THE FUTURE	LINEAR NETWORK	INTEGER DRIVER ASSIGNMENTS AND LINK FLOWS	NETWORK SIMPLEX
11: SET PARTITIONING	30,000 TOURS 2,000 CON- STRAINTS	DRIVERS/LOADS MODELLED EXPLICITLY THROUGHOUT	HEURISTIC GENERATION OF FORECASTED LOADS	LINEAR PROGRAM	DRIVER TOURS	LAGRANGIAN RELAXATION

1: Problem sizes are approximately estimated based on 60 regions, 10 day explicit planning horizon (for transshipment networks) and assuming there are 300 drivers and loads actively being assigned.

2: Characterized by an assignment problem front-end and a deterministic transshipment network with integer forecasts.

3: Based on a 600 vehicle fleet.

specific drivers and loads. These details include equipment type, the precise deadhead distance from a driver to a load, precise time of availability of a driver and the precise pickup and delivery times of each load. The assignment model (2) is the only network flow formulation where individual drivers and loads are represented explicitly, allowing most details to be incorporated. The key limitation is its inability to perform these assignments in the context of any forecasting. On the other hand, the inability of network flow models to handle driver and load details is a major impediment to their practical implementation in the real-world.

Cape [6] combined the assignment model and the network flow model into a single network where each driver and load is explicitly represented by a node in the first time period. From a driver node, links might extend to five or ten known loads (which are within a reasonable distance and which are compatible in terms of equipment type and driver arrival and pickup times). In addition, there will be a link from a driver node to a node associated with the driver's region from the first time period, out of which forecasted opportunities are modelled. Flow from a driver node to a known load node flow forward in time ending in a node corresponding for the region and time period where the load terminates. Thus, details about an individual driver and load are retained for the first assignment and are then lost as the flows move into the future. This combined assignment/transshipment model is represented by formulations 5 and 10 in Table 2.

The last formulation, based on set partitioning, was not discussed in this paper but was included in the table for completeness. This approach, which has received widespread attention both in the research literature and in practice, requires generating feasible tours for each driver and then choosing the best set of tours so that each load is covered by one driver. This approach allows for a very high level of detail in representing drivers and loads both in the present as well as the future. The difficulty here is that an extremely large number of tours must often be generated, resulting in a very large integer programming problem, generally restricting their use to smaller private fleet operations. In addition, this approach does not lend itself well to forecasting uncertainties.

Among the 11 formulations, the Markov decision process approach is at this time restricted to toy problems, as would be exact N-period stochastic programming formulations. Aside from the MDP approach and the set partitioning approach, which will comfortably handle fleets of several hundred trucks, all the remaining formulations will easily handle fleets of several thousand trucks. This is true even of the simple and null recourse models, although they are somewhat slower due to their nonlinear nature. However, the fractional solutions produced by the nonlinear models, as well as the transshipment model with fractional forecasts, require some method to "integerize" the solution prior to implementation.

The remainder of this section reviews implementation issues involved with DVA models in general and the state of implementation in the industry. Finally, we review major research issues still facing the dynamic vehicle allocation problem.

6.1 Implementation of DVA models

Real-world implementations of dynamic fleet dispatching models are still quite few. The models are still evolving in terms of their ability to properly handle forecasting while also coping with the high level of detail that is required to manage a real operation. There are two applications of DVA models. The first is in a batch mode which might be run once or twice a day to determine repositioning strategies. This mode places less emphasis on specific real-time details and instead focuses attention on the broader pattern of surpluses and deficits. The second mode is for real-time driver dispatching, which places a very high emphasis on driver and load details so as not to make infeasible assignments. Using a network model for real-time driver assignments places much higher demands on the carrier's MIS system, which must be up-to-date at all times, and on the network model, which must be capable of detecting changes in the drivers or loads and then reoptimizing from a previous optimal solution within a few seconds.

The data requirements for all the DVA models are effectively the same, with the only exception being the assignment or transportation formulations which may not require any forecast information. Regardless of whether the model is being run in batch or real-time, a network model requires real-time data on drivers, trucks and loads pending. In addition, there is a set of base files which are used to forecast future activities as well as to provide information for calculating empty distances. When the model is run, it is necessary first to extract the status of each driver, including his estimated time of arrival, his destination and his equipment type (in some applications it is necessary to know his domicile and recent dispatching history). If there are different compensation rates for drivers, this will also be needed. Information about current pending loads includes origin and destination, pickup and delivery dates, equipment restrictions and compensation rates.

Base historical files required for forecasting include region to region average empty and loaded distances and travel times, empty movement costs and average contribution per load. Also needed is a set of models for forecasting loads over a 10 or 20 day planning horizon. These models typically work on a one year base of data supplemented by recent activities. Historical loaded contributions (revenue minus direct operating costs) and travel times reflect actual delivery costs (including special handling charges and extra drop-off costs) and additional times resulting from multiple stops.

The biggest hurdle facing most carriers is the lack of an up to date MIS system that both retains the necessary historical data as well as being able to provide current extracts of drivers and loads. Second to this is the traditional hesitation of management to accept help in day to day operations. Just the same, three carriers (to this author's knowledge) are actively using network models for fleet management. The first to do so uses a dynamic transshipment model (formulation 3 in Table 2) each night to plan general fleet movements. Then, an assignment model is used in real-time to perform detailed assignments of drivers to loads. This application is not reported in the research literature and no quantified estimates of impacts are available. However, the system has been in use for over six years suggesting that management is quite happy with the results.

More recently, the author was involved in the installation of a package called LOADMAP, which is an algorithm based on the N-stage stochastic programming heuristic, at North American Van Lines' Commercial Transport division. This implementation is described in Powell *et al.* [24]. To estimate the value of the package, a simulated game was conducted where two teams of six dispatchers each, made up of upper management from North American Van Lines, competed against each other. The performance of these teams was then compared to the performance of the network model. The model outperformed the best of the two teams by 12 percent, with 43 percent fewer refused loads (loads the carrier was unable to carry), 15 percent fewer empty miles and 6 percent higher revenues. This package is being used by two motor carriers and is run in batch approximately three to six times per day.

6.2 Directions for further research

The challenge facing the dynamic vehicle allocation problem is one of developing a computationally feasible algorithm which incorporates planning uncertainties. For the most part stochastic considerations have been largely ignored within the vehicle routing literature. At the same time, the stochastic optimization literature has not progressed very far in terms of handling large problems.

Some of the research directions that are of highest priority include the following:

1. We do not have a rigorous formulation of the stochastic DVA. Sections 4 and 5 of this paper provide a foundation for the structure of the decision variables, but this presentation needs to be firmed up considerably.
2. Can the special structure of the problem be exploited to provide a computationally feasible, optimal solution to the MDP or stochastic programming formulations? The one-vehicle MDP and the two stage stochastic program with network recourse provide glimpses of what is possible here.
3. Does a planning horizon exist, where the optimal policy for time period $t = 0$ based on a P period horizon is optimal as $P \rightarrow \infty$? Recent research in this area has established conditions for planning horizons, and these should be investigated.
4. Can bounds be developed to evaluate the efficiency of heuristics? It is likely that a bound for a medium to large problem will be more useful than an optimal solution for a very small problem.

Separate from basic theoretical issues are a range of more experimental research topics. These include:

5. The development of the software to rigorously test alternative heuristics. Research *is* progressing in this area and has exposed a variety of important experimental design questions.

6. The most obvious question is, of course, how well do the different formulations actually perform in a rigorous test environment? It is possible, for example, that the simple transportation network in Figure 2 will perform adequately. Actual performance may easily depend on the degree of uncertainty.

APPENDIX

Both the deterministic transshipment networks and the N-stage stochastic programming heuristic make use of regional salvage values, $p_i(t)$, giving the expected net contribution of a truck in region i at time t until the end of a secondary planning horizon, P_s. This can be accomplished through a simple backward recursion. Assume we have available the following:

$g_{ij}(t)$ = forecasted number of trucks moving loaded from i to j at time t,

$e_{ij}(t)$ = forecasted number of trucks moving empty from i to j at time t.

This information can be obtained in two ways. First it is possible to use six months of actual historical activities which are worked into a set of weekly averages. The weekly cycle is then assumed to repeat itself indefinitely. This approach has actually been applied in practice but suffers from some important limitations. First of course is the fact that a six month rolling average does not necessarily forecast the future. Second, and actually more significant, is that from a practical perspective a carrier's database on empty activities, $e_{ij}(t)$, can be of low quality. Often there is no record of activities of trucks holding in a region, $e_{ii}(t)$, which must then be inferred from flow conservation equations. The principal advantage of the use of historical activities is that they represent actual activities and as such may provide a better prediction of actual future costs and revenues.

The second method for estimating loaded and empty activities is to develop a deterministic transshipment network model with a planning horizon P_s' that is substantially longer than P_s. The activities $g_{ij}(t)$ and $e_{ij}(t)$ are then just the optimal loaded and empty flows off this network. Given that integer solutions are not really necessary here, it is best to use fractional upper bounds on the loaded movement links. The advantage of this approach is that the loaded and especially the empty activities become true forecasts. The disadvantage is that the network is very large, since $P_s > P$, and $P_s' > P_s$, and because the use of fractional forecasts greatly expands the number of links required. Also, one is never sure that the network model is actually predicting future activities. Note that since this network model cannot use any salvage values, it will be necessary to use $P_s' \gg P_s$ to mitigate truncation effects. Once the loaded and empty activities are estimated, salvage values can be calculated as follows. Let

$$p_i(P_s) = 0 \quad \forall i . \tag{A.1}$$

Then, beginning with $t = P_s - 1$ and working backward in time, let

$$P_i(t) = \sum_{j \in R} \theta_{ij}(t) \cdot w_{ij}(t) - \alpha_{ij}(t) \cdot \overline{w}_{ij}(t) \tag{A.2}$$

where

$$w_{ij}(t) = \begin{cases} r_{ij}(P_s - t)/t_{ij} & \text{if } P_s - t \leq t_{ij} \\ r_{ij} + p_j(t + t_{ij}) & \text{otherwise} \end{cases} \tag{A.3}$$

and where $\overline{w}_{ij}(t)$ is defined similarly using the empty cost c_{ij} instead of the load contribution. Note that we are allowing the travel times, t_{ij}, to differ from unity. The fractions θ_{ij} and α_{ij} are the fraction of trucks moving loaded and empty, given by

$$\theta_{ij}(t) = \frac{g_{ij}(t)}{\sum_{l \in R} (g_{ie}(t) + e_{ie}(t))} \tag{A.4}$$

and

$$\alpha_{ij}(t) = \frac{e_{ij}(t)}{\sum_{l \in R} (g_{il}(t) + e_{il}(t))} \,. \tag{A.5}$$

Equation A.2 defines a backward recursion that is exceptionally fast and provides salvage values that are fairly robust with respect to errors in the estimates of the activity variables.

REFERENCES

[1] Baker, L., "Overview of Computer Based Models Applicable to Freight Car Utilization", Report prepared for U.S. Department of Transportation, NTIS, Springfield, VA (1977).

[2] Bean, J. C. and Smith, R. L., "Conditions for the Existence of Planning Horizons", *Mathematics of Operations Research* 9 (1984) 391-401.

[3] Bhat, U. N. *Elements of Applied Stochastic Processes* (John Wiley and Sons, New York, 1984).

[4] Birge, J. R. and Wets, R. J-B., "Designing Approximation Schemes for Stochastic Optimization Problems, in Particular for Stochastic Programs with Recourse", *Mathematical Programming Study* 27 (1986) 54-102.

[5] Birge, J. R. and Wallace, S. W., "A Separable Piecewise Linear Upper Bound for Stochastic Linear Programs", University of Michigan, Department of Industrial and Operations Engineering Working Paper, (1987).

[6] Cape, D. J., "MICROMAP: A Dynamic Dispatch and Planning Model for Truckload Motor Carriers", Bachelor's thesis, Department of Civil Engineering, Princeton University, (1987).

[7] Chih, K. C-K., "A Real Time Dynamic Optimal Freight Car Management Simulation Model of the Multiple Railroad, Multicommodity Temporal Spatial Network Flow Problem", Ph.D. dissertation, Department of Civil Engineering and Operations Research, Princeton University, (1986).

[8] Dantzig, G. B., "Linear Programming Under Uncertainty", *Management Science* 1 (1955) 197-206.

[9] Dejax, P. J., and Crainic, T. G., "Models for Empty Freight Vehicle Transportation Logistics", *Transportation Science* to appear (1987).

[10] Grinold, R., "Finite Horizon Approximations of Infinite Horizon Linear Programs", *Mathematical Programming* 12 (1977) 1-17.

[11] Grinold, R., "Convex Infinite Horizon Programs", *Mathematical Programming* 15 (1983) 64-82.

[12] Grinold, R., "Model Building Techniques for the Correction of End Effects in Multistage Convex Programs", *Operations Research* 31 (1983) 407-431.

[13] Grinold, R. and Hopkins, D. S. P., "Computing Optimal Solutions for Infinite-Horizon Mathematical Programs with a Transient Stage", *Operations Research* 21 (1973) 179-187.

[14] Heyman, D. and Sobel, M. J., *Stochastic Models in Operations Research Vol. 11*, (McGraw-Hill, New York, 1984).

[15] Hughes, R. E. and Powell, W. B., "Mitigating End Effects in the Dynamic Vehicle Allocation Model", forthcoming in *Management Science* (1985).

[16] Jordan, W. C., "The Impact of Uncertain Demand and Supply on Empty Rail Car Distribution", Ph.D. dissertation, Cornell University, Ithaca, New York, (1982).

[17] Jordan, W. C. and Turnquist, M. A., A "Stochastic, Dynamic Model for Railroad Car Distribution", *Transportation Science* 17 (1983) 123-145.

[18] Misra, S. C., "Linear Programming of Empty Wagon Disposition", *Rail International* 3 (1972) 151-158.

[19] Ouimet, G. P., "Empty Freight Car Distribution", Master Thesis, Queen's University, Kingston, Ontario (1972).

[20] Philip, C. E. and Sussman, J. M., "Inventory Model of the Railroad Empty Car Distribution Process", *Transportation Research Record* 656 Transportation Research Board (1977) 52-60.

[21] Powell, W. B., Sheffi, Y. and Thiriez, S., "The Dynamic Vehicle Allocation Problem with Uncertain Demands", *Ninth International Symposium on Transportation and Traffic Theory*, (1984).

[22] Powell, W. B., "A Stochastic Model of the Dynamic Vehicle Allocation Problem", *Transportation Science* 20 (1986) 117-129.

[23] Powell, W. B., "An Operational Planning Model for the Dynamic Vehicle Allocation Problem with Uncertain Demands", *Transportation Research* 21B (1987) 217-232.

[24] Powell, W. B., Sheffi, Y., Nickerson, K., Butterbaugh, K. and Atherton, S., "Maximizing Profits for Truckload Motor Carriers: A New Framework for Pricing and Operations", *Interfaces* to appear (1987).

[25] Shan, Yen-Shwin, "A Dynamic Multicommodity Network Flow Model for Real Time Optimal Rail Freight Car Management", Ph.D. dissertation, Department of Civil Engineering and Operations Research, Princeton University, (1985).

[26] Wallace, S. W., "Solving Stochastic Programs with Network Recourse", *Networks* 16 (1986) 295-317.

[27] Wets, R. J-B., "Solving Stochastic Programs with Simple Recourse", *J. Stochastics* 10 (1983) 219-242.

[28] Wets, R. J-B., "Stochastic Programming: Solution Techniques and Approximation Schemes" in: Bachem, A., Grotschel M., and Kurte, B. (eds.), *Mathematical Programming: The State of the Art* (Springer-Verlag, Berlin, 1983) 565-603.

[29] Wets, R. J-B., "Large Scale Programming Techniques in Stochastic Programming", Working Paper WP-84-90, IIASA, Austria, in: Ermoliev Y., and Wets, R. J-B. (eds.), *Numerical Methods in Stochastic Programming*, (1984).

[30] White, W. W. and Bomberault, A. M., "A Network Algorithm for Empty Freight Car Allocation", *IBM System Journal* 8 (1969) 147-169.

Vehicle Routing: Methods and Studies
B.L. Golden and A.A. Assad (Editors)

THE PROBABILISTIC VEHICLE ROUTING PROBLEM

PATRICK JAILLET

CERMA - Ecole Nationale des Ponts et Chaussees
La Courtine, B.P. 105
93194 Noisy-le-grand, France

AMEDEO R. ODONI

Operations Research Center, Massachusetts Institute of Technology
Room 33-404, M.I.T.
Cambridge, MA. 02139, U.S.A.

Probabilistic vehicle routing problems (PVRPs) and probabilistic traveling salesman problems (PTSPs) are important variations of their classical counterparts in which only a subset of potential customers need to be visited on any given instance of the problem. This subset is determined according to some given probability law. After describing a number of applications, several interesting properties of PTSPs and PVRPs are presented. The paper also includes brief discussions of heuristics for solving PTSPs, bounds and asymptotic results.

1. INTRODUCTION

The scholarly literature devoted to vehicle routing problems in a deterministic context has been growing rapidly over the last several years, as this volume attests (see also Bodin et al. [1983] with around 700 references!). By a deterministic context, we mean situations in which the number of customers, their locations and the size of their demands are known with certainty before the routes are designed. One can identify, however, a practically endless variety of problems in which one or more of these parameters are random variables, i.e. subject to uncertainty in accordance with some probability distribution. In fact, these problems, specified as they are in a probabilistic context, are often more applicable than their deterministic counterparts.

In this paper, we shall discuss a family of vehicle routing problems whose probabilistic aspect is a very simple and fundamental one: only a subset of all the potential customers need to be visited on any given instance of the problem, the subset being determined according to some known "probability law".

More specifically, consider a problem in which a package delivery company wishes to design a tour through n commercial customers and is only interested in minimizing route length. Assume that this tour is to be used for a prolonged period of time and that, over this time horizon, the set of customers to be visited on a daily basis varies. For instance, the probability of having to make a delivery to customer i $(1 \leq i \leq n)$ on a random day may be equal to p_i and independent of the probability of a delivery to any other customer. (This is an example of a "probability law".) We shall examine here the case in which the company will not be redesigning the tour every day (once the subset of customers to be visited that day is known) but will instead always follow the customer sequence as it appears in the predesigned tour that contains all n potential customers. In other words, if on a given day only $m (0 \leq m \leq n)$ customers must be visited, the delivery vehicle will simply skip the missing n-m customers and visit the other m in the same order in which they appear in the predesigned tour. (This is illustrated in Figure 1 for n=10.)

We are interested in finding a predesigned tour of minimum expected length through the n potential customers, where the expectation is computed over all possible instances of the problem. That is, given an a priori tour t, if problem instance k will occur with probability α_k and will require covering a total distance $r_{t,k}$ to visit the associated subset of customers, that problem instance will receive a "weight" of $\alpha_k r_{t,k}$ in the computation of the expected length. If we denote the length of the tour t by L_t (a random variable), then our problem is to find an a priori tour t*, through all n potential customers, which minimizes the quantity

$$E[L_t] = \sum_k \alpha_k r_{t,k} \tag{1}$$

with the summation being over all possible instances, k, of the tour.

Note that the traveling salesman tour (TST) through the n potential customers may not be the solution to our problem -- and, in fact, as will be seen later, will probably not be. The reason is that, not only must the predesigned tour be a "good" one (small route length) when all customers are present, but it must also remain "well-behaved" when some, possibly many, customers are skipped. There is no guarantee that a TST through all n potential points will have this desirable property.

By analogy to deterministic problems (where all customers are always visited), we shall refer to the problem described above as a probabilistic traveling salesman problem (PTSP) whenever the delivery vehicle's capacity imposes no constraints and as a probabilistic vehicle routing problem (PVRP) otherwise. Our objective is to review informally and with a minimum of mathematical formalism, some recently-obtained results on PTSPs and PVRPs. We begin in Section 2 by describing a number of applications of these

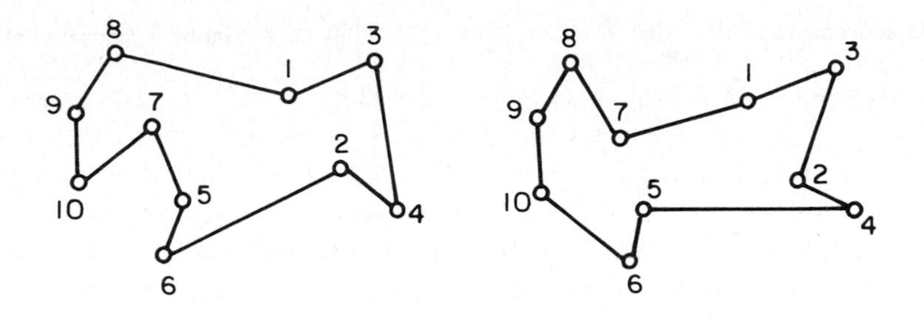

1.1 Two a-priori tours through the same set of points.

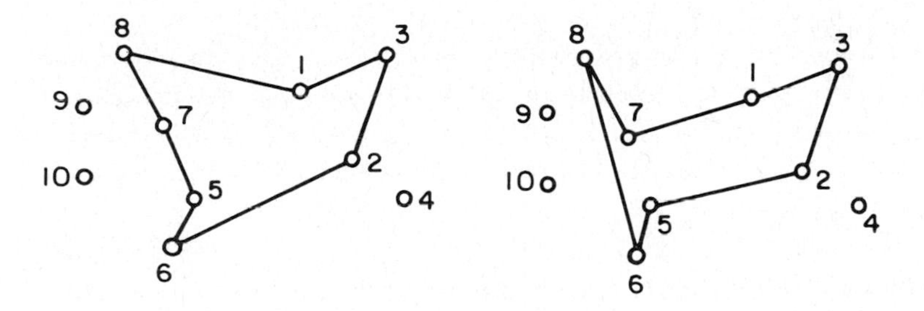

1.2 The two resulting tours when the points 4, 9, and 10 need not be visited.

Figure 1 Simple graphical example of a PTSP

problems in both "strategic" and "tactical" route planning. We then present our principal analytic results for PTSPs, namely how to compute efficiently their expected lengths and how to use such expressions to improve our understanding of the occasionally surprising characteristics of these problems and of their solutions (Section 3). Section 4 extends the results and observations of Section 3 to the case of capacitated vehicles (PVRP). The development of algorithms for solving non-trivial instances of PTSPs and PVRPs is still in its early stages. Some heuristics for this purpose will be discussed in Section 5, including the potential application of PTSP-oriented algorithms to solving standard traveling salesman problems (TSPs). Finally, Section 6 briefly reviews some bounds and some "asymptotic" results, i.e., results that, in practice, could be useful with problems that involve large numbers of customers under certain conditions. In addition to the value of the problem-specific results presented, we believe that this paper serves the more general purpose of alerting the reader to the rich possibilities of probabilistic routing problems, as well as to the new and hitherto unexplored theoretical and computational issues that such problems raise for future research.

2. EXAMPLES OF APPLICATIONS

The applications of PVRPs are many and varied. Of particular importance are those in the area of preliminary ("strategic") planning for collection, delivery and distribution services. We begin this section with two examples of this type.

2.1. Resource Planning

Consider once again a small package delivery-and-collection company XYZ which has decided to begin service in a particular area. XYZ has carried out a market survey and identified a number n of potential major customers who, during any given collection/distribution period, have a significant probability (e.g., more than 10%) of requiring a visit. XYZ now wishes to estimate the resources (vehicles, drivers, dispatching staff, etc.) necessary to serve these customers. Assume for simplicity that each potential customer i is assigned a probability p_i of requiring a visit on any given day, independently of any other customer. (At this stage of planning, it is most likely that all customers will be assigned the same probability p or that they will be classified into a small number of groups, with customers in group 1 having probability p_1 of requiring a visit, those in group 2 p_2, etc.; however, the techniques to be discussed in this chapter can be applied to problems with distinct p_i's for each customer, as well as to cases where the independence assumption does not hold.)

To address this planning problem, XYZ will wish to estimate approximately E[L], the expected amount of travel (e.g., in vehicle miles) that will be necessary on a typical day to serve the subset of the n customers and locations that will require a visit. (XYZ may, in addition, be interested in some related statistics, such as the probability that travel distance on a given day will exceed M miles.)

The techniques available in the VRP literature to date for addressing this "generic" problem are not particularly satisfactory. The exhaustive enumeration approach, for example, would call for designing minimum length tours for <u>all possible instances</u> that might occur and making the corresponding decisions. This, of course, makes little sense: in the example we described there are 2^n such instances (each customer may or may not be present) and, even for 40 potential customers, this would mean solving more than one trillion VRPs.

A second approach might be to use the celebrated limit law of Beardwood et al. [1959] which, informally stated, says that the expected length of a TSP tour through m points, uniformly and independently located in a square of area A, tends to $\beta\sqrt{mA}$ for large m, where β is a constant that depends on the travel metric ($\beta \approx 0.76$ and $\beta \approx 0.97$ for the Euclidean and right-angle metrics respectively). Since, in our case, the expected number of customers to be visited on any given day is equal to

$$W = \sum_{i=1}^{n} p_i \quad (= np, \text{ for } p_i = p \quad \forall\, i)$$

we could use $\beta\sqrt{WA}$ as our estimate of expected travel distance, where A is the area of the region that XYZ will service.

Unfortunately, <u>all</u> of the assumptions underlying this approximation may not be valid in this case and thus the estimate itself may be a very poor one indeed: the locations of the n potential customers may be far from uniformly and independently distributed within the area of interest -- and, in fact, there will likely be regions with a high concentration of customers and others which are very "sparse"; the travel metric will certainly not be purely Euclidean or right-angle or of any other single type; and, finally, the expected number of customers present, W, may not be sufficiently large for the limit theorem to hold in the first place (this is especially true when some of the p_i are small).

The recent results to be presented in this chapter, on the other hand, do offer a satisfactory way to deal with this problem. Let t denote the predesigned tour through all n potential customers which constitutes our solution to the PTSP for this case, i.e., t minimizes the expected length of the tour over all instances of the problem, if the order in which present customers are visited is maintained unchanged. Using the approach to be described in Section 3, we can then compute efficiently the expected length $E[L_t]$, in the PTSP sense, of the tour t. It should then be clear that $E[L_t]$ is a tight upper bound on the quantity of interest $E[L]$: for, if XYZ decides to always visit those customers present each day in the same predetermined order, then $E[L]=E[L_t]$; and, if instead XYZ wants to re-optimize its tour every day (after it is known which customers and locations must be visited that day), then $E[L]\leq E[L_t]$.

It is important to realize that the approach just described requires <u>no special assumptions</u> regarding the locations of customers, independence, travel metric or the number of actual or potential customers. The true locations, distances between any pair of customers, and probabilities of customer presence -- including possible dependences among customers -- can be used.

The weakest link in the approach is, in fact, its first step, i.e., finding the tour t which constitutes the solution to the n-point PTSP. At this time, we have no efficient exact algorithms for solving to optimality PTSPs for even quite modest values of n. However, we can use any one of a number of good heuristics (see Section 5), which will yield tours whose expected length $E[L_t]$ will provide the required upper bound for $E[L]$.

Once such an estimate of $E[L]$ is obtained, any further limitations imposed by vehicle capacity constraints (maximum distance that a vehicle can cover between successive returns to the depot, maximum number of stops allowable per vehicle tour, etc.) can be taken into consideration through what would essentially be a route-first-cluster-second procedure (see, for example, Bodin et al. [1983]).

Finally, it should be mentioned that questions such as the one concerning the probability that the distance traveled on a given day will exceed some critical value of M miles, can be addressed, at least approximately, using the PTSP and PVRP approach. However, any details on this topic are beyond the scope of this paper.

2.2. Facility Location

This is another "strategic" question that naturally arises within the scenario just described: Where should XYZ locate one or more depots to minimize $E[L]$? We assume, as before, that the locations of the potential customers and the corresponding probability law that governs the pattern of customer visit requirements are known "inputs".

2.3. Design of Actual Tours

There are a number of environments which are consistent with our PVRP model, even at the "tactical" level of everyday tour design. In such environments, tours are not "re-optimized" on a daily basis, even when information is available on which customers must actually be visited on any given day. Instead, those customers who are present are always visited in the same pre-determined sequence, i.e., on the basis of a tour designed a priori. The reason for not re-optimizing the tour on every problem instance could be that the system's operator does not have the resources for doing so; or, it may be decided that such re-design of tours is not sufficiently important to justify the required effort and cost, or the operator may have other priorities that could best be attained by following a PVRP-like strategy in vehicle routing -- such as achieving regularity and personalization of service by having the same vehicle and driver visit a particular customer every time. Examples in this category that have been described in the literature include a "hot meals" delivery system (Platzman and Bartholdi [1983]) routing of forklifts in a cargo terminal or in a warehouse and, interestingly, the daily delivery of mail to homes and businesses by Post Office mail carriers everywhere.

2.4. Non-Routing Contexts

PTSP-like models can also be of interest in many situations in which a sequence (ordering) of entities has to be found and that sequence has to be preserved even when some of the entities may be absent. One such example can be given from the area of job-shop scheduling: Consider the problem of loading n jobs on a machine in which a changeover cost is incurred whenever a new job is loaded. With any given ordering of the n jobs on the machine, we can then associate a total changeover cost. Any given ordering of the n jobs may also impose specific long-term requirements on the job shop, such as a set of tasks to be performed before and after the processing of the jobs on the machine. These requirements often are difficult to modify on a daily basis so that, if on a given day some jobs need not be processed, the relative ordering previously specified is nonetheless left unmodified. The models discussed in this chapter are relevant in analyzing such situations, as well.

3. THE UNCAPACITATED PROBLEM (PTSP)

To facilitate the presentation, we shall begin with a discussion of some results on the uncapacitated routing problem, i.e., the PTSP. For the same reason, we shall limit ourselves throughout this paper to the case in which each potential customer i has probability p_i $(0<p_i\leq 1)$ of requiring a visit

on any given instance of the problem (and $1-p_i$ of not requiring a visit) <u>independently</u> of all other customers. Cases in which some types of statistical dependence among customers may exist are discussed in Jaillet [1985].

3.1. The Expected Length of a Tour

Consider $n+1$ points of which point "0" denotes the depot where all routes originate and terminate and the other n are the locations of the customers. Let $d(i,j)$ denote the distance between points i and j. (It will be easiest to think of the travel environment as the Euclidean plane, but many of our results apply as well to all metrics as well as to cases where the points are located on a network with positive arc lengths.)

Let t represent a given tour through the $n+1$ points and let us re-index the customers in the order of their appearance along t by writing t as follows:

$$t = (0, 1, 2, \ldots, n, 0)$$

In the most general case, when $p_0=1$ and $p_i<1$ for all $i\neq 0$, the distance L_t (covered in traversing the set of points actually present in the order in which they appear in the <u>a priori</u> tour t) can take 2^n different values, the same as the number of different instances involving "present" and "absent" customers. For each such instance k, we would require $O(n)$ additions to compute the value $r_{t,k}$ of L_t ($r_{t,k}$ is the sum of the distances between the points present, visited in the order in which they appear in t). Thus, were we to use an enumeration approach, the computational effort would be $O(n\cdot 2^n)$, in order to compute $E[L_t]$ for a <u>given</u> tour t.

Fortunately, a much more efficient alternative exists.

Theorem 1: The expected length of t is equal to:

$$E[L_t] = \sum_{i=0}^{n} \sum_{j=0}^{n} \alpha_{ij}\, d(i,j) \tag{2}$$

where

$$\alpha_{ii} = 0 \qquad \forall\, i \in [0\ldots n] \tag{3a}$$

$$\alpha_{i,i+1} = p_i\, p_{i+1} \qquad \forall\, i \in [0\ldots n] \tag{3b}$$

$$\alpha_{ij} = p_i p_j \left(\prod_{k=i+1}^{j-1} (1-p_k) \right) \qquad \text{otherwise} \tag{3c}$$

Sketch of proof:

This result follows directly from the following argument: the instances of the problem for which the arc (i,j) is in the resulting tour [or equivalently $d(i,j)$ makes a contribution to the specific value $r_{t,k}$ then taken by L_t] are those for which the nodes i and j are present, while the nodes $i+1,\ldots,j-1$ are absent and thus skipped. The probabilities of such instances are simply the α_{ij}'s. #

Theorem 1 shows that $E[L_t]$ can be computed in $O(n^3)$ time under very general conditions. We have $O(n^2)$ terms α_{ij} and for each we need $O(n)$ elementary operations. For special cases, the computational effort can be even less. For example, when $p_i=p$ for all customers $(i=1,\ldots,n)$, then $\alpha_{ij}=\alpha_{gh}$ for all quadruplets (i,j,g,h) such that $j-i=h-g$. This means that we can save $O(n)$ effort in computing $E[L_t]$, i.e., the expected tour length can be obtained in $O(n^2)$ time. Other special cases of this type are discussed in Jaillet [1985].

3.2. Relationship Between PTSP and TSP

Theorem 1 provides an efficient method for computing $E[L_t]$ for any given tour t. Even more important, however, is the fact that (2) gives an explicit expression for the objective function in the PTSP. By analyzing and appropriately grouping together the terms of this expression, it is possible to derive certain properties (see Jaillet [1985] for a very extensive discussion of these properties) which, in turn, lead to an understanding of many interesting characteristics of PTSPs and PVRPs.

For example, it is natural to inquire about the links between solutions to the PTSP and to the TSP. In other words, how well would a TSP tour through the $n+1$ points in our problem do as a solution to the true problem, i.e. the PTSP? The answer is: potentially very poorly. Under some very special conditions the solutions to the PTSP and the TSP are identical (for example, when the $n+1$ points lie at the $n+1$ corner points of a convex polygon). In general, however, no assurance can be offered that this will be the case, except for trivially small problems, as indicated by the following:

Theorem 2: When the matrix of distances between points is symmetric, the optimum TSP tour is guaranteed to solve optimally the PTSP for problems with 4 or fewer points; when the matrix of distances is asymmetric, the optimum TSP tour is guaranteed to solve optimally the PTSP for problems with 3 or fewer points.

This theorem (for proof and examples see Jaillet [1985]) suggests that it may be necessary to devise new solution procedures specially designed for PTSPs and PVRPs, when addressing these and similar problems. The importance of doing so is underscored by our next result:

Let us denote by s the optimal PTSP tour for our $n+1$ point problem and by v the optimal TSP tour for the same $n+1$ points. Let us also define the quantity $W = \sum_{i=1}^{n} p_i$, i.e., W is the expected number of customers present on any given day (or "instance of the tour"). Then we have:

Theorem 3:

$$\frac{E[L_v] - E[L_s]}{E[L_s]} \leq \frac{n - W}{1 + W} \tag{4}$$

The left-hand side in this result (Jaillet [1985]) represents the error (as a fraction of the value of $E[L_s]$) that would result if the TSP solution, v, were used to solve the PTSP, instead of the true solution, s. The right-hand side gives an upper bound to this error. Note that when $p_i=1$ for all n customers, i.e., we have a TSP over n+1 points, then W=n and Theorem 3 states, correctly, that v is the optimal solution to the problem. When, however, the p_i's are generally small, i.e., the average number of customers present on any given day is small relative to the number n of potential customers, $(n-W)/(1+W)$ can be arbitrarily large, i.e., v can be a <u>very poor</u> solution to the PTSP. This can be best seen if we assume that $p_i=p$ for all i, in which case $(n-W)/(1+W) = n(1-p)/(np+1)$ and tends to n as p tends to zero. Jaillet [1986] has shown examples in which the TSP tour can indeed become an arbitrarily bad solution to the PTSP as p tends to zero. However, no examples exist so far to show that the bound in (4) is tight.

Example: These points can be illustrated through the 24-point example shown in Figure 2(i). In it we assume that all 24 points, representing the locations of the customers, have identical p_i's (i.e., $p_i=p$ for $i=1,...,24$) and we have not shown a location for the depot (i.e., the tours shown cover only the customer locations). For this configuration -- customers are located 12 each on two concentric circles and are evenly spaced on each circle with an offset of 15° between circles -- Figure 2(ii) shows on the left (tour v) the solution of the TSP and on the right an alternative star-shaped tour b. In a specific numerical example, the following were obtained:

For p=1(TSP): $E[L_v] = 19.561$; $E[L_b] = 19.704$.
For p=0.9: $E[L_v] = 19.193$; $E[L_b] = 17.846$.
For p=0.5: $E[L_v] = 16.110$; $E[L_b] = 12.286$.

Thus, while for p=1, the TSP solution, v, is slightly better than tour b, for p=0.9 and p=0.5, $E[L_v]$ is 8% and 31% greater than $E[L_b]$. Some thought will convince the reader that this is because tour b "takes better advantage" of missing points than tour v to reduce the distance that must be traveled on any given day. We could say that b is a good <u>a priori</u> tour for a wider range of circumstances (= probabilities of having to "cover" a customer) than tour v. Indeed, for problems in the Euclidean plane, tours

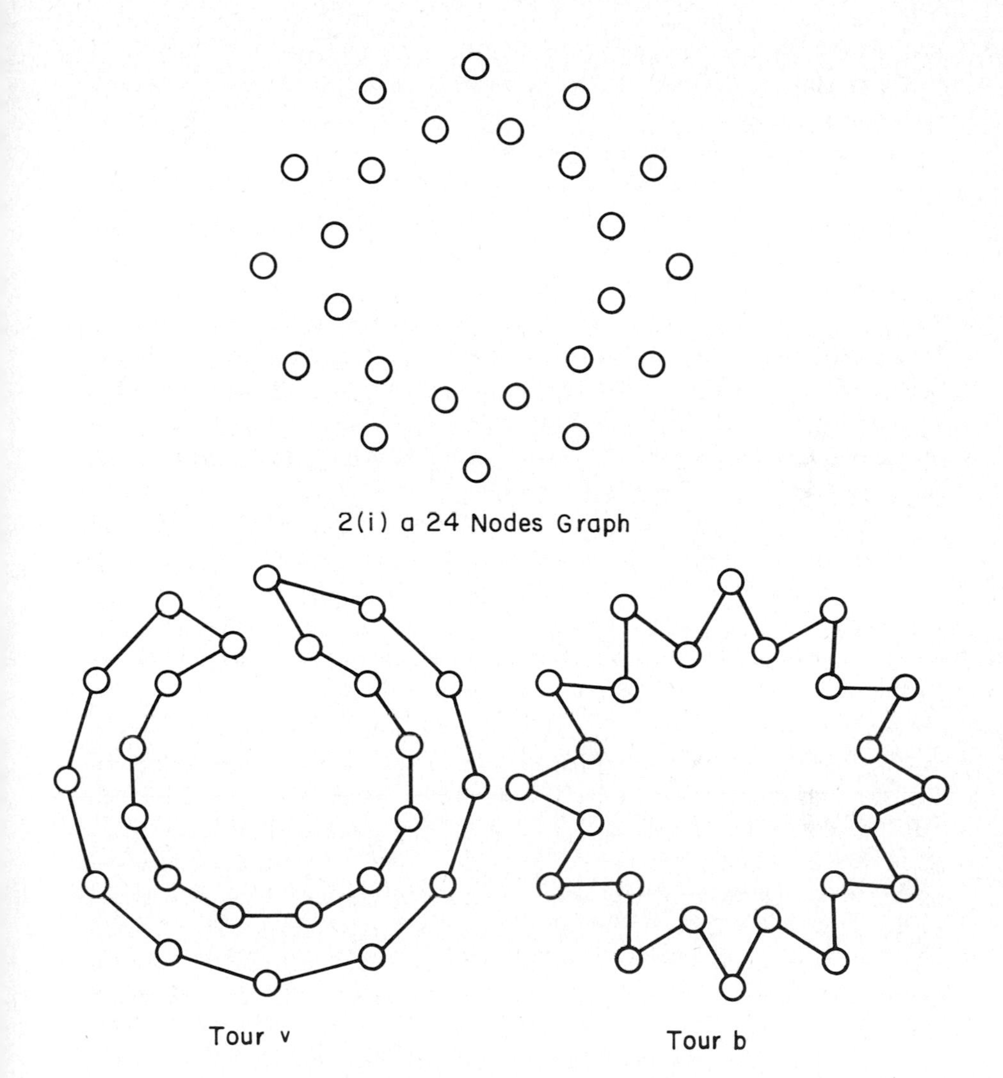

2(ii) Two Tours of the 24 Nodes Graph

Figure 2 Graph and Tours for the Numerical Example

that solve PTSPs are more and more likely to contain zig-zag and star-shaped patterns, like that of tour b, as the probability p of visiting individual customers decreases.

3.3. Two Other Observations on the PTSP

A couple of additional observations on the PTSP illustrate further how its probabilistic aspects induce some characteristics which are distinctly different from those of the TSP.

First, it is true that an optimal PTSP tour *may intersect itself* in the Euclidean plane! This, of course, is in sharp contrast with one of the first known properties of optimal TSP tours, namely that in a Euclidean metric the optimal tour does not intersect itself. As this property is often exploited by heuristic algorithms for solving the TSP in the plane (e.g.,see Larson and Odoni [1981] or Golden et al. [1980]), one must be careful when using (modified versions of) such algorithms with PTSPs.

Example: Consider the 5-point example of Figure 3(i). Points 3 and 5 are always present ($p_3=p_5=1$) while points 2,3,4 are "probabilistic", being present independently, each with the same probability $p (0<p\leq 1)$. For a particular set of Euclidean distances among them, it turns out that the tour v on the left in Figure 3(ii), which is also the TSP solution for the 5-point problem, is optimal for values of p greater than 0.25. However, for values of p less than 0.25, tour b on the right is optimal. To understand why one should recognize that there are exactly two instances for which the two *a priori* tours v and b result in different actual tours. One is the instance in which all 5 points are present. The probability of this happening is p^3 and the actual tours resulting from the *a priori* tours v and b are those shown in Figure 3(ii). For this instance v clearly enjoys an advantage over b. The other instance of interest is the one shown in Figure 3(iii). Now the points present are 2,3,4 and 5 and the probability of this instance is $p^2(1-p)$. But, in this case, the actual tours resulting from the *a priori* tours v and b are such that b enjoys an advantage over v, as can be seen from Figure 3(iii). (Remember that points are always visited in the same order as in the *a priori* tours).

For small values of p, the probability of the instance of Figure 3(iii) is much greater than that of Figure 3(ii) (for instance, 9 times greater for $p=0.1$). Thus, when averaged over all possible instances of the problem, i.e., over all possible actual tours, b turns out to be a better solution to the PTSP than v for small values of p. (The reader may wish to verify that for the other 6 possible instances of our example, i.e., for the ones not covered by Figures 3(ii) and 3(iii), the actual tours resulting from *a priori* tours v and b are identical.) #

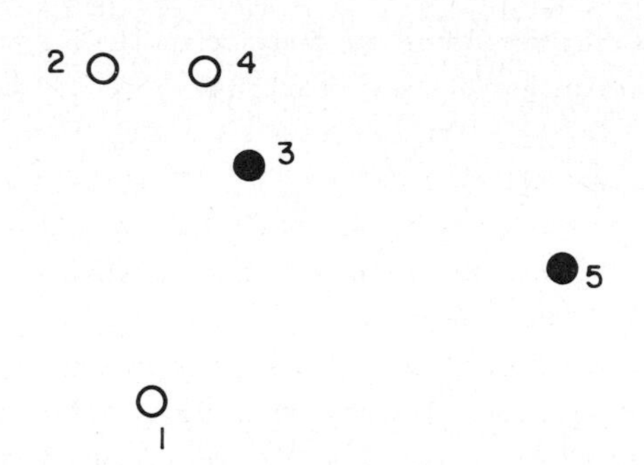

3(i) The set of five points

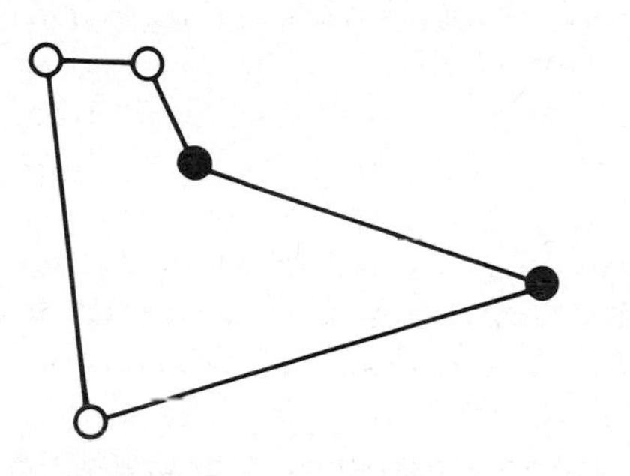

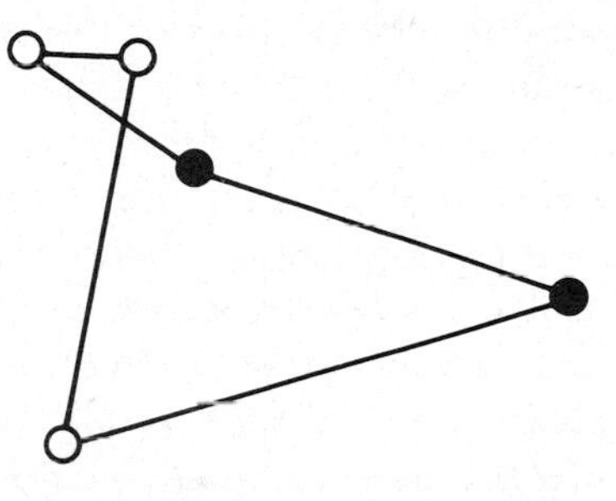

Tour v

Tour b

3(ii) Tour when all points are present

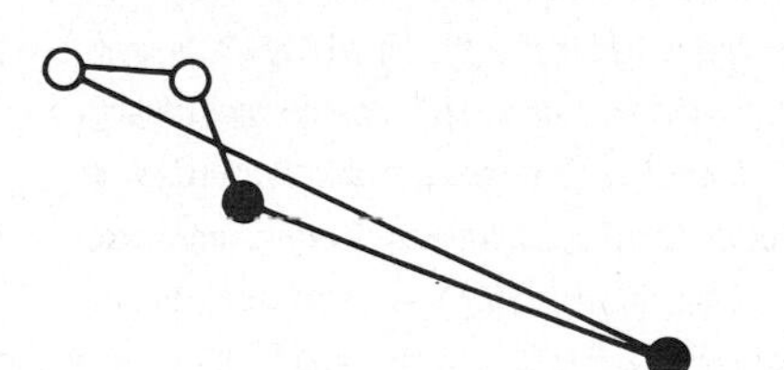

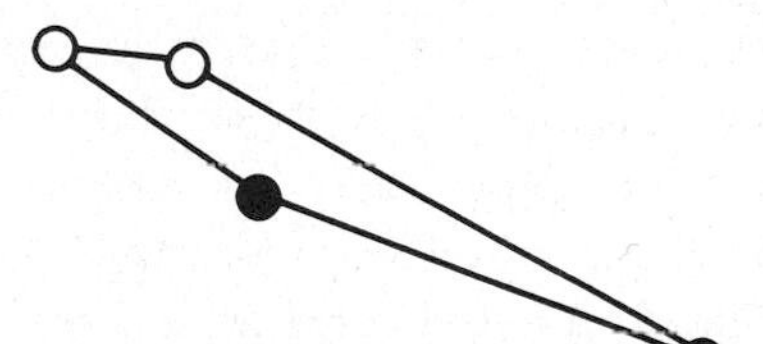

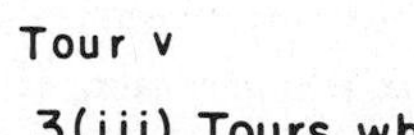

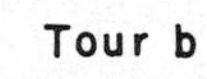

3(iii) Tours when points 2, 3, 4, 5 are present

Figure 3 Intersection of the optimal PTSP tour

The second observation concerns the classical dynamic programming (DP) formulation and solution of the TSP (Held and Karp [1962]). This is probably the most widely-known (but far from the most efficient) approach for solving the TSP optimally. Given the fact that the PTSP can be formulated in exactly the same manner as the TSP, apart from a different objective function (see also Section 5), one might reasonably expect that the DP approach could also be used to solve (small-size) PTSPs. Indeed, it is not difficult to write a set of DP-like recursive relationships, which would seem to implement such an approach. Unfortunately, it turns out that this straightforward "translation" from the TSP is not a valid method for solving the PTSP optimally, the reason being that the PTSP cannot be decomposed into stages (as required by Bellman's principle of optimality on which the DP approach is based) in a straightforward fashion. A similar observation can be made for other routing problems with stochastic elements (see, e.g., Andreatta et al. [1988]).

4. THE CAPACITATED PROBLEM (PVRP)

The probabilistic vehicle routing problem (PVRP) which is examined in this section is a direct extension of the probabilistic traveling salesman problem (PTSP) discussed in the previous section. Specifically, we still consider demands which are probabilistic in nature and our problem is to determine an a priori route of minimal expected length for a single vehicle with finite capacity. The complications introduced by the finite capacity of the vehicle is a major point of interest.

Our approach will be to design a giant a priori vehicle tour through all the demand points. While covering this tour the vehicle may run out of capacity and, in such an event, it will have to return to the depot -- for instance, in order to unload the packages it has picked up at the points it has already visited. Thus, the expected tour length to be minimized must also include any additional distance traveled to and from the depot whenever the vehicle reaches its capacity.

There is, of course, an alternative interpretation under which the very same problem can be viewed as a multi-vehicle PVRP. This can be seen best if one sets p_i, the probability of visiting point i, equal to 1 for all i. Then our approach is identical to one of the two standard approaches to multi-vehicle VRPs, namely "route first, cluster second" -- see, for example, Bodin et al [1983]. Under this interpretation, the returns of the vehicle to the depot result in multiple tours, so that we are dealing with multiple-VRP tours as solutions to the overall problem. However, in the general case when some of the p_i are strictly less than 1, a "probabilistic" criterion must be used in order to break up the giant a priori tour into clusters of customers -- with each cluster served by a different vehicle. An example of such a criterion might be that "the probability of a vehicle having to return to the depot more than once while serving its cluster of customers should not exceed $\delta(0<\delta<1)$", Stewart and Golden [1983].

To date we have not dealt much with the clustering aspects of PVRPs. For this reason we shall limit the discussion in the rest of this section to the single-vehicle interpretation of the PVRP. We shall examine the PVRP under two different strategies for serving demands and dealing with cases where the vehicle may have to return to the depot prematurely. These are summarized in Figure 4. Under Strategy A the vehicle visits all the points in the same fixed order as under the a priori tour, but serves only customers requiring service that day. For example, in the instance shown on the left-hand side of Figure 4(ii), all six potential customers will be visited but a load will be picked up -- assuming this is a pick-up service -- only from customers 2, 3 and 5. In practice, Strategy A may correspond to cases -- e.g.,

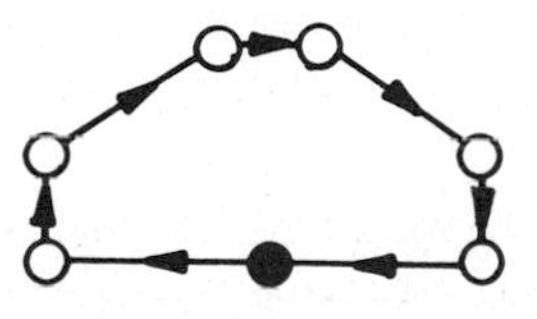

4(i) An a priori route through 6 customers (each with a demand of zero or one unit) by a vehicle of capacity 2

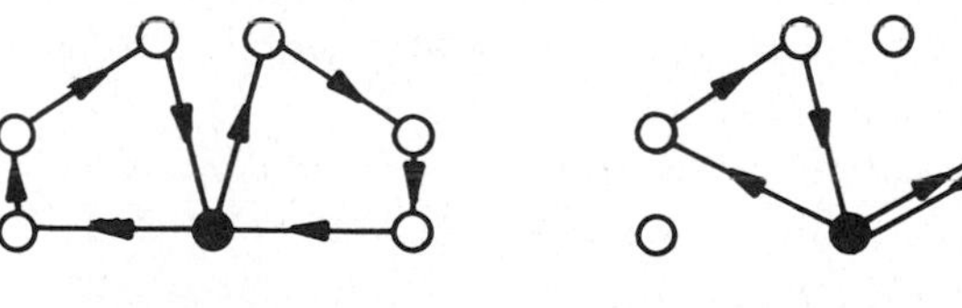

Class A Class B

4(ii) The two strategies when only the second, third and fifth customers have a non-zero demand.

Figure 4 Illustration of the two strategies

distribution of a product to many small retailer shops -- in which no advance information is available on the needs of a customer prior to visiting the customer. The total expected distance traveled corresponds to the fixed length of the a priori tour plus the expected value of the additional distance that must be covered whenever the demand on the route exceeds vehicle capacity and forces the vehicle to go back to the depot to deliver its load before continuing on its route. This class of problems is similar in description to the stochastic VRPs described in Stewart and Golden [1983]. However, our interest and approach from this point on diverges from theirs, as we are concerned here only with routing costs and do not introduce any additional parameters as they do.

Strategy B is defined similarly to A, with the sole difference that customers with no demand on a particular instance of the vehicle tour are simply skipped, as shown on the right-hand side of Figure 4(ii).

To maintain analytical tractability and derive results which are conceptually useful and free of unnecessary "clutter", we shall examine from here on cases for which each customer has a demand of one unit with probability p, independently of all other customers, and no demand with probability $1-p$. Even under this simplifying assumption, the analysis of the capacitated vehicle routing problem turns out to be a difficult one (Jaillet [1986]).

Let us first try to obtain results analogous to Theorem 1 for the expected length of a given PVRP tour. Consider an a priori vehicle tour $t=(0,1,2,\ldots,n,0)$ through n potential customers who are re-indexed according to the order of their appearance in t. The case of interest is when the capacity of the vehicle, Q, is strictly less than n. Otherwise, the vehicle will never run out of capacity and the problem reduces to a TSP under Strategy A and PTSP under Strategy B. Assuming then that $\lceil n/Q \rceil > 1$ ($\lceil x \rceil$ denotes the smallest integer greater than or equal to x), we have:

Theorem 4: Under Strategy A the expected length of the vehicle route is given by:

$$E^a[L_t] = \sum_{i=0}^{n} d(i,i+1) + \sum_{i=1}^{n-1} \beta_i \, s(i,i+1) \tag{5}$$

where:
$$s(i,i+1) = d(i,0) + d(0,i+1) - d(i,i+1) \tag{6}$$

$$\beta_i = 0 \qquad \forall\, i \in [1,\ldots,Q-1] \tag{7a}$$

$$\beta_i = \sum_{k=1}^{\lfloor i/Q \rfloor} p^{kQ}(1-p)^{i-kQ} \cdot \binom{i-1}{kQ-1} \qquad \forall\, i \in [Q,\ldots,n-1] \tag{7b}$$

Sketch of proof:

The expected length of the route is a summation of two terms: the length of the a priori tour $(\Sigma\ d(i,i+1))$ plus the expected value of the extra distance when the vehicle reaches its capacity. To evaluate this second term, let i be a node on the route where the vehicle reaches its capacity. The vehicle will then go to the depot before going back to the following node on its route, which is $i+1$ under Strategy A (even if node $i+1$ has no demand on a particular instance of the problem). The extra distance is then $s(i,i+1)=d(i,0)+d(0,i-1)-d(i,i+1)$. In Theorem 4, β_i represents the probability that the vehicle reaches its capacity Q at node i. The summation over k is due to conditioning on the number of times the vehicle reaches capacity. #

The computation of each of the β_i requires at most $O(n)$ steps; the overall complexity of (5) is thus $O(n^2)$.

Let us now consider Strategy B, under which customers with no demand are skipped.

Theorem 5: Under Strategy B the expected length of the vehicle route is given by

$$E^b[L_t] = \sum_{i=0}^{n} \sum_{j=0}^{n} \alpha_{ij}\, d(i,j) + \sum_{i=1}^{n-1} \sum_{j=i+1}^{n} \gamma_{ij}\, s(i,j) \qquad (8)$$

where α_{ij} is defined in Theorem 1

and $\gamma_{ij} = 0 \qquad \forall\, i \in [1,\ldots,Q-1] \qquad$ (9a)

$$\gamma_{ij} = \sum_{k=1}^{\lfloor i/Q \rfloor} p^{kQ+1}\,(1-p)^{j-kQ-1} \cdot \binom{i-1}{kQ-1} \quad \text{otherwise} \qquad (9b)$$

Sketch of proof: Identical to that of Theorem 4 except that when the vehicle reaches its capacity at node i, it goes back (after a visit to the depot) to the first node j with a non-zero demand (skipping nodes $i+1, i+2,\ldots,j-1$ with no demand). Obviously, $i+1 \leq j \leq n$, and it is easy to see that:

$$\gamma_{ij} = p(1-p)^{j-i-1}\,\beta_i. \qquad \#$$

$E^b[L_t]$ can then be obtained in $O(n^3)$ steps.

As in the case of Theorem 1, Theorems 4 and 5 not only provide expressions for computing efficiently the expected length of vehicle routes, but also can be used as starting points for deriving many interesting properties of PVRP tours. Using (5) and (8), one can show that the same observations that we made in the previous section for the PTSP also apply to the PVRP -- although, occasionally, in somewhat modified form. For example, the optimal VRP (deterministic version) tour may be a very poor solution to the corresponding PVRP. Some thought, with the aid of Figure 4, will also convince the reader that the direction in which a PVRP tour is traveled, i.e., $t=(0,1,2,\ldots,n,0)$ vs. $t'=(0,n,n-1,\ldots,2,1,0)$, may affect its expected length, even when distances between potential customers are symmetric.

It should also be mentioned that (5) and (8) make it possible to analyze the influence of the capacity of the vehicle on the expected length of routes. For example, one can find examples in which an increase in vehicle capacity leads to an increase in the expected length of a given PVRP route t! The reason is that vehicle capacity obviously affects the locations on the tour where the vehicle is most likely to run out of capacity. This, in turn, affects the expected length of the extra distance required to return to the depot before resuming the tour.

Finally, we note that the above results can be generalized in two directions. First, to cases in which, instead of independence among customer demands, we allow certain types of dependence. Second, and more significantly from the practical viewpoint, to situations where the demands of individual customers are mutually independent random variables of a specific but quite flexible form, namely $Y_j=a_jX_j+b_j$ where Y_j is the demand of customer j, a_j and b_j are constants and X_j is a binomial random variable, $B(k_j,p)$, i.e. the sum of k_j Bernoulli random variables with parameter p. Obviously this is much less restrictive than requiring passenger demands to be either 0 or 1. The analysis for the binomial random variable becomes tractable because a customer with binomially distributed demand is equivalent to k_j collocated customers with Bernoulli (0 or 1) demand.

5. ALGORITHMIC APPROACHES

The last two sections suggested strongly that PTSPs and PVRPs have sufficiently different features from their famous counterparts, the TSP and the VRP, to warrant an investigation ab initio into specially designed solution approaches. One can also observe at the outset that, since the PTSP is at least as "hard" as the TSP, there is little hope of developing exact optimization methods that could solve more than modest-scale PTSPs. It is indeed possible to formulate the PTSP as an integer optimization problem with 0-1 variables (as in the TSP) and a set of linear constraints identical to those used in the standard assignment-based formulation of the TSP (see, e.g., Bodin et al [1983]). Unfortunately, the objective function is nonlinear, namely a polynomial of degree equal to n, the number of customers. This renders the problem impractical to solve exactly for any significant number of customers. The problem can also be transformed into a mixed integer linear program, as well as into a pure integer linear program (Jaillet [1985]), but at the cost of a very large increase in the number of variables and of constraints. At this point, the most promising exact solution procedure for PTSPs seems to be a branch-and-bound approach described by Jaillet [1985] with which, however, we have no computational experience to date.

One must, consequently, turn to heuristics for practically viable PTSP and PVRP algorithms. In doing so, one must give special consideration to the complications raised by the probabilistic nature of customer demands. Even the simplest standard heuristics for capacitated and uncapacitated routing problems must be appropriately modified to take these considerations into account. This point can be illustrated with reference to one such simple heuristic, the "nearest neighbor algorithm" (NNA).

The NNA is a "greedy" tour-construction heuristic: one begins at the depot and builds a tour by going to the yet-unvisited demand point which is the nearest to the last visited point. When all points have been visited the tour is completed by returning to the origin. Consider now the application of the NNA to a PVRP with n customers, each of whom has probability p of being visited independently of all others (the case of different individual probabilities, p_i, is a simple extension). Suppose that we are at a stage in the algorithm where k customers have already been included in the (*a priori*) PVRP tour, and there are still $n-k$ customers who are not in the tour. Which customer should be included next? Note that the proper interpretation of "nearest neighbor" is now to "add the least possible expected length to the tour". If the tour so far consists of $(0,g,h,\ldots,t,i)$ then the extra expected length added to the tour by including point j, as the next one in the tour is equal to

$$E[j] = p^2 d(i,j) + p^2(1-p)\, d(t,j) + \ldots + p^2 \cdot (1-p)^{k-2}\, d(h,j) + p^2(1-p)^{k-1}\, d(g,j) + p(1-p)^k\, d(0,j)$$

This is because, depending on which customers are missing, one could visit j immediately after customer i or customer t, etc., or the depot. Moreover, the probability that customer j will have to be visited at all is p. Thus, to select the next point on the tour, one should compute $E[j]$ for all $n-k$ customers not presently in the tour. This means that, even for a simple heuristic like the NNA, the computational effort increases by $O(n^2)$ relative to the deterministic VRP (or TSP), since for each of the n points to be added to the tour $O(n)$ simple operations must be carried out.

One possibility for reducing the computational effort in this case is to perform "limited depth" searches. This leads to a set of "Almost Nearest

Neighbor Algorithms" (ANNA). For example, a "level-0" ANNA uses only $p^2d(i,j)$ as a proxy for $E[j]$ (i.e., disregards all possibilities for arriving at j other than from the very last point i currently in the tour); a level-1 ANNA uses $p^2d(i,j)+p^2(1-p)d(t,j)$, i.e., looks at the last two points in the tour; and so on for level-2, etc. Clearly the most desirable level depends on the trade-offs between computational effort and desired degree of approximation to the spirit of the heuristic -- the latter also being a function of p, with, generally, a large p favoring a low "depth" of search.

The preceding is a typical, albeit the simplest, example of the types of modifications we have proposed to several well-known TSP and VRP heuristics in order to solve PTSPs and PVRPs. These include variations of tour-construction heuristics (such as the Clark-Wright savings algorithm, tour-merging algorithms, and insertion algorithms), as well as tour-improvement heuristics (such as 2-opt and 3-opt exchanges). The modifications require varying degrees of difficulty and computational effort. The reader is referred to Golden et al. [1980] for a fine description of these heuristics as they apply to deterministic problems and to Jaillet [1985] for a discussion of their adaptation to the probabilistic setting. Another recently-proposed heuristic, the space-filling curve (Platzman and Bartholdi [1983]) is also particularly well-suited for the PTSP.

Some computational experience with several of these heuristics has been obtained recently by Jezequel [1986]. He solved eight test problems with one depot and n ranging from 9 to 75, assuming a binomial probability law (independent customer demands with identical probabilities p) and solving each problem for values of p ranging from 0.1 to 1.0 at intervals of 0.1. He obtained consistently good performance, by comparison to tours produced by other heuristics, from a variation of the Clarke-Wright savings algorithm that he called NEWSAVE. Figure 5, based on Jezequel's work, illustrates how the value of p may affect dramatically the configuration of tours for any given set of points. The tour on the left, for $p=0.1$, emphasizes our earlier observation to the effect that small probabilities of visiting individual customers induce the use of zig-zag-type local patterns within tours.

One very interesting insight gained from these computational experiments was the following: Changes in the value of p are helpful in identifying significantly different alternative tour configurations for any given set of points, as shown in Figure 5. In several instances, it was observed that a tour which was discovered while solving a PTSP for a specific value of p, turned out to be a very good solution (and in a few cases the "best"

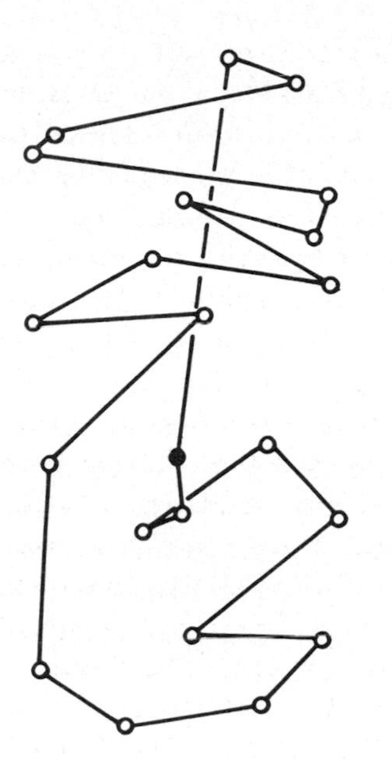

ANNAO with p=0.1

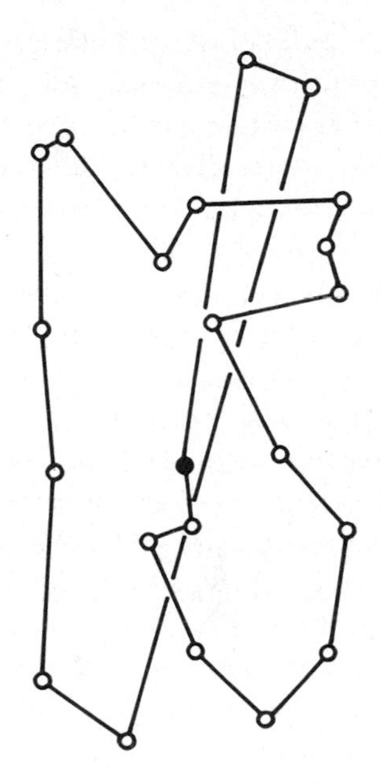

ANNAO with p=0.5

Figure 5 Examples of heuristic solutions for a 22-point problem

solution) for other values of p, as well. For example, one might solve a given problem for, e.g., $p=0.4$, and develop a tour t_1. Then, it may turn out that t_1 is also a better solution for, e.g., $p=0.6$ than the best tour, t_2, that had been found when the heuristics were previously used to solve the same problem with $p=0.6$! In other words, $E[L_{t1}] < E[L_{t2}]$ for $p=0.6$ (as well as for $p=0.4$).

This suggests that, in varying the value of p, one obtains another "dimension" along which to search for improved tours. This new dimension may be helpful in avoiding being "trapped" into a locally optimal tour -- a problem common to all local-search heuristics. Stated differently, the "probabilistic" dimension may permit local-search heuristics to explore possibilities for tour configurations that would not otherwise have been examined. In this respect, there are some analogies here with the simulated annealing approach that has attracted much attention recently.

In concluding this section we would note that, while some progress has obviously been made in developing algorithms for PTSPs and PVRPs, as we have just described, the topic is still open for much further work.

6. BOUNDS AND ASYMPTOTIC ANALYSIS FOR THE PTSP

In vehicle routing and scheduling, especially in situations involving long-term planning, it is often useful to have on hand simple expressions that provide upper and lower bounds on route lengths. The derivation of such bounds is usually achieved by "construction", i.e., by devising graphs or tours which are guaranteed to have lengths that bound from above or below the optimal solutions to the routing problem at hand. In addition to bounds, researchers in recent years have often attempted to understand the properties of some routing and network problems "at the limit", i.e., as the number of points to be visited (or of network nodes or of some other input parameter) increases. In section 2, we mentioned informally the limit theorem on the TSP due to Beardwood et al. [1959], which is probably the best-known result of this type. Such results are very useful in a number of situations, such as obtaining approximations for large-scale instances of routing problems, in the sense suggested by our example of Section 2; or analyzing the properties of some heuristic algorithms -- as, for example, in the case of the fundamental "partitioning" algorithm proposed by Karp [1977] for the TSP.

In this section we give a survey of such results obtained for the PTSP in the plane. Let us first summarize the notation and assumptions used throughout this section. We consider a set of points in R^2 and we assume the distances between points to be the ordinary Euclidean distance; $x=\{x_1,x_2,\ldots\}$ represents an infinite sequence of points in R^2 and $x^{(n)}$ indicates the first n points of x. If the positions of the points are random, the sequence will be denoted by upper-case letters, i.e., $X=\{X_1,X_2,\ldots\}$. We assume that each point of the sequence requires a visit with a probability p (the coverage probability), independently of the others. Let the functional $\phi_p(x^{(n)})$ define the expected length, in the PTSP sense, of an optimal PTSP tour through $x^{(n)}$. When the positions of the points are random, then $\phi_p(X^{(n)})$ is a random variable and its expectation (with respect to the position of the points) is written $E\phi_p(X^{(n)})$.

6.1. Bounds for Sequence of Points in $[0,r]^2$

We consider a sequence of n points lying in a square of side r. Our interest is to obtain bounds on $\phi_p(x^{(n)})$. The results are summarized in the following theorem:

Theorem 6:

(1) Let x be an <u>arbitrary</u> sequence of points in $[0,r]^2$ and p be the coverage probability. Then,

$$\phi_p(x^{(n)}) \le \begin{cases} \left(\sqrt{2(np-2)} + \frac{13}{4}\right) r & \text{if } np \ge 2.5 \\ \left(\frac{np}{2} + 3\right) r & \text{if } np \le 2.5 \end{cases} \tag{10}$$

(2) Let X be a sequence of points independently and uniformly distributed over $[0,r]^2$ and p be the coverage probability. Then,

$$E\phi_p(X^{(n)}) \leq \begin{cases} \left(\sqrt{\frac{4}{3}(np-3)} + \frac{11}{2} + \sqrt{2}\right) r & \text{if } np \geq 3.75 \\ \left(\frac{np}{3} + \frac{2}{3} + \sqrt{2}\right) r & \text{if } np \leq 3.75 \end{cases} \tag{11}$$

$$E\phi_p(X^{(n)}) \geq \frac{5}{8}\left(p\sqrt{n} - n(1-p)^{n-1}\right) r . \tag{12}$$

Comments:

(i) The proofs of (10) and (11) are simple and are both obtained via constructions of tours whose values are at most equal to the upper bounds; the constructions are respectively similar to the ones presented for the TSP in Few [1955] and Beardwood et al. [1959] (Lemma 4). The proof of (12) is slightly more involved and proceeds along the techniques given in the proof of Lemma 3 in Beardwood et al. [1959]. The main idea (applied also to other problems, see for example Papadimitriou [1978]) is that, as the shortest tour through all points does not visit any point twice, the sum of the lengths of the two segments of this tour that terminate at a given point is at least as large as the sum of the distances from this point to its first and second nearest points. A lower bound on the latter quantity leads to the result of (12). The interested reader is referred to Jaillet [1985] for details on the proof of Theorem 6. One may also note that, for $p=1$, the bounds of Theorem 6 reduce to well-known results for the TSP.

(ii) The results summarized in this theorem are valid for any finite size n. For large n, the lower and upper bounds are all of order $O(\sqrt{n})$. In the next theorem, we will see that the value of an optimal PTSP tour through n points drawn from a uniform distribution in the unit square is almost surely asymptotic to $c(p)\cdot\sqrt{n}$ where $c(p)$ is a positive constant depending only on the coverage probability p.

6.2. Asymptotic Analysis

We turn next to the limiting behavior, as n goes to infinity of the random variable $\phi_p(x^{(n)})$; this is described in the following theorem:

Theorem 7:

Let X be an infinite sequence of points independently and uniformly distributed over $[0,1]^2$ and p be the coverage probability. Then there exists a constant $c(p)$ such that:

$$\forall\ p \in [0,1]: \lim_{n\to\infty} \frac{\phi_p(X^{(n)})}{\sqrt{n}} = c(p) \quad \text{(a.s.)} \tag{13}$$

with $c(p) > 0$ for $p \in (0,1]$.

Comments:

(i) The proof of this theorem is based on a very interesting result due to Steele [1981] who uses the theory of independent subadditive processes to obtain strong limit laws for a class of problems in geometrical probability that exhibit nonlinear growth. We prove that the PTSP is one member of this class (the main difficulty is to show that the functional $\phi_p(x^{(n)})$ is subadditive, see Jaillet [1985]).

(ii) Theorem 7, together with Theorem 6, implies (dominated convergence theorem) that

$$\lim_{n\to\infty} \frac{E\phi_p(X^{(n)})}{\sqrt{n}} = c(p) \tag{14}$$

(iii) We have proved the existence of a constant c(p) without giving details on its value. In fact, it is interesting to note that, for all similar asymptotic results (TSP, matching, spanning tree), the respective limiting constants are unknown and only bounds have been established for them (see Beardwood et al. [1959], Papadimitriou [1978], and Gilbert [1965] respectively). Our problem is no exception and one can already infer from Theorem 6 and (6.4) that

$$\frac{5}{8}p \leq c(p) \leq \sqrt{\frac{4}{3}}\sqrt{p}$$

In fact our best bounds for c(p) are expressed in the following:

Lemma:

The constant c(p) in the result given in Theorem 7 is bounded as follows:

$$\beta\sqrt{p} \leq c(p) \leq \min\{\beta,\ 0.9204\sqrt{p}\}\quad \forall\, p \in [0,1] \tag{13}$$

where β is the "TSP-constant"

Note that, for p=1, the bounds give c(1)=β and if β=0.76, as estimated by Stein [1977], then the upper bound in (13) is equal to β for values of p greater than approximately 0.83 and equal to $0.9204\sqrt{p}$, otherwise.

Finally, we note that results presented in this section can be generalized in several directions such as: cases involving n probabilistic points (only present with a probability p) and m deterministic points (always present); and cases involving any bounded Lebesgue measurable set of a d-dimensioned Euclidean space under more general metrics.

Besides algorithmic applications such as presented in the introduction to this section, Theorem 7, together with an estimator of the constant c(p), provides an important practical "by-product" (take β=0.76): $(c(p)-0.76)\sqrt{p}]\sqrt{nA}$ represents an approximation for the penalty one has to pay when n customers (each of them present only with fixed probability p) are served within a region of area A and the route is not reoptimized for each instance of the problem.

7. CONCLUSIONS

Probabilistic vehicle routing problems such as the ones discussed here represent an emerging and exciting new area of study. While it is fair to state that the area is still in its infancy, there is little question, as suggested by this chapter, that PVRPs and related problems present distinct and challenging opportunities for both methodological and algorithmic work which can be highly applicable.

We anticipate that much new material will begin appearing soon in this area, given that some fundamental concepts have now been established regarding problem definition and properties. For example, very recently the "mini-tour" facility location problem in the PTSP context which was described in Section 2 (example b) was addressed very successfully [see Berman and Simchi-Levi [1986] and, especially, Simchi-Levi [1986]). Much research on algorithms, including acquisition of computational experience specific to PVRPs, also remains to be done. Finally, the co-authors of this chapter are currently working on related problems with similar probabilistic notions, such as a probabilistic shortest path problem and a probabilistic minimum spanning tree problem.

References

Andreatta, G. and L. Romeo (1988) "Stochastic Shortest Paths with Recourse", Networks, 18, to appear.

Beardwood, J., J. Halton and J. Hammersley (1959) "The Shortest Path through Many Points", Proc. Cambridge Philosophical Society, 55, 299-327.

Berman, O. and D. Simchi-Levi (1986) "Minisum Location of a Traveling Salesman", Networks, 16, no. 3, 239-254.

Bodin, L., B. Golden, A. Assad and M. Ball (1983) "Routing and Scheduling of Vehicles and Crews: The State of the Art", Computers and Operations Research, 10, no. 2, 69-211.

Few, L. (1955), The Shortest Path and the Shortest Road through N points, Mathematika, 2, 141-144.

Gilbert, E. (1965), Random Minimal Trees, J. Soc. Ind. Appl. Math., 13, 376-387.

Golden, B., L. Bodin, T. Doyle and W. Stewart (1980) "Approximate Traveling Salesman Algorithms", Operations Research, 28, no. 3, 694-711.

Held, M. and R.M. Karp (1962) "A Dynamic Programming Approach to Sequencing Problems", SIAM REVIEW, 10, no. 2, 196-210.

Jaillet, P. (1986), "Stochastic Routing Problems", in Proceedings of CISM Advanced School on Stochastics and Optimization, G. Andreatta and P. Serafini (eds.), Springer-Verlag, Berlin (in press).

Jaillet, P. (1985), Probabilistic Traveling Salesman Problems (Ph.D. Thesis) Technical Report no. 185, Operations Research Center, Massachusetts Institute of Technology, Cambridge, MA. 02139.

Jezequel, A. (1986), Probabilistic Vehicle Routing Problems (S.M. Thesis) Department of Civil Engineering, Massachusetts Institute of Technology, Cambridge, MA 02139 (unpublished).

Karp, R. (1977), "Probabilistic Analysis of Partitioning Algorithms for the Traveling Salesman Problem in the Plane", Math. of Opers. Research, 2, no. 2, 209-224.

Larson, R.C. and A.R. Odoni (1981) Urban Operations Research, Prentice-Hall, Englewood-Cliffs, NJ.

Papadimitriou, C. (1978), The Probabilistic Analysis of Matching Heuristics, Proc. 15th Annual Conf. Comm. Contr. Computing, 368-378.

Platzmann, L.K. and J.J. Bartholdi (1983) Spacefiling Curves and the Planar Traveling Salesman Problem, Report no. PDRC 83-02, School of Industrial and Systems Engineering, Georgia Institute of Technology, Atlanta, GA 30332.

Simchi-Levi, D. (1936) The Multistop Location Problems (Ph.D. Thesis) Tel-Aviv University, Israel.

Steele, J. (1981), Subadditive Euclidean Functionals and Nonlinear Growth in Geometric Probability, Ann. Prob. 9, 365-376.

Stein, D. (1977), Scheduling Dial-A-Ride Transportation Systems (Ph.D. Thesis), Harvard University, Cambridge, MA 02138.

Stewart, W. and B. Golden (1983) "Stochastic Vehicle Routing: A Comprehensive Approach", Eur. J. Opers. Res., 14, no. 3, 371-385.

Vehicle Routing: Methods and Studies
B.L. Golden and A.A. Assad (Editors)

AN ADDITIVE APPROACH FOR THE OPTIMAL SOLUTION OF THE PRIZE-COLLECTING TRAVELLING SALESMAN PROBLEM*

Matteo Fischetti and Paolo Toth
D.E.I.S., University of Bologna, Italy

Consider the following generalization of the well-known Travelling Salesman Problem. Given a depot, at which a vehicle is stationed, and a set of cities, each having an associated non-negative *prize* p_i, let $c_{i,j}$ be the cost of routing city j just after city i and γ_i be the cost of leaving city i unrouted. The *Prize-Collecting Travelling Salesman Problem* (*PC-TSP*) is to find a minimum-cost route for the vehicle, visiting each city at most once and collecting a total prize not less than a given *goal* g. Such a problem arises in several routing and scheduling applications and belongs to the class of $\mathcal{NP}$-hard problems. In this paper, we introduce several mathematical models of the problem and point out its main substructures. Additive bounding procedures are then designed yielding sequences of increasing lower bounds on the optimum value of the problem. Extensive computational results on randomly generated test problems are reported, comparing the performances of the proposed bounding procedures. A branch and bound algorithm for the optimal solution of *PC-TSP* is finally described and computationally analyzed.

1. Introduction

Consider the following routing problem, which generalizes the well-known *Travelling Salesman Problem* (TSP). Given a depot at which a vehicle is stationed, and a set of cities, each having an associated non-negative *prize* p_i, let $c_{i,j}$ be the cost of routing city j just after city i and γ_i be the cost of leaving city i unrouted. The *Prize-Collecting Travelling Salesman Problem* (*PC-TSP*) is to find a minimum-cost route for the vehicle, visiting each city at most once and collecting a total prize not less than a given *goal* g.

Such a problem arises, for instance, when a factory (located at city 1) needs a given amount g of a product, which can be provided by a set of suppliers (located at cities $2, \cdots, n$).

* Work supported by the Ministero della Pubblica Istruzione, Italy

Let p_i be the (indivisible) amount supplied at city i, β_i the corresponding cost ($i = 2, \cdots, n$), and $c_{i,j}$ the transportation cost from city i to city j ($i,j = 1, \cdots, n$). Assuming that only one trip is required, such a problem can be formulated as an instance of *PC-TSP* in which $\gamma_i = -\beta_i$ is the saving from city i not supplying the product (the total cost being $\sum_{i=2}^{n} \beta_i +$ optimal value of the *PC-TSP* instance).

The problem also arises in several scheduling problems. Balas and Martin (1985) introduced *PC-TSP* as a model for scheduling the daily operation of a steel rolling mill. A rolling mill produces steel sheets from slabs by hot or cold rolling. Let $c_{i,j}$ be the "cost" of processing order j just after order i, and p_i the weight of the slab assigned to order i. Scheduling the daily operation consists of selecting a subset of orders that satisfies a given lower bound g on the total weight, and of sequencing them so as to minimize the global cost.

PC-TSP can be formulated through a graph theory model as follows. Let $G = (V, A)$ be a directed complete graph, where $V = \{1, 2, \cdots, n\}$ is the vertex set (vertex 1 corresponding to the depot) and A the arc set. For each $(i,j) \in A$ let $c_{i,j}$ be the cost of arc (i,j) (with $c_{i,i} = \infty$ for each $i \in V$) and, for each $i \in V$, let γ_i and p_i be respectively the cost and the prize associated with vertex i (with $\gamma_1 = \infty$ and $p_1 = 0$). A vertex subset $S \subseteq V$ is *feasible* iff $\sum_{i \in S} p_i \geq g$ and $1 \in S$. The *PC-TSP* is to find a Hamiltonian circuit in the subgraph induced by a feasible vertex subset S, so as to minimize the sum of the costs of the arcs in the circuit plus the sum of the costs associated with vertices in $V \setminus S$.

PC-TSP contains as a particular case the *TSP* obtained when $\sum_{i \in V} p_i = g$ (or when $\gamma_i = \infty$ for each $i \in V$), and hence it belongs to the class of $\mathcal{NP}$-hard problems (in the strong sense).

PC-TSP can also be viewed as a generalization of the minimization form of the *0-1 Single Knapsack Problem*, arising when $\gamma_i = 0$ for each $i \in V \setminus \{1\}$, and when all the arcs entering the same vertex have the same cost.

Without loss of generality we assume:

$$c_{i,j} \geq 0 \quad \text{for each } (i,j) \in A, \text{ and } \gamma_i \geq 0 \quad \text{for each } i \in V\,.$$

In fact, for each $i \in V$ the addition of any constant α_i to γ_i and $c_{i,j}$ ($j \in V$) does not alter the relative ranking among the feasible solutions to *PC-TSP*. We also assume the feasibility condition $\sum_{i \in V} p_i \geq g$.

We do not assume that the triangle inequality ($c_{i,k} + c_{k,j} \geq c_{i,j}$ for $i,j,k \in V$ and $i \neq j$) holds. If, however, the triangle inequality applies and $\gamma_i = 0$ for $i \in V \setminus \{1\}$, as in several practical applications, we can also assume that $p_i > 0$ for $i \in V \setminus \{1\}$, since vertices with zero prize are not worth routing; for the same reason, only minimal (with respect to deletion of one element) feasible vertex subsets S can lead to optimal solutions.

To our knowledge, no optimal algorithm has been proposed for *PC-TSP*. Heuristic methods and structural properties have been discussed in Balas and Martin (1985) and in Balas (1987), respectively. A related problem is the *Orienteering Problem*, in which the vehicle has to collect the maximum possible prize p^* through a route whose total cost does not exceed a given bound $\bar{c}$. Such a problem could be optimally solved through binary search for the maximum feasible prize p^*, exploiting the property that for each total prize p we have $p^* \geq p$ iff there exists a feasible solution to the instance of *PC-TSP* with goal $g = p$, whose cost does not exceed $\bar{c}$. The existence of such a solution can be checked by optimally solving the corresponding *PC-TSP* instance. Heuristic algorithms for the Orienteering Problem and some generalizations have been proposed by Tsiligirides (1984), Golden, Levy and Vohra (1985, 1987), Golden, Storchi and Levy (1986) and Golden, Wang and Liu (1987), while optimal methods are given in Laporte and Martello (1987).

In Section 2, integer linear programming models are given, pointing out different substructures of the problem. In Section 3 we propose several lower bounds, and combine them to obtain additive procedures yielding a combined bound which is generally superior to the individual ones used to produce it. Section 4 experimentally analyzes the performances of the proposed bounds on randomly generated test problems. A branch and bound algorithm is then described and computationally evaluated in Section 5.

2. Mathematical models

Several integer linear programming models can be given for *PC-TSP*. For each $i, j \in V$, $i \neq j$, let $x_{i,j}$ be a binary variable set to 1 if arc (i,j) is in the optimal solution, set to 0 otherwise. For each vertex $i \in V$, let y_i be a binary variable set to 1 if vertex i is routed, set to 0 otherwise. A first model is

$$(PC-TSP) \quad v(PC-TSP) = \min \sum_{i \in V} \sum_{j \in V \setminus \{i\}} c_{i,j}\, x_{i,j} + \sum_{i \in V} \gamma_i (1 - y_i) \tag{1}$$

subject to

$$y_i = \sum_{h \in V \setminus \{i\}} x_{h,i}\,, \qquad \text{for each } i \in V \tag{2}$$

$$\sum_{h \in V \setminus \{i\}} x_{i,h} = \sum_{h \in V \setminus \{i\}} x_{h,i}\,, \qquad \text{for each } i \in V \tag{3}$$

$$\sum_{i \in V} p_i\, y_i \geq g \tag{4}$$

$$\sum_{i \in S} \sum_{j \in V \setminus S} x_{i,j} \geq y_h\,, \qquad \begin{array}{c}\text{for each } h \in V \setminus \{1\} \text{ and for each} \\ S \subset V : 1 \in S, h \in V \setminus S\end{array} \tag{5}$$

$$\left.\begin{array}{ll} x_{i,j} \in \{0,1\}\,, & \text{for each } i,j \in V, i \neq j \\ y_i \leq 1\,, & \text{for each } i \in V \end{array}\right\} . \tag{6}$$

Constraints (2), (3) and (6) ensure that each vertex i is either unrouted (in which case no arc enters or leaves vertex i) or routed once. Constraints (4) ensure that the sum of the prizes of the routed vertices is not less than goal g. Any feasible solution satisfying constraints (2), (3),(4) and (6) can be viewed as a family of disjoint subtours (each of cardinality at least 2) visiting a subset of vertices whose total prize is at least g. Constraints (5) ensure the "connectivity" of the solution, in the sense that each routed vertex h can be reached from vertex 1.

Constraints (5) can be replaced by the "subtour elimination" constraints

$$\sum_{i \in S} \sum_{j \in S \setminus \{i\}} x_{i,j} \leq |S| - 1, \qquad \text{for each } S \subset V : 1 \in V \setminus S \tag{7}$$

forbidding subtours not visiting vertex 1.

A tighter formulation of the problem could be obtained by replacing constraint (4) with constraints

$$\sum_{i \in S} \sum_{j \in S \setminus \{i\}} x_{i,j} \leq |S| - 1, \qquad \text{for each } S \subset V : 1 \in S, \sum_{j \in S} p_j < g \tag{8}$$

which eliminate subtours visiting vertex 1 but not collecting a sufficient prize. On the assumption that $\gamma_i = 0$ for $i \in V \setminus \{1\}$, constraints (5) (or (7)) become redundant, and then (1), (2), (3), (6) and (8) give a valid formulation of *PC-TSP*.

Model (1) – (6) can be rearranged so as to point out the assignment problem substructure of *PC-TSP*. To this end, for each $i \in V$ we define $x_{i,i} = 1$ if vertex i is left unrouted, and $x_{i,i} = 0$ otherwise (i.e., $x_{i,i} = 1 - y_i$). Consequently, for each $i \in V$ we set $c_{i,i} = \gamma_i$ (the cost incurred if vertex i is left unrouted). With these definitions, a valid formulation of the problem is

$$(PC-TSP) \quad v(PC-TSP) = \min \sum_{i \in V} \sum_{j \in V} c_{i,j}\, x_{i,j} \tag{9}$$

subject to

$$\sum_{i \in V} x_{i,j} = 1, \qquad \text{for each } j \in V \tag{10}$$

$$\sum_{j \in V} x_{i,j} = 1, \qquad \text{for each } i \in V \tag{11}$$

$$\sum_{i \in V} p_i\, x_{i,i} \leq \sum_{j \in V} p_j - g \tag{12}$$

$$\sum_{i \in V} \sum_{j \in V \setminus S} x_{i,j} \geq 1 - x_{h,h}, \qquad \begin{array}{c}\text{for each } h \in V \setminus \{1\} \text{ and for each} \\ S \subset V : 1 \in S, h \in V \setminus S\end{array} \tag{13}$$

$$x_{i,j} \in \{0,1\}, \qquad \text{for each } i, j \in V. \tag{14}$$

Constraints (10), (11), and (14) – with objective function (9) – define the well-known *Linear Min-sum Assignment Problem* (*AP*). Any feasible solution to *AP* gives a family of disjoint subtours (possibly visiting only one vertex) covering all the vertices. Constraints (12) impose an upper bound on the total prize of the unrouted vertices (i.e., of the vertices covered by loops in the *AP* solution), while constraints (13) ensure that the routed vertices can be reached from vertex 1. An example of a feasible solution to model (9) – (14) is given in Figure 1.

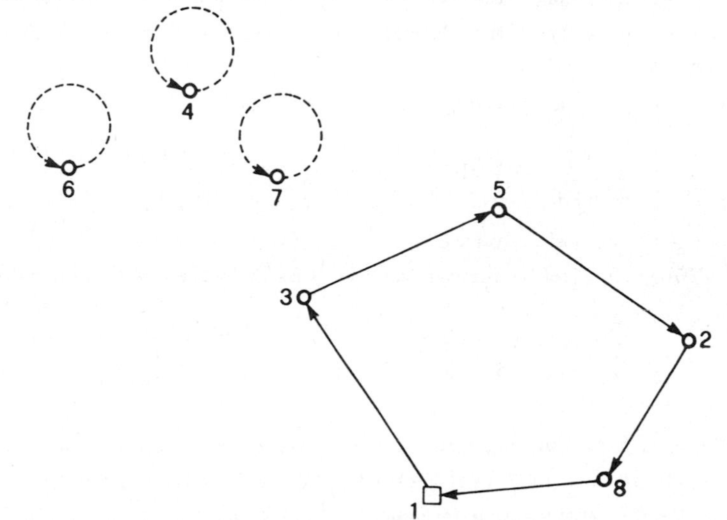

Figure 1. Feasible solution to model (9) – (14), assuming $(p_j)=(0,8,5,2,4,5,3,6)$, $g = 22$.

It is also possible to point out the underlying arborescence problem substructure of *PC-TSP* by rearranging model (1) – (6) as follows. Let us consider an augmented complete graph $G' = (V', A')$, where $V' = V \cup \{0\}$, and for each $i \in V$ define $x_{0,i} = 1$ if vertex i is left unrouted, and $x_{0,i} = 0$ otherwise (i.e., $x_{0,i} = 1 - y_i$). Consequently, for each $i \in V$ we set $c_{0,i} = \gamma_i$ (the cost incurred if vertex i is left unrouted). In addition, we set $c_{i,0} = 0$ for each $i \in V \setminus \{1\}$, and $c_{0,0} = c_{1,0} = \infty$. *PC-TSP* can now be formulated as

$$(PC-TSP) \quad v(PC-TSP) = \min \sum_{i \in V'} \sum_{j \in V'} c_{i,j}\, x_{i,j} \tag{15}$$

subject to

$$\sum_{h \in V'} x_{h,i} = 1, \qquad \text{for each } i \in V' \tag{16}$$

$$\sum_{i \in S} \sum_{j \in V' \setminus S} x_{i,j} \geq 1, \qquad \text{for each } S \subset V' : 1 \in S \tag{17}$$

$$\sum_{h \in V} x_{i,h} = \sum_{h \in V} x_{h,i}, \qquad \text{for each } i \in V \tag{18}$$

$$\sum_{i \in V} p_i\, x_{0,i} \leq \sum_{j \in V} p_j - g \tag{19}$$

$$x_{i,j} \in \{0,1\}, \qquad \text{for each } i,j \subset V'. \tag{20}$$

Constraints (16), (17), and (20) – with objective function (15) – give the well-known *Shortest Spanning 1-Arborescence Problem* (*1-SSAP*). Any solution to *1-SSAP* is a collection of $|V'| - 1$ arcs defining a directed spanning tree rooted at vertex 1, plus an arc entering vertex 1. Constraints (18) ensure that each vertex $i \in V$ is either unrouted (when $x_{0,i} = 1$ and hence, from (16), $\sum_{h \in V} x_{h,i} = 0$) or routed once. Constraint (19) imposes an upper bound on the total prize of the unrouted vertices. An example of a feasible solution to model (15) – (20) is given in Figure 2.

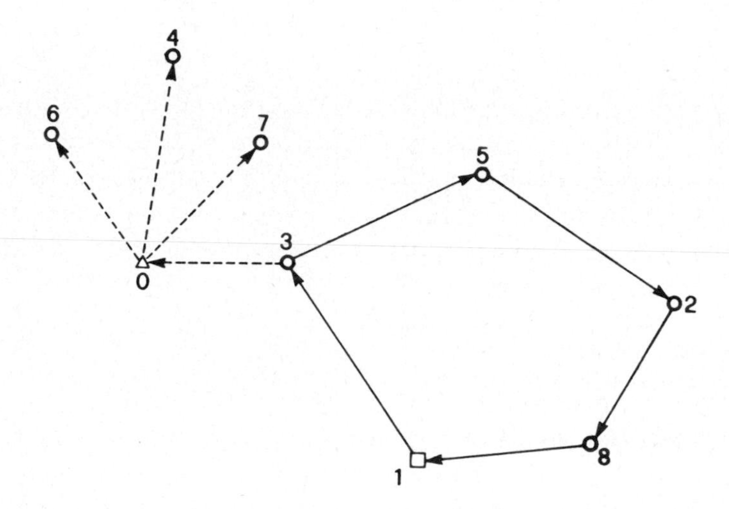

Figure 2. Feasible solution to model (15) – (20), assuming (p_j)= (0,8,5,2,4,5,3,6), $g = 22$.

3. Lower bounds

Different lower bounds for *PC-TSP* can be obtained by exploiting different relaxations of the problem as, for example, those based on the assignment problem or on the arborescence problem substructures pointed out in the previous section. Since none of these bounds dominates the others, a possible way to obtain a strengthened bound is to compute the maximum among them. In this way, however, only one substructure is taken into account, while all the others are completely lost. To partially overcome this drawback and to exploit the complementarity of the available bounds, we use the *additive approach* proposed by Fischetti and Toth (1986), leading to a sequence of increasing lower bounds.

3.1. An additive lower bounding procedure

Let us suppose r bounding procedures $\mathcal{L}^{(1)}, \mathcal{L}^{(2)}, \cdots, \mathcal{L}^{(r)}$ are available for *PC-TSP*, each based on a different relaxation. Let $\tilde{I} = [\tilde{n}, (\tilde{c}_{i,j}), (\tilde{\gamma}_i), (\tilde{p}_i), \tilde{g}]$ be a given *instance* of *PC-TSP*, defined through input data $\tilde{n}$, $(\tilde{c}_{i,j})$, $(\tilde{\gamma}_i)$, $(\tilde{p}_i)$, and $\tilde{g}$, and let $v(\tilde{I})$ be the corresponding optimal solution value. Let us also suppose that, for $h = 1, 2, \cdots, r$ and for any instance $\tilde{I}$, procedure $\mathcal{L}^{(h)}(\tilde{I})$, when applied to $\tilde{I}$, returns a lower bound $\delta^{(h)}$ as well as a *residual instance* $I^{(h)} = [n^{(h)}, (c_{i,j}^{(h)}), (\gamma_i^{(h)}), (p_i^{(h)}), g^{(h)}]$ such that:

i) $v(I^h) \geq 0$;

ii) $\delta^{(h)} + v(I^h) \leq v(\tilde{I})$.

(For example, a residual instance $I^{(h)}$ associated with bound $\delta^{(h)}$ could be obtained by taking $n^{(h)} = \tilde{n}$, $(p_i^{(h)}) = (\tilde{p}_i)$, $g^{(h)} = \tilde{g}$, and by conveniently "reducing" the input costs $(\tilde{c}_{i,j})$ and $(\tilde{\gamma}_i)$ so as to produce non-negative *residual costs* $(c_{i,j}^{(h)})$ and $(\gamma_i^{(h)})$ satisfying condition ii)).

The additive approach generates a sequence of instances of problem *PC-TSP*, each obtained from the previous one by applying a different bounding procedure. A Pascal-like outline of the approach follows.

ALGORITHM ADD-PCTSP:

1. **input** instance $I = [n, (c_{i,j}), (\gamma_i), (p_i), g]$;
2. **output** lower bound δ;

 begin

3. initialize $I^{(0)} := I$, $\delta := 0$;
4. **for** $h := 1$ **to** r **do**

 begin

5. apply $\mathcal{L}^{(h)}(I^{(h-1)})$, thus obtaining value $\delta^{(h)}$ and the residual instance $I^{(h)}$;
6. $\delta := \delta + \delta^{(h)}$

 end

 end.

In order to show that each value δ computed at step 6 is a valid lower bound for *PC-TSP*, we prove by induction on h that

$$\sum_{l=1}^{h} \delta^{(l)} + v(I^{(h)}) \leq v(I) \tag{21}$$

holds for $h = 1, 2, \cdots, r$. The basis, $h = 1$, is trivially true. Suppose now that inequality (21) holds for $h = \overline{h} - 1$. From conditions ii) on the residual instance, we have

$$\sum_{l=1}^{\overline{h}-1} \delta^{(l)} + \delta^{(\overline{h})} + v(I^{(\overline{h})}) \leq \sum_{l=1}^{\overline{h}-1} \delta^{(l)} + v(I^{(\overline{h}-1)})$$

so inequality (21) holds for $h = \overline{h}$ as well.

Each value δ computed at step 6 is then a valid lower bound for *PC-TSP* since, because of condition i), $v(I^{(h)}) \geq 0$. The sequence of values δ is non-decreasing, since increments $\delta^{(h)}$ are clearly non-negative for $h = 2, \cdots, r$.

More details on the additive approach are given in Fischetti and Toth (1986). Applications to the Asymmetric Travelling Salesman Problem, the Multiple Depot Vehicle Scheduling Problem and the Symmetric Travelling Salesman Problem are reported, respectively, in Fischetti and Toth (1987a), Carpaneto, Dell'Amico, Fischetti and Toth (1987), and Carpaneto, Fischetti and Toth (1987).

Algorithm *ADD-PCTSP* requires the computation of valid residual instances corresponding to all the available lower bounds, except the last (since residual instance $I^{(r)}$ is not used to increase the final value of δ). In the following, we propose different relaxations of *PC-TSP*, and describe how to obtain the corresponding lower bounds and residual instances.

3.2. A bound based on the assignment problem substructure

Let us consider instance $I = [n, (c_{i,j}), (\gamma_i), (p_i), g]$, and the formulation of *PC-TSP* given by (9) – (14). Removing connectivity constraints (13) produces a relaxed problem in which one calls for a minimum cost collection of subtours covering all the vertices and such that the sum of the prizes of the vertices not covered by loops is not less than goal g. Such a *Prize-Collecting Assignment Problem* (*PC-AP*) turns out to be $\mathcal{NP}$-hard, since it contains the *0-1 Knapsack Problem* (formulated as a minimization problem) as a particular case, arising when $c_{i,j} = b_j$ (for $i, j \in V, i \neq j$), $c_{j,j} = 0$ (for $j \in V \setminus \{1\}$), $c_{1,1} = \infty$, where $b_j > 0$ (for $j \in V \setminus \{1\}$), and $b_1 = 0$.

A lower bound for *PC-AP*, and hence for *PC-TSP*, could be obtained by solving its continuous relaxation, $\overline{PC\text{-}AP}$, given by (9) – (12) and

$$x_{i,j} \geq 0, \qquad \text{for each } i, j \in V.$$

A different approach (leading to the same value of the lower bound as $\overline{PC\text{-}AP}$) is to relax *PC-AP* in a Lagrangean fashion by imbedding constraint (12) in the objective function (9), and to solve the corresponding Lagrangean dual problem:

$$(LD - AP) \quad v(LD - AP) = \max_{\lambda \geq 0} L(\lambda)$$

where

$$L(\lambda) = -\lambda \Delta + \min \sum_{i \in V} \sum_{j \in V} \tilde{c}_{i,j}(\lambda) x_{i,j}$$

subject to (10), (11) and (14),

with

$$\Delta = \sum_{j \in V} p_j - g,$$
$$\tilde{c}_{i,j}(\lambda) = c_{i,j}, \qquad \text{for } i, j \in V, i \neq j,$$
$$\tilde{c}_{i,i}(\lambda) = c_{i,i} + \lambda p_i, \qquad \text{for } i \in V.$$

$L(\lambda)$ can be efficiently computed by solving the instance of the assignment problem $AP(\lambda)$ defined by cost matrix $(\tilde{c}_{i,j}(\lambda))$. As for the solution of *LD-AP*, the maximization of $L(\lambda)$ over $\lambda \geq 0$ is obtained in an effective way by applying an iterative subgradient optimization technique specialized for the scalar multiplier case, yielding a sequence $\lambda_1, \lambda_2, \cdots$ of multipliers. A "good" value for λ_1 can be heuristically computed through the solution of the following continuous *Knapsack Problem*:

$$\min \sum_{i=1}^{n} a_i y_i$$

subject to

$$\sum_{i=1}^{n} p_i y_i \geq g$$
$$0 \leq y_i \leq 1, \qquad \text{for } i = 1, \cdots, n,$$

where $a_i = \min\{c_{h,i} : h = 1, \cdots, n, h \neq i\} - \gamma_i$ is a lower bound on the extra cost incurred for visiting vertex i instead of leaving it unvisited ($i = 1, \cdots, n$). Value λ_1 is set equal to the optimal dual multiplier associated with constraint $\sum_{i=1}^{n} p_i y_i \geq g$.

Computing $L(\lambda_1)$ requires $O(n^3)$ time, e.g., through the Hungarian algorithm (see, e.g., Lawler (1976)). Computation of $L(\lambda_2), L(\lambda_3), \cdots$ can be speeded-up through parametric techniques which take advantage of the fact that, at each subgradient iteration, only coefficients $\tilde{c}_{i,i}(\lambda)$ over the main diagonal are affected by the change of λ with respect to the previous iteration.

Let $\lambda^* \geq 0$ be an optimal (or near optimal) solution of *LD-AP*. Consider the assignment problem $AP(\lambda^*)$ defined by cost matrix $(\tilde{c}_{i,j}(\lambda^*))$, and let $(u_i^*) - (v_j^*)$ be the optimal solution of the associated linear programming dual problem. It is well known that $\sum_{i \in V}(u_i^* + v_i^*)$ gives the optimal solution value of $AP(\lambda^*)$. Hence $L(\lambda^*) = -\lambda^* \Delta + \sum_{i \in V}(u_i^* + v_i^*)$ is a valid lower bound for *PC-AP* and for *PC-TSP* as well. As for the corresponding residual instance $\overline{I}$, we define $\overline{I} = [\overline{n}, (\overline{c}_{i,j}), (\overline{\gamma}_i), (\overline{p}_i), \overline{g}]$, where $\overline{n} = n$, $(\overline{\gamma}_i) = (\overline{c}_{i,i})$, $(\overline{p}_i) = (p_i)$, $\overline{g} = g$, and

$$\overline{c}_{i,j} = \tilde{c}_{i,j}(\lambda^*) - u_i^* - v_j^* \geq 0, \qquad \text{for each } i, j \in V, \tag{22}$$

i.e., the ***residual-cost matrix*** $(\overline{c}_{i,j})$ is given by the non-negative ***reduced-cost matrix*** associated with the optimal solution to $AP(\lambda^*)$. In fact, consider an optimal solution to instance I and let $(x_{i,j})$ be the corresponding incidence matrix in formulation (9) – (14). Because of (10) and (11), we have:

$$L(\lambda^*) + \sum_{i \in V} \sum_{j \in V} \overline{c}_{i,j} x_{i,j} =$$

$$= -\lambda^* \Delta + \sum_{i \in V} (u_i^* + v_i^*) + \sum_{i \in V} \sum_{j \in V \setminus \{i\}} (c_{i,j} - u_i^* - v_j^*) x_{i,j} +$$

$$\sum_{i \in V} (c_{i,i} + \lambda^* p_i - u_i^* - v_i^*) x_{i,i} =$$

$$= \lambda^* (\sum_{i \in V} p_i \, x_{i,i} - \Delta) + \sum_{i \in V} u_i^* (1 - \sum_{j \in V} x_{i,j}) + \sum_{j \in V} v_j^* (1 - \sum_{i \in V} x_{i,j}) +$$

$$\sum_{i \in V} \sum_{j \in V} c_{i,j} \, x_{i,j} = \lambda^* (\sum_{i \in V} p_i \, x_{i,i} - \Delta) + \sum_{i \in V} \sum_{j \in V} c_{i,j} \, x_{i,j} \, .$$

Hence, because of (12) and since $\lambda^* \geq 0$,

$$L(\lambda^*) + \sum_{i \in V} \sum_{j \in V} \overline{c}_{i,j} \, x_{i,j} \leq \sum_{i \in V} \sum_{j \in V} c_{i,j} \, x_{i,j} \tag{23}$$

holds. Since $(x_{i,j})$ is clearly feasible for reduced instance $\overline{I}$, we have

$$L(\lambda^*) + v(\overline{I}) \leq L(\lambda^*) + \sum_{i \in V} \sum_{j \in V} \overline{c}_{i,j} \, x_{i,j} \leq \sum_{i \in V} \sum_{j \in V} c_{i,j} \, x_{i,j} = v(I) \, .$$

In addition, $v(\overline{I}) \geq 0$ follows from (22), so both conditions i) and ii) on the residual instance $\overline{I}$ are satisfied.

It is worth noting that, from (23), residual costs $\overline{c}_{i,j}$ and $\overline{\gamma}_i = \overline{c}_{i,i}$ give a lower bound on the extra cost incurred if, respectively, arc (i, j) is included in the solution and vertex i is left unrouted ($\overline{\gamma}_i$ can be greater than 0 even when $\gamma_i = 0$ in the original instance).

Also note that the solution of $AP(\lambda^*)$ is generally infeasible for *PC-TSP* both because it can contain more than one subtour of cardinality greater than one, and because the total prize "collected" can be less than g.

3.3. A bound based on the arborescence problem substructure

Let us consider instance $I = [n, (c_{i,j}), (\gamma_i), (p_i), g]$ and the formulation of *PC-TSP* given by (15) – (20). Removing equilibrium constraints (18) produces a relaxed problem in which one calls for a shortest spanning 1-arborescence in the augmented graph G' such that the sum of the prizes of vertices j for which arc $(0, j)$ is not in the solution, is greater or equal to goal g. Such a problem is $\mathcal{NP}$-hard, since the 0-1 Knapsack Problem easily transforms to it. Note that the relaxed problem cannot be viewed as a *Prize-Collecting 1-Arborescence Problem*, since removing artificial vertex 0 could produce a solution given by a family of disconnected branchings. Due to the lack of reachability from vertex 1 of the vertices whose prize has been "collected", this problem generally leads to a poor lower bound.

A second approach is to relax in a Lagrangean fashion the formulation of *PC-TSP* given by (15) – (20) by imbedding constraints (18), with $i \in V \setminus \{1\}$, and (19) in the objective function. The corresponding Lagrangean dual problem is

$$(LD - SSAP) \quad v(LD - SSAP) = \max\{L(\lambda, u) : \lambda \geq 0, u \in \Re^n, u_1 = 0\}$$

where

$$L(\lambda, u) = -\lambda\,\Delta + \min \sum_{i \in V'} \sum_{j \in V'} \tilde{c}_{i,j}(\lambda, u) x_{i,j} \tag{24}$$

subject to (16), (17), (20) and

$$\sum_{h \in V'} x_{1,h} = 1 \tag{25}$$

with

$$\Delta = \sum_{j \in V} p_j - g\,,$$

$$\tilde{c}_{i,j}(\lambda, u) = c_{i,j} + u_i - u_j\,, \qquad \text{for } i, j \in V\,,$$

$$\tilde{c}_{0,j}(\lambda, u) = c_{0,j} + \lambda\, p_j\,, \qquad \text{for } j \in V\,,$$

$$\tilde{c}_{i,0}(\lambda, u) = c_{i,0}\,, \qquad \text{for } i \in V'\,.$$

It is worth noting that u_1 is set to 0, since the corresponding constraint (18) (with $i = 1$) has not been removed, but transformed to (25) (recall $c_{1,0} = c_{0,1} = \infty$). $L(\lambda, u)$ can be efficiently computed in $O(n^2)$ time by solving an equivalent instance of the shortest-spanning 1-arborescence on G' defined by cost matrix $(c'_{i,j}(\lambda, u))$, where

$$c'_{i,j}(\lambda, u) = \tilde{c}_{i,j}(\lambda, u)\,, \qquad \text{for } i, j \in V', i \neq 1\,,$$

$$c'_{1,j}(\lambda, u) = \tilde{c}_{1,j}(\lambda, u) + M\,, \qquad \text{for } j \in V'\,,$$

with M a sufficiently large positive number (so as to ensure out-degree one for vertex 1).

The maximization of $L(\lambda, u)$ can be obtained by applying standard subgradient optimization techniques. Let $\lambda^* \geq 0$ and $u^* \in \Re^n$ be the optimal (or near optimal) solution of *LD-SSAP*. Hence, $L(\lambda^*, u^*)$ is a valid lower bound for *PC-TSP*.

The corresponding residual instance $\overline{I}$ can be defined as follows. It is well known that the problem given by (24), (16), (17), (25), and (20) is equivalent to its linear programming relaxation (see Edmonds (1967) and Fulkerson (1974)). Let (v_i^*), $(\mu^*(S))$ and α^* $(= M)$ be the optimal linear programming dual variables associated, respectively, with constraints (16), (17), and (25), when $\lambda = \lambda^*$ and $u = u^*$. Hence $L(\lambda^*, u^*) = -\lambda^*\,\Delta + (\sum_{i \in V'} v_i^* + \sum_{S \subset V' : 1 \in S} \mu^*(S) - \alpha^*)$. We can now define $\overline{I} = [\overline{n}, (\overline{c}_{i,j}), (\overline{\gamma}_i), (\overline{p}_i), \overline{g}]$, where $\overline{n} = n$, $(\overline{\gamma}_i) = (\overline{c}_{0,i})$, $(\overline{p}_i) = (p_i)$, $\overline{g} = g$, and

$$\overline{c}_{i,j} = c'_{i,j}(\lambda^*, u^*) - v_j^* - \sum_{\substack{S \in V' : 1 \in S, \\ i \in S, j \in V' \setminus S}} \mu^*(S) \geq 0\,, \qquad \text{for } i \in V', j \in V\,.$$

So the residual-cost matrix $(\bar{c}_{i,j})$ is given by the non-negative reduced-cost matrix associated with the optimal solution of the instance of the shortest-spanning 1-arborescence defined by cost matrix $(c'_{i,j}(\lambda^*, u^*))$. This reduced cost matrix can be computed in $O(n^2)$ time (see Fischetti and Toth (1987b)). The proof of correctness, based on standard LP manipulations of the objective function, is similar to that of the previous subsection and is hence omitted.

3.4. A bound based on disjunction

Let us consider instance $I = [n, (c_{i,j}), (\gamma_i), (p_i), g]$ of *PC-TSP*, and let σ be any feasible solution to I. For each vertex subset $S \subset V$ with $1 \in S$, two cases can occur:

1) solution σ routes only vertices in S;
2) solution σ routes at least one vertex in $V \setminus S$.

A valid lower bound $\delta(S)$ for instance I is then given by the minimum between the lower bounds associated with the two problems, say P_1 and P_2, obtained by imposing condition 1) or 2).

A lower bound $\delta_1(S)$ for problem P_1 can be obtained by defining instance $I^{(1)}(S)$ derived from I by setting $c_{i,j} = \infty$ if $i \in V \setminus S$ or $j \in V \setminus S$, with $i \neq j$ (thus imposing as unrouted all the vertices in $V \setminus S$), and by solving the corresponding *LD-AP*, as described in Section 3.2 (obviously, $\delta_1(S) = \infty$ if $\sum_{i \in S} p_i < g$).

As for problem P_2, let f_h (resp. b_h) be the cost of the shortest path from vertex 1 to vertex h (resp. from vertex h to vertex 1), for each $h \in V$. A lower bound $\delta_2(S)$ for P_2 is then given by:

$$\delta_2(S) = \min\{f_h + b_h : h \in V \setminus S\}.$$

In order to obtain the best lower bound δ, one has to choose subset S so as to maximize $\delta(S) = \min\{\delta_1(S), \delta_2(S)\}$. To this end, we note that for any vertex subset S' with $1 \in S'$ and $S' \subseteq S$, conditions $\delta_1(S') \geq \delta_1(S)$ and $\delta_2(S') \leq \delta_2(S)$ hold. Now, for any given subset S define $S' = \{j \in V : f_j + b_j < \delta_2(S)\}$. Clearly, $\delta_2(S') = \delta_2(S)$ (from the definition of S'), while $S' \subseteq S$ (since no vertex $j \in V \setminus S$ can have $f_j + b_j < \delta_2(S)$), and then $\delta_1(S') \geq \delta_1(S)$. It follows that $\delta(S') \geq \delta(S)$, and therefore the maximization of $\delta(S)$ requires that we explicitly consider only subsets $S_2, S_3, \cdots, S_n$, with $S_h = \{j \in V : f_j + b_j < f_h + b_h\}$ for $h = 2, 3, \cdots, n$ (since $\delta_2(S)$ can attain only values $f_2 + b_2$, $f_3 + b_3$, $\cdots$, $f_n + b_n$). Let S_{h^*} $(2 \leq h^* \leq n)$ be the subset maximizing $\delta(S_h)$. Since $S_2 \subseteq S_3 \subseteq \cdots \subseteq S_n$, we have $\delta_1(S_2) \geq \delta_1(S_3) \geq \cdots \geq \delta_1(S_n)$ and $\delta_2(S_2) \leq \delta_2(S_3) \leq \cdots \leq \delta_2(S_n)$. Index h^* can then be determined through binary search by exploiting the property that, for each h $(2 \leq h \leq n)$, $h^* \geq h$ if $\delta_1(S_h) \geq \delta_2(S_h)$, while $h^* \leq h$ if $\delta_1(S_h) \leq \delta_2(S_h)$.

A residual instance $\bar{I} = [\bar{n}, (\bar{c}_{i,j}), (\bar{\gamma}_i), (\bar{p}_i), \bar{g}]$ corresponding to lower bound $\delta(S_{h^*})$

is obtained as follows. Define $\overline{n} = n$, $(\overline{p}_i) = (p_i)$, and $\overline{g} = g$. Now let $\overline{I}^{(1)} = [\overline{n}, (\overline{c}^{(1)}_{i,j}), (\overline{\gamma}^{(1)}_i), (\overline{p}_i), \overline{g}]$ be the residual instance associated with $\delta_1(S_{h^*})$ computed over instance $I^{(1)}(S_{h^*})$ as described in Subsection 3.2 (with $\overline{c}^{(1)}_{i,j} = \infty$ and $\tilde{\gamma}^{(1)}_i = \infty$ for $i,j \in V$ when $\sum_{i \in S_{h^*}} p_i < g$ and hence $\delta_1(S_{h^*}) = \infty$). So, $\delta_1(S_{h^*}) + v(\overline{I}^{(1)}) \leq v(I^{(1)}(S_{h^*}))$. Define $\overline{I}^{(2)} = [\overline{n}, (\overline{c}^{(2)}_{i,j}), (\overline{\gamma}^{(2)}_i), (\overline{p}_i), \overline{g}]$, where $\overline{\gamma}^{(2)}_i = \gamma_i$ for $i \in V$, and $\overline{c}^{(2)}_{i,j} = \min\{c_{i,j} + f_i - f_j, c_{i,j} - b_i + b_j\}$ for each $i,j \in V$. Values $\overline{c}^{(2)}_{i,j}$ are non-negative because of the definition of f_h and b_h as shortest path costs. Residual costs $(\overline{c}_{i,j})$ and $(\overline{\gamma}_i)$ of instance $\overline{I}$ can now be computed as: $\overline{c}_{i,j} = \min\{\overline{c}^{(1)}_{i,j}, \overline{c}^{(2)}_{i,j}\}$, $\overline{\gamma}_i = \min\{\overline{\gamma}^{(1)}_i, \overline{\gamma}^{(2)}_i\}$ for $i,j \in V$. In fact, $v(\overline{I}) \geq 0$ clearly holds. To prove that $\delta(S_{h^*}) + v(\overline{I}) \leq v(I)$, consider an optimal solution σ of instance I defined through arc set $A(\sigma)$ and the set of unrouted vertices $U(\sigma)$ (hence $v(I) = \sum_{(i,j) \in A(\sigma)} c_{i,j} + \sum_{i \in U(\sigma)} \gamma_i$). If solution σ routes only vertices in S_{h^*}, then $\delta(S_{h^*}) + v(\overline{I}) \leq \delta_1(S_{h^*}) + v(\overline{I}^{(1)}) \leq v(I^{(1)}(S_{h^*})) = v(I)$. If at least one vertex $k \in V \setminus S_{h^*}$ is routed in solution σ (i.e. $k \notin U(\sigma)$), a more complex analysis is needed. Partition arc set $A(\sigma)$ into arc sets A_1 and A_2, A_1 defining a path from vertex 1 to vertex k, and A_2 defining a path from vertex k to vertex 1. We have $\sum_{(i,j) \in A_1} (c_{i,j} + f_i - f_j) = f_1 - f_k + \sum_{(i,j) \in A_1} c_{i,j}$, and $\sum_{(i,j) \in A_2} (c_{i,j} - b_i + b_j) = b_1 - b_k + \sum_{(i,j) \in A_2} c_{i,j}$, where $f_1 = b_1 = 0$, and $f_k + b_k \geq \delta_2(S_{h^*})$ (since $k \in V \setminus S_{h^*}$). Hence $\delta(S_{h^*}) + v(\overline{I}) \leq \delta_2(S_{h^*}) + \sum_{(i,j) \in A(\sigma)} \tilde{c}^{(2)}_{i,j} + \sum_{i \in U(\sigma)} \overline{\gamma}^{(2)}_i \leq f_k + b_k + \sum_{(i,j) \in A_1} (c_{i,j} + f_i - f_j) + \sum_{(i,j) \in A_2} (c_{i,j} - b_i + b_j) + \sum_{i \in U(\sigma)} \gamma_i = \sum_{(i,j) \in A_1} c_{i,j} + \sum_{(i,j) \in A_2} c_{i,j} + \sum_{i \in U(\sigma)} \gamma_i = v(I)$.

3.5. Instance transformation

In this subsection we introduce a technique to transform an instance $I = [n, (c_{i,j}), (\gamma_i), (p_i), g]$ of *PC-TSP* to a residual one, obtaining a lower bound of value 0, but allowing the following procedures of the additive scheme to produce better bounds. Let $S_1, S_2, \cdots, S_m$ be a partition of V with $1 \in S_1$ and $\sum_{j \in S_1} p_j < g$. A "residual" instance $\overline{I} = [\overline{n}, (\overline{c}_{i,j}), (\overline{\gamma}_i), (\overline{p}_i), \overline{g}]$ of *PC-TSP*, defined on a reduced complete graph $\overline{G} = (\overline{V}, \overline{A})$ with $|\overline{V}| = \overline{n} = m$, can be built up as follows. Vertices $1, 2, \cdots, \overline{n}$ of $\overline{V}$ correspond, in graph G, to subsets $S_1, S_2, \cdots, S_m$, respectively. We define $\overline{g} = g$, $\overline{p}_i = \sum_{j \in S_i} p_j$ and $\overline{\gamma}_i = \sum_{j \in S_i} \gamma_j$ for $i = 1, 2, \cdots, \overline{n}$. The residual costs $\overline{c}_{i,j}$ $(i,j \in \overline{V})$ are obtained in two steps.

Step 1 (*Shrinking*): define $c'_{i,j} = \min\{c_{h,k} : h \in S_i, k \in S_j\}$ for $i,j \in \overline{V}$, $i \neq j$;

Step 2 (*Compression*): let $L = \{i \in \overline{V} : |S_i| \geq 2\}$, and define $\overline{c}_{i,j}$ = cost of the shortest path from vertex i to vertex j, computed with respect to costs $c'_{i,j}$ and allowing as intermediate vertices only those in L, for $i,j \in \overline{V}$, $i \neq j$.

Step 1 requires $O(n^2)$ time, while step 2 requires $O(\overline{n}^2 \, |L|)$ time through a straightforward modification of the Floyd-Warshall shortest-path algorithm (see Floyd (1962) and Warshall (1962)).

Residual costs $\overline{c}_{i,j}$ satisfy the triangle inequality $\overline{c}_{i,k} + \overline{c}_{k,j} \geq \overline{c}_{i,j}$ for each $i, j \in \overline{V}$, $i \neq j$, and for each $k \in L$.

We now prove that $\overline{I}$ is a valid residual instance associated with lower bound $\delta = 0$. Condition $v(\overline{I}) \geq 0$ trivially follows from hypotheses $c_{i,j} \geq 0$ and $\gamma_i \geq 0$ for each $i, j \in V$, which imply $\overline{c}_{i,j} \geq 0$ and $\overline{\gamma}_i \geq 0$ for each $i, j \in \overline{V}$. As for condition $v(\overline{I}) \leq v(I)$, it is enough to show that any feasible solution σ of instance I corresponds to a feasible solution $\overline{\sigma}$ of instance $\overline{I}$ such that the cost of $\overline{\sigma}$ does not exceed the cost of σ. Feasible solution $\overline{\sigma}$ can be obtained in two steps as follows. First, an intermediate solution $\tilde{\sigma}$ of instance $\overline{I}$ is built up, containing an arc $(i, j) \in \overline{A}$ for each arc (h, k) in σ with $h \in S_i$, $k \in S_j$ and $i \neq j$ (see Figure 3). Because of the definition of $(\overline{p}_i)$, the sum of the prizes of the vertices in $\overline{V}$ visited (one or more times) by solution $\tilde{\sigma}$ is not less than $\overline{g} = g$. Note however that solution $\tilde{\sigma}$ is generally infeasible, since some vertices could be visited more than once. The cost of solution $\tilde{\sigma}$ is less or equal to the cost of solution σ, because of the definition of $(\overline{\gamma}_i)$ and $(\overline{c}_{i,j})$ (recall that $\overline{c}_{i,j} \leq c'_{i,j}$ for each $i, j \in \overline{V}$; see the compression step). A feasible solution $\overline{\sigma}$ of instance $\overline{I}$ is then obtained from solution $\tilde{\sigma}$ by using *shortcuts*, i.e., by replacing the pair of arcs (i, k) and (k, j) of $\tilde{\sigma}$ with arc (i, j) so as to avoid visiting vertex k more than once (see Figure 3b). Because of the compression step, the cost of $\overline{\sigma}$ does not exceed the cost of $\tilde{\sigma}$, and hence that of σ.

3.6. An additive bounding procedure based on instance transformation

We now combine the bound based on the assignment problem substructure (Subsection 3.2) with the instance transformation of the previous subsection to obtain a first additive bounding procedure.

LD-AP is first solved, obtaining a residual instance $\overline{I}$ in which a family of zero-cost subtours exists. A further solution of *LD-AP* on $\overline{I}$ is clearly useless, since no increase of the bound can occur. However, applying the instance transformation (with an appropriate partition of the vertex set) on $\overline{I}$ generally yields a residual instance $\overline{\overline{I}}$ in which no family of zero-cost subtours collecting the goal exists. So, the solution of *LD-AP* on instance $\overline{\overline{I}}$ generally leads to an increase of the lower bound. The two steps (*LD-AP* solution – instance transformation) can be iterated until the assignment problem solution found on the current instance contains only one subtour collecting the goal, thus solving the current *PC-TSP*. The corresponding bounding procedure is given below.

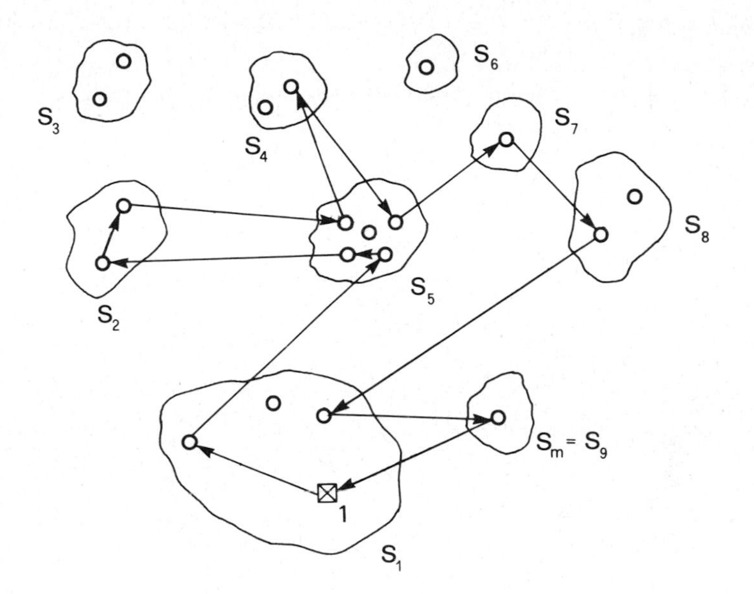

Figure 3a. A feasible solution σ of instance I.

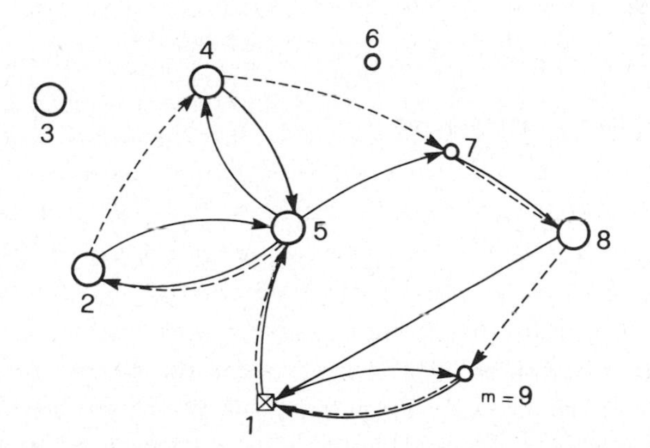

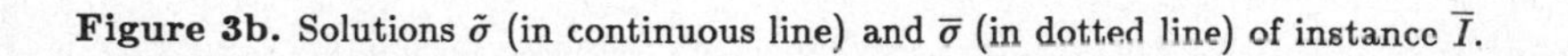

Figure 3b. Solutions $\tilde{\sigma}$ (in continuous line) and $\overline{\sigma}$ (in dotted line) of instance $\overline{I}$.

Procedure ADD1:

1. **input** instance $I = [n, (c_{i,j}), (\gamma_i), (p_i), g]$;
2. **output** lower bound $\delta 1$, residual instance $\overline{I} = [\overline{n}, (\overline{c}_{i,j}), (\overline{\gamma}_i), (\overline{p}_i), \overline{g}]$;

   begin

3. initialize $\overline{I} := I$, $\delta 1 := 0$;

   repeat

4. (**comment** step 1: *LD-AP* solution)
5. solve *LD-AP* over instance $\overline{I}$, thus obtaining lower bound $L(\lambda^*)$, the new residual instance $\overline{I}$ and the optimal solution σ^* of $AP(\lambda^*)$ (see Subsection 3.2);
6. $\delta 1 := \delta 1 + L(\lambda^*)$;
7. let $R_1, R_2, \cdots, R_q$ (with $1 \in R_1$ and $|R_h| \geq 2$ for $h = 1, 2, \cdots, q$) be the subtours in solution σ^* found at step 5, and let $1, v_1, \cdots, v_t$ be the vertices visited by subtour R_1, and $\pi = p_1 + \sum_{k=1}^{t} p_{v_k}$ be the total prize collected by R_1;
8. **if** $(q > 1)$ **or** $(\pi < g)$ **then**
9. **begin** (**comment** step 2: instance transformation)
10. let $u_1, \cdots, u_r$ be the vertices not visited by any subtour R_h $(h = 1, \cdots, q)$;
11. **if** $q > 1$ **then**

    define $S_1 := \{1\}$; $S_k := \{\text{vertices in } R_k\}$ for $k = 2, \cdots, q$; $S_{q+k} := \{v_k\}$ for $k = 1, \cdots, t$; $S_{q+t+k} := \{u_k\}$ for $k = 1, \cdots, r$; $m := q + t + r$

12. **else** define $S_1 := \{\text{ vertices in } R_1\}$; $S_{1+k} := \{u_k\}$ for $k = 1, \cdots, r$;

    $m := r + 1$;

13. apply the instance transformation on $\overline{I}$ with respect to partition $S_1, \cdots, S_m$, obtaining the new residual instance $\overline{I}$

    end

14. **until** $(q = 1)$ **and** $(\pi \geq g)$

    end.

Lines 9 to 13 define an appropriate partition $S_1, S_2, \cdots, S_m$ of the vertex set of the current instance $\overline{I}$ (see Figure 4a) and the corresponding transformed instance (see Figure 4b). Each subtour R_k is collapsed into a single vertex, for which both the extra costs to route it and to leave it unrouted are generally greater than zero. However, this transformation leads to a "loss of information" both because of the shrinking step (degree one constraint is neglected for the vertices in R_k) and the compression step (costs are decreased). Hence we choose not to collapse subtour R_1 (if possible) so as to avoid a loss of information in the "neighbourhood" of vertex 1 (which probably contains the optimal solution).

Execution of step 5 can be accelerated through parametric techniques exploiting the closeness between two consecutive residual instances.

The approach can be viewed as a generalization of that proposed by Christofides (1972) for the Asymmetric Travelling Salesman Problem, both approaches using the shrinking and

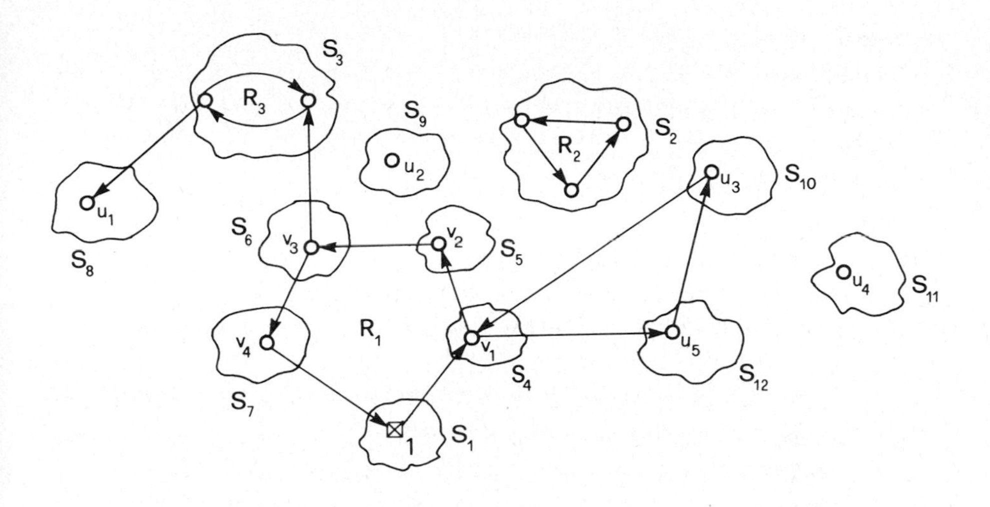

Figure 4a. Current instance $\overline{I}$ and the corresponding partition $S_1, \cdots, S_m$ (with $m = q + t + r = 3 + 4 + 5 = 12$); only arcs having zero residual-cost are drawn.

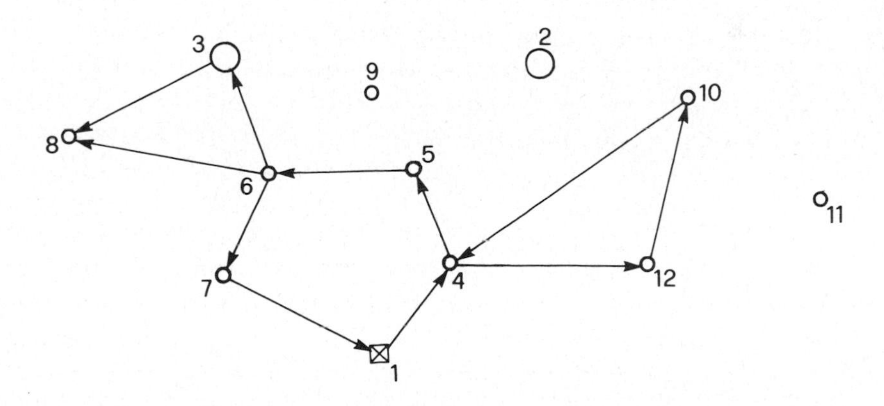

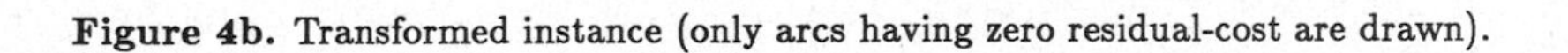

Figure 4b. Transformed instance (only arcs having zero residual-cost are drawn).

compression steps to produce an instance leading to an increase of the bound. A main difference is the relaxation used, at each iteration, to compute the extra bound (LD-AP instead of AP). In addition, the proof of correctness given by Christofides cannot be straightforwardly extended to prove the correctness of procedure $ADD1$ which, instead, immediately follows from the general additive approach.

4. An experimental analysis of lower bounds

The lower bounds proposed in the previous section have been computationally evaluated on different classes of randomly generated test problems.

We report an analysis of the following bounding procedures:

LB1, computing the lower bound based on the assignment problem substructure (see Subsection 3.2);

LB2, computing the lower bound based on the arborescence problem substructure (see Subsection 3.3), and imposing at most 30 subgradient iterations;

LB3, computing the lower bound based on disjunction (see Subsection 3.4);

LB4, computing the lower bound based on instance transformation (see procedure $ADD1$ of Subsection 3.6);

LB5, applying in sequence procedures LB1 and LB3 according to the additive approach.

Analysis of other possible sequences of bounding procedures is not reported, since the higher computing times are generally not rewarded by significant improvements in the bound quality. All procedures have been implemented in FORTRAN ANSI, and run on a Digital VAX 11/780. Solution of the assignment problems has been obtained through procedure APC given in Carpaneto, Martello and Toth (1987), while solution of the arborescence problems was accomplished via the procedure given in Fischetti and Toth (1987b).

Test problems have been randomly generated according to the following distributions:

p_i = uniformly random integer in range (1,100) for $i = 2, \cdots, n$;

$g = \lfloor \alpha \sum_{i=2}^{n} p_i \rfloor$ (with α =0.2, 0.5, 0.8, 1.0).

We have considered $\gamma_i = 0$ for $i = 2, \cdots, n$, as in several practical applications.

Three different generations have been considered for costs $c_{i,j}$ $(i, j = 1, \cdots, n; i \neq j)$:

Class A (pure asymmetric): $c_{i,j}$ = uniformly random integer in range (1,1000);

Class B (triangularized asymmetric): $c_{i,j}$ = shortest path cost from i to j with respect to a pure asymmetric cost matrix;

Class C (Euclidean symmetric): $c_{i,j} = \lfloor \sqrt{(x_i - x_j)^2 + (y_i - y_j)^2} \rfloor$, with (x_i), (y_i) uniformly random integers in range (1,100).

For each class (A,B,C) and each goal (α =0.2, 0.5, 0.8, 1.0), 5 different values of n

(n = 20,40,60,80,100) have been considered. Note that when α = 1.0, *PC-TSP* reduces to *TSP*.

Tables 1 and 2 compare bounding procedures LB1, LB2, LB3, LB4 and LB5 on test problems of classes A and C, respectively (results corresponding to test problems of class B are not given since they are equivalent to those of class A). Each table gives, for each bounding procedure, the average values (computed over five instances) of the ratios (lower bound)/(lower bound computed by LB1), and of the computing times (in VAX 11/780 seconds).

Table 1

Class A (pure asymmetric cost matrices)

Lower bounds comparison (time is given in VAX 11/780 seconds)

α	n	LB1		LB2		LB3		LB4		LB5	
		ratio	time	ratio	time	ratio	time	ratio	time	ratio	time
0.2	20	1.000	0.05	0.949	0.49	1.345	0.03	1.137	0.09	1.142	0.08
	40	1.000	0.20	0.985	1.32	1.160	0.15	1.125	0.38	1.109	0.27
	60	1.000	0.37	0.899	2.58	1.033	0.52	1.007	0.55	1.020	0.51
	80	1.000	1.00	0.853	4.43	1.002	2.59	1.021	1.63	1.018	1.25
	100	1.000	1.65	0.850	6.89	0.982	3.44	1.015	3.36	1.009	2.07
0.5	20	1.000	0.06	1.022	0.49	1.110	0.07	1.097	0.11	1.076	0.08
	40	1.000	0.25	0.983	1.32	0.854	0.32	1.053	0.34	1.034	0.35
	60	1.000	0.67	0.954	2.88	0.619	0.68	1.006	0.79	1.001	0.89
	80	1.000	1.32	0.928	4.70	0.518	1.31	1.005	1.91	1.004	1.62
	100	1.000	2.56	0.942	6.90	0.430	2.20	1.006	3.37	1.001	3.49
0.8	20	1.000	0.06	1.027	0.54	0.736	0.05	1.049	0.12	1.049	0.09
	40	1.000	0.31	0.987	1.34	0.491	0.18	1.013	0.41	1.011	0.42
	60	1.000	0.88	0.977	3.00	0.294	0.60	1.004	1.11	1.001	1.20
	80	1.000	1.89	0.982	5.01	0.258	1.43	1.005	2.67	1.001	2.47
	100	1.000	3.02	0.977	7.05	0.204	1.98	1.003	3.90	1.001	3.82
1.0	20	1.000	0.02	0.938	0.67	0.382	0.02	1.002	0.04	1.002	0.03
	40	1.000	0.06	0.941	1.68	0.255	0.04	1.000	0.11	1.000	0.10
	60	1.000	0.19	0.942	3.36	0.153	0.08	1.000	0.26	1.000	0.26
	80	1.000	0.45	0.972	5.46	0.141	0.14	1.001	0.63	1.001	0.59
	100	1.000	0.74	0.924	7.82	0.102	0.21	1.000	0.93	1.000	0.93

The computing times of procedure LB1, and hence those of LB3, LB4 and LB5, are considerably shorter when α = 1.0 since, in this case, no unidimensional subgradient iteration is required (the optimal value of λ being ∞).

Table 2

Class C (Euclidean symmetric cost matrices)

Lower bounds comparison (time is given in VAX 11/780 seconds)

α	n	LB1		LB2		LB3		LB4		LB5	
		ratio	time	ratio	time	ratio	time	ratio	time	ratio	time
0.2	20	1.000	0.04	1.101	0.35	2.368	0.03	2.133	0.18	2.250	0.05
	40	1.000	0.08	1.181	1.50	2.421	0.07	2.859	0.94	2.523	0.14
	60	1.000	0.16	1.250	3.05	3.517	0.15	4.267	2.13	3.665	0.32
	80	1.000	0.29	1.136	5.53	1.966	0.29	2.669	4.18	2.186	0.55
	100	1.000	0.43	1.253	9.18	1.669	0.42	2.498	7.44	1.893	0.82
0.5	20	1.000	0.04	1.148	0.55	1.603	0.05	1.772	0.18	1.707	0.07
	40	1.000	0.11	1.285	1.63	1.388	0.16	2.010	0.59	1.576	0.21
	60	1.000	0.23	1.343	3.42	1.424	0.50	2.119	1.32	1.743	0.50
	80	1.000	0.41	1.263	5.83	1.146	0.96	1.948	2.63	1.462	0.84
	100	1.000	0.63	1.184	10.73	1.046	1.56	1.788	4.29	1.301	1.25
0.8	20	1.000	0.05	1.150	0.70	1.198	0.07	1.433	0.14	1.322	0.08
	40	1.000	0.13	1.209	1.90	0.840	0.14	1.446	0.37	1.288	0.23
	60	1.000	0.29	1.280	4.13	0.880	0.41	1.510	0.90	1.274	0.60
	80	1.000	0.52	1.239	6.92	0.611	0.54	1.470	1.70	1.191	1.02
	100	1.000	0.84	1.255	9.78	0.568	0.86	1.480	3.05	1.148	1.71
1.0	20	1.000	0.02	1.039	0.93	0.637	0.02	1.188	0.06	1.116	0.03
	40	1.000	0.04	1.204	2.93	0.462	0.04	1.190	0.18	1.107	0.09
	60	1.000	0.11	1.199	6.57	0.433	0.08	1.167	0.40	1.084	0.21
	80	1.000	0.17	1.211	10.33	0.326	0.14	1.184	0.79	1.082	0.35
	100	1.000	0.29	1.238	15.89	0.296	0.21	1.204	1.45	1.065	0.56

Procedure LB2 provides lower bound values worse than those of LB4, with higher computing times.

Procedure LB3 shows good performance for problems with small values of α and n, but gives poor results in the other cases. This behavior can be explained by considering that shortest paths give a reliable approximation of the optimal tour only when both α and n are small.

Procedures LB4 and LB5 exhibit the best performances. In particular, LB4 requires longer computing times but provides tighter bounds, mainly for problems of class C.

Note that the improvements on the value of the bound obtained through procedures LB4 and LB5, with respect to LB1, are remarkable for problems of class C, mainly for small values of α. On the other hand, for problems of classes A and B, the extra computational effort with respect to LB1 is generally small (about 2 to 1 in the worst case). The bound improvement given by LB4 and LB5 decreases when α tends to 1. This is mainly due to the decrease of the gap between the optimal solution value and the bound computed by LB1 (see Table 3 of Section 5). In addition, procedures LB4 and LB5 have been designed for *PC-TSP* , rather than for the pure TSP (more effective additive bounding procedures for both the Asymmetric and the Symmetric TSP have been proposed by Fischetti and Toth (1987a) and Carpaneto, Fischetti and Toth (1987), respectively).

According to the computational results above, a good overall bounding procedure to be imbedded within a branch and bound algorithm can be obtained by applying LB5 for small values of both n and the ratio $g/\sum_{i=1}^{n} p_i$, and LB4 otherwise.

5. A branch and bound algorithm

We now describe a simple branch and bound algorithm for the optimal solution of *PC-TSP*, based on bounding procedures LB4 and LB5 (see Section 4). As for branching, we do not propose an "ad hoc" scheme, but use a straightforward adaptation of the subtour elimination scheme for the Asymmetric *TSP*, as implemented by Carpaneto and Toth (1980).

At each node ν of the branch decision tree, arcs in $I^{(\nu)}$ and $F^{(\nu)}$ are, respectively, imposed and forbidden in the current solution. For each arc $(i,j) \in F^{(\nu)}$, cost $c_{i,j}$ is set to ∞; for each arc $(i,j) \in I^{(\nu)}$, costs $c_{i,h}$ for $h \in V \setminus \{j\}$, $c_{h,j}$ for $h \in V \setminus \{i\}$, γ_i and γ_j are set to ∞. We apply bounding procedure LB5 if $g \leq 0.3 \sum_{i=1}^{n} p_i$ and $n \leq 50$, bounding procedure LB4 otherwise. Let σ^* be the solution of the assignment problem $AP(\lambda^*)$ (see Subsection 3.2), found by LB1 if procedure LB5 is applied, or by the first execution of Step 1 of $ADD1$ if procedure LB4 is applied. If the subtour of σ^* passing through vertex 1 collects a total prize not less than g, we evaluate (using the original costs $(c_{i,j})$ and (γ_i)) the cost of the corresponding feasible solution to *PC-TSP* and possibly update the incumbent optimal solution. In any case, we compare the current lower bound δ_ν with the cost of the incumbent optimal solution z^*. If $\delta_\nu \geq z^*$, node ν is fathomed and a backtracking step is performed. Otherwise, we choose in σ^* a subtour $(v_1, v_2, \cdots, v_h)$, with $h \geq 2$, having the minimum number of non-imposed arcs, and generate h descendent nodes $\nu_1, \nu_2, \cdots, \nu_h$ defined by:

$$\left.\begin{array}{ll} I^{(\nu_i)} = & I^{(\nu)} \cup \{(v_1,v_2), \cdots, (v_{i-1},v_i)\} \\ F^{(\nu_i)} = & F^{(\nu)} \cup \{(v_i, v_{i+1})\} \end{array}\right\} \quad i = 1, \cdots, h,$$

where $v_{h+1} = v_1$. Nodes ν_i for which $I^{(\nu_i)} \cap F^{(\nu_i)} \neq \emptyset$ are clearly not generated.

The branch and bound algorithm has been implemented in FORTRAN ANSI, and run on a Digital VAX 11/780 by considering the test problems of Section 4. Table 3 gives the average values (computed over five instances) of the global computing time, of the number of nodes generated in the branch decision tree, and of the ratio (optimum value)/(lower bound computed by LB1 at the root node). Instances of class C with $n \geq 60$ have not been tried because of excessive computing time.

Table 3

Branch and bound algorithm (time is given in VAX 11/780 seconds)

α	n	Class A			Class B			Class C		
		time	nodes	ratio	time	nodes	ratio	time	nodes	ratio
0.2	20	0.4	7	1.360	0.9	17	1.362	5.2	75	2.969
	40	3.6	25	1.300	21.8	144	1.300	166.0	645	3.589
	60	8.5	39	1.140	69.7	238	1.140	—	—	—
	80	44.9	77	1.102	267.5	386	1.098	—	—	—
	100	94.6	75	1.073	256.8	217	1.073	—	—	—
0.5	20	1.7	38	1.303	9.4	214	1.237	88.0	773	2.245
	40	17.7	154	1.122	116.6	1116	1.120	1377.1	2237	2.291
	60	67.4	365	1.059	485.1	2148	1.051	—	—	—
	80	174.3	412	1.032	677.7	1318	1.028	—	—	—
	100	283.7	461	1.040	753.0	897	1.034	—	—	—
0.8	20	3.6	115	1.153	22.6	859	1.109	216.1	2429	1.697
	40	40.4	457	1.057	102.6	1533	1.037	1193.6	2971	1.597
	60	164.6	1060	1.036	795.7	5045	1.024	—	—	—
	80	307.2	855	1.028	1079.8	5583	1.023	—	—	—
	100	253.8	468	1.015	932.3	2149	1.014	—	—	—
1.0	20	0.4	16	1.057	0.3	9	1.028	31.6	803	1.277
	40	3.7	65	1.017	2.2	30	1.015	463.8	2420	1.278
	60	7.0	76	1.012	1.2	10	1.002	—	—	—
	80	7.7	20	1.009	1.7	4	1.004	—	—	—
	100	41.3	138	1.006	5.7	12	1.001	—	—	—

Problems of class A can be solved within small computing times even for large instances. The computational effort increases with α, except for $\alpha = 1.0$ (the pure TSP case).

Problems of class B are harder than those of class A, although the gap between the optimal solution value and the lower bound at the root node is almost the same. This behaviour can be explained by considering that many solutions exist for these instances, with approximately the same cost, and that the branching rule used is not effective in dealing with such situations.

Problems of class C can be solved only for small values of n, due to the well-known difficulty of solving symmetric and Euclidean travelling salesman problems through AP-

based branch and bound algorithms.

Additional computational experience has been performed, for large-size instances of class A, on the Digital VAX 8650 computer of the Carnegie Mellon University of Pittsburgh (this computer turned out to be about 6 times faster than Digital VAX 11/780). For each goal ($\alpha = 0.2, 0.5, 0.8$), 3 different values of n ($n = 120, 160, 200$) have been considered. Table 4 gives the average results computed over 5 problems.

Table 4

Branch and bound algorithm (time is given in VAX 8650 seconds)

Large-size instances of class A

n	$\alpha = 0.2$			$\alpha = 0.5$			$\alpha = 0.8$		
	time	nodes	ratio	time	nodes	ratio	time	nodes	ratio
120	27.8	125	1.076	93.9	739	1.023	80.6	567	1.023
160	132.6	365	1.077	616.1	1994	1.020	309.9	1147	1.011
200	408.9	489	1.061	794.1	1531	1.023	664.5	1347	1.006

The performance of the branch and bound algorithm could be improved upon considerably, mainly for instances of classes B and C, by designing an "ad hoc" branching scheme and introducing dominance criteria to deal with almost-equivalent solutions, and reduction procedures to impose or exclude subsets of vertices from the optimal solution.

Acknowledgment

We thank the Graduate School of Industrial Administration of the Carnegie Mellon University, Pittsburgh, for the use of Digital VAX 8650.

References

E. Balas (1987), "The Prize-Collecting Traveling Salesman Problem", Research Report No. MSRR-539, GSIA, Carnegie Mellon University, Pittsburgh.

E. Balas and G. Martin (1985), "ROLL-A-ROUND: Software Package for Scheduling the Rounds of a Rolling Mill", Copyright Balas and Martin Associates, 104 Maple Heights Road, Pittsburgh.

G. Carpaneto, M. Dell'Amico, M. Fischetti and P. Toth (1987), "A Branch and Bound Algorithm for the Multiple Depot Vehicle Scheduling Problem", Working Paper OR/87/2, DEIS, University of Bologna, Italy.

G. Carpaneto, M. Fischetti and P. Toth (1987), "New Lower Bounds for the Symmetric Travelling Salesman Problem", Working Paper OR/87/7, DEIS, University of Bologna, Italy.

G. Carpaneto, S. Martello and P. Toth (1987), "Algorithms and Codes for the Assignment Problem", forthcoming in G. Gallo, F. Maffioli, S. Pallottino, B. Simeone and P. Toth, eds., *Fortran Codes for Network Optimization*, Annals of Operational Research, J.C. Baltzer AG, Basel.

G. Carpaneto and P. Toth (1980), "Some New Branching and Bounding Criteria for the Asymmetric Travelling Salesman Problem", *Management Science* 26, 736-743.

N. Christofides (1972), "Bounds for the Traveling Salesman Problem", *Operations Research*, 20, 1044-1055.

J. Edmonds (1967), "Optimum Branchings", *Journal of Research of the National Bureau of Standards-B.Mathematics and Mathematical Physics* 71B, 233-240.

M. Fischetti and P. Toth (1986), "An Additive Bounding Procedure for Combinatorial Optimization Problems", Working Paper OR/86/4, DEIS, University of Bologna, Italy, forthcoming in *Operations Research.*

M. Fischetti and P. Toth (1987a), "An Additive Bounding Procedure for the Asymmetric Travelling Salesman Problem", Working Paper OR/87/5, DEIS, University of Bologna, Italy.

M. Fischetti and P. Toth (1987b), "An Efficient Algorithm for the Min-sum Arborescence Problem", Working Paper OR/87/6, DEIS, University of Bologna, Italy.

R.W. Floyd (1962), "Algorithm 97: Shortest Path", *Communications of ACM*, 5, 345.

D.R. Fulkerson (1974), "Packing Rooted Cuts in a Weighted Directed Graph", *Mathematical Programming* 6, 1-13.

B. Golden, L. Levy and R. Vohra (1985), "Some Heuristics for the Generalized Traveling Salesman Problem", *Proceedings of 1985 Southeast TIMS Conference* (J. Hammesfahr, ed.), Myrtle Beach, 168-170.

B. Golden, L. Levy and R. Vohra (1987), "The Orienteering Problem", *Naval Research Logistics* 34 (3), 307-318.

B. Golden, G. Storchi and L. Levy (1986), "A Time-Relaxed Version of the Orienteering Problem", *Proceedings of 1986 Southeast TIMS Conference* (J.A. Pope and A. Ardalar, eds.), Myrtle Beach, 35-37.

B. Golden, Q. Wang and L. Liu (1987), "A Multi-Faceted Heuristic for the Orienteering

Problem", forthcoming in *Naval Research Logistics.*

G. Laporte and S. Martello (1987), "The Selective Travelling Salesman Problem", Working Paper, EHEC, Universitè de Montreal, Quebec, Canada (presented at the ORSA/TIMS Meeting, New Orleans, May 1987).

E.L. Lawler (1976), *Combinatorial Optimization: Networks and Matroids*, Holt, Rinehart and Winston, New York.

T. Tsiligirides (1984), "Heuristic Methods Applied to Orienteering", *Journal of the Operational Research Society* 35, 797-809.

S. Warshall (1962), "A Theorem on Boolean Matrices", *Journal of ACM* 9, 11-12.

PRACTICAL APPLICATIONS

Vehicle Routing: Methods and Studies
B.L. Golden and A.A. Assad (Editors)

VEHICLE ROUTING AND SCHEDULING IN THE NEWSPAPER INDUSTRY

J. N. Holt and A. M. Watts

Department of Mathematics, University of Queensland
St. Lucia, Queensland 4067, Australia

In this chapter, the vehicle routing and scheduling problem as found in the newspaper industry is addressed. The authors' experience in designing delivery truck routes for the distribution of morning newspapers at three different newspaper companies in different cities of Australia is summarized. The intention is to convey some of the unique features of routing and scheduling in the newspaper industry as well as to discuss more general aspects of route design and maintenance.

1. INTRODUCTION

In this chapter, the vehicle routing and scheduling problem as found in the newspaper industry is addressed. The material presented here summarizes the authors' experience in designing delivery truck routes for the distribution of morning newspapers at three different newspaper companies in different cities of Australia. The intention is to give some of the flavour of the application as well as discuss the technical aspects of route design and maintenance.

For newspapers, the actual routing problem for delivery trucks is only part of a complex production and distribution exercise involving several departments within the company. In order to produce a set of vehicle routes which is acceptable to the company, it is essential to understand first the sometimes conflicting priorities of the different departments, and to decide the relative importance of different data which are provided.

Although newspaper companies like to keep the routes static as far as possible, demand is constantly changing, as are paper size and newsagents. Hence it is necessary to refine the routes regularly. An important part of our work is to provide the means for such refinement in an on-going way. A successful route-editing computer package designed for this purpose is described.

2. PRODUCTION AND DISTRIBUTION OF NEWSPAPERS

Attention is restricted to morning newspapers, where production and distribution normally take place in the early hours of the morning. Such papers are usually produced in a number of different editions. For example, there may be an Air Edition for distribution to remote places by airfreight, a Country Edition for distribution to surrounding rural areas, a Home Edition for metropolitan consumers and perhaps a Late City Edition. Technically then, the problem is one of multi-commodity distribution. However, in practice it is almost totally decoupled. Early editions are printed and despatched before the

Home Edition is started. The routing problem usually only applies to the Home Edition, which accounts for the bulk of the distribution in terms of numbers of papers.

Figure 1 depicts the stages that a newspaper goes through from its production on the printing press to its final destination, the consumer.

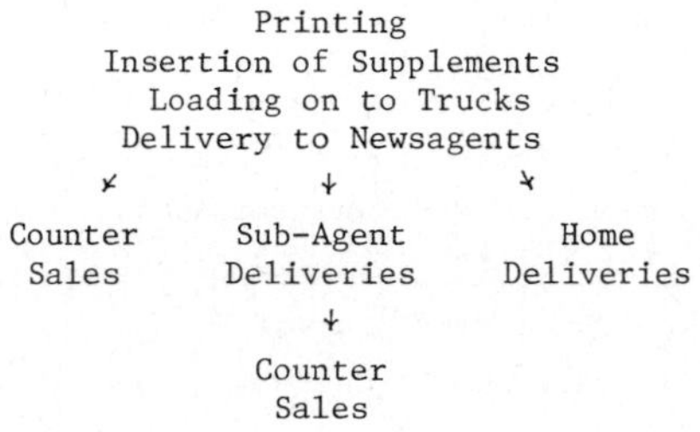

Printing
Insertion of Supplements
Loading on to Trucks
Delivery to Newsagents

↙ ↓ ↘

Counter Sales	Sub-Agent Deliveries	Home Deliveries
	↓	
	Counter Sales	

Figure 1

Newspapers come from the presses on conveyors, and advertising supplements are often inserted either manually, or increasingly, by mechanical inserters. From there, after tying into standard weight bundles, the papers are fed to a number of loading docks into which delivery trucks are reversed. These trucks perform Stage 1 of the distribution exercise, the stage that this chapter is concerned with. They carry the bundles to the city and suburban newsagents and perhaps other contractors throughout the circulation area. On receipt of the papers, the newsagents perform Stage 2 of the distribution by making home deliveries to individual residences, by delivering to their sub-agents which are usually smaller shops, and by making over-the-counter sales when their own shops open. Prior to home deliveries, it is common for agents to have to wrap papers in paper sleeves or plastic wraps. They normally have machines to do this at their premises.

It can be seen from this brief overview that the design of delivery routes in Stage 1 is a sub-problem which is constrained by a range of factors from other components of the total system. In the next section, some of these constraints are discussed in more detail, as is the objective of the route design problem.

3. THE ROUTE DESIGN PROBLEM--CONSTRAINTS AND OBJECTIVES

The aim is to despatch a fleet of trucks from the depot in order to deliver the newspapers to meet specified service standards, usually expressed as a latest time by which the home consumer must have his paper delivered. Typically, this might be 6:30 am during the week and a little later on weekends. The effect of this ultimate time constraint flows back through the system and translates into deadlines by which the newsagents must receive their papers in time to wrap and deliver them. Normally, there is no earliest time for deliveries to agents. Hence their time windows are one-sided. Unfortunately, these deadlines, which are crucial for the design of routes, are not usually known very accurately by the Circulation Department of the newspaper company, and a central part of the exercise is to get good estimates of them. There are several factors which can impact the agent deadlines, such as

wrapping time,
length of agent's delivery run,
number of sub-agents serviced,
number of vehicles used to deliver,
opening time of shops,
the need to meet trains,
etc.

The Circulation staff are usually very sensitive to the needs of the agents to get their papers on time, and it is necessary to convince them to take a hard look at the data they have. Once they are shown the effects on the computed delivery routes of imposing very early deadlines, they better understand the problem.

Of course, the agents' deadlines would be no problem if the trucks could leave early enough. It has been said that if the Home Edition went to press an hour earlier, all the distribution problems would be solved immediately. The starting time of the presses is determined by Editorial policy. The later it is, the greater the probability of catching a story which breaks late, so beating their competition. The effect is that the production and distribution processes are squeezed as tightly as possible. This is a critical fact of life in the newspaper industry.

The marginal cost associated with delaying the press starting times arises from the need to increase the fleet size so that the delivery times can be achieved. For any practical starting time, there is a minimum cost associated with an optimal fleet.

A procedure used in practice when it is known that publishing start time will be delayed, is to hire extra trucks and to split the loads on some existing routes. This can be expensive, and there is always a compromise between cost and delivery times. To our knowledge, no systematic study of this marginal cost has been carried out.

The presses, once started, typically roll for several hours for a large circulation metropolitan daily. During that time, trucks are progressively loaded and despatched. Consequently, superimposed on the routing problem is a scheduling problem for the despatch of trucks. The feasibility of a set of routes will depend on the order in which the trucks are despatched. An algorithm for designing the routes must also contain a scheduling module.

Variability in demand further complicates the route design problem. The numbers of newspapers required by agents exhibit seasonal trends, and the number of pages and hence the weight varies from day to day through the week. Typically, the weekend papers are much larger than those on weekdays, except possibly for Wednesday, when the mid-week advertising occurs. It is common practice to have one set of delivery routes for the weekdays, another for Saturday and another for Sunday. Sometimes a separate set for Wednesday might be warranted. A compromise must be reached between efficiency and ease of implementation. Too many sets of routes, such as one for each day, or different routes daily depending on the loads is an unacceptable solution from the Transport Section's viewpoint. Staff there are concerned about driver training and prefer stability where possible.

In designing routes for any given day, a paper size can be settled on which covers all but a small number of outlier paper sizes for that day based on historical data. Once the routes have been devised, the effect of a larger paper size can be evaluated using the route editing system to be described later.

The existence of capacity constraints on the delivery vehicles is another feature of the problem. The existing fleet mix can be used in the computations, or alternatively the optimization process can be used to recommend the best fleet size mix by performing an analysis with effectively no capacity restrictions on the vehicles. The time constraints then determine the loads and hence the required capacities.

It is common practice in large newspaper companies for some of the delivery trucks to carry out more than one route on the nights corresponding to big papers. The first route takes mainly papers for home delivery, while the second or bulk route delivers the remainder of the papers which are for sale from the agent's shop and so are not needed as early. Thus we have a problem of multiple routes per vehicle with variable route start times. This can be handled by first producing routes for the loads and deadlines appropriate to the first drops, and then dealing with the remainder. Not all agents receive second drops, so that the second set of routes is not a replica of the first. A useful facility in this situation is one for computing the allocation of trucks to routes, re-using trucks where possible, so as to minimize the number of drivers and trucks needed. Such a program has been developed and results in the need to hire fewer rental trucks to supplement the existing fleet. It has been used to reduce the number of trucks by up to three in about thirty.

The objective of any delivery route study is to improve the efficiency of the operation subject to the various sets of constraints discussed above. It is usually assumed that the principal objective is to minimize the number of drivers and vehicles. As a secondary objective, the total distance travelled is to be minimized.

4. DATA REQUIRED

The discussion in the previous section has implied most of the data which needs to be collected. As indicated, some of this data may not be available readily and some investment of effort on the part of newspaper staff would then need to be made in order to enhance the value of the study. The data required for the design of each set of routes is summarized here for completeness.

a) Newsagent Information

1. The name and address of each agent
2. The number of papers required, split into home deliveries and other sales
3. The deadline for delivery (for early and late drops if necessary)
4. Special information such as different addresses for different drops

b) Production Data

1. Starting time of presses
2. Rate of production (papers per minute)
3. Number of pages and weight per page
4. Number of loading docks in use.

c) Publisher's Report and Circulation Manifest

As will be seen in the next section, it is necessary to develop a vehicle travel time model for use in the analysis. This model typically contains a small number of parameters whose values must be determined. By knowing the departure and return times as well as the loads for the routes existing before the analysis, a calibration can be made to choose the parameter values. The Publisher's Report contains the times and loads and the Circulation Manifest contains the list of agents on each of the existing routes.

It should be understood that the data provided will be average-case data. There is variability in almost all aspects of the problem. Factors ranging from the quality of newsprint to the weather can affect the result.

5. THE CALCULATION OF DISTANCES

A variety of approximations and approaches are possible for estimating the distances between pairs of agents and between agents and the depot.

The simplest, and yet often highly effective, approach is to measure the position of each agent by using coordinates on a map relative to some arbitrary origin and to use the Euclidean distance between pairs of points on the map. This straight-line distance should then be multiplied by a factor between one and two to obtain an estimate of actual distance. The factor is found to be around 1.4. For most pairs of agents, this may be adequate. However, there are usually cases where it is not. For example, a river may pass through the city, dividing the delivery region into two parts connected only by bridges. In this case, the coordinates of all the bridges may be added to the locations database. For two agents on opposite sides of the river, the distance between them can then be computed as the minimum over the bridges of the sum of the two straight-line segments connecting the agents to the bridge. If the river has significant meanders, it may also occur that there are agents near, and on the same side of the river for which the actual road distance differs markedly from the approximation based on straight lines. Actual distances can be measured from the map for such cases and entered into the distance matrix using an editing facility. In practice, such cases account for a very small percentage of agents.

If the city has a significant freeway network or the division by waterways is not simply into two disconnected parts as above, a more elaborate means of computing distances may be called for.

A network representation of the freeways may be introduced and used in conjunction with the straight-line distance calculations as follows. Let the nodes of the network be the on-ramps and off-ramps and the arcs the segments of freeway between them, of known lengths. To a first approximation, the main effect of the freeways is in the travel times between the depot and the first drop on a route, and between the last drop and the depot. A shortest path calculation can be made for the distances between the depot and each agent. This is done by extending the network to include arcs from the depot to each on-ramp, and from each off-ramp to each agent. Some selectivity can obviously be invoked here if desired. The arc lengths for these included arcs can be taken as 1.4 times the Euclidean distance. It is also necessary to weight the freeway arc lengths according to an estimated ratio of travel speed on non-freeway roads to travel speed on freeways. The calculation of shortest distance via the network is then compared with the calculation from the straight-line model described earlier, and the minimum is chosen. When computing travel times, non-freeway speeds are then used because of the way the freeway arc lengths have been adjusted. Inter-agent distances are taken from the straight-line model.

Finally, in a complex urban region marked by a dense freeway network and several coastal inlets, rivers and harbours, both of the methods described so far can be expected to be inadequate. In such a case, it may be necessary to construct a detailed approximation to the road network and allow travel only on the approximate network, dispensing altogether with the straight-line approach.

Since the exercise of entering the network may involve thousands of nodes (typically the intersections of roads), and tens of thousands of arcs, it is essential to make good use of modern technology in carrying out the data entry in order to make the approach viable. It has been found that a digitizing palette used in conjunction with a personal computer forms an adequate hardware basis for the exercise. Custom written software for interfacing the digitizer has been written which also interfaces with a Relational Database Management System such as UNIFY, allowing map, node and arc information to be entered to the database as quickly as it takes to reposition the cursor over nodes, and to display the road network quickly.

As well as defining road intersections and freeway ramps as nodes, other nodes are created for the newsagents. Extra arcs connecting these nodes to nearby road-network nodes via actual street distances are added, as well as arcs connecting close agents. Shortest distances between pairs of agents and between agents and the depot can then be computed without the need to consider bridges, railway lines, waterways, etc. An efficient implementation of the Dijkstra shortest path algorithm such as that of Dial [2] is required for the scale of problem dealt with. The network is normally stored in forward star representation.

The feasibility of keeping the resulting distance matrix in RAM during the optimization computations depends on the number of agents, the available memory and the mode of representation. In a large city, it is not unusual to find nearly 1000 agents. In this case, it is desirable to use sparse matrix techniques. Only a small percentage, around 5 to 10 percent, of the entries in the distance matrix need be considered as non-infinite, because of the relative locations of the corresponding agents.

6. THE TRAVEL TIME MODEL

Once the distance matrix is available, the travel times can be estimated. Data on actual travel times between agents is seldom available. However, departure and return times for vehicles on the existing routes are often collected, or can be if so requested. These can be used to fit a simple model with parameters representing average travel speed, a set-up time per stop and an unloading rate. Such a model has been very successful in predicting the mean behaviour of existing routes. Routes of around two hours duration are typically predicted to within 10 - 15 minutes of their mean durations which is within the observed variability caused by weather, different drivers on the route, etc.

7. THE DESPATCH ORDER OF ROUTES

Because of time constraints imposed by the production and loading rates on the one hand and by the newsagent deadlines on the other hand, the feasibility of a given set of routes depends, as stated earlier, on the despatch order, which determines route start times. The development of good routes cannot be considered in isolation from the determination of despatch order. In the route generation method to be described later, these interrelated problems are solved as follows. For each partially developed set of routes at each stage, a heuristic is used to find a despatch order and the feasibility of the routes using this despatch order is checked. A change to the routes is only accepted in the method if it leads to feasibility.

The heuristic used is based on the EDD (earliest due date) rule which minimizes maximum job tardiness in the context of scheduling jobs with due dates on a single machine. See, for example, Baker [2]. In the present context, for the due date of a given route or partial route, we compute a critical time which is the latest time the loading of the vehicle for the route can start without causing a late delivery to an agent on the route. This computation uses the known agent deadlines and the distances and travel time model discussed earlier. Vehicles are normally loaded in parallel on a number of docks. The heuristic schedules the truck with the earliest critical time to start loading whenever a dock is released.

The number of docks used is a parameter of the problem. In our model, it is fixed for the entire production period. In practice, it may be varied a little from time to time according to changes in the rate of production due to equipment malfunction, etc. This is not a serious problem. There is a practical limit to the rate at which drivers can load their vehicles. An estimate of the number of docks to use is given by the ratio of the total production rate to this loading rate. Using fewer docks than this will obviously cause a backup of papers on the docks. Using significantly more than this will delay the early departures, which is usually an undesirable effect.

8. VEHICLE CAPACITY CONSTRAINTS

When checking the feasibility of a set of routes or partial routes, part of the exercise is to ensure that the vehicles are not forced to be despatched with loads above their legal capacities. The variation in vehicle capacities is often large. This requires a procedure which allocates trucks to routes in descending order of load, each time choosing the available truck with the smallest spare capacity. The vehicle allocation program mentioned earlier for use when trucks are to be re-used after returning from their first route can be employed here. Trucks can be re-entered into the available fleet list at appropriate times computed as their return times plus a suitable buffer to allow for variability in route times and for reloading.

9. THE DESIGN OF THE ROUTES

The time-constrained vehicle routing problem is a complex one. The addition of the despatch scheduling makes the problem described here even more so. A variety of heuristic methods for various vehicle routing problems have been developed over recent years, and continue to be developed as reported elsewhere in this volume. It is not our intention to review them here. Rather, a single method is briefly outlined which has proved to be useful in practice.

For a simple one dimensional problem, where all customers are on a straight line passing through the depot, and in which customers on a single route are assigned to delivery routes without time constraints, it can be shown that the minimum distance is achieved by assigning customers to routes according to their distances from the depot. At any stage, the most distant unassigned customer is assigned to the longest feasible route. It might be expected then that in two dimensions, and with time constraints, it is best to give primary consideration to customers at long distances. Any error with them is likely to lead to large increases in the distance travelled and, because of the time constraints, produce a need for extra routes.

A method described in the literature which directly uses this principle is the proximity priority method of Williams [5]. The method, as originally presented, does not consider time constraints but can be easily modified to do so. It also has the virtue of simplicity and is reported to perform about as well as other heuristics on the time constraint free problems.

The method develops routes in parallel. Initially each customer is assigned to a separate route. The algorithm then attempts to combine or modify the routes iteratively to reduce their number and to reduce total distance. Each partial route existing at any iteration has two free route ends which may be candidates for linking to other route ends.

At the start of each iteration, the free route end furthest from the depot, say i, is considered. The route end closest to it, say j, is then found. If it is feasible to do so, the routes to which these ends belong are joined. If this is not feasible, due to time or capacity constraint restriction, an attempt is made to swap the route end j onto the end of the route containing i. The swap is carried out if j is closer to i than to its current neighbour and if the feasibility tests are passed. If neither of these route adjustments are possible, i is tagged and is not considered again. The time and capacity constraints are not used in deciding which agents to consider in order to modify the routes. They are used only in determining the feasibility of the resulting routes. Both possible directional orientations through the extended route are considered when checking time feasibility. In the early stages of the algorithm, the number of partial routes greatly exceeds the number of vehicles, say M, and the capacity feasibility check should only look at the M largest capacity routes.

The basic algorithm as outlined above can be improved upon by the application of a travelling salesman heuristic to each of the constructed routes. The 2-opt procedure of Lin and Kernighan [4] is useful here. In evaluating rearranged routes, a suitable criterion is the critical time for starting load as defined in the section Despatch Order of Routes. The later it is the better, since this will free up valuable time early in the despatch process.

A further possible improvement which is currently being explored is to apply simulated annealing (Kirkpatrick et al. [3]) to the routes, with the interchanges obtained by 2-opt applied across routes as well as within routes. No indication of the success of this technique is available yet.

At the conclusion of the automatic route generation process, there are invariably ways in which the routes can be polished by an experienced route designer. For example, a "side of road" constraint may be violated. That is, a route may be inconvenient because it passes an agent on the wrong side of the road at a place where a turn cannot be made. Because of its limitations, the algorithm sometimes leaves agents by themselves on routes when an opportunity for a join exists. Consequently, final tuning of the routes is usually done manually, with decisions made in conjunction with the staff of the Circulation and Transport Departments of the company. Such joint exercises are useful in promoting better understanding and cooperation between the two groups.

10. ROUTE EDITING SOFTWARE (RUNED)

To facilitate the changes to the routes and to provide a way to update the routes in the future, a route editing and evaluation program is essential. The authors have developed such a system, RUNED, for the newspaper industry. It is a tree structured menu-driven system with entry screens written in standard FORTRAN-77 which is able to be implemented on any computer from a PC to a mainframe.

The selections from the RUNED system main menu are as follows.

1. Exit
2. Agents or Drops
3. Routes
4. Trucks
5. Distances
6. Reports
7. System Parameters

Selection of Agents or Drops displays the menu of that name containing the following entries.

0. Quit without saving changes
1. Exit saving changes
2. Display Agent Information
3. Create an Agent
4. Change Agent Information
5. Delete an Agent
6. Change Drop Information

Selection of one of these items causes a further screen to be displayed through which the requested function can be carried out on the database. Agent information consists of name, address, account code and location. Drop information consists of agent code for the drop, number of papers required, deadline for delivery and route number if required.

Selection of Routes from the Main Menu displays the RUNS menu whose selections are below.

0. Quit without saving changes
1. Exit saving changes
2. Display a Route
3. Create a Route
4. Change a Route Description
5. Delete a Route
6. Move Drops
7. Check Feasibility
8. Change a Drop Requirement or Deadline
9. File a Full Report
10. Report Preordered Routes
11. Optimize

Further entry screens are generated where necessary when selections are made from the list. The functions on this menu allow complete freedom to rearrange the current set of routes. The full report generated may be printed on the line printer. It contains a complete list of routes, showing despatch times, loads and a list of drops with requirements and estimated arrival time at each drop. A version of this without times can be produced if required for driver manifests. The Optimize option allows the application of 2-opt to a set of routes. There is also a version of RUNED which contains the heuristic route generation method as a menu item on the RUNS screen.

Selections 4, 5 and 7 from the Main menu are for maintenance of data. The distance matrix is, however, automatically updated whenever an agent is added or deleted. The system parameters include those concerning production and the travel time model. The Reports available in addition to the Full Report are complete lists of agents, trucks and a short list of routes.

RUNED has been installed at two newspaper companies to date, and has proved to be a valuable tool in two respects. The first is for minor on-going modifications to agent and route lists. The second is for planning exercises preceding significant changes such as newspaper size.

11. DISCUSSION AND CONCLUSIONS

Using the methods discussed here, it has been possible to effect significant savings in terms of numbers of vehicles used and kilometres travelled. In one case, the number of vehicles and drivers was reduced from 35 to 30 on weekends and from 28 to 23 during the week. In another case, a reduction from 19 to 14 was achieved. In this latter case, the original routes were so inefficient that Circulation was able to reduce their number to 15 independently during the course of the work. Nevertheless, a savings of one route and around 8% in total distance was achieved. In another exercise, we showed how to cope with a significantly increased paper size, from 100 pages to about 120 pages, using no extra trucks than were currently being used. A further study is in progress.

The reaction of managers to the analyses we have performed has been very encouraging. In the early stages, there was some apprehension probably due to the fact that we were outsiders to the industry. Driver acceptance has been no problem to date after the initial transition from the existing routes to the new ones. One of the most pleasing side effects of the work is that management now understands the importance of obtaining good data on delivery deadlines, since these crucially affect the quality of the routes devised.

Furthermore, with studies such as these, it is easy to explain to management the effect on the distribution system of the press starting times and the production rate. Certainly the routes are very sensitive to both of these parameters, but as has been stated, there are other considerations which must be taken into account. In summary, this chapter has attempted to discuss vehicle routing and scheduling as applicable to the newspaper industry. The application of the methods described have proven fruitful in all cases the authors have been involved with. The methods use a combination of optimization techniques and judicious use of the knowledge of experts implemented through an easy to use computer software system.

There is always room for improvement, and it is felt that this will come through improved heuristics for the optimization and through the use of evolving technology such as expert systems.

REFERENCES

[1] Baker, K. R., Introduction to Sequencing and Scheduling (John Wiley and Sons, N.Y., 1974).
[2] Dial, R. B., Algorithm 360 Shortest-Path Forest with Topological Ordering, Communications of the ACM 12 (1969), 632-633.
[3] Kirkpatrick, S., Gelatt, C. D. Jr. and Vecchi, M. P., Optimization by Simulated Annealing, Science 220 (1983), 671-680.
[4] Lin, S. and Kernighan, B. W., An Effective Heuristic Algorithm for the TSP, Operations Research 21 (1973), 498-516.
[5] Williams, B. W., Vehicle Scheduling: Proximity Priority Scheduling, J. of the Operational Res. Soc. 33 (1982), 961-966.

Vehicle Routing: Methods and Studies
B.L. Golden and A.A. Assad (Editors)

Scheduling the Postal Carriers for the United States Postal Service: An Application of Arc Partitioning and Routing

Laurence Levy

DISTINCT Management Consultants, 10705 Charter Drive
Suite 440, Columbia, Maryland 21044

Lawrence Bodin

College of Business and Management, University of Maryland, College Park, Maryland 20742

In the Walking Line of Travel Problem, we are given a set of postal carriers who must deliver mail to a specified area (such as a zip code area) while walking their routes. Two algorithms are proposed for solving this problem. These algorithms are tested on a variety of problems based on a portion of a zip code area in Ft. Worth, Texas. It is shown in this paper that both algorithms work well with respect to the multiple criteria proposed for this problem. Furthermore, it is shown that these algorithms are computationally efficient.

1. Introduction

A problem facing the United States Postal Service (USPS) is the scheduling of their postal carriers. Postal carriers deliver mail to residences, industries, and commercial enterprises. Part of this service is carried out in a vehicle while another part of this service is carried out by the carriers walking a route. A different problem evolves depending upon the mix of service made on foot or by vehicle. The problem considered in this paper is called the Walking Line of Travel Problem (WLT) of the USPS. In the WLT, we are given a set of postal carriers who must deliver mail to a given area (for example, a zip code area) by walking their routes. The only driving that the carriers do is between parking locations and the depot. The problem is to develop a procedure which will generate daily work schedules for the carriers.

In the WLT, the daily work schedule for a postal carrier looks like the following. He drives from the depot to an intersection and parks his vehicle on one of the streets adjacent to the intersection. He then walks a cycle of approximately one hour in duration, while delivering mail. After returning to the vehicle at the end of the cycle, the carrier either begins another walking cycle with the vehicle still parked or moves the vehicle to another intersection, where he parks his vehicle and continues as above. After completing his workday (of about five hours in duration), the carrier returns to the depot in his vehicle.

All of the streets in the area where the deliveries take place comprise the arc set A of the Travel Network(TN). Some streets in the arc set A may not require service but all streets in the arc set A can be used by the carrier for deadheading (i.e., walking a street but not making deliveries on the street). The TN is assumed to be a connected network. The arcs in A are undirected (since a carrier can walk these streets in either direction). The nodes of the Travel Network represent all the intersections in the area. In Figure 1a, a Travel Network is displayed. The entire set of arcs in Figure 1a comprise the arc set A.

Two arcs are referred to as undirected counterpart arcs if these two arcs are undirected and oriented between the same two nodes. As an example, a street network where each side of the street is explicitly represented as an arc would have an arc set made up solely of counterpart arcs. It is possible that only one of the two arcs in a counterpart pair requires service. For the WLT, all of the arcs in the Travel Network A occur in counterpart pairs.

Let R be the subset of the arcs in A which actually requires delivery. The Delivery Network (DN) is defined to be the network with arc set R. For the WLT, virtually all of the arcs in the DN occur in counterpart pairs. The time to walk each street in R while delivering that street is assumed known and the Delivery Network need not be connected. In Figure 1b, the Delivery Network extracted from the TN in Figure 1a is shown. In comparing Figures 1a and 1b, note that there are arcs in A which are not in R. For example, only one of the two arcs between nodes 4 and 9 needs to be covered but its counterpart arc does not require delivery.

A partition is defined to be a connected network consisting of a subset of the arcs from the DN and possibly some deadhead arcs from the TN. A walking cycle is assumed to be a sequencing of the arcs in a partition such that every arc in the partition is

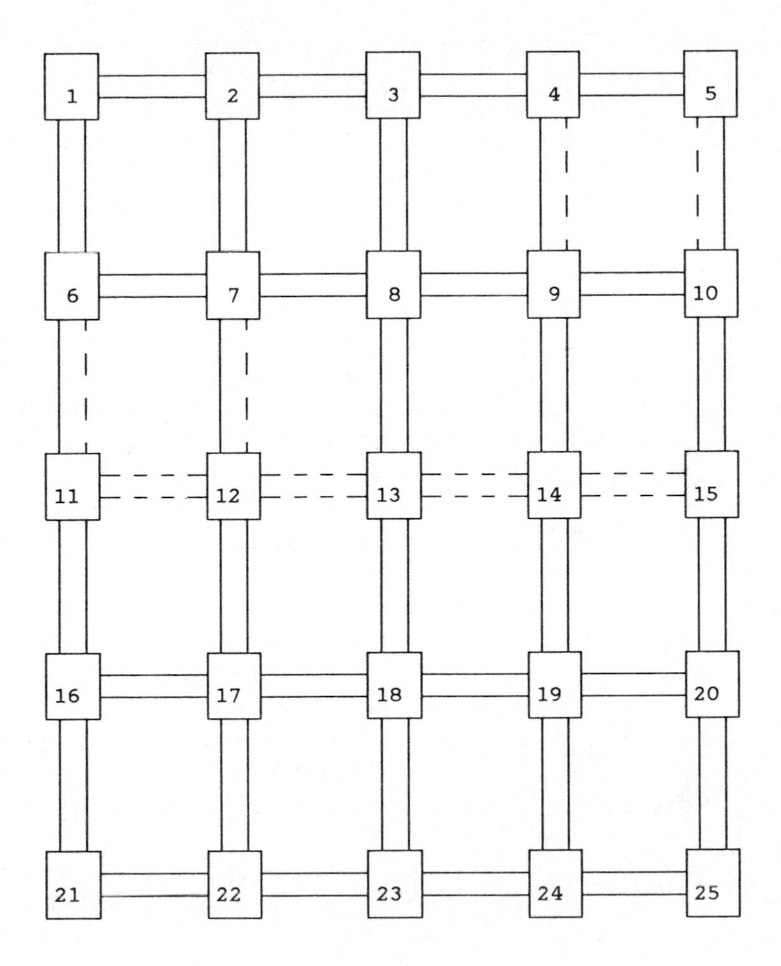

Figure 1a

A Travel Network

Solid arcs are required, dashed arcs are not required.

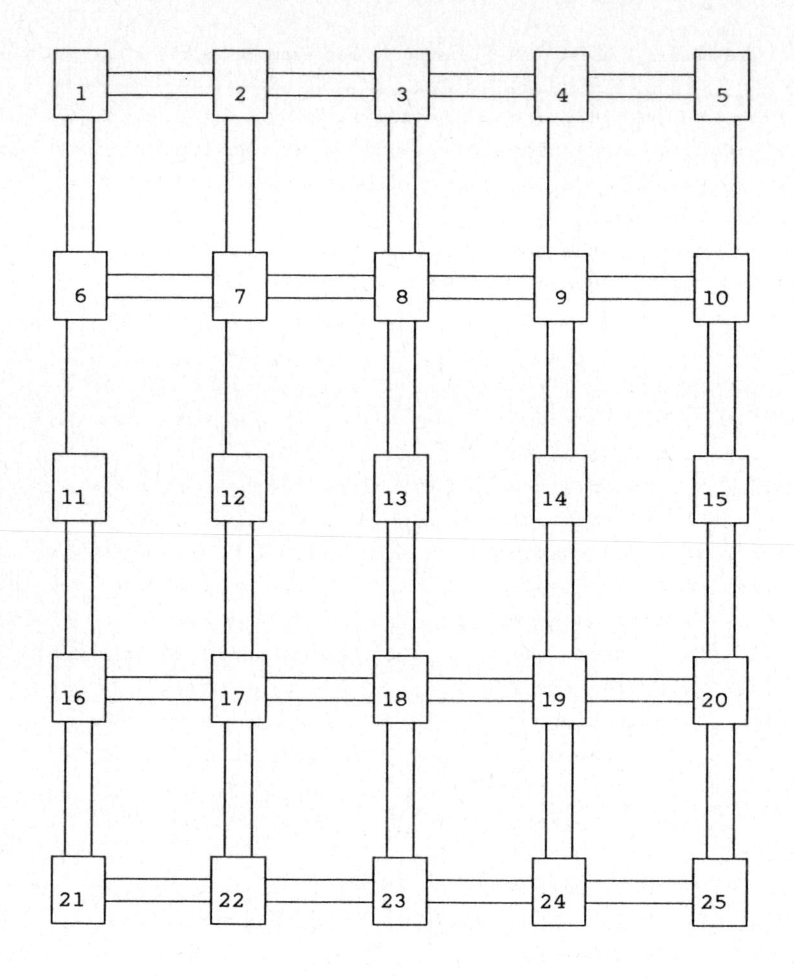

Figure 1b

A Delivery Network

The subset of required arcs from the TN in Figure 1a.

on the cycle and the cycle begins and ends at the same intersection (parking location). The streets to be walked in a daily work schedule are a set of walking cycles. In practice, a walking cycle corresponds to a one hour delivery route and a daily work schedule is roughly five hours of walking time (i.e., 5 walking cycles).

A solution to the WLT is a partitioning of the arcs in R into a set of walking cycles. These cycles are combined into a set of daily work schedules. The constraints on the problem are the following:

A. Every arc in R is covered at least once;

B. Each walking cycle is no longer in duration than the specified upper bound;

C. Both arcs of a counterpart pair receive service within time T of each other, if both of these arcs are in R.

In most arc routing problems, the single objective is to minimize total deadheading time. While this is certainly a concern in forming the walking cycles for the WLT, other criteria on which the set of cycles can be evaluated include ensuring that the walking cycles and the daily work schedules are balanced with respect to duration. Criteria of secondary importance are minimizing the number of returns to the vehicle by the carriers and minimizing the number of parking locations. In Section 2, the criteria, constraints and assumptions of the WLT are described in more detail.

In this paper, we have developed two basic approaches for solving the WLT. The Composite Algorithm initially constructs the daily work schedules and then creates the actual walking cycles within each daily work schedule. The Conventional Algorithm builds the walking cycles first and then aggregates these walking cycles into daily work schedules. The Composite Algorithm derives its name from the use of two separate arc partitioning heuristics. The Conventional Algorithm is named for its similarity to the current manual system used by the USPS.

The Composite Algorithm and the Conventional Algorithm are described in Section 3. The two arc partitioning algorithms fundamental to both the Composite and the Conventional Algorithms are the Arc Begin Partitioning Algorithm (ABPA) and the Node Begin Partitioning Algorithm (NBPA). These algorithms are described in considerable detail in the doctoral dissertation of Levy [16] and are sketched in Section 4.

In Section 5, results which compare the Composite Algorithm and the Conventional Algorithm on six data sets are presented and in Section 6, conclusions based on this research are given.

The problem presented in this paper can be considered a variant of an arc routing and partitioning problem. In [1]-[15] and [17] - [22], other variants of arc routing and partitioning problems are described. In [16], a description of how the WLT is different from the problems considered in these papers can be found.

2. A Description of the WLT

In this section, a more formal description of the WLT is presented. Included in this description are the assumptions, constraints and evaluation criteria of the WLT.

2.1. Assumptions

A. Driving Time

In both the Composite and Conventional Algorithms, the driving route that the carrier is to follow (and, hence, the driving time to and from the depot) is determined after the walking cycles and daily work schedules are formed. The carrier drives from the depot to his first parking location, between parking locations, and from his last parking location to the depot. In the remainder of his day, the carrier walks while delivering the mail. Since the carrier only spends a few minutes of his daily work schedule driving and the rest of his day walking, it is assumed that the driving time can be disregarded when forming the daily work schedules. If the problem had a larger ratio of driving time to walking time, this assumption would not be reasonable.

B. Graphs with Counterpart Arcs

As described in Section 1, all arcs in the Travel Network and a majority of the arcs in the Delivery Network are assumed to occur in counterpart pairs. The algorithms for solving the WLT were developed with this assumption in mind.

C. Arc Weights

The weight on each arc in the TN is the time to deadhead the arc and the weight on each arc in the DN is the time to walk the arc while making deliveries. The cost of a walking cycle is the total time for traversing the cycle while delivering and deadheading the arcs in the cycle.

2.2. Constraints

A. Delivery Time Between Counterpart Arcs

Whenever both arcs in a counterpart pair require delivery, both of these arcs must receive service within some specified time T of each other. In the WLT, T is approximately one hour. This constraint rules out the situation where one side of a street gets service early in the day and the other side does not receive delivery until hours later.

B. Lower and Upper Bounds on Cycle Time and Daily Work Schedule

There exists both a target duration and a lower bound and upper bound on the duration of a walking cycle. The target time for a walking cycle is the ideal length of the cycle. The lower and upper bounds on the duration of the walking cycle provide some flexibility in terms of forming a feasible set of cycles.

A hard bound is a constraint which may not be violated, whereas a soft bound is more of a guideline than a constraint for considering a cycle feasible. In the WLT, the lower bound on the length of a walking cycle is soft, whereas the upper bound is hard. The hard upper bound reflects the amount of weight that can be carried in a mail bag by a postal carrier. The soft lower bound is present to force the cycles created to be of about the same duration. This is done to ensure that the postal carrier does not return to his vehicle an excessive number of times since excessive returns to the vehicle lead to wasted time in refilling the mail bag, inefficient daily work schedules, and potentially increased vehicle movements.

In addition to the bounds on cycle duration, there exist bounds on the walking time for the complete daily work schedule. In the WLT, both the lower and upper bounds on the daily work schedule are soft. If the upper bound is violated, the carrier's work day may be longer than desired so that the carrier might earn some overtime pay. The lower bound on the daily work schedule exists to ensure that a carrier has a realistic minimum amount of time associated with his work day. Even if the individual walking cycles fail to all satisfy the lower bound on cycle travel time, the daily work schedules can often be constructed in such a way that they all fall between the upper and lower bounds in length.

2.3. Criteria for a Good Solution

A. Minimize Deadheading

If all the delivery arcs in the WLT were counterpart pairs, then every node in the DN would be of even degree and it would be possible to find a partitioning of the DN with no deadhead walking time. Because of the other criteria described below, this partitioning might not represent a good solution. On the other hand, in networks where the DN is disconnected, the issue of minimizing deadheading time becomes very important because each carrier might spend a considerable amount of time deadheading between parts of his walking cycles.

B. Balance

A set of walking cycles are balanced when all the cycle times fall between pre-specified lower and upper bounds. For the same amount of deadheading, a solution in which all the cycle times satisfy the soft lower bound is preferred.

The daily work schedules must also be balanced. A balanced set of daily work schedules would have the walking time on each daily work schedule also fall between pre-specified lower and upper bounds. If all the daily work schedules satisfied these bounds, the set of carriers would have approximately equal work days. This concept of balanced work days is extremely important in the acceptance by the workers and management of the schedules generated by the algorithm.

C. Minimize Number of Park Locations

Although the travel time to and from the depot and the travel time between intersections serving as start/end nodes for cycles are not considered in the development of algorithms for the WLT, there is an obvious incentive to reduce the vehicle movements for each carrier. A secondary criteria in the WLT is to keep the number of vehicle movements to a minimum in order to reduce the total driving time. Thus, if two solutions appeared equal with respect to deadheading and balance, the solution with fewer park locations would be preferred.

D. Minimize Returns to the Vehicle

There is a setup time involved every time the carrier extracts mail from his vehicle to walk a cycle. This time is not explicitly modeled in the algorithms but there is a desire to minimize the number of returns to the vehicle the carrier makes during his work day. To accomplish this in the algorithms, the number of walking cycles formed is forced to be approximately equal to the total pairs of arcs incident to all of the parking locations. Variations occur as a result of satisfying the upper bound on walking cycle duration.

2.4. Glossary of Terms

The terms used throughout this paper (all of which have been defined previously) are defined again in a reference list.

Counterpart Arcs - two arcs between the same two nodes.

Daily Work Schedule - a set of walking cycles that constitute a full delivery work day for a carrier.

Deadhead - an arc which is walked by no deliveries are made.

Delivery Network - the subset of the arcs in the Travel Network which require walking delivery service.

Partition - a connected network consisting of a subset of the arcs from the Partition Network.

Partition Network - the arcs of the Delivery Network are supplemented with a dummy counterpart arc for every arc which does not already have a counterpart arc.

Travel Network - a street network where every street is represented by a counterpart arc pair.

Walking Cycle - a sequencing of the arcs in a partition such that every arc in the partition is in the cycle and the cycle begins and ends at the same intersection.

Walking Line of Travel - the problem of scheduling the walking delivery routes for the United States Postal Service.

3. Approaches for Solving the WLT

The two algorithms for solving the WLT, call the Composite Algorithm and the Conventional Algorithm, are now presented. Imbedded within these algorithms are two arc partitioning heuristics, called the ABPA and the NBPA. The ABPA and the NBPA are designed to partition a network which consists of all counterpart arcs (i.e., every node has even degree). As noted previously, the Delivery Network may not consist of all counterpart arcs. The DN is modified to produce the Partition Network. The Partition Network (PN) is a network consisting of all arcs from the Delivery Network with the addition of a dummy counterpart arc for every arc which does not already have a counterpart arc present in the DN. The time associated with a dummy counterpart arc is the minimum deadhead time of the corresponding pair of arcs from the TN. The PN for the DN in Figure 1b is shown in Figure 1c.

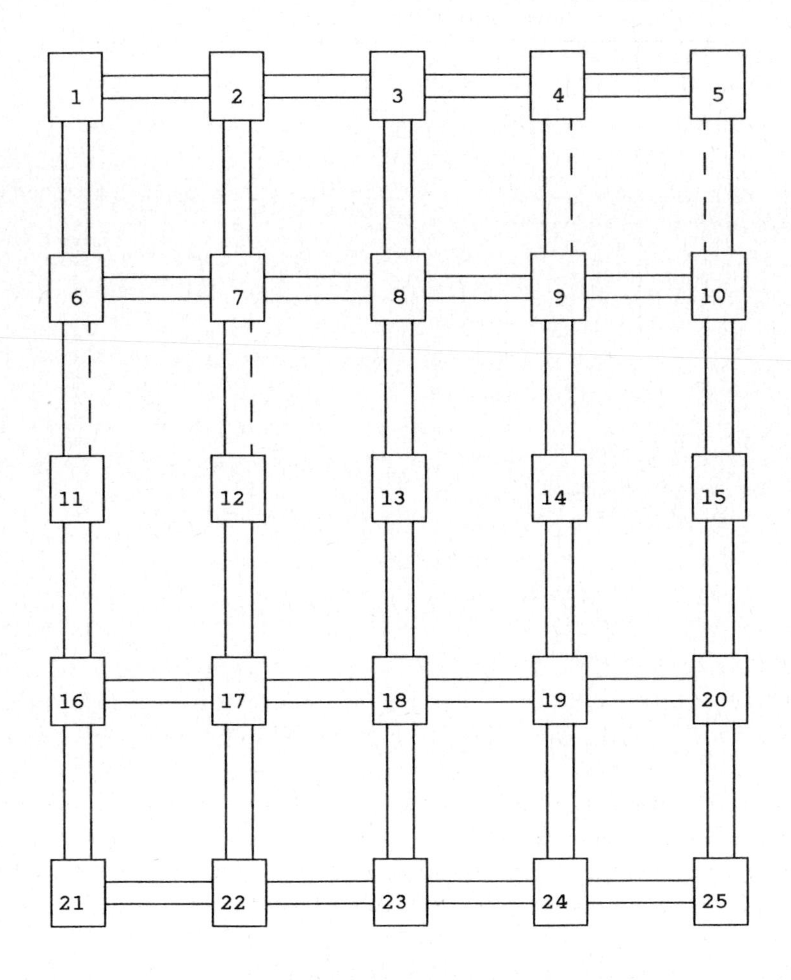

Figure 1c

A Partition Network

Solid arcs are required, dashed arcs are the dummy counterpart arcs added into the DN in Figure 1b.

3.1. The Composite Algorithm

The steps of the Composite Algorithm are shown in Figure 2. These steps are described below.

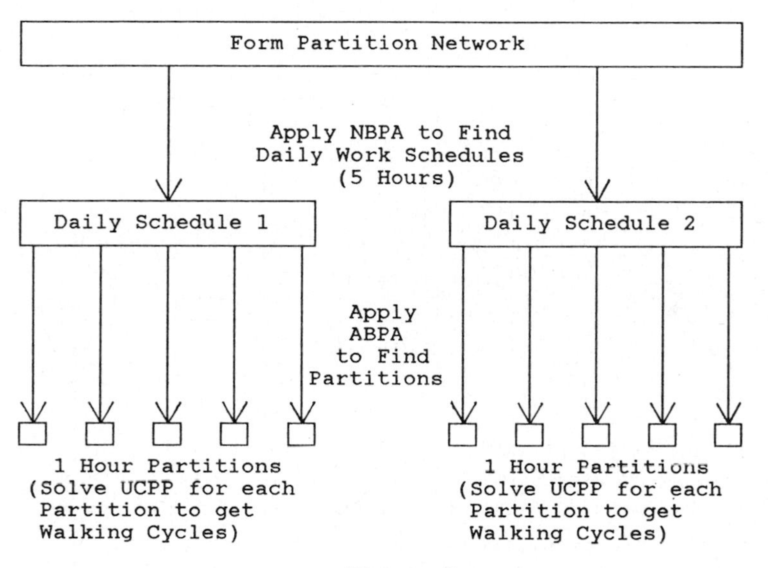

Figure 2

Steps of the Composite Algorithm

3.1.1. Divide the PN into Daily Work Schedules

In the first step of the Composite Algorithm, the Partition Network is broken down into subnetworks representing the daily work schedules for the carriers (in this step, the walking cycles for the carriers are not created). These subnetworks have the characteristics that they are connected and are approximately balanced in terms of required workload. The algorithm used to accomplish this network partitioning is the NBPA and is described in Section 4.

3.1.2. Partition the Daily Work Schedules

The second step of the Composite Algorithm forms the partitions representing the walking cycles. This partitioning is done separately for each carrier's daily work schedule. In this step, the intersections where a carrier will park the vehicle are determined. The algorithm used to carry out this partitioning is the ABPA and is described in Section 4.

3.1.3. Create Walking Cycles from the Partitions

The third step in the Composite Algorithm forms walking cycles over each of the partitions found in Step 2. For each partition, an undirected Chinese Postman problem (UCPP) is solved in order to form the walking cycle while minimizing the total deadhead walking time (the deadhead arcs added in previously to ensure that all arcs in the PN occur in counterpart pairs are removed at the beginning of this step). A procedure for determining the actual walking cycles for each partition is carried out, based on a procedure suggested by Bodin and Kursh [4,5].

3.1.4. Form the Driving Tour

The final step in the Composite Algorithm is to find a minimum length driving tour which joins the depot with the parking locations for each of the partitions. Since there are generally no more than 2 or 3 parking locations within each daily work schedule, this problem is a traveling salesman problem with 3 or 4 nodes and the optimal solution is trivial to find.

3.1.5. Example

The following example illustrates the Composite Algorithm. The TN, DN and PN are shown in Figures 1a, 1b and 1c, respectively. Each required arc in the PN is assumed to have a delivery time of 2 and each dummy counterpart arc is assumed to have a deadhead walking time of 1. Thus, the PN consists of 140 units to partition -- 136 on required arcs and 4 on dummy counterpart arcs. The driving time on each arc is .25. The depot is node 13.

For illustration purposes, we will create two daily work schedules. Ideally, to be balanced, each daily work schedule would have 70 units. A seed point is a node in the PN for beginning the partitioning. (Seed points are defined more formally in Section 4.1). Here, we chose two seed points, one for each daily work schedule. Selecting nodes 11 and 15, the NBPA is able to generate two 70 unit daily work schedules in the PN. The daily work schedules are labelled I and II in Figure 3a.

In Step 2, we divide each daily work schedule into two partitions (two is simply for this example). Daily schedule I is partitioned into two 35 unit partitions emanating from node 3. Daily schedule II is partitioned into two 35 unit partitions emanating from node 18. These four partitions, labelled A, B, C and D, are illustrated in Figure 3b.

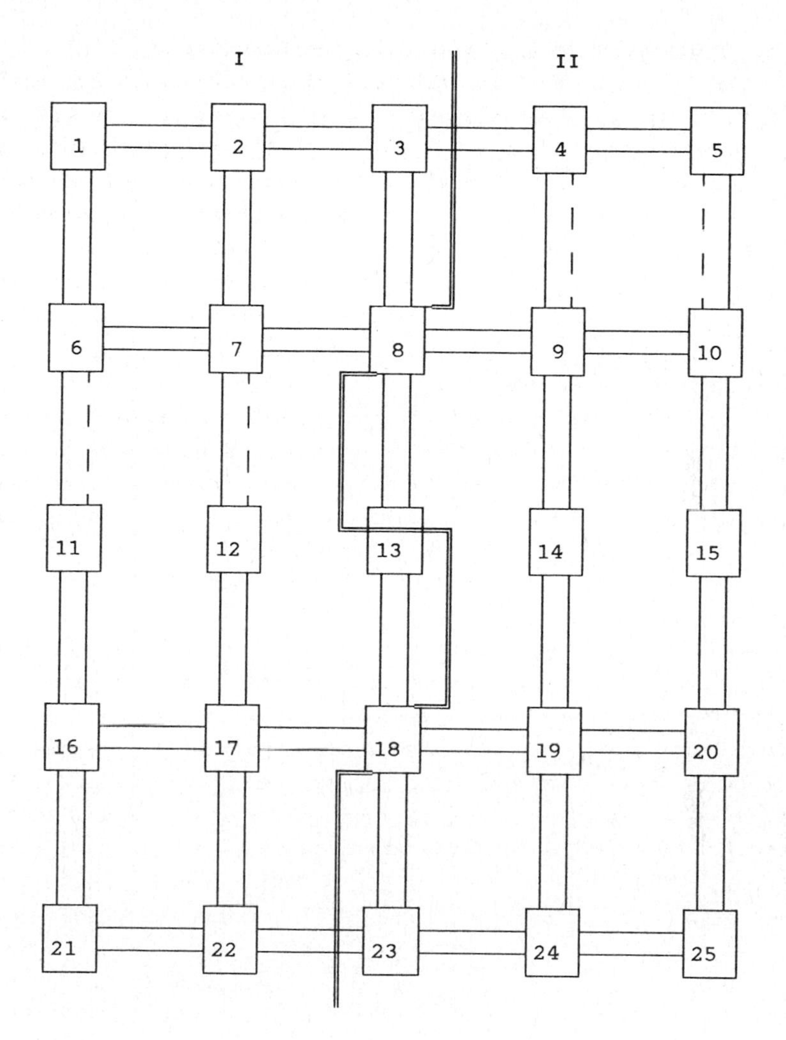

Figure 3a

Two Daily Work Schedules from the Composite Algorithm over the PN from Figure 1c

The seed points are nodes 11 and 5.

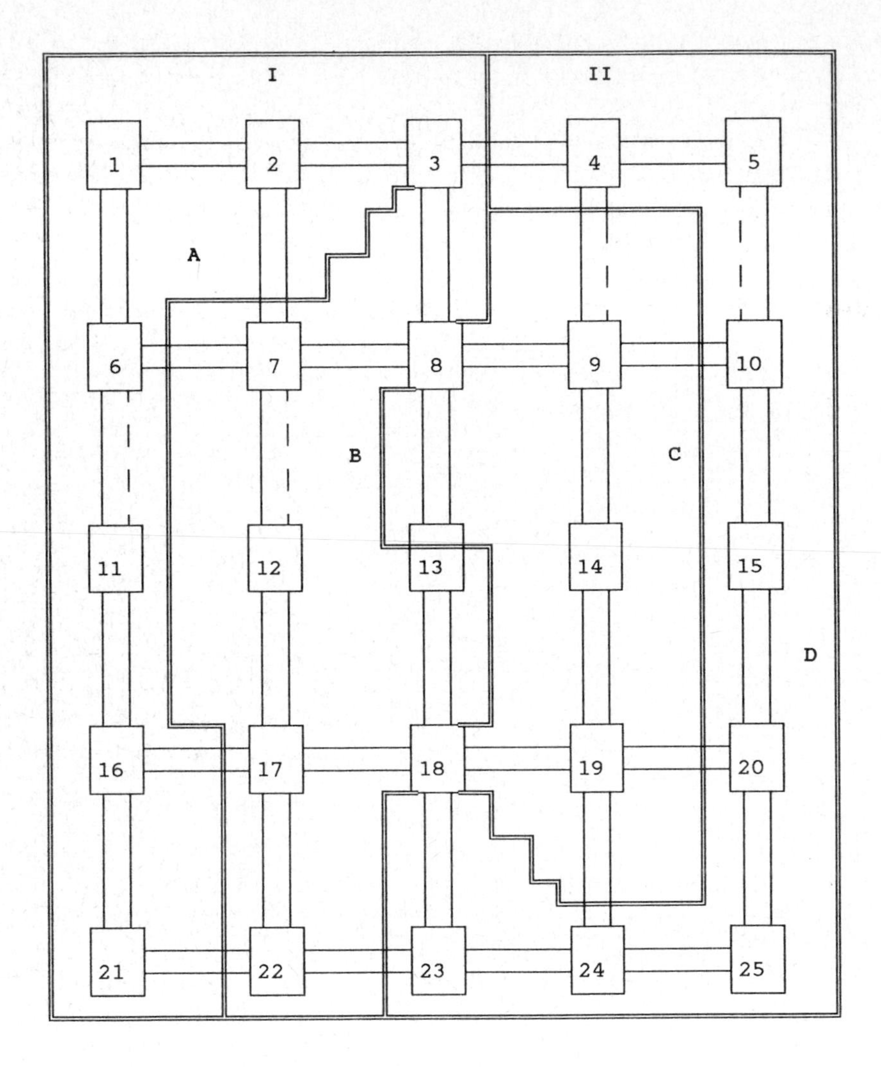

Figure 3b

Four Partitions from the Composite Algorithm
Based on the Daily Work Schedules in Figure 3a

The seed points are nodes 3 and 18.

In Step 3, the minimum deadhead solution over each partition is found. Thus, the actual walking cycles which begin and end at the respective park locations are created. In this example, the best deadhead solution is identical with the partitions already produced above in Step 2 (Figure 3b). The last step in the Composite Algorithm is to find the driving tour for each carrier. For daily schedule I, this tour is node 13 to node 3 to node 13 at a total cost of 1.0. For daily schedule II, this tour is node 13 to node 18 to node 13 with a total driving time of 0.5 units.

3.2. The Conventional Algorithm

The steps of the Conventional Algorithm are shown in Figure 4. These steps are described next.

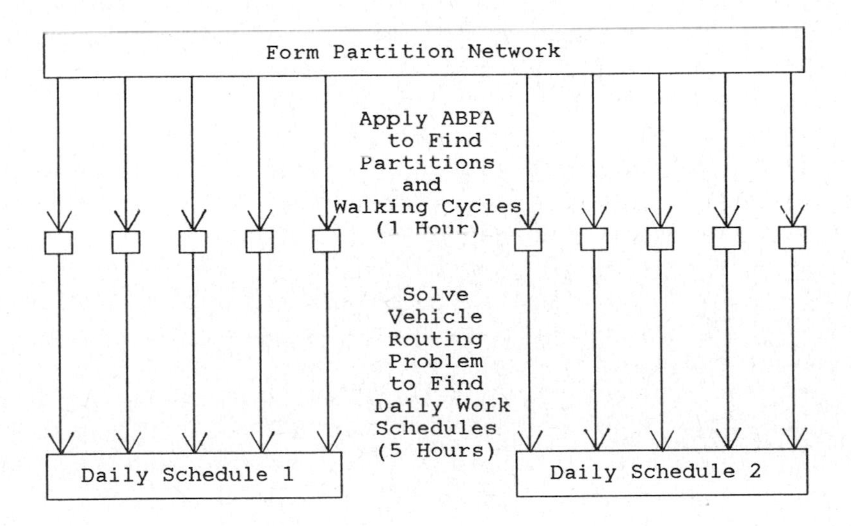

Figure 4

Steps of the Conventional Algorithm

3.2.1. Divide the PN into Partitions Representing Walking Cycles

The Partition Network is broken down into connected partitions representing one hour walking cycles. These partitions are approximately balanced in terms of required workload. The parking locations to be utilized in the region are determined and each pair of arcs incident to each of the parking locations is required to be in a different partition. The algorithm used to accomplish this network partitioning is the ABPA. This is the same algorithm utilized in Step 2 of the Composite Algorithm.

3.2.2. Create the Walking Cycles from the Partitions

This step is identical to Step 3 of the Composite Algorithm.

3.2.3. Form the Daily Work Schedules and the Driving Tours

In this step, the walking cycles found in Step 2 are aggregated into daily work schedules and the driving tours for each of the carriers are found. Each of the daily work schedules contains about 5 hours of walking time and each driving tour begins and ends at the depot. This problem can be formulated as the following single depot vehicle routing problem.

Each walking cycle is represented by a node whose location corresponds to the parking location for the walking cycle. The demand on each of the nodes is 0 and the delivery time for each of the nodes is the time to walk the walking cycle. The upper bound on a vehicle route is the upper bound on the length of a workday. A vehicle route identifies the walking cycles in the daily work schedule for a carrier and the driving route that the carrier should follow. All vehicles are assumed to begin and end their route at the depot and the objective is to minimize the total driving time for the fleet. As the ABPA forms several walking cycles emanating from each parking location, more than one vehicle route can pass through the same parking location.

3.2.4. Example

The same example used previously to illustrate the Composite Algorithm is used to illustrate the Conventional Algorithm. The 140 unit Partition Network is to be broken down into four partitions where each partition contains about 35 units of work. Step 1 of the Conventional Algorithm chooses a single seed point, node 18, from which to begin all four partitions, each emanating from a different pair of counterpart arcs incident to that node 18. The four partitions are labelled I, II, III, and IV in Figure 5 and each contains 35 units.

For each of the four partitions individually, the second step removes the dummy counterpart arcs and computes the minimum deadhead walking cycles by solving the undirected Chinese Postman problem. In this example, the partitions obtained in the previous step cannot be improved upon and the walking cycles are the same as shown in Figure 5. The last step of the Conventional

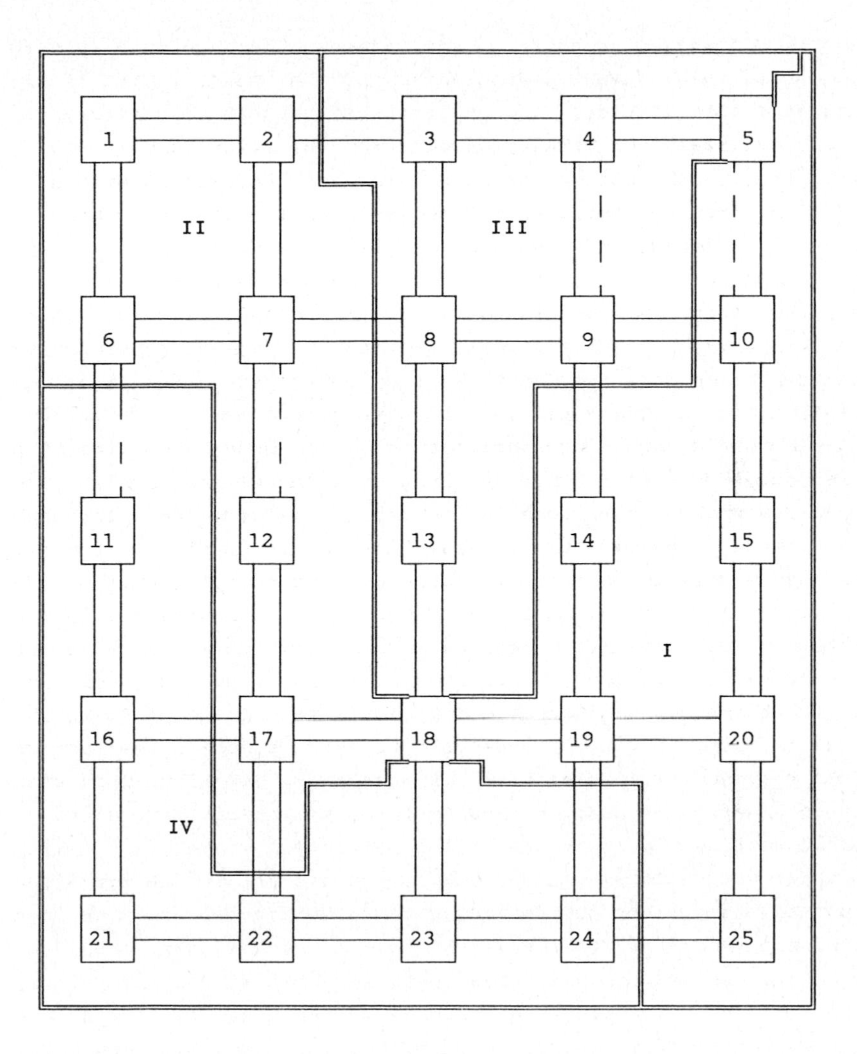

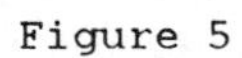

Figure 5

Four Partitions from the Conventional Algorithm

The seed point is node 18.

Algorithm creates the daily work schedules and the actual driving tours. In this example, one daily work schedule would be the aggregation of cycles I and II with total work time of 70 units. The second daily work schedule would consist of cycles III and IV with total work time of 70 units. Each carrier would drive from node 13 to node 18 and then return to node 13. The total driving time for each carrier would be .5 units.

4. Description of the ABPA and the NBPA

Both the ABPA and the NBPA are complicated arc partitioning and routing procedures. They are developed in considerable detail by Levy [16]. Furthermore, in [16], numerous computational tests are carried out in order to determine reasonable parameter settings and to compare the ABPA and the NBPA with regard to the quality of results. These two procedures are outlined next. It is important to note that the ABPA is a one pass algorithm whose performance is linked to the seed points selected whereas the NBPA is an iterative algorithm whose performance is somewhat independent of the initial seed point selection.

4.1. The Arc Begin Partitioning Algorithm

A carrier is said to return to his vehicle if the carrier walks through the intersection at which his vehicle is parked. In the WLT, the carrier returns to his vehicle several times during the day. Each return to the vehicle involves a delay in time at the vehicle as the employee organizes the mail for the next walking cycle to be delivered. The USPS believes that if a carrier passes his vehicle in the middle of a walking cycle, he will only carry the mail for the first portion of the walking cycle. He will then organize his mail for delivery to the remaining portion of the walking cycle and incur a second delay on this walking cycle. Thus, in developing a set of daily work schedules, the USPS wishes to minimize the number of returns to the vehicle by placing each pair of counterpart arcs emanating out of a parking location on a different partition. Since the ABPA is used in both the Composite and the Conventional Algorithms to create the one hour partitions and parking locations, the ABPA was developed to ensure that the total number of returns to the vehicles be as small as possible.

The flow of the ABPA is shown in Figure 6. These steps are now discussed.

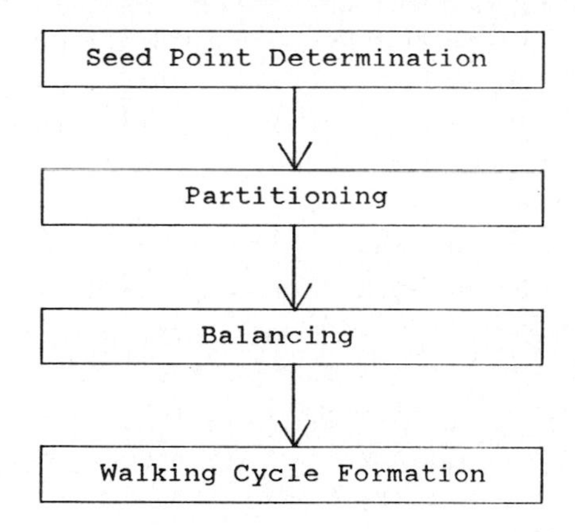

Figure 6

Steps of the ABPA

4.1.1. Seed Point Determination

As noted earlier, a seed point is a node in the PN out of which partitions are built and corresponds to a parking location. Each pair of arcs emanating from a seed point becomes the root of a partition (or the first pair of arcs in the partition). In the ABPA, each pair of arcs incident to a seed point serve as the root of a different partition. The number of partitions to be formed is estimated as the total workload in the DN divided by the target size of the partition. The seed point determination step must find an appropriate set of seed points (or nodes) in the Partition Network so that the desired number of partitions can be formed about these seed points.

A node is called a <u>border node</u> of the PN if all arcs emanating out of that node are connected to just one other node of the PN. In Figure 7, nodes 3, 13, and 23 are border nodes. If a seed point is selected too close to a border node, then the partition rooted in the direction of the border node may not have any room to expand during the partitioning step. For example, if node 2 became a seed point, the partition rooted on the arcs between node 2 and node 3 would be unable to expand beyond node 3. In general, nodes which are close to border nodes are eliminated

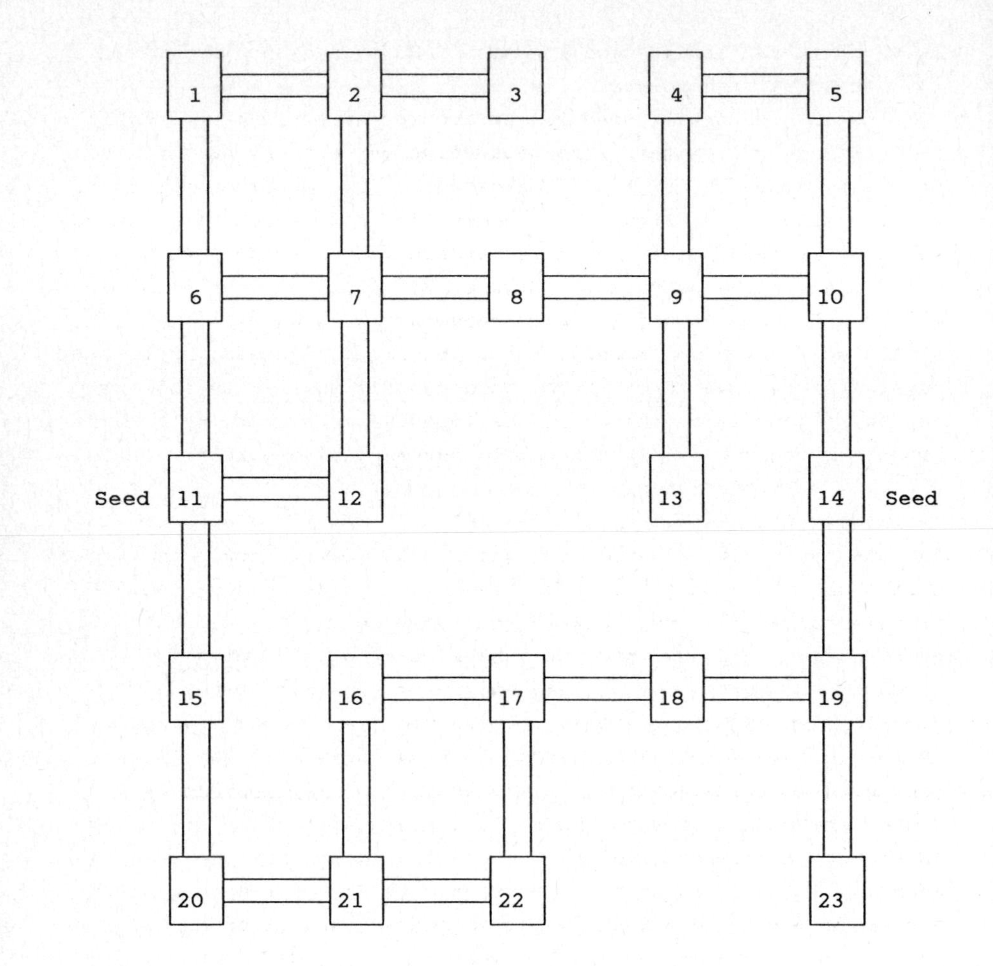

Figure 7

Border Nodes and Seed Point Selection

Border Nodes are 3, 13 and 23. Seed Points are 11 and 14.

from consideration as seed points. However, some border nodes can make good seed points.

Also, a good set of seed points are spread out over the entire PN. To accomplish this, the separation between any pair of seed points selected is required to exceed some minimum number of arcs. Thus, in Figure 7, nodes 11 and 20 would not both be selected as seed points because they are too close to each other.

Assume for the network in Figure 7, that 5 partitions are required. Nodes 11 and 14 could serve as two seed points for the ABPA since they have a total of 5 pairs of incident arcs, are several arcs away from the border nodes and are sufficiently separated from each other. It is important to note that there are other pairs of nodes that could serve as seed points.

4.1.2. Partitioning

The partitioning step of the ABPA assigns every pair of arcs in the PN (not already designated as root arcs) to some partition. The partition with the least amount of time assigned is selected as the next partition to expand. Of all pairs of arcs incident to any node in the partition, the pair with the largest delivery time is added into the partition. This algorithm ensures that each partition remains connected.

This procedure is illustrated in Figures 8a-8c. The PN given has delivery times adjacent to the arcs. The seed point is node G and the roots of three partitions are displayed in Figure 8a. In Figure 8b, three iterations of the partitioning step are shown. In order, the first three arc pairs added to partitions are as follows: the arc pair between nodes F and E added to the partition rooted on the arcs between nodes G and F, the arc pair between nodes H and I added to the partition rooted on the arcs between nodes G and H, and the arc pair between nodes B and A is added to the partition rooted on the arcs between nodes G and B. The complete partitioning is displayed in Figure 8c.

4.1.3. Balancing

At the completion of the ABPA, some of the partitions may violate the lower and upper bounds on the specified duration of a walking cycle. The purpose of the balancing step is to transfer arcs among partitions to ensure that the duration of each partition is between the lower and upper bound. To completely evaluate all possible exchanges of arcs would be computationally prohibitive. Several kinds of swaps are examined in [16]. Here,

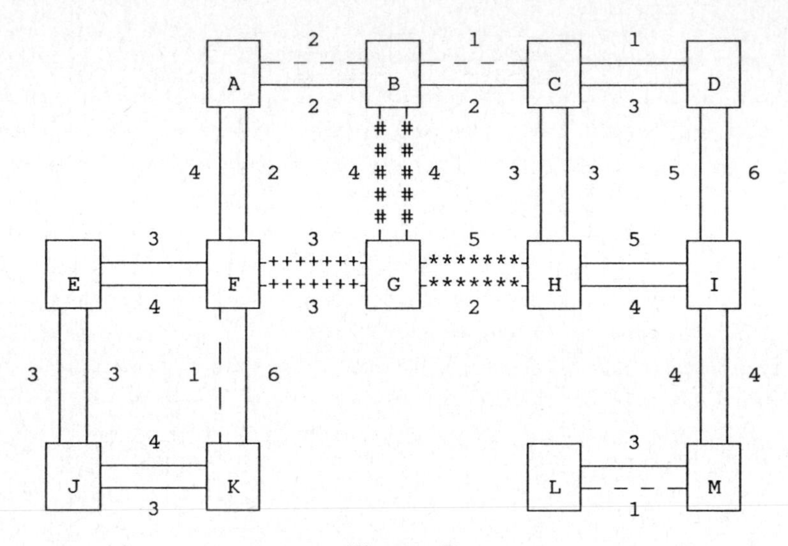

Figure 8a

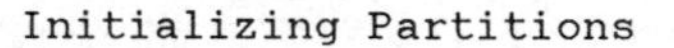
Initializing Partitions

The seed point is node G. Dashed arcs are dummy counterpart arcs.

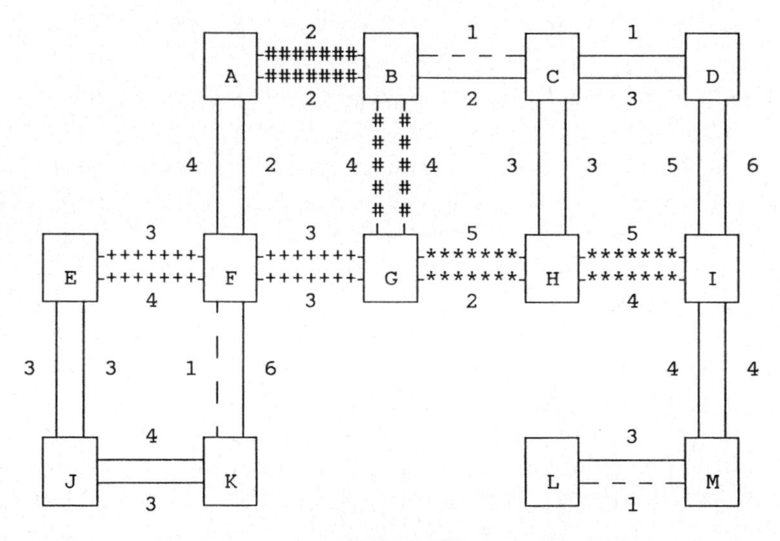

Figure 8b

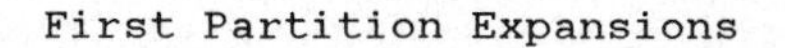
First Partition Expansions

The seed point is node G. Dashed arcs are dummy counterpart arcs.
First pair F-E, second pair H-I and third pair B-A.

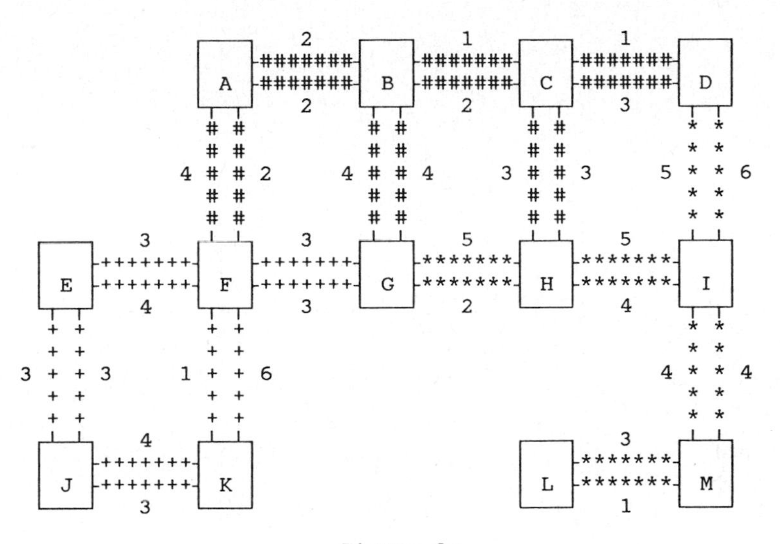

Figure 8c

Final Partitions

The seed point is node G.

we illustrate a branch swap. In Figure 9a two partitions are shown, both emanating from seed point I. The arrows indicate the order in which the arcs of each partition are stored. Assume we wish to transfer the arcs between nodes Q and B from their current partition. We observe that the arcs between nodes A and B as well as node B and C would become disconnected. Hence, to carry out the swap and maintain the connectivity of the partitions, these arc pairs would also have to be transferred. The effect of swapping the entire branch (arc pairs between nodes Q and B, B and C, and B and A) is illustrated in Figure 9b.

4.1.4. Cycle Formation

A cycle always exists in each partition using only the arcs in the partition. However, the deadhead walking time in a partition may be greater than the minimum deadhead walking time which would actually be required to form the walking cycle. Solving an undirected Chinese Postman problem over the required arcs of a partition provides the minimum deadhead walking time and subsequently the walking can be found.

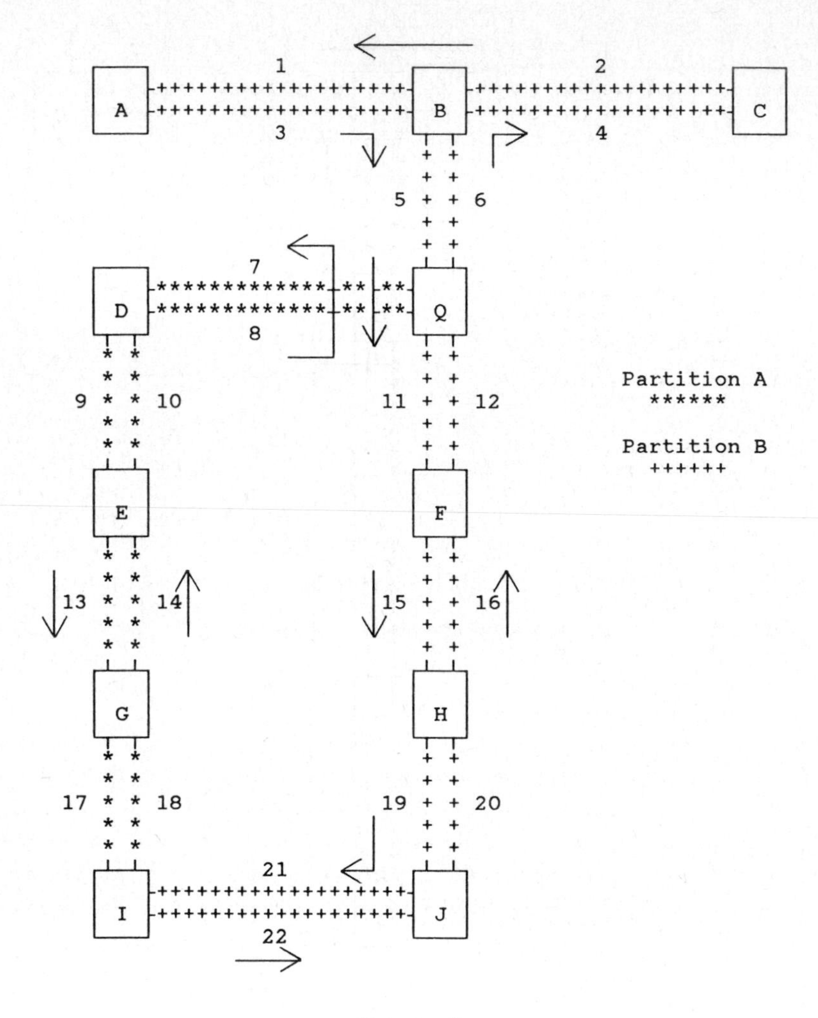

Figure 9a

Potential Branch Swap

The seed point is node I.
Numbers are arc numbers, not delivery times.

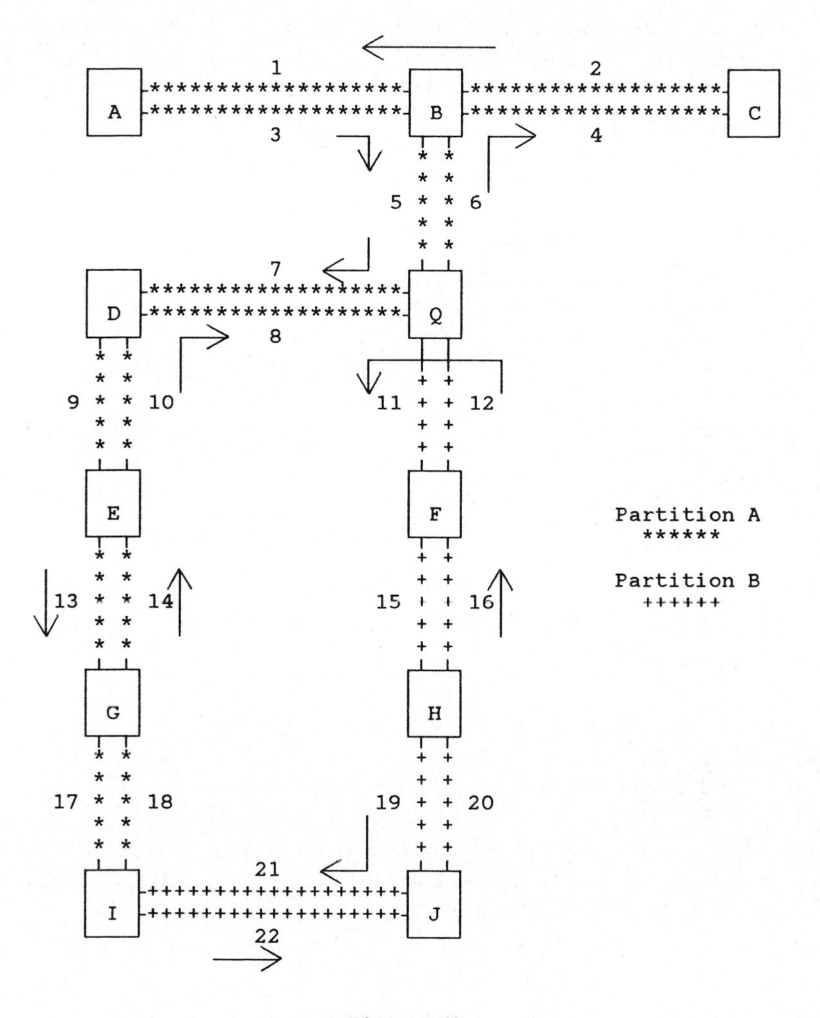

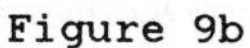

Figure 9b

Completed Branch Swap

The seed point is node I.
Numbers are arc numbers, not delivery times.

In Figure 10, we show a partition with a total time of 13 units, including 3 dummy counterpart arcs. We remove the dummy counterpart arcs and solve the Chinese Postman problem, producing a total time of 11.

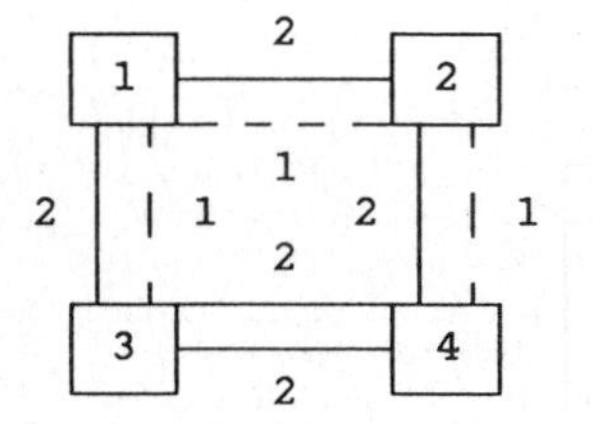

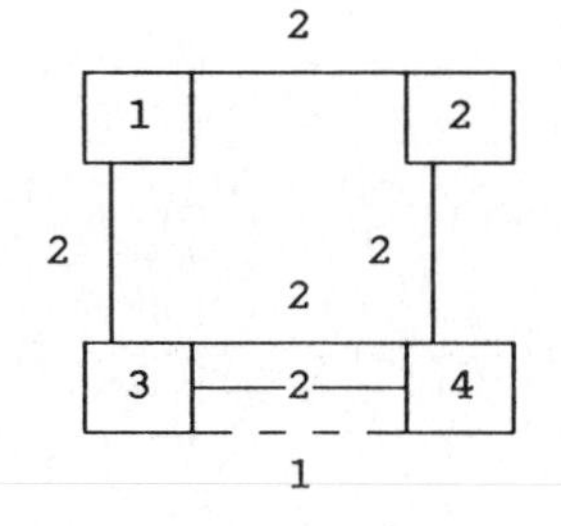

Total Time = 13 Total Time = 11

Figure 10

Reduction in Deadheading from the Chinese Postman Problem

Seed point is node 1.

4.2. The Node Begin Partitioning Algorithm

The Node Begin Partitioning Algorithm builds a single partition out of a seed point rather than a partition out of each pair of arcs incident to a seed point as in the ABPA. Thus, in the NBPA, all arcs incident to a seed point can be in the same partition, although this is not a requirement. The NBPA is used in the Composite Algorithm to construct the daily work schedules. Thus, the seed points do not correspond to parking locations. The parking locations will be determined by the ABPA for each daily work schedule. Thus, the seed points are not fixed once they are chosen and the NBPA is free to establish revised seed points. As a result, the NBPA can iterate whereas the ABPA is a one pass algorithm. This is characterized in Figure 11, which shows the steps of the NBPA. These steps are discussed next.

4.2.1. Initial Seed Point Determination

The initial seed point determination step selects a set of seed points equal to the specified number of daily work schedules to be built.

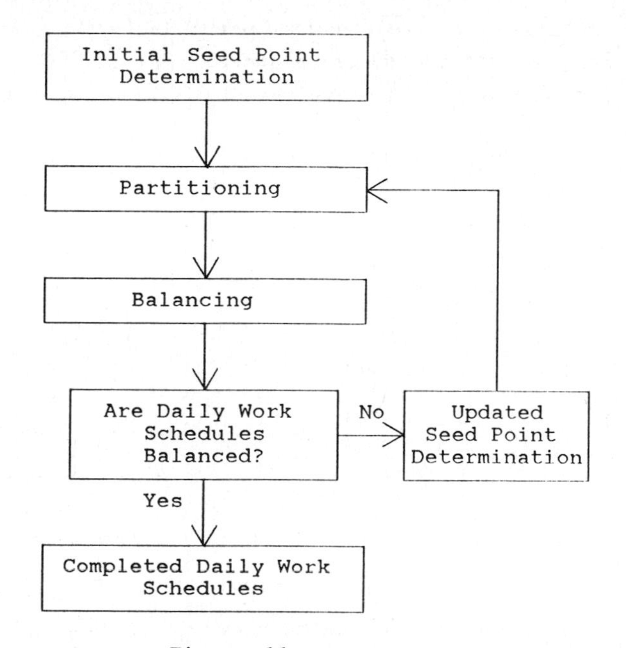

Figure 11

Steps of the NBPA

4.2.2. Partitioning

The same data structure and partitioning rules used in the partitioning step of the ABPA are used in the NBPA. To avoid the constraint imposed on the ABPA that every pair of arcs incident to a seed point be the root of a different partition, we modify the PN by adding a supernode with a pair of dummy counterpart arcs of duration 0 to each designated seed point. This is illustrated in Figure 12. By rooting each partition along the pair of dummy counterpart arcs from the supernode to the seed point, we are able to utilize the ABPA partition procedure to do the partitioning in the NBPA.

4.2.3. Balancing

The swaps considered in balancing the NBPA are the same as those examined in the ABPA.

The balancing step continues until no more swaps are possible. A decision is then made as to whether to repeat the NBPA with a new set of seed points. If all the daily work schedules are

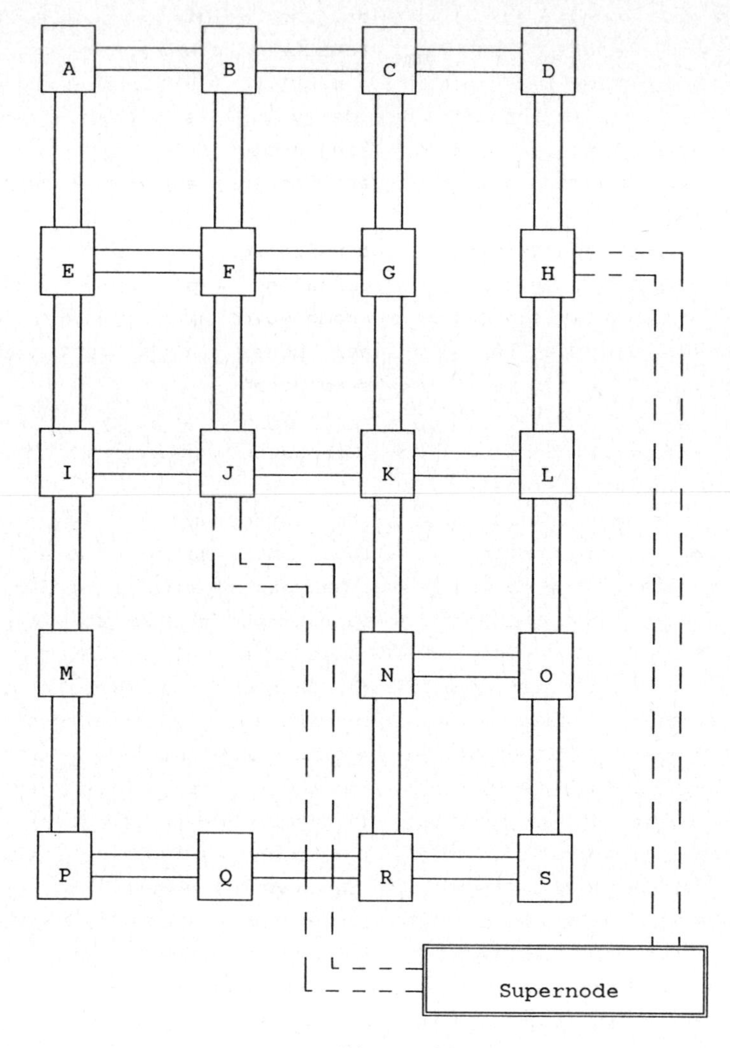

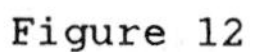
Figure 12

NBPA Initial Partitions Using the Supernode

Two initial partitions rooted from the Supernode to node H and to node J, shown with dashed arcs.

within the specified bounds or the solution on the current iteration is worse than the previous iteration (so that the previous solution is considered best), the algorithm ceases and the subnetworks representing the daily work schedules are passed on to the ABPA to create the walking cycles. Otherwise, the seed points are updated and the partitioning and balancing steps repeated.

4.2.4. Updated Seed Point Determination

The updated seed point determination step chooses, for each current daily work schedule, the node which minimizes the maximum Euclidean distance to all other nodes in the subnetwork and designates this node as the new seed point.

5. Computational Testing of the Composite Algorithm and the Conventional Algorithm

In this section, we analyze the performance of the Composite Algorithm and the Conventional Algorithm in solving the WLT. The USPS provided us with one Travel Network consisting of 206 nodes and 564 arcs and representing approximately 20% of a zip code area in Ft. Worth, Texas. Although a natural problem test size would be an entire zip code region, this data set was considered sufficient for testing the algorithms. We created six test problems (i.e., Delivery Networks with different characteristics) from this Travel Network by varying the set of required arcs.

A summary of the problems and their characteristics is given in Table 1. Problems 1 and 2 are connected Delivery Networks. The noncounterpart arcs refer to one side of the street receiving delivery while the other side is not required. Problems 3A-3D have the same disconnected Delivery Network. The problem differences arise in the setting of the amount of deadhead walking time allowed between disconnected pieces. (A complete discussion of the extension of the algorithms for the disconnected case in given in Levy [16].)

In carrying out these computational tests, we attempted to reflect the real-world constraints of the WLT as closely as possible by setting the bounds on the daily work schedule to be around five hours and the target walking cycle time to be about one hour. These targets varied slightly for each problem and were a function of the total walking time contained in the problem. We first determined the number of daily work schedules

Table 1

Problem Characteristics

	Test Problems					
	1	2	3A	3B	3C	3D
# of Nodes	206	167	154	154	154	154
# of Arcs Required	564	425	319	319	319	319
# of Non-Counter-part arcs	0	35	28	28	28	28
Total Delivery Hours	25.460	19.337	16.373	16.373	16.373	16.373
Estimated Deadhead Hours	0.000	1.006	0.362	0.455	0.566	0.676
Daily Schedule Target	5.1	5.0	5.6	5.6	5.6	5.6
Upper and Lower Bounds	5.05-5.15	4.95-5.05	5.55-5.65	5.55-5.65	5.55-5.65	5.55-5.65
Walking Cycle Target	1.02	1.0	0.93	0.93	0.93	0.93
Upper and Lower Bounds	0.82-1.22	0.80-1.20	0.73-1.13	0.73-1.13	0.73-1.13	0.73-1.13

to construct and then estimated the target workload in a daily work schedule as the total workload in the problem divided by the number of daily work schedules to form. In an actual situation, there would be a tradeoff between the target walking time in a daily work schedule and the number of daily work schedules to be formed. One could vary the number of daily work schedules to form, examine the work schedules created, see if any overtime was dictated by these schedules, and then decide if overtime should be built into the schedules.

The lower and upper bound on the duration of the daily work schedules was set to be 3 minutes on each side of the target workload. Moreover, the range for each walking cycle was always set to 12 minutes on each side of the target duration of a walking cycle. These ranges are reflective of the desired ranges for walking cycles and daily work schedules of the USPS.

In Tables 2 and 3, we present a comparison of the Composite Algorithm and the Conventional Algorithm on the six test problems. In these tables, a <u>cycle time deviation for a walking cycle</u> is defined to be the time that a walking cycle falls below its lower bound (in hours) and the <u>sum of the time deviations of the cycles</u> is defined to be the sum of cycle time deviations over all the walking cycles (in hours). Also, in Tables 2 and 3, we record over all the daily work schedules the total deadheading in hours, the final number of walking cycles and the number of parking locations that the algorithm created for each of the problems.

Furthermore, in these tables, the <u>time deviation for a daily work schedule</u> is defined as the absolute value of the deviation of the length of the work schedule from either its lower bound or its upper bound (in hours). The <u>sum of the time deviations for the work schedules</u> is defined as the sum of time deviations over all the daily work schedules (in hours). Finally, in Tables 2 and 3, vehicle miles is the total number of miles travelled by the fleet of vehicles in order to service all the daily work schedules is reported. This quantity is computed assuming the depot is located in the middle of the Travel Network.

The results presented in Tables 2 and 3 indicate that both the Composite Algorithm and the Conventional Algorithm produce very reasonable results. In every case for both algorithms, the sum of cycle deviations is less than two hours, implying that most of

Table 2

Composite Algorithm (Time in hours)

	Test Problems					
	1	2	3A	3B	3C	3D
Sum Cycle Deviations	0.000	0.023	1.427	1.249	0.739	0.472
Total Deadhead	---	0.548	0.336	0.429	0.578	0.508
Final # Cycles	25	20	21	21	19	19
Final # Parking Locations	10	8	9	10	9	7
Sum Daily Schedule Deviations	0.000	0.106	0.011	0.000	0.312	0.050
Vehicle Miles	6.365	6.559	4.205	4.937	4.714	4.351

Table 3

Conventional Algorithm (Time in hours)

	Test Problems					
	1	2	3A	3B	3C	3D
Sum Cycle Deviations	0.438	0.050	1.439	1.439	1.670	1.290
Total Deadhead	---	0.586	0.336	0.429	0.539	0.649
Final # Cycles	27	20	20	20	21	21
Final # Parking Locations	9	8	9	9	10	10
Sum Daily Schedule Deviations	0.639	0.415	0.123	0.011	0.181	0.209
Vehicle Miles	5.397	5.992	5.726	5.213	4.902	5.988

the walking cycles are within or very close to the desired range. In every case, the sum of cycle deviations is less with the Composite Algorithm than with the Conventional Algorithm. In most cases, however, this difference is not large.

For the Composite Algorithm, the total amount by which the daily work schedules violate the lower or upper bounds is less than six minutes in every case except Problem 3C, where the violation is 20 minutes. This result indicates that by using the Composite Algorithm the total walking time in each worker's schedule is about the same, even if there exists a slight variation in walking time among the individual walking cycles. The Conventional Algorithm also does well with respect to this measure but its performance is not as favorable as the Composite Algorithm. The Conventional Algorithm has a 40 minute daily work schedule time deviation on Problem 1 and exceeds 6 minutes on all problems except 3B.

Other comparisons between the two algorithms are as follows:

a. The total amount of deadhead walking time does not appear to be significantly different between the two algorithms and extremely small (generally about 30 minutes out of a total workday of 15-25 hours).

b. There is very little difference with respect to the final number of cycles and the number of parking locations. It is worth noting that on Problems 1 and 2 both algorithms produced roughly 2 parking locations per daily work schedule. This is considered to be as good as possible by the USPS since virtually all intersections in a region have no more than 4 street segments incident to it. Thus, if five walking cycles are to be contained in a daily work schedule, the daily work schedule has to contain at least two parking locations.

The disconnected problems 3A-3D have approximately 3 parking locations per daily work schedule, with the only exception being Problem 3D with the Composite Algorithm. This behavior is not unexpected since in a disconnected problem the carrier may have to move the vehicle between disconnected parts of the network.

c. In all cases, the total number of vehicle miles to be driven is extremely small. If one assumes that the carrier drives the vehicle at 15 miles/hour, then the total driving time in all cases for both algorithms is less than 30 minutes. Most of this driving time is associated with the vehicle traveling to and from the depot. Obviously, if the depot is in a more remote location than we assumed, then the total driving time would increase. However, it does not appear that driving time is a serious concern in this problem. Thus, the assumption we made in Section 2.1.A in disregarding driving time until the last step of the Composite and Conventional Algorithms seems reasonable.

d. The algorithms described have been developed in FORTRAN on a UNIVAC 1100/90. The CPU time for the Conventional Algorithm ranged from 5 to 17 seconds depending on the problem. The Composite Algorithm, on the other hand, took considerably longer. The CPU time ranged from 1.5 minutes to 16 minutes. While the running time for the Composite Algorithm is significantly larger than the Conventional Algorithm, it is not prohibitive for conducting "what-if" analyses. Furthermore, once a set of daily work schedules are produced, they can be used for months or longer. Thus, the algorithms do not need to be executed on a daily basis.

It is important to realize that the running times reported are for a network 1/5 the average size zip code. The running time for the Conventional Algorithm is very fast and we would not expect it to increase significantly for a zip code size network. We did, however, examine the growth in running time for the Composite Algorithm. We found that over 80% of the running time for the Composite Algorithm was devoted to the ABPA. But, the ABPA is applied to subnetworks the size of daily work schedules. Thus, as the problem size grows we would expect some growth in running time attributable to the NBPA and a growth in running time for the ABPA which is a linear function of the number of daily work schedules. As such, we would not expect the Composite Algorithm to become computationally prohibitive as the problem size grows.

6. Conclusions

In this paper, the USPS´s WLT problem has been presented as an application of arc routing and partitioning. Two algorithms, the Composite and the Conventional, have been developed and tested to show their effectiveness in solving the WLT. Furthermore, two arc partitioning heuristics, the ABPA and the NBPA, essential in the Composite and Conventional Algorithms have been described.

The results of the Composite and the Conventional Algorithms indicate that both approaches are very successful in solving the

WLT with respect to the multiple criteria which we specified for evaluating these solutions. While either algorithm performs well, the Composite Algorithm has an advantage because of its excellent ability in balancing both the daily work schedules and the individual walking cycles.

These results have been presented to the USPS. The USPS has awarded us a contract to incorporate these procedures into a highly interactive user friendly computerized system for scheduling their postal carriers.

Acknowledgments

We wish to thank Jim Bailey of the United States Postal Service, George Fagan and Ron Welebny of Bowne Management Systems and Russ Tredway of Electronic Data Systems for their support, advice and guidance in this project. Furthermore, we want to thank Bruce Golden, Arjang Assad, Michael Ball and Roy Dahl for their comments on this paper and their assistance in the design of the algorithms.

References

[1] Assad, A., Pearn, W.L. and Golden, B., Working Paper #85-032, University of Maryland (1985).
[2] Beltrami, E. and Bodin, L., Networks 4 (1974) 65.
[3] Bodin, L., Golden, B., Assad, A. and Ball, M., Computers and Operations Research 10 (1983) 63.
[4] Bodin, L. and Kursh, S., Operations Research 26 (1978) 525.
[5] Bodin, L. and Kursh, S., Computers and Operations Research 6 (1979) 181.
[6] Chapleau, L., Ferland, J., Lapalme, G. and Rousseau, J.M., Operations Research Letters 3 (1984) 95.
[7] Christofides, N., Omega 1 (1973) 719.
[8] Christofides, N., Campos, V., Corberan, A. and Mota, E., Imperial College Technical Report ICOR 81-5, London (1981).
[9] Christofides, N., Campos, V., Corberan, A. and Mota, E., Imperial College Technical Report ICOR 82-4, London (1982).
[10] Christofides, N., Benavent, E., Campos, V., Corberan, A. and Mota, E., Imperial College Technical Report ICOR 83-6, London (1983).
[11] Edmonds, J. and Johnson, E., Mathematical Programming 5 (1973) 88.
[12] Euler, L., Scientific American 189 (1953) 66.
[13] Golden, B. and Wong, R., Networks 11 (1981) 305.
[14] Golden, B., DeArmon, J. and Baker, E., Computers and Operations Research 10 (1983) 47.
[15] Kwan, M.K., Chinese Mathematics 1 (1962) 273.
[16] Levy, L., "The Walking Line of Travel Problem: An Application of Arc Routing and Partitioning," Ph.D. Dissertation, University of Maryland (1987).
[17] McBride, R., Computer and Operations Research 9 (1982) 145.
[18] Orloff, C., Networks 4 (1974) 35.
[19] Orloff, C., Networks 6 (1976) 281.
[20] Papadimitriou, C., Journal of A.C.M. 23 (1976) 544.

[21] Pearn, W. L., "The Capacitated Chinese Postman Problem," Ph.D. Dissertation, University of Maryland (1984).
[22] Stern, H. and Dror, M., Computers and Operations Research 6 (1979) 209.

Vehicle Routing: Methods and Studies
B.L. Golden and A.A. Assad (Editors)

FLEET SIZING AND DISPATCHING FOR THE MARINE DIVISION OF THE NEW YORK CITY DEPARTMENT OF SANITATION

Richard C. Larson, Massachusetts Institute of Technology, Cambridge, MA 02139

Alan Minkoff, IBM Corporation, P.O. Box 950, Poughkeepsie, NY 12602

Paul Gregory, Office of Resource Recovery, Department of Sanitation, New York, NY 10007

New York City's Department of Sanitation (DOS) operates a marine-based system comprising barges, tugboats and various fixed facilities to transport refuse ("garbage") from various parts of the City to the world's largest landfill on Staten Island. The purpose of this chapter is to report on a modeling project whose purpose was to provide DOS with a planning tool for sizing the fleet of barges and tugs in the face of a projected doubling of transported load. Vehicle routing in this context consisted of procedures for dispatching the tugs towing barges from point to point within the harbor. The developed model has been used for fleet sizing and other strategic (capital planning) decisions and has resulted in savings to the City of New York of at least $6 million.

1. INTRODUCTION

The Marine Division of the New York City Department of Sanitation (MDDOS) operates a fleet of tugboats and barges for the transport of refuse from New York City to Fresh Kills Landfill on Staten Island. In the early 1980's, MDDOS was faced with a situation in which many of its barges were aged and decrepit, yet workload was scheduled to increase as another major landfill was shut down. With major barge investment decisions on the horizon, MDDOS commissioned the development of BOSS, the *B*arge *O*perations *S*ystems *S*imulator, for assistance in the task of fleet sizing. Although BOSS simulates MDDOS's operations on a strategic scale, it had to incorporate the sophisticated tactical decisions that represent good dispatching practice. This chapter reports on the development and use of the BOSS model.

"Vehicle routing" in this chapter corresponds to directing movements of tugboats and barges throughout New York City's harborways. The individual that performs the routing is the tugboat dispatcher who communicates via radio with each of the tugboat captains. As will become clear from the detailed description that follows, vehicle routing is a complex state-dependent activity that must respond appropriately to time-dependent probabilistically changing conditions throughout the refuse transport system. One consequence of bad routing is a lack of empty barge capacity at locations where refuse trucks dump their contents into barges; such lack of dumping capability can create severe land-based congestion and significantly increase overall system costs.

While "vehicle routing" is an important and complex feature of overall MDDOS operations, it is not the primary focus of the work reported in this chapter. Rather, the emphasis here is on the development of an original model whose primary purpose is to aid fleet sizing and related strategic decisions. Vehicle routing is an integral component of this model, indicating by example how vehicle routing methodologies can be important "modules" in larger modeling and analysis efforts.

In summary, the chapter differs somewhat from many of the other chapters in the book in that (1) its focus is strategic rather than tactical; (2) emphasis is given more to modeling than to optimization; (3) it is developed within the context of a specific application; (4) it reports cost savings and "return on investment" obtained from implementation results of the model.

2. DESCRIBING SYSTEM OPERATIONS

A. The Strategic General Setting

In 1983, the New York City Department of Sanitation (DOS) operated three landfills, three incinerators and nine *m*arine *t*ransfer *s*tations (MTS's) for loading barges to transport refuse to the largest of the landfills, the Fresh Kills Landfill (FKL). The entire DOS refuse transportation system handled between 22,000 and 24,000 tons of refuse per day. Two of the landfills handled the majority of the refuse: Fountain Avenue (in Brooklyn) took in 7,000 to 9,000 tons per day and FKL took in approximately 11,000 tons per day. Of the 11,000 tons per day at FKL, approximately 9,600 tons were delivered by barge.

Because of the impending shutdown of the Fountain Avenue Landfill, DOS was confronted with the problem of moving up to 9,000 tons of refuse per day to some destination other than Fountain Avenue. The logical (and only feasible) choice was to FKL via the marine transport system. Thus by 1986 DOS was projecting a near *doubling* of demand on the marine system, up to approximately 18,000 tons per day. Prior to 1983, the rule of thumb for fleet sizing was strictly linear in the number of tons carried per day. Apparently the logic for the "linear thinking" was that for every barge being unloaded (at FKL), one barge would either be in transit to an MTS or returning from an MTS, and one barge would be at an MTS being filled. A doubling of load would have meant a doubling of fleet size under the "straight line" rule.

The approximate fleet size in 1983 was 60 barges. At that time new barges cost approximately $1 million each, so DOS was confronted with a capital investment decision of up to $60 million. Not willing to trust the linear rule of thumb for such a large acquisition, DOS issued a request for proposals to develop a strategic planning model to help in making the fleet sizing decision. This chapter reports on the model that was ultimately developed and used for fleet sizing.

B. Elements of System Operations

Assuming the reader is familiar with transportation network terms such as nodes, arcs, and "vehicles," we introduce these aspects of the model as we describe system operations.

2.1. The "Vehicles" - Barges and Tugboats

The mobile units or vehicles in the MDDOS transportation system are the barges used for carrying refuse and the tugboats used for towing the barges. The dependence of the barges on the motorized power of the tugboats is analogous to the dependence of rail cars on a locomotive or of a trailer on its tractor. Each barge can handle approximately 550 to 600 tons of ordinary (nonincinerated) refuse. Tugboats haul barges in barge trains with one to four barges per train. To avoid barge hull deterioration, all barges in a train must be in the same state, either empty (or, in MDDOS venacular, "light") or full ("heavy"). The MDDOS operates only with a homogeneous fleet of (identical) barges.

While traveling, a barge train can be considered to be moving along a link of a transportation network, under direction of a radio dispatcher (Figure 1). Link travel times depend on barge train size and state, tides, and weather. For instance, a four barge train travels approximately 30% more slowly than a one barge train; a heavy barge train travels 20% more slowly than a light barge train; traveling against the tide can increase travel times from 10 to 30%.

2.2. The "Source Nodes" - Marine Transfer Stations

A marine transfer station (MTS) is a two-storey pier that facilitates transfer of the refuse from trucks ("garbage trucks)" to the barges (Figure 2). Each truck enters the upper storey, is weighed, and is directed to one of several "holes in the pier floor" to dump its load into the barge below.

Input to an MTS is measured in "tons per hour." The average value of this quantity varies greatly by MTS and hour of the week. Most MTS's allow for two barges simultaneously to be in "receiving position," below the pier's upper storey. Each MTS also has a finite capacity for accommodating other barges (either light or heavy) tied up along side the MTS. This additional holding capacity, ranging from one to four barges, allows delivery of several light barges at one time and thereby lengthens the time interval between visits by a tugboat to remove heavy barges and replenish with light barges. Barges are entered into and removed from receiving position by a "rope and pulley" process known as shifting. Tidal conditions at some MTS's mandate that a tug be present during the shifting process.

A major operational problem occurs whenever all barges resident at an MTS are "heavy". In that case, there exists no additional capacity to accept refuse from arriving trucks and the MTS is said to be "blocked." In practice whenever an MTS is blocked arriving trucks may be queued (if the anticipated blockage time is short) or they may be sent to another MTS, resulting in considerable additional driver time and

BOSS: Waterway Link

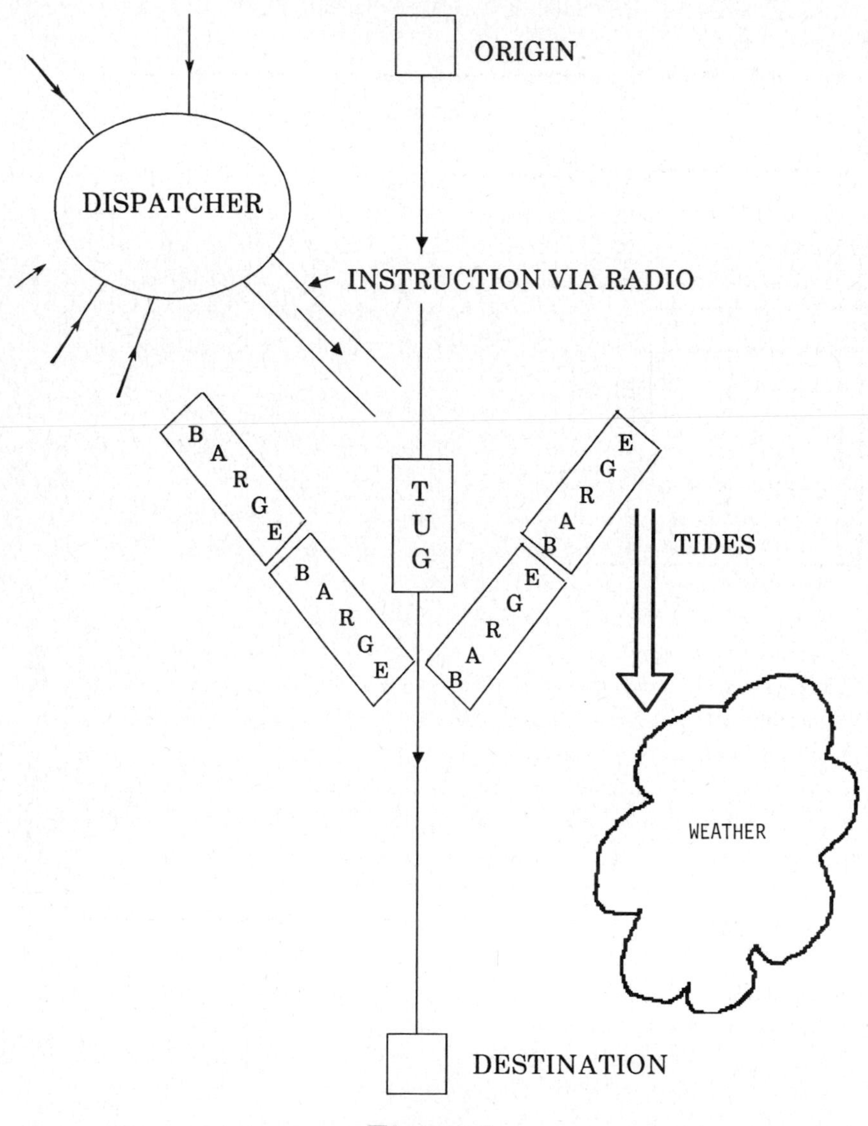

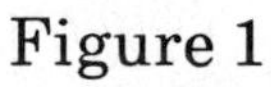

Figure 1

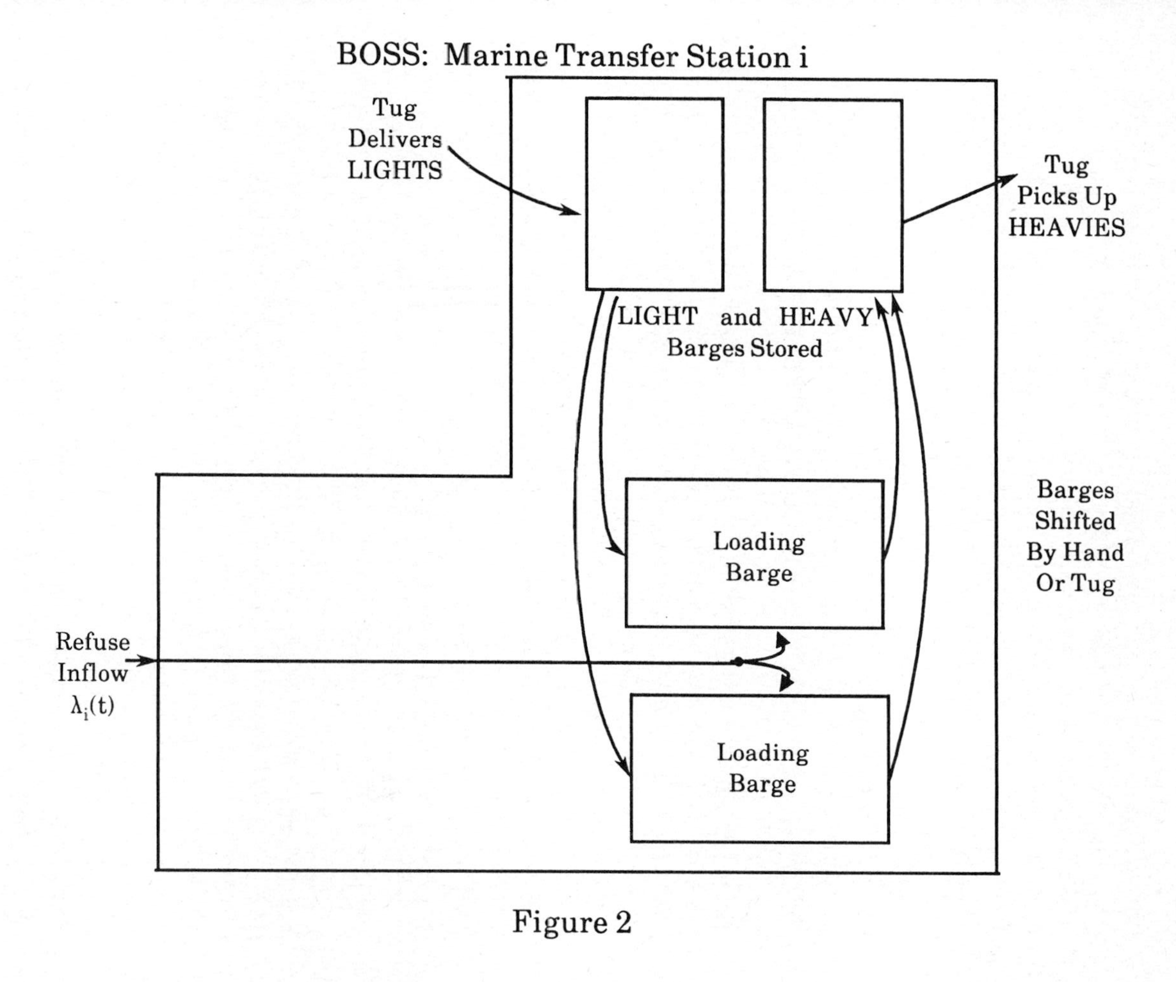
BOSS: Marine Transfer Station i
Tug Delivers LIGHTS
Tug Picks Up HEAVIES
LIGHT and HEAVY Barges Stored
Loading Barge
Loading Barge
Barges Shifted By Hand Or Tug
Refuse Inflow $\lambda_i(t)$

Figure 2

fuel costs. Since blockage is a condition that system operators try to avoid, we will discuss it further (within the context of the "dispatcher's problem").

2.3. The "Sink" Node - Fresh Kills Landfill

Each heavy barge that is picked up at an MTS is eventually joined to a barge train and towed to FKL. Heavy barges arriving at FKL are entered into a queue of other "heavies," each such queue awaiting the "service" of a "digger," namely a crane that digs out or empties a barge of this refuse load (Figure 3). After a barge has been transformed from a heavy to a light, it is moved to a queue of other lights, awaiting transport back to an MTS.

FKL is in many ways a classic multi-server queueing process with very little practical limitation on queueing capacity, of either lights or heavies. The "service time" per barge, that is, the time to empty a barge of its contents, is approximately two hours. However, as we discovered after several months of study, there is an unwritten rule called the "three-barge-per-shift-rule" that allows a digger's crew to end their workday after the third barge on an eight hour shift is "dug out;" our model, to be realistic, had to include the three-barge-per-shift-rule.

The queueing paradigm was appealing, and one modeling approach that is embodied in our final model is to treat the entire system as a "closed cyclic queue," with "customers" (barges in this case) cycling through the system in alternating states of "fullness" and "emptiness." Certain complications (e.g., tides, customer input statistics, the three-barge-per-shift-rule) prevented us from developing a strictly analytical model within this framework.

2.4. Staging Areas and the Hierarchical Transportation Network

Suppose a train of four lights has left FKL. Rarely can one (or should one) MTS accept all four lights. Rather, more typically, two lights would be delivered to one MTS and the remaining two would be delivered to another. A typical tugboat visit to an MTS would also result in one or more heavies being towed from the MTS. Because of the constraint that all barges in a train be in the same state (heavy or light), it is necessary for the tugboat operator to "drop off" two of the lights at some intermediate location prior to visiting the first MTS. Such an intermediate location is known as a *staging area* (SA). In practice staging areas are (often dilapidated) piers in proximity to one or two MTS's (Figure 4).

The following "guided tour" illustrates the use of the staging areas.

1. A tugboat leaves FKL with one or more light barges, heading for a staging area.
2. The tug arrives at the staging area and drops off some of the lights.
3. It heads for the first MTS of the tour with the remaining lights.
4. It arrives at the MTS, drops off the lights, and picks up one or more heavy barges.
5. The tug returns to the staging area, and drops off the heavies.

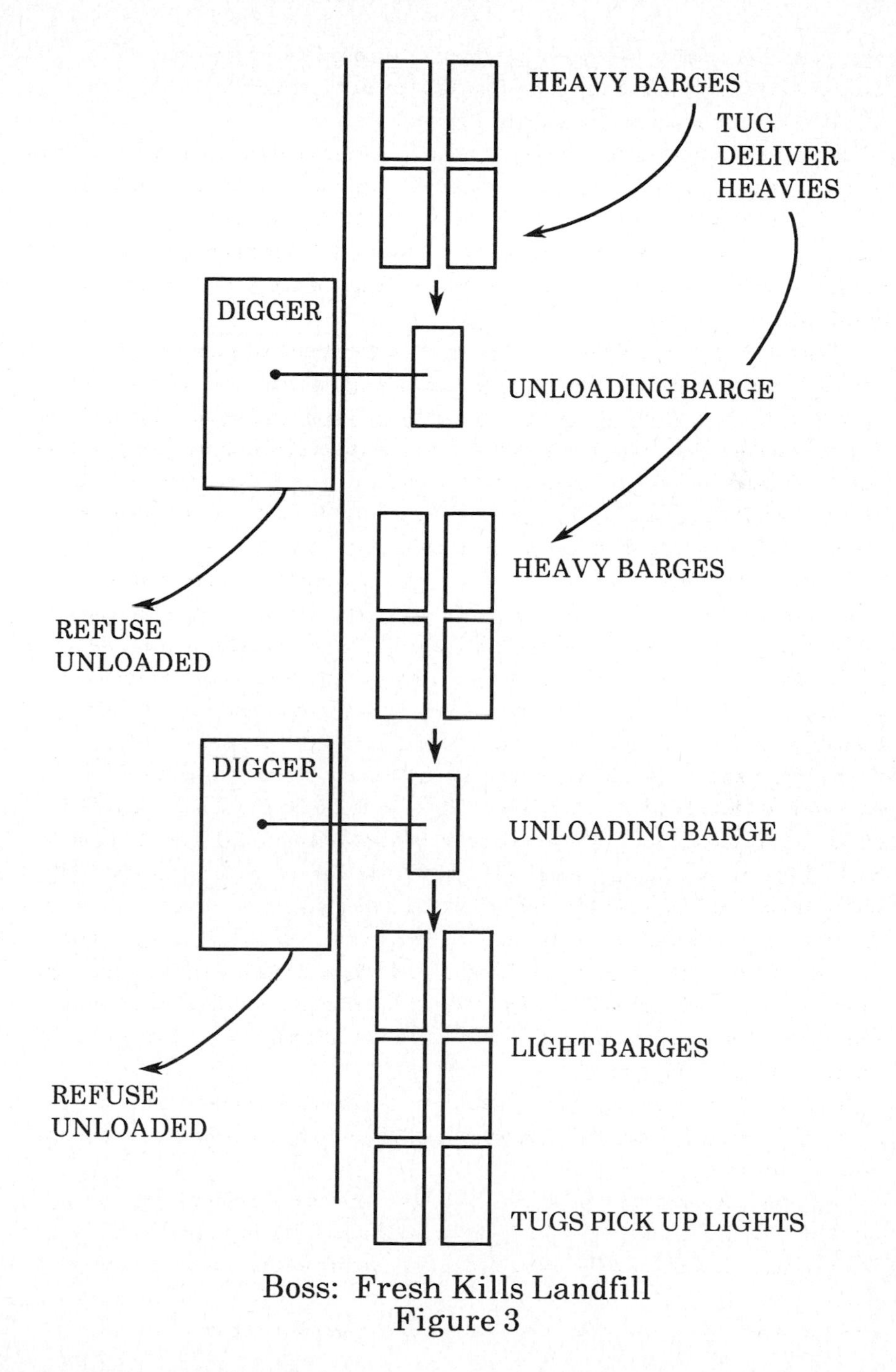

Boss: Fresh Kills Landfill
Figure 3

BOSS: Staging Area

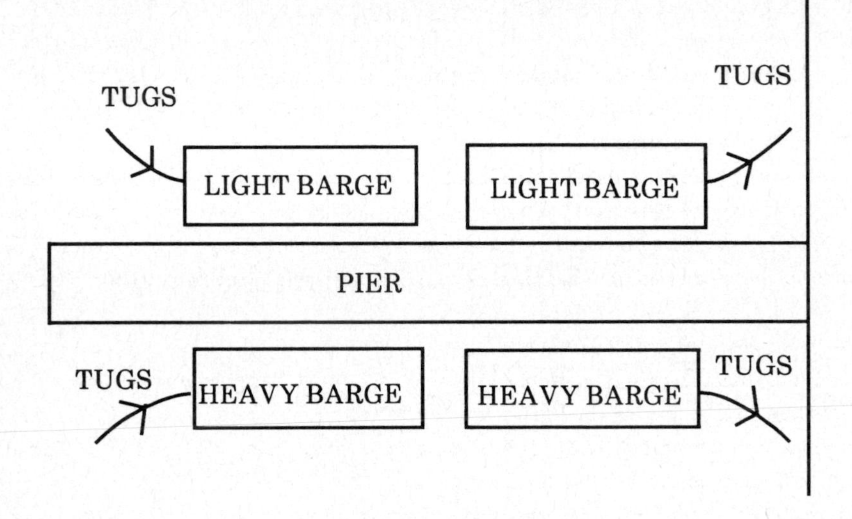

Figure 4

6. It picks up the lights it had stowed there previously and proceeds to the second MTS.
7. After arriving at the MTS, the tug deposits the lights and collects the heavies.
8. The tug visits the staging area one last time. It adds the heavies from the staging area to those from the last MTS (to a maximum of four heavies).
9. The tug returns to FKL.

The process just described implies a rather simple transportation network on which the tugboats and barges travel. The network has an arbitrary number of source nodes (the MTS's), a smaller number of intermediate nodes (the staging areas) and one sink or destination node (FKL). This hierarchical network structure is depicted in Figure 5.

C. The Dispatcher's Problem

Each tugboat captain is in radio communication with the *tugboat dispatcher* who directs the captain around the harborway with his "cargo" of heavy and light barges. The major decisions made by the dispatcher on any given tugboat tour commencing from FKL relate to operational details of that tour: the time of commencement (constrained by availability of lights and a tugboat); the number of lights to be towed from FKL; and the sequence of "network nodes" (i.e., staging areas and MTS's) to visit.

Modeling this aspect of operations proved to be the most difficult part of our assignment. Experienced dispatchers seemed to have a sound intuitive sense for these decisions, but were unable to articulate the objectives motivating their decisions. Their approach to dispatching, as related in answers to interview questions, tended to be situation-specific, that is, clinical rather than statistical (or should we now say "rule-based" rather than "optimum-seeking?").

Dispatch targets (number of lights to deliver to each MTS over the course of a day) for any given day are drawn up the day before by people in "downtown planning." These targets are based on informal analyses of recent and seasonal barge consumption rates, as well as the current deployment of barges throughout the system. As the day progresses, the variety of random events that inevitably impact the system (outages, tides, refuse arrival patterns) influence the order in which MTS's are replenished and often cause dispatch targets to be changed. The dispatchers felt that they were operating a system under stress, where their job was to identify potential emergencies at MTS's and to avert them if at all possible.

1. Sources of Time-Dependence and Randomness

a. Time Varying Gaussian Inputs

We were unable to develop a strictly analytical model of system operations because of the "twin complexities" of system time-dependence and randomness of input flows. Regarding time-dependence, we found it necessary to characterize the mean "input load" at each MTS by a vector having 168 elements, one for each hour of the week.

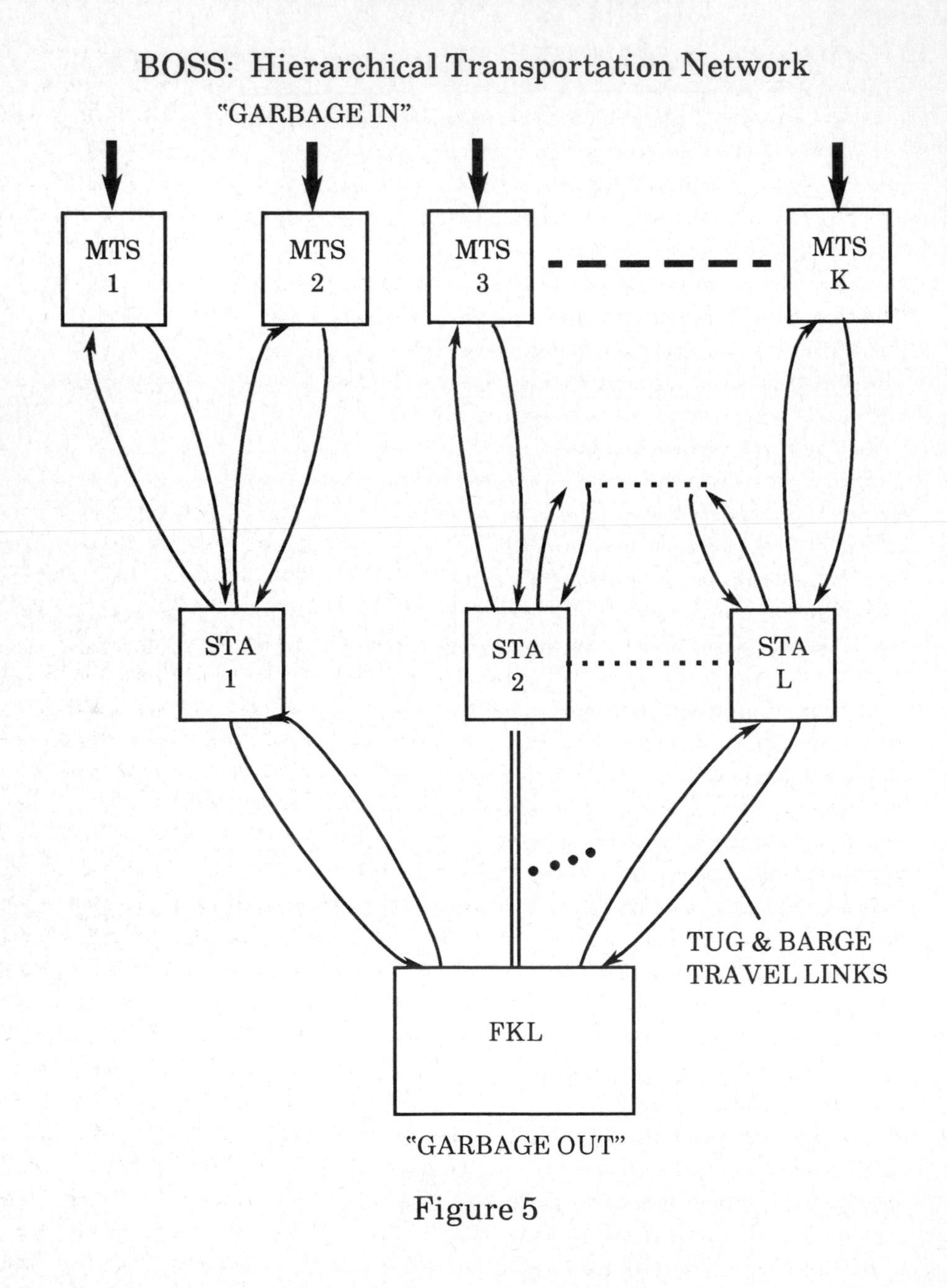
BOSS: Hierarchical Transportation Network
"GARBAGE IN"
MTS 1
MTS 2
MTS 3
MTS K
STA 1
STA 2
STA L
FKL
TUG & BARGE TRAVEL LINKS
"GARBAGE OUT"

Figure 5

Regarding randomness, the daily number of tons of refuse delivered seemed to follow a Gaussian distribution, after seasonal effects were factored in, with a coefficient of variation of approximately 0.10. While this level of variation may seem relatively small, the hourly deviations from the projected means can be quite large. This was particularly troublesome during the mid- to late-morning hours (of weekdays), when an MTS's hourly load tended to be highest. Without any planned staggering of the start times of truck routes, the trucks tended to arrive with their first load of the day within short intervals of each other. For several MTS's it was not unusual for an entire barge load of refuse to be delivered to it within the time period 9:00 a.m. to noon. The number of extra lights stored adjacent to the MTS at the beginning of the day would range from 0 to 2. In the case of 0 or 1, it was important that a "delivery" of additional lights occur sometime during the morning, prior to all barges at the MTS becoming heavy, in which case MTS blockage would occur. Positive Gaussian deviations above the mean load could move up the blockage time by 30 minutes or even an hour or more. Because of this stochastically changing situation at each of the MTS's, the dispatchers had to monitor continuously the loading status of all barges at each MTS and make their dispatching decisions dynamically and adaptively.

b. The "Integer Roundoff" Phenomenon

Another source of randomness appeared as a surprise. The typical MTS "produced" 2 to 3 heavies per day, implying that it required 2 to 3 lights per day as replenishment. Consider an MTS that generates on average, say, 2.6 heavies per day. For the sake of simplicity assume that there is not Gaussian randomness in the loads, i.e., the input load is deterministic. Then, for 60% of the days of MTS operation, precisely 3 heavies will be produced; for 40% of the days, precisely 2 heavies will be produced. If replenishment with lights were to occur daily on a one-for-one basis with heavies, then 40% of the days require only 2 lights, whereas 60% of the days require 3, a 50% increase from the point of view of the dispatcher. Moreover, 2 lights can usually be supplied in one trip by a tugboat, whereas 3 usually require 2 trips (because most trips from FKL entail visits to at least two MTS's). While this complication may seem relatively minor when considering just one MTS, the same complication occurs daily with all eight MTS's. A moment's reflection will reveal that for N MTS's there are 2^N different combinations of daily system replenishment requirements, even in a totally deterministic system (!), due to what we call the "integer roundoff phenomenon" of heavy barge production. With 8 MTS's (the correct number in 1984), $2^8 = 256$, or nearly one combination for each working day of the year! This we believe is one major reason why the dispatchers viewed every day as somewhat unique and why they developed contingency rules for "getting through" each day.

Finally, if the mean daily load to an MTS could be changed to be an integer number of barges (or approximately integer), then the negative effects of integer roundoff would be greatly diminished; the dispatchers of the marine transport system would then enjoy a much more even daily pattern of barge replenishment requirements. This could be accomplished by redesigning the "MTS catchment zones" that assigned truck routes to (typically) the closest MTS. Thus we see a strong linkage between land-based and marine-based logistics system design. While previous to our work each system had been designed separately, MDDOS planners now understand that deviations from "land-based optimality" (by assigning some trucks to, say, the second closest MTS) may in fact improve entire system operation.

2. Performance Measures and The Concept of Deferred Tonnage

At any given time a barge is either (i) in transit *and* light or heavy or (ii) in queue *and* light or heavy or (iii) in the process of having is state changed (either from light to heavy or heavy to light). As might be expected many key performance measures of system operations are time related.

- Time spent heavy (or light) at an MTS or at a staging area or at FKL.
- Time between filling of a barge and the emptying of that barge.
- Fraction of work time that a working tugboat has i barges in tow, where $i = 0$ (the case of dead-heading), 1, 2, 3, or 4.
- Fraction of time that an MTS is "blocked" (i.e., cannot accept refuse from arriving trucks).

Our major problem with modeling the dispatch decisions, even for a strategic model (which could be much more approximate than a tactical model), was quantifying the dispatcher's objectives. None of the "standard measures" listed above seemed to capture what the dispatchers were attempting to do, namely to avoid MTS blockages. One reasonable objective would be to dispatch in order to minimize the probability of blockage anywhere in the system. However, such a criterion would give equal weight to blockage at a high load MTS and a low load MTS; it would also give equal weight to a 15 minute blockage or a four hour blockage.

We thus "invented" a new performance measure called "deferred tonnage." A ton of refuse delivered to an MTS is deferred if upon arrival in the truck the MTS cannot immediately accommodate the refuse because of MTS blockage. We then offered as a dispatcher objective the minimization of the expected (average) number of deferred tonnage (system-wide) per day. Even though no statistics had ever been kept on deferred tonnage, the dispatchers seemed comfortable with the dispatch recommendations implied by its minimization.

In addition to operational performance measures, there are numerous cost- related performance measures, identified at an aggregate level as either capital or operating costs.

D. Data Sources

The operations of MDDOS were not automated and no data were machine readable. In fact virtually no data describing system operations were routinely collected in a deliberate fashion. When we started our analyses, we fully expected to employ research assistants "with clipboards and stop-watches" to sample data from various points of system operations.

One day while visiting the dispatcher's room at FKL, we noticed a stack of dusty boxes piled behind the dispatcher. We discovered that those boxes contained completed "daily condition logs" describing the last several years of operation. One such sheet is used each day (24 hours) by the dispatchers to monitor the movement and status of barges and tugboats. Once an entry became obsolete, it was simply crossed out with one line drawn through it and a new entry was written in an appropriate (unused) box on the sheet. We found that with diligence the entire pattern of vessel movements and history of barge status could be recovered from these sheets! We thus had an excellent data base: the entire detailed operational history of MDDOS for the past several years. The data were reasonably accurate, since the primary purpose of collection was accurate status knowledge for the dispatchers and (not unlike a number of public agencies) historical accountability "in case anything ever went wrong." Among other things, the data (when machine coded) revealed point-to-point travel times throughout the harbor, queueing delays at all key points, tugboat deadheading statistics, etc. The potential positive impacts of such data were in fact so great that MDDOS later decided to implement a data base management system that incorporates the types of data routinely recorded in the daily condition logs.

3. THE MODEL

When it became time to develop the full model, beyond simply identifying the nodes and links of its transportation network, we had reached the following conclusions.

- Due to the requirements of the end user (i.e., DOS planners and managers), the modelers could not sacrifice reality for analytical tractability.
- Randomness of input loads, time-of-week variability and the effects of tides could not be ignored.
- No uncontested objective function or set of constraints existed to develop an "optimization model," even assuming such a model would be computationally practicable.

We decided to develop a descriptive (or "what if") model which was required by its complexity to include elements of Monte Carlo simulation. In fact, as to be discussed below, the resulting model was a hybrid, incorporating simulation, transportation network concepts, ideas drawn from closed cyclic queues, and rule-based concepts motivated from Markovian decision theory. We decided also that the model should be "validated" in at least three ways: (1) with a "back-of-the-envelope" model that includes many of the key features of system operations; (2) with operational data; and (3) with feedback from operations personnel.

A. General Structure

The simulation depicted the movements and status of each tugboat and barge in the system. The status of each fixed facility (MTS, FKL, SA) was also maintained within the event-paced simulation.

All travel throughout the harborway was depicted as occurring on a transportation network having the simple hierarchical structure shown previously in Figure 5. When traveling on any given link (see Figure 1), the tugboat with its barge train was under direction of the radio dispatcher and its link travel time was affected by tides, train size and train status (light or heavy), as discussed earlier. An alternative summary depiction of the model is shown in Figure 6.

The only probabilistic feature of operation was the time varying "Gauss-Markov" process generating input tonnages at each of the MTS's. Using the random number generator of the Monte Carlo simulation, we obtained exact samples of the appropriate Gaussian distribution by using the well-known Rayleigh transformation technique (Larson and Odoni, p. 526).

The entire marine transport system was assumed to operate in a time cyclic manner, with period seven days (168 hours). To obtain reliable (i.e., statistically stable) statistics regarding system performance and costs, the DOS user typically ran the simulation for the equivalent of one year (52 weeks) of operation. To avoid counting unusual behavior in the "start-up transient phase" of the simulation, the user usually omitted the first one or two weeks of data from the statistical tabulations. Since the MDDOS experiences strong seasonality in its load demand patterns, the user had the option of choosing the season of the year that he wished to simulate.

B. The Dispatch Algorithm

Since the reader of this book is most interested in routing and scheduling, we focus our technical description of the model on the routing and scheduling component of the BOSS model, namely the dispatch algorithm.

3.1. An Inventory Replenishment System

We have already discussed many of the complexities of dispatching. More generally, the activities of a MDDOS tugboat dispatcher closely resemble more traditional vehicle routing and multi-location inventory paradigms. In such a paradigm, a dispatcher routes and/or schedules a fixed fleet of delivery vehicles from a central depot or distribution point to spatially dispersed customers to replenish their inventories before they run out of stock. Real-life operations to which this paradigm has been applied include industrial gas and beverage distribution. In the case of MDDOS, the "inventory" that is replenished is emptiness! That is, one may view the tugboats as delivering a barge load of emptiness with each light barge that is deposited at an MTS. An MTS suffering blockage is a customer that has run out of stock. The consequences of the "stockout" are felt each time an arriving truck is denied immediate dumping; the magnitude of the penalty is the number of tons of refuse the truck is carrying. The concept of emptiness as a replenishable inventory item is not new, extending back to some important work by Hardgrave [1966] in the routing and scheduling of coin collectors for public pay telephones. The same concept has been used recently by Larson [1987] in designing a new logistics system to transport New York City's sewage sludge to an ocean destination.

Summary Diagram Describing a Simulation BOSS

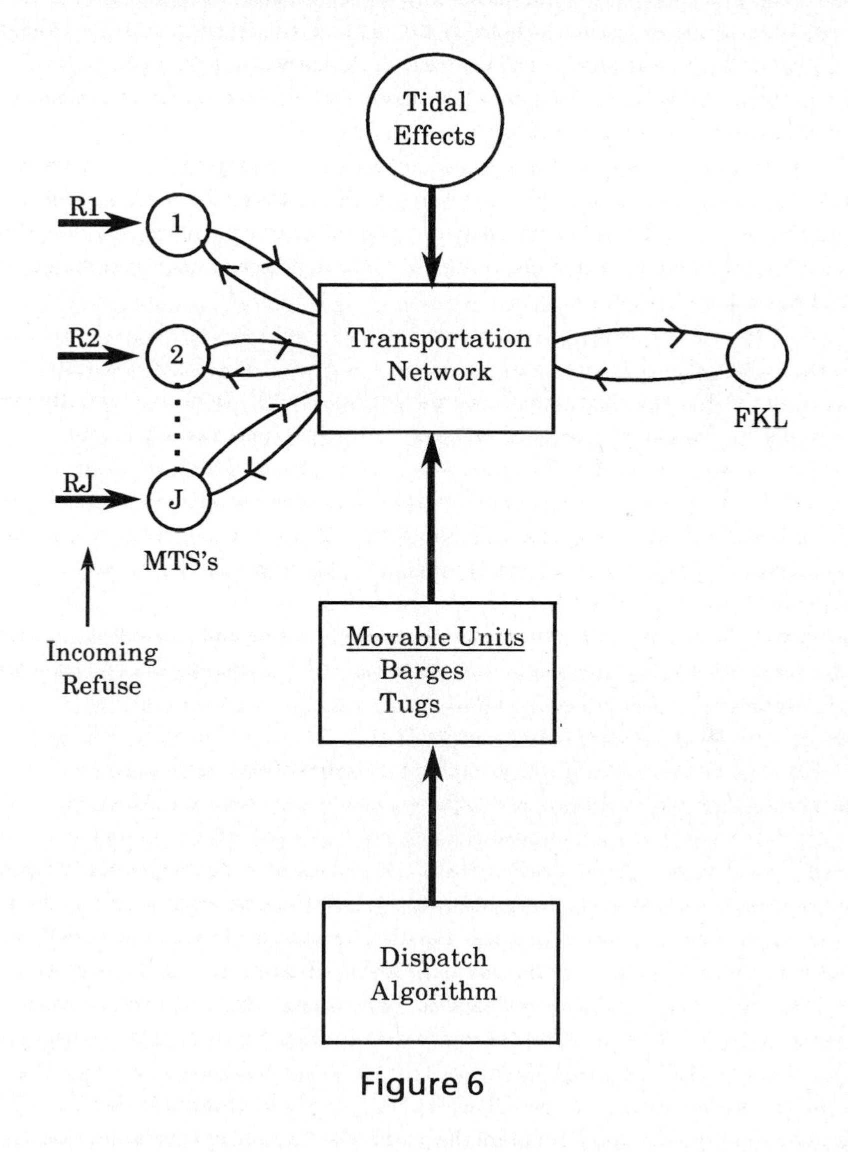

Figure 6

Completing the routing/inventory paradigm, the central depot from which vehicles are routed is Fresh Kills Landfill, a "production facility" that "produces" replenishment stock, namely empty barges. The focus of the BOSS dispatch algorithm is vehicle scheduling and routing from FKL, that is the determination of tugboat dispatch times, barge train sizes, the sequences of nodes (staging areas and MTS's) to visit on a tour, and the transactions to be undertaken at each node.

3.2. Concepts of Markovian Decision Theory

The BOSS dispatch algorithm is a rule-based procedure motivated by ideas of Markovian decision theory (see, for instance, Bertsekas [1987]). We have a probabilistic system that is evolving dynamically over time, and decisions must be made at particular epochs (the dispatch times) using all information available about current system states and projected states resulting from any given dispatch decision. Given that the number of system states is finite (or at least countable) and given a Markovian (or "limited memory") transition behavior, the dispatch process can be formulated as a Markovian decision problem. While these ideas can be formalized in elegant and compact notation (Minkoff, 1985), our emphasis here is to provide a technically complete and intuitive narrative description of the algorithm.

The optimal decision in a given state, according to classical Markov decision theory, is the one that maximizes the sum of immediate reward for the decision in that particular state, *plus* the expectation of the value of the next state that is entered. "Value" here is a relative quantity and measures how much more or less desirable it is to be in a given state as opposed to a reference state. In some sense, the expected value term "corrects" for the myopic policy of basing decisions on a greatest immediate reward criterion. Minkoff (1985) shows that it is equivalent to calculate expected reward collected over the next n decisions and to apply the expected state value correction to the state entered at the n + 1st decision. Moreover, the greater the number of decisions considered before applying the expected value correction, the smaller the influence of the correction in the optimal decision. It is upon this last idea that the BOSS dispatch algorithm's effectiveness relies.

As will be seen below, there is one fundamental property of the optimal policy of a Markovian decision problem that surfaces within the dispatching context: the optimal decision at any given time is often not the one that minimizes the immediate costs associated with the decision, but rather one that minimizes the sum of the costs associated with the immediate decision and the costs associated with the "state trajectory" associated with following that decision. Often it is preferable to choose a decision having greater than the minimum immediate costs in order to leave the system in a set of states which then reduces subsequent costs. Translated into operational terms, "putting out fires" is not necessarily an optimal policy.

3.3. A Narrative Description

The BOSS dispatch algorithm, as applied to the routing of tugboats leaving FKL, seeks to minimize expected deferred tonnage (working in terms of costs rather than rewards) throughout the simulated marine system over a user-specified time horizon. Expected deferred tonnage is calculated via scaled-down "expected value" simulations of the marine system over the horizon. The remainder of this section describes the algorithm in finer detail.

When a tug leaves FKL a "dispatch decision" is made. A dispatch decision is in fact an integrated set of decisions, constituting the time of tug departure, the number of light barges in the train, routing (i.e., the sequence of nodes to visit), and the transactions to be undertaken at each visited node. The success of the BOSS dispatch algorithm, and consequently of the entire BOSS simulation, hinges on the quality of such "decision sets" for tugs as they leave FKL. Although the complexity of the simulated system prevented standard routing algorithms from filling this role, a new dispatching algorithm inspired by Markov decision theory concepts did meet the challenge.

If we were to consider the dispatch decision-making aspect of the BOSS simulation as residing outside the simulation proper, we could classify the dispatching-plus-simulation process as a Markov decision process, and in principle, apply one of various decision-optimizing procedures to obtain the dispatch decisions. This is clearly infeasible in the case of BOSS, due to the dimensionality of the state space and the need to calculate the "values" of the states that the process could enter at the next decision point. But, by the idea advanced in the previous section, the detrimental effects of not including the next state's expected value can be mitigated by estimating the costs incurred over a longer horizon. Hence, the BOSS dispatch algorithm functions by seeking the dispatch decision leading to optimum system performance over a user-specified time horizon (known as the "look-ahead window") (also see Bertsekas [1987]).

In the BOSS dispatch algorithm, the "dispatcher" conceives of the marine operations system in terms of "zones." (In the remainder of this discussion, "dispatcher" refers to the algorithm's dispatch procedure and not the individual that actually dispatches tugboats). A zone consists of one staging area and the MTS's it services. That is, each MTS is in one and only one zone. Dispatched tugs never make pickups and deliveries in more than one zone before returning to FKL. Barge needs are thus classified by zone. The dispatcher determines in which zone it would be most appropriate for the next dispatched tug to operate. If the MTS's in this zone cannot accommodate a number of barges equal to or greater than the BOSS-user-prescribed lower bound on barge train size for tugs leaving FKL, the "need" of this zone is not considered substantial and dispatch of this tug is delayed. Otherwise, the dispatcher determines how many light (heavy) barges should be delivered to (removed from) each MTS, and the tug departs for the zone's SA.

When any tug is ready to be dispatched, the dispatcher begins its job by predicting for each MTS the next time it will experience a blockage that can only be alleviated by the presence of a tug. This is called the "tug block time." Then, for each zone, the dispatcher finds the minimum of the tug block times among all MTS's in the zone. If, and only if, this time falls within the BOSS-user-specified dispatch window, and it is predicted that the zone can accommodate at least one barge, the dispatcher will consider this zone as a possible dispatch destination. If, after all such zonal block times have been found, no zones are under consideration, the dispatch is delayed by half the predicted length of time until at least one zone would come under consideration, or by one hour, whichever is greater. (Half the interval is used to guard against inaccuracies in the prediction.) Otherwise, the dispatcher continues by sorting the zonal block times and eliminating those zones from consideration whose block times are furthest into the future, until there are no more than four zones to deal with.

The dispatching algorithm then proceeds by generating a sequence of *expected value simulations*. (An expected value simulation is a simulation in which random variables have been replaced by their expected values.) The dispatcher maintains a list of predicted times when each tug will be dispatched next from FKL. Using this information, appropriately updated when necessary, the dispatcher simulates all possible $4! = 24$ sequences of dispatches and obtains an estimate of the severity of system-wide blockage that would occur under each sequence. In no sequence are tugs dispatched to the same zone more than once.

The first task of each expected value simulation consists of finding the next time of arrival at each MTS of tugs which have been dispatched in the simulation (denoted as the "simulated visit time"). This is a function of the zone and order of MTS visits within the zone, predicted time of dispatch, and predicted light barge availability. The second task entails estimating the number of tons deferred (or cost of deferral at the user's option) at each MTS from its tug block time until the simulated visit time. If simulated visit time occurs before the tug block time, the dispatcher assumes no tons are deferred at this MTS. If the simulated visit time occurs beyond the dispatch window, the dispatcher only measures the number of tons deferred up to the end of the window. The deferral measures are summed to form a total for the sequence.

At the conclusion of the expected value simulation series, the dispatcher recalls the zone which began the sequence having minimum expected total deferred tons over the duration of the tug dispatch window. This becomes the zone to which the dispatcher intends to dispatch. The dispatcher checks whether a "block alarm" should be put into effect in this zone. A block alarm signifies that an MTS will be blocked before the tug is scheduled to arrive there. Next, the dispatcher attempts to "commit" as many barges as possible to the zone's MTS's, being limited, of course, by the number of light barges currently at FKL and the capacity of the tug (four barges). A barge is committed to an MTS if the dispatcher predicts that this barge and all barges previously committed to the MTS can be accommodated. The dispatcher chooses which MTS to try to commit the next barge to on the basis of minimum tug block time, updated for all newly committed barges. If the barge that the dispatcher is trying to commit to a particular MTS cannot be accommodated there, this MTS will be noted as having received its full possible commitment, and will not be tried again. After all MTS's in the zone are marked in this manner, or there are no barges left to commit, the dispatcher notes how many barges were just committed. If the number is zero, the tug cannot be immediately dispatched. If the number is positive but less than the lower bound on departing barge train size, the tug may only be immediately dispatched when the BOSS user has specified that lower bound violations are allowed when a block alarm is in effect. If the lower bound is equalled or exceeded, the tug is immediately dispatched. Tugs which may not be immediately dispatched wait until the time the dispatcher projects that the tug will be needed, but not for more than six hours. These tugs are considered to be "waiting for need." Otherwise, the tug departs FKL for the SA of the selected zone on its just-established assignment.

The dispatch algorithm, admittedly, is quite complex; a short numerical example will clarify some of the procedures that the dispatcher performs in determining tug destination and assignments. Figure 7 depicts the marine operations system as the dispatching algorithm might look at it. A tug is waiting to leave FKL with a number of light barges (three, in this case) and it needs a barge delivery assignment. The dispatcher goes to work. First, it predicts the (expected) time for each MTS (numbered 1 through 8 in this example) at which it becomes blocked in such a way that only a tug visit can unblock it (tug block time). Each MTS must have one such time, because each MTS will run out of barges at one time or another and will require a tug towing light barges to visit; however, an MTS can block sooner if a tug is required to maneuver (shift) heavies out of and lights into receiving position at an MTS. The blockage times appear in parentheses above each MTS in Figure 7. The minimum of these times for all MTS's served by each SA (1 through 4) is recorded in parentheses next to the SA, and can be called the "zonal block time."

A simple dispatching scheme might send the tug to SA 1, because an MTS served by this SA blocks soonest of all. The dispatcher is wary of such schemes. In addition to tug block time, the dispatcher takes into account the rate of refuse inflow at each MTS and the times of subsequent dispatches. The dispatcher looks at all possible sequences of dispatches to the SA's, where no two dispatches in any sequence go to the same SA. Two such sequences are considered in the exhibit. The first, 1-2-3-4, dispatches tugs to SA's by the rule of "soonest zonal block time first." The times of arrival of dispatched tugs at the MTS's for this sequence, which appear directly above the tug block times, show that each MTS is visited before blockage can occur, except for SA 2's two MTS's. The total predicted deferred tonnage for this sequence is then the sum of the mean amounts of refuse delivered by trucks between times 6 and 10 at MTS 3 and between times 7 and 11 at MTS 4. The other sequence changes the order of dispatch for the first two tugs, yielding the sequence 2-1-3-4. Predicted tug arrival times are given in the top row. Under this sequence, blockage is only expected to occur from times 5 to 9 at MTS 1, and total predicted deferred tonnage corresponds to the amount delivered by trucks at that MTS between those hours.

Suppose that MTS 1 received a heavier inflow of refuse than MTS's 3 and 4 combined. Under these circumstances, and considering only the two indicated sequences, sequence 1-2-3-4 would be chosen as the best dispatching scheme (four hours of deferral at each of MTS's 3 and 4 from sequence 1-2-3-4 produces fewer deferred tons than four hours at MTS 1 from sequence 2-1-3-4). If, on the other hand, MTS 1 received little tonnage compared to 3 and 4, the sequence 2-1-3-4 would be selected. This example shows that, given the latter set of MTS refuse inflow characteristics, tugs are not necessarily dispatched to the MTS which will block soonest.

The best sequence only serves to identify the destination of the *current dispatch*, and that is SA 2 in the second case above. At the time of the next dispatch decision, system conditions may have changed so that expected tug blockage times and related system quantities may assume different (updated) values, perhaps resulting in a dispatch decision different from the second dispatch decision in the current best sequence.

The role of the dispatch window can be illustrated here. Suppose the window extends only until time 20. Then SA 4, whose MTS's are predicted to block no earlier than time 21, would be removed from consideration in the sequences--its critical time lies too far into the future. If the window were further shortened to 10 hours, the dispatcher would modify its "thinking" in two ways. First, only SA's 1 and 2 would be included in sequences; only the sequences 1-2 and 2-1 would be considered. Second, deferred tonnage at MTS 4 in sequence 1-2 would only be measured to the end of the window, time 10 (rather than from 7 to 11).

Given that the next tug is to be dispatched to SA 2 and its MTS's, the tug needs an assignment listing the order to visit MTS's and the number of light barges to deliver. The process that the dispatcher employs is displayed in Figure 8. First, the dispatcher arbitrarily sends the tug to the MTS's in the order they are predicted to become blocked, soonest first. Then each barge is committed in turn by the dispatcher to the MTS that will block soonest. After each assignment, the tug block time for the assigned-to MTS is updated, given the current barge assignment. The updated tug block times appear in Figure 8. The numbered arrows between tug block times in the exhibit indicate which barge assignment caused the update. In the example, the first barge was assigned to MTS 3, and the new tug block time, given that one barge was to be delivered to MTS 3 at time 4 (see Figure 7), increased to time 12. That time being later than MTS 4's tug block time of 7, barge 2 was assigned to MTS 4. MTS 4's new tug block time grew to 10, but since that was less than MTS 3's tug block time, the third and final tug was also committed to MTS 4. The dispatched tug would, upon arriving at the zone, attempt to fulfill this assignment.

Suppose that the assignment of the third barge to MTS 4 left its tug block time at 10. This suggests that the second barge would not be of any use to MTS 4, presumably because it cannot be accommodated at the MTS. In such an instance, the dispatcher

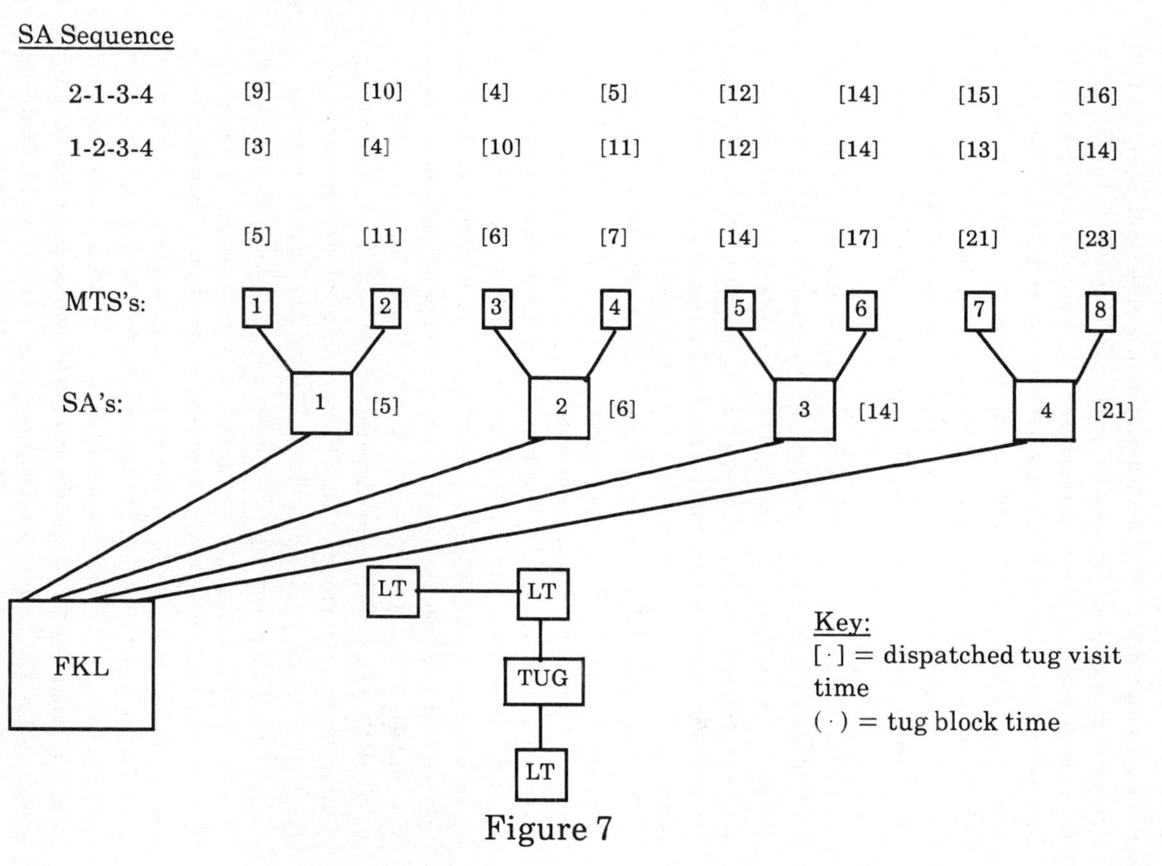
System in the Eyes of the Dispatcher
SA Sequence
2-1-3-4 [9] [10] [4] [5] [12] [14] [15] [16]
1-2-3-4 [3] [4] [10] [11] [12] [14] [13] [14]
[5] [11] [6] [7] [14] [17] [21] [23]
MTS's:
1 2 3 4 5 6 7 8
SA's:
1 [5] 2 [6] 3 [14] 4 [21]
FKL
LT
LT
TUG
LT
Key:
[·] = dispatched tug visit time
(·) = tug block time

Figure 7

Barge-to-MTS Assignment

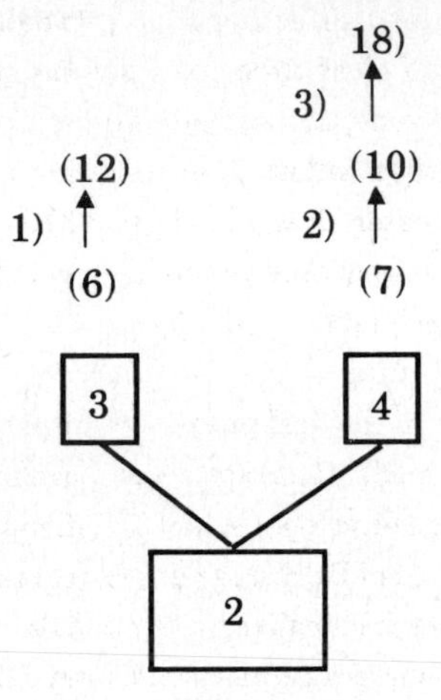

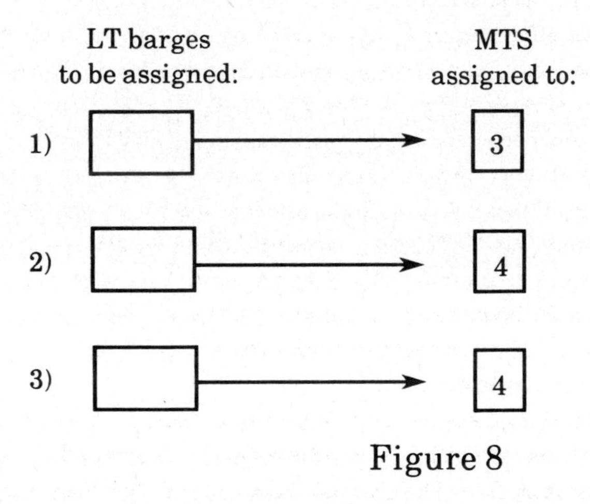

Figure 8

would not bother trying to assign any more barges to MTS 4. It would try to commit barge 3 to MTS 3. If MTS 3's tug block time increased as a result, the barge would be committed there instead, directing the tug to deliver two barges to MTS 3 and one to MTS 4. If MTS 3's tug block time remained the same, the third barge would be of no use anywhere in the zone, and so would remain at FKL. The tug would be dispatched with just the two barges, unless the lower bound on barge train size for barges leaving FKL was three or greater. A tug held up at FKL on these grounds would be considered "waiting for need." A new time would be selected by the dispatcher for attempting to dispatch this tug once more.

C. Model Validation

In 1983 (the year of the first production runs) the model was validated with a separate analytical model that required approximately 25 pages of analysis (or the back of one very large envelope), which verified the accuracy of important aggregate performance measures of BOSS; the analytical model was not capable of computing the detailed performance measures of each fixed facility within the system. In addition, input data were carefully calibrated using statistics derived from the Daily Condition Logs, from earlier consultant's studies, and from other MDDOS reports. Finally, BOSS outputs were checked by operations personnel for "reasonableness."

In 1986, while preparing to design a *c*omputer *a*ided *d*ispatch (CAD) system for the MDDOS tug dispatcher, a much more extensive BOSS validation and evaluation effort was undertaken. This effort, requiring approximately one person year of effort, included tracing detailed tug and barge movements for one week, merging that week's operations will all data available from MTS's, FKL, and elsewhere within the system, and running BOSS under comparable circumstances to calibrate performance. It also required the development of rather complex indirect ways for measuring "deferred tonnage," since MDDOS does not yet measure directly this critical quantity. In virtually all tests comparing BOSS outputs against the real system's performance, as reflected by the one week data sample, BOSS was an accurate predictor of system performance when the BOSS dispatch look-ahead window was set to 24 hours and the minimum barge train size (leaving FKL) was set to two barges.

D. Boss Computer Implementation

The BOSS model was implemented in an interactive (time-sharing) mode, utilizing approximately 7,000 lines of FORTRAN code for the model and an additional 2,000 lines of codes for the cost module. The user interface was designed to be user friendly, menu driven, and error forgiving. The model is run from MDDOS offices in downtown Manhattan, using a modem that connects (via telephone lines) to a New York City-owned mainframe computer. More details are found in Minkoff and Larson (1983).

4. USE OF THE BOSS MODEL

The Department of Sanitation contracted for development of the BOSS model as a tool to evaluate the fleet size required to move refuse expected at present and in the future at the marine transfer stations. In addition, the model was required to have the flexibility of structure to allow adding or taking out components of the marine transportation system in order to assess the effects of such actions.

A. Sizing the Barge Fleet

Several objectives were set up for use of BOSS to size the barge fleet. First, MDDOS was not able to predict exactly how much refuse would be moved by barge. The expected refuse load for the entire system was between 17,000 and 18,000 tons per day. This expectation was tempered by the knowledge that ultimately MDDOS could limit the amount to whatever level it could handle. As a result, one of the objectives of the fleet size determination was to obtain a barge fleet large enough to handle the expected refuse loads and yet flexible enough to handle peaks and troughs around this average.

Since the purchase price of each barge was expected to be approximately \$1,000,000, the second objective was to produce a justifiable expenditure. The third objective was to keep the expected number of deferred tons as small as possible.

Production runs performed to assess fleet sizes were performed on a configuration of the model which described the marine transportation system expected in the late 1980's. This configuration is depicted in Figure 9.

Because MDDOS operates a closed system - each barge loaded is unloaded and sent out again for filling - the number of barges required is related directly to the number of tugs assigned. The fewer tugs assigned, the more barges required to obtain the same performance in the system. To account for this, the fleet sizes of barges and tugs assigned were assessed in tandem. The cost per ton moved, the tons deferred, and the incremental costs of the fleet sizes were assessed by varying the number of barges in the fleet and the number of tugs assigned. Integrated sets of production BOSS simulation runs were executed, from which charts were produced to depict the relationship of cost per ton moved for each fleet size (Figure 10). Other figures were produced from the production runs to depict, for instance, number of deferred tons as a function of fleet size. Considering deferred tonnage performance as well as costs, DOS determined that fleet sizes ranging from 65 to 75 barges exhibit acceptable performance and costs.

In terms of the objectives of the study, seventy barges is neither the least nor the most expensive barge fleet size, in terms of cost per ton moved, but does allow the system to achieve good performance in the measure of deferred tons. It is large enough to handle over 18,000 tons per day average without resorting to a seventh tug, but not so large that the costs become excessive. As a result, a 70 barge fleet meets the objectives of the fleet sizing study.

MDDOS Configuration for Fleet Sizing Production Runs

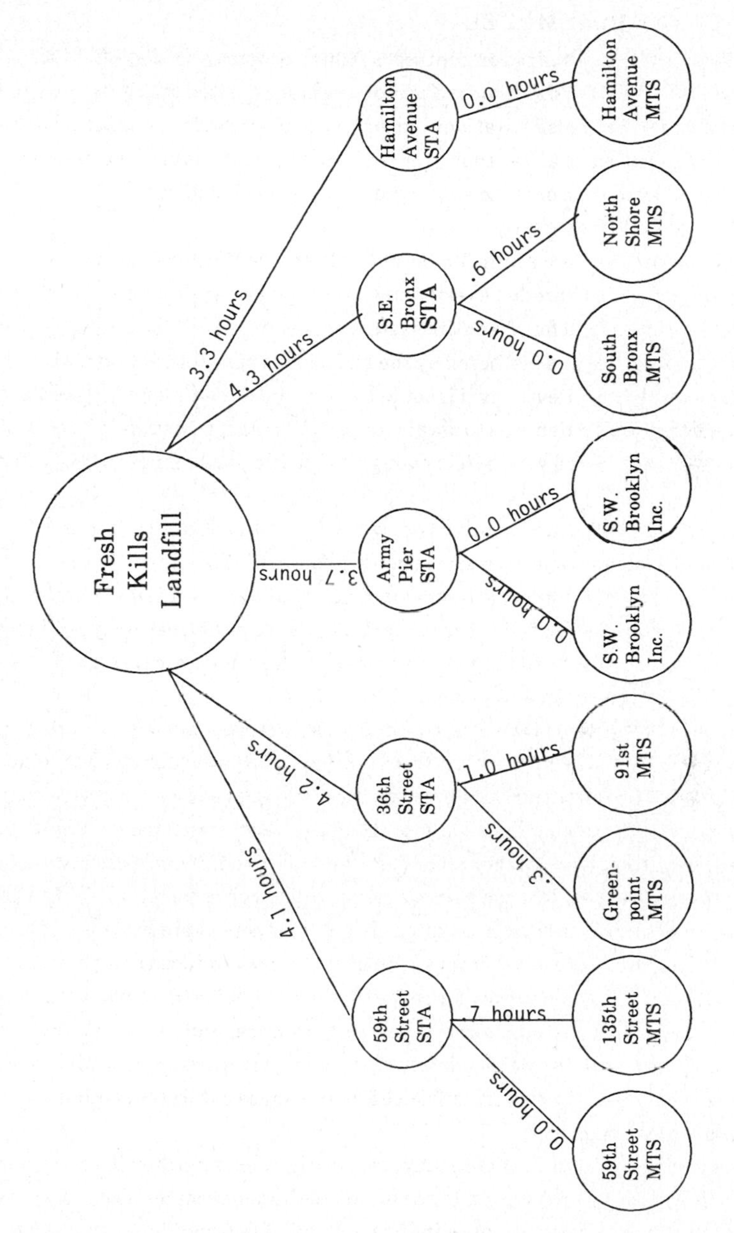

Figure 9

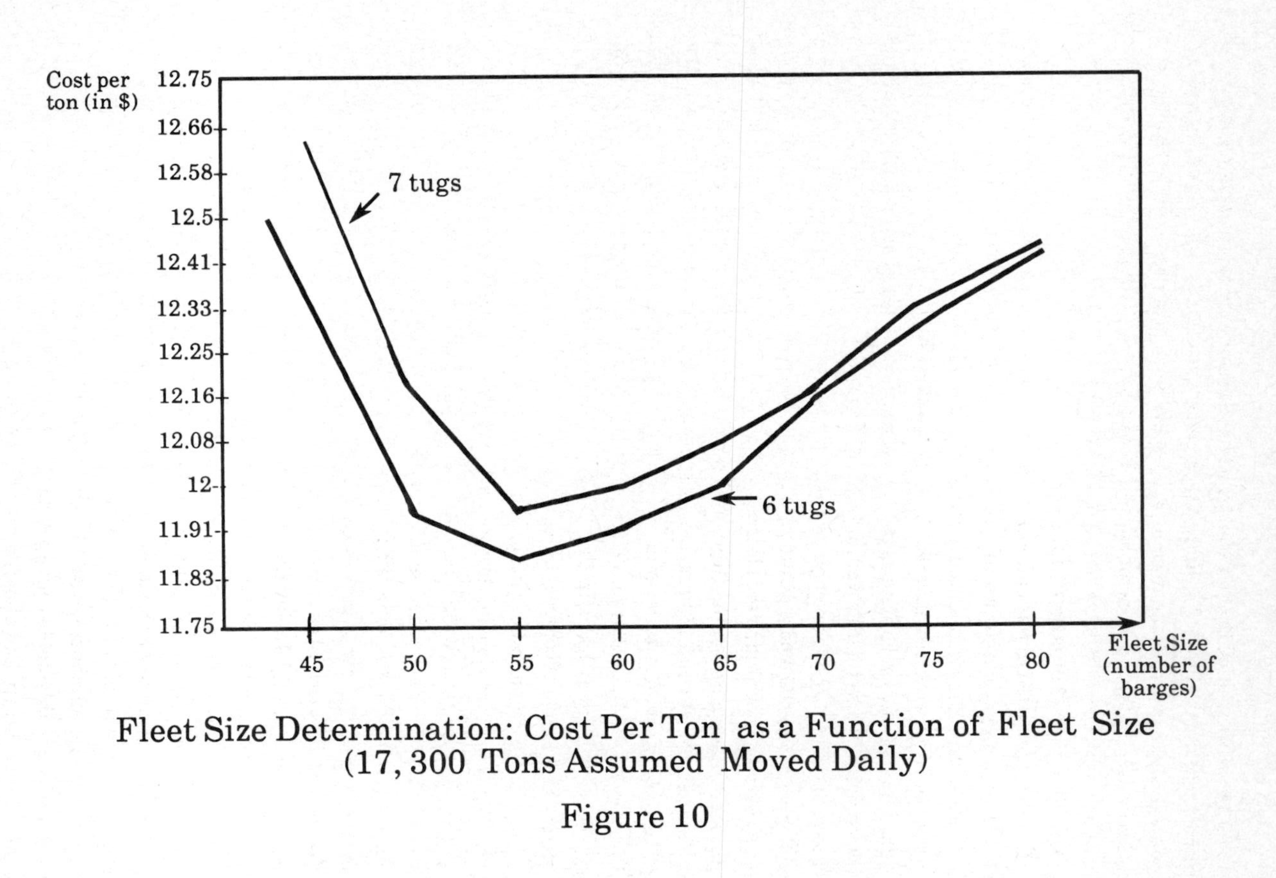

Fleet Size Determination: Cost Per Ton as a Function of Fleet Size (17, 300 Tons Assumed Moved Daily)

Figure 10

This process revealed that a fleet size of 70 barges was probably best for the expected refuse deliveries. To this number, the Department would normally add 25% for outage and downtime due to emergency repair, rehabilitation, and alternative service or use. This outage factor added 18 barges to the 70 barges required for moving the refuse, which results in a fleet size of 88 barges.

4.1. Fleet Size Actually Purchased

In 1983, the MDDOS was operating a fleet of 63 barges. Of these, 40 were over 20 years of age and beyond the useful life for barges in the fleet. The MDDOS in opting to purchase 70 new barges, planned to relinquish 30 of these 40 old barges as soon as the replacement barges were delivered. The remaining older barges were repaired and planned for use until such time as the fleet could reasonably expect to get along without them. This strategy was adopted so MDDOS would have 103 barges in its fleet for a period of about seven years, later dropping to 93. Though 103 barges is 15 barges more than the model would recommend, this would occur during the time that MDDOS was responding to the closure of two of its major landfills and when work was underway to completely rehabilitate and reconstruct the marine transfer stations. It was felt these additional 15 barges would provide a measure of reliability to the fleet during these most uncertain years. The cost of this acquisition was approximately $42 million, or $600,000 per barge. (The figure of $600,000 per barge was $400,000 less than anticipated by DOS, reflecting the depressed state of the ship and barge building industry at the time of contract bid quotations.)

4.2. Comparison to Original Fleet Sizing Strategy

As mentioned previously, the rule of thumb for fleet sizing used by MDDOS prior to implementation of the BOSS model stated that MDDOS would add three barges for every 600 tons of waste added to the marine transport system. Had his rule of thumb been used, MDDOS would have determined that a fleet size of 90 barges for refuse movement, plus 25% for outage was required. This calculation would have resulted in a fleet size of 113 barges, or 10 barges more than the fleet size which exists now (and currently is operating) during the period when the 15 additional older barges are still in the fleet. This "rule of thumb" fleet size is 25 barges more than the model would justify. In order to assure that there would be sufficient barges to handle the service interruptions caused by rehabilitating the marine transfer stations and to allow the MDDOS to respond to Fountain Avenue's closing, MDDOS decided to avoid the purchase of 10 barges through use of this fleet sizing tool, saving approximately $6 million in barge purchases, with the additional 10 barges to be relinquished in the early 1990's possibly not being replaced. This strategy allows some flexibility during a period of uncertainty and allows for obtaining nearly all of the savings calculated by BOSS if the early 1990's system does not warrant the purchase of new barges to account for unforeseen changes in operations.

4.3. Summary of Fleet Sizing Study

The flexibility of BOSS allowed the users to simulate the number of marine transfer stations expected in the future, to try various arrival patterns of refuse at each marine transfer station, and to look for the most sensitive elements in the system. The outcomes of each run were plotted on several graphs depicting total tons deferred, total marine transportation system costs, and percent tons deferred. These graphs were compared and assessed to determine a fleet size which responded best to the variety of total tons moved in a day in each season and to compare the marginal costs of additional barges to the cost of not having immediate transfer capability if the barge fleet were smaller.

The "cost" to MDDOS for creating and implementing BOSS was $100,000. If ultimately only 10 barges are "saved" by BOSS-generated analyses, the return on investment for MDDOS is $6,000,000: 100,000 or 60:1. Considering that up to 10 additional barges may be saved and that BOSS has been used for other purposes (see next section), the figure of 60:1 may be considered a lower bound on New York City's return on investment.

B. Other Uses of BOSS

In addition to fleet sizing uses, BOSS was designed to be used to simulate the system under a variety of system configurations. This capability has been exploited to determine the budget impacts on the towing contract of marine transfer station shut downs for rehabilitation. The number of months required to rehabilitate each MTS was simulated in periods of shut down slips or closed MTS's, and the number of tugs required was determined and budgeted. These runs were used to justify larger towing contracts to allow MDDOS a measure of flexibility in its towing practices during service interruptions.

BOSS has also been used to measure the impacts of different marine transfer and staging area locations and numbers on the towing contract. These runs were used to justify the request for permanent assignment of City-owned piers on the waterfront in Brooklyn for MDDOS barge staging areas, thereby putting to an end the borrowing and renting of piers. Because the waterfront areas in the City are being developed for more recreation uses, piers for this type of use are becoming scarce. The results of these runs allowed MDDOS to attach a dollar figure to not receiving piers for its use. MDDOS has been performing major rehabilitation projects at marine transfer stations. As a result, the possibility for changing the storage and barge movement capabilities at those stations has been considered during the design phases of these projects. In most cases, the existing storage is more than adequate to the needs of MDDOS. However, in one case, BOSS was used to simulate the effect of adding storage at a fairly remote MTS, Southwest Brooklyn. The results of this simulation were used to justify a change in barge storage capability. Another newly designed MTS, Hamilton Avenue, was not provided with storage in the original project. The BOSS model was used to justify locating a staging area adjacent to this station.

Though cost savings are difficult to assess for these uses of the model, these changes (which were justified in large part through use of the model) have resulted in a more efficient marine transport system.

ACKNOWLEDGMENTS

BOSS was implemented under contract no. 2P01141 from the City of New York to ENFORTH Corporation, Cambridge, Massachusetts. We thank Michael F. Cahn and James M. Tien at ENFORTH for their helpful contributions. We also thank in MDDOS Walter Czwartacky and the members of the "User's Committee" all of whom offered suggestions leading to a more useful model. While the BOSS model is property of the City of New York, views and opinions expressed in this chapter are those of the authors and do not necessarily reflect the official positions of either the City of New York or its Department of Sanitation.

REFERENCES

[1] Bertsekas, D.P., Dynamic Programming, Deterministic and Stochastic Models, (Prentice Hall, Englewood Cliffs, New Jersey, 1987).

[2] ENFORTH Corporation, A Computer-Aided Dispatch System for the New York City Department of Sanitation Marine Transport System, Phase I, Final Report, Cambridge, MA, 1986.

[3] Hardgrave, W.W., Coin Telephone Collection Operations, *Bell Laboratories Record*, (December 1966), pp. 359-363.

[4] Larson, R.C., Transporting Sludge to the 106 Mile Site: An Inventory/Routing Model for Fleet Sizing and Logistics Systems Design, to appear in *Transportation Science*, 1988.

[5] Larson, R.C., and Odoni, A.R., Urban Operations Research, (Prentice Hall, Englewood Cliffs, New Jersey, 1981).

[6] Minkoff, A.S., Real-Time Dispatching of Delivery Vehicles, *Ph.D. Thesis in Operations Research*, Massachusetts Institute of Technology, Cambridge, MA, 1985.

[7] Minkoff, A.S., and Larson, R.C., Barge Operations System Simulator, A Planning Model Developed for the New York City Department of Sanitation, ENFORTH Corporation, 929 Massachusetts Avenue, Cambridge, MA, 1983.

Vehicle Routing: Methods and Studies
B.L. Golden and A.A. Assad (Editors)
Elsevier Science Publishers B.V. (North-Holland), 1988

CAR TRANSPORTATION BY TRUCK

Uwe Pape

Technische Universitaet
Berlin, W. Germany

Automobile carriers haul cars from manufacturers to dealerships. Since a standard truck can hold approximately 8 cars, depending on the type of truck and the size of the cars, it is often necessary for the carrier to combine cars of various sizes in the same shipment to optimize the utilization of trucks. A single shipment may visit more than one dealer, and shipments to the same dealer from different manufacturers may be combined. Moreover, several shipments which don't fill a whole truck themselves can be arranged as a series of smaller shipments (for up to five days) to achieve continuous utilization of the vehicle. The main objective of an auto carrier operation is generally the minimization of cost per car transported. This is strongly related to reducing the number of empty miles traveled. Another goal is to restructure the scheduling of vehicles in order to achieve more economic assignments of cars to trucks, and to acquire and fit new orders into the present schedule easily. Very little work has been devoted to this problem. In this paper, the problem and a simple heuristic solution procedure will be discussed.

1. INTRODUCTION

Transportation is a major factor in society, in general, and the automotive industry in particular. In Europe and the U.S., the automotive industries:

- account for a notable percentage of gross national product,
- provide necessary mobility to individuals and industries, and
- create employment in major industries such as vehicle manufacturing (e.g., auto, trucks, etc.), operating companies (e.g., trucking companies, railroads, etc.), and service industries (e.g., terminals, rail ramps, airports, etc.).

The impact of fluctuating markets, new technological opportunities, and energy considerations are dramatically changing the types of companies, systems, and services that are competitive in the current transportation marketplace. In particular, the impact of computers on the operations of manufacturers has begun to influence the development of transportation divisions, operating companies, and service industries. Manufacturers are treating transportation more and more as an extension of the manufacturing process, and thus they are beginning to require carriers to provide transfer between plants and carrier terminals.

Transportation divisions and common carriers have themselves shown increasing interest in the use of computers to improve their productivity in applications such as:

- development of computer networks between manufacturers and transportation divisions (or associated terminals of carriers) for fast data transfer,
- development of computer networks between terminals and headquarters of carriers for centralized route planning, and
- use of computers in the transportation divisions of terminals for automated load make-up.

One major transportation problem is the cost-effective distribution of new cars.

2. PROBLEM DESCRIPTION

Automobile carriers haul cars from assembly plants or import locations to dealerships or export locations. Since the standard truck can hold approximately 8 cars, depending on the type of truck and the size of the cars, it is necessary for the carrier to combine several different car sizes in one shipment in order to make economic use of a truck. A single truck load may be shipped to more than one destination, and shipments from different origins to one destination may be combined. In Germany, for example, one company may serve about 30 origins and about 3000 destinations. In the U.S., the numbers are about twice as large.

If the demand for each origin-destination pair is not a full truck load, to achieve continuous utilization of the truck the load may be arranged as a sequence of smaller shipments over several days. A typical such route might be: deadhead, pick up 4 cars, deadhead, pick up 4 more cars, deadhead, deliver 3 cars, pick up 2 cars, deadhead, deliver 6 cars, deadhead, deliver 1 car, return to the depot.

Of course, the volume of demand differs from company to company. In Europe the monthly volume at one of the bigger carriers is about 50,000 cars; in the U.S. it may be ten times as large. A European carrier may operate with 200 trucks, while the largest American carrier operates with nearly 5000.

In the U.S., the ratio of "empty" to "loaded" miles is very high; more than 40% of driven miles are empty ones. This is a relic of times when automobile manufacturing was concentrated in Detroit and backhauling was much less important. Today the situation has changed, and assembly plants and importers are located all over the U.S. Automobile carriers are aware of the backhaul opportunity but haven't yet obtained much improvement through manual operations. Dispatchers still think in terms of haul and backhaul, not flexible round-trip operations, as they should.

In Europe the situation is different, mainly because of smaller distances and the historic growth of manufacturers in different countries or different states in the same country. For example, in Germany we see Volkswagen in

Lower Saxony, BMW in Bavaria, Mercedes-Benz in Baden-Wurtemberg, and so on. An analysis of driven miles in one month for a typical selection of trucks in a North German company shows the following distribution of loads:

load	% of mileage
8 to 11 cars (full load)	62.2%
5 to 7 cars	21.4%
0 to 4 cars	16.0% .

Another issue is the redefinition of the business to get more economic assignments of cars to trucks and to acquire new orders which might fit into the existing schedule. Examples include transportation of used cars, rent-a-cars, driveaway company cars, and even cars of competitors who are unable to meet time requirements or are trying to eliminate costly trips.

Each order is composed of from 1 to 11 cars and has to be performed within a specified amount of time. In Europe the constraint is very tight--dealerships want to receive their cars within two days after they've been manufactured. In the U.S. the situation is somewhat different; manufactureres want to dispose of cars within five days.

Additional constraints are crew work rules (such as maximum number of hours per interval that a driver/crew can drive) and break-time restrictions. Some examples of German union regulations that serve to demonstrate the complexity of timing constraints follows:

- 118 hours maximal driving per two weeks (one or two drivers);
- 34 hours maximal cabin time (two drivers, truck with sleeping cabin);
- 12 hours maximal length of shift (one driver);
- 17 hours maximal length of shift (two drivers);
- 22 hours maximal length of shift (two drivers, truck with sleeping cabin);
- 8 hours driving time per shift;
- 4 hours maximal continuous driving time;
- 30 minutes minimal rest between two continuous driving periods;
- 11 hours rest between two shifts (one driver);
- 10 hours rest between two shifts (two drivers);
- 8 hours rest between two shifts (two drivers, truck with sleeping cabin);
- 29 hours additional rest per week.

American and European union regulations are sometimes even more complicated and demand excessive work and much adroitness in system design.

3. THE MODEL

3.1. Outline

The automobile routing and scheduling problem can be characterized by task precedence, mixed loads, and time window constraints. In this combined routing and scheduling problem we are given:

- goods or people to be transported;
- a transportation network;
- a set of terminals at certain locations;
- a set of vehicles at each terminal;
- a set of pickup locations;
- a set of demand locations;
- a set of customer service levels (frequency of deliveries, time windows, etc.).

Goods to be transported are passenger cars. Although car sizes and weights are (in reality) different, we simplify the problem by assuming equal car size and neglecting the weight constraints. The transportation network is the underlying road and highway system, with real distances measured in miles or kilometers. Terminals are courtyards of transportation companies, mostly located near assembly plants, rail ramps, or ports.

A certain number of trucks are assigned to each of the terminals, which in turn serve as depots. In general, each truck has to return to its depot within 5 or 6 days. These time windows are not absolute constraints, rather they represent a desirable limit. Each truck may be driven by one or two drivers. Examples of time constraints for drivers have already been mentioned in Section 2. Any routing algorithm will be highly inefficient if all union regulations are included, so we model only the most important ones. Even these may sometimes be relaxed, however, because carriers often assign a third driver to each truck in order to continuously utilize the vehicles. We assume that each truck can hold 8 cars (if orders of less than 8 cars are combined). Orders with more than 8 cars will be handled manually and assigned to trucks as full loads.

Pickup locations are assembly plants, rail ramps, and ports. Other locations in the transportation network may also serve as pickup locations, if dealers and rent-a-car companies are involved. Demand locations are dealerships and ports. Orders consist mainly of the following information: pickup location, demand location, number of cars, earliest pickup time, and latest delivery time (in Europe) or latest pickup time (in the U.S.).

Pickup and delivery activities are called tasks. Task precedence relationships force the pickup activity to precede the delivery activity. Pickup and delivery tasks must be on the same vehicle. Service constraints involve the servicing of tasks within specific time windows. A time window on a service task requires that the task be executed within a specific time interval. Thus, any potential route that involves this particular task must ensure that the pickup time and delivery time fall within the appropriate time bounds. Time windows may be neglected if drivers have access to the courtyards of pickup or delivery locations.

3.2. Mathematical Model

A section is a segment of the transportation network and is denoted by (i,j) where i is the beginning of the segment (origin) and j is the end of the segment (destination). An order o(i,j) is the demand for transportation of a specified number of cars n(o(i,j)) from an origin i to a destination j. The cost for the partial execution of an order over a section is represented by c(i,j). There are a variety of cost functions used in practice, but we shall adopt a simple model--the shortest distance traveled by the truck. Each truck has a capacity of CP cars. The load cp(i,j) on each section is limited to CP. We let t(i) denote the arrival time at location i. The time it takes to load cars at origins and unload them at destinations is represented by tl(i) and tu(j). The duration of the journey tj(i,j) on a section (i,j) is the traveling time from i to j. A trip f(i,j) is a productive journey carried out by a single truck without becoming empty on the way from i to j. A trip covers at least one order, is characterized by a sequence of origins and destinations, and takes into account all information about duration, distance, and time windows. The duration of a trip tt(i,j) is the shortest time to travel from i to j (including all loading and unloading). The cost c(f(i,j)) is the sum of the distances of all sections traveled. Time windows at pickup and demand locations are represented by (w1(i), w2(i)). An intertrip g(i,j) is an unproductive (i.e., empty) journey carried out by a truck. The intertrip g(i,j) goes from the end of one trip to the beginning of another trip. The cost is the distance of the section traveled. A route is a sequence of trips and intertrips beginning and ending at the depot which is carried out by a single vehicle. A tour is a sequence of origins and destinations beginning and ending at the depot.

The problem is to create trips by combining orders and to determine routes and schedules so as to minimize travel costs while obeying network and scheduling constraints. Travel costs, in our case, are the sum of all distances traveled. Let R be the set of orders, O the set of origins, and D the set of destinations, with $O \cap D = \emptyset$. We shall describe the problem with a single depot intersection where each truck leaves it only once. The node p represents the depot. The network used is defined by the set of nodes $N = O \cup D \cup \{p\}$ and the set of arcs $A = N \times N$. Other versions of the problem with more than one depot or with several exits per truck can be handled in a similar manner.

Two types of variables are used: flow variables x(i,j), $(i,j) \in A$, and continuous time variables t(i), $i \in N$, associated with the beginning of the task at each node.

Optimal routes satisfying the scheduling constraints solve the following problem:

$$\min \sum_{(i,j) \varepsilon A} c(i,j)\, x(i,j) \tag{1}$$

s.t.

$$\sum_{j \varepsilon N} x(i,j) = \sum_{j \varepsilon N} x(j,i) = 1 \qquad i \varepsilon N \backslash p \tag{2}$$

$$x(i,j) \geq 0 \qquad (i,j) \varepsilon A \tag{3}$$

$$x(i,j) > 0 \Longrightarrow cp(i,j) \leq CP \qquad (i,j) \varepsilon A \tag{4}$$

$$wl(i) \leq t(i),\ t(i) + tl(i) \leq w2(i) \qquad i \varepsilon O \tag{5}$$

$$wl(j) \leq t(j),\ t(j) + tu(j) \leq w2(j) \qquad i \varepsilon D \tag{6}$$

$$x(i,j) > 0 \Longrightarrow t(i) + tl(i) + tj(i,j) \leq t(j) \qquad i \varepsilon O,\ j \varepsilon N \tag{7}$$

$$x(i,j) > 0 \Longrightarrow t(i) + tu(i) + tj(i,j) \leq t(j) \qquad i \varepsilon D,\ j \varepsilon N \tag{8}$$

Additional time window constraints can be set up to handle union regulations. Finally, each order has to be part of a route and tour, respectively.

As the whole dispatch process has to be carried out within a few hours every morning and routes will be changed several times during the rest of the day, an efficient procedure is an absolute requirement, even if the solution is not optimal.

4. THE SYSTEM

This section gives the description of an integrated information and control system supporting car transportation by truck, implemented on a Siemens mainframe (1985-86) and on an IBM-AT (1986-87). The AT version uses a special windowing technique and will be used as a prototype for future research. We refer to this implementation in the following paragraphs.

4.1. Data Files

The system is comprised of two kinds of data files--object files and scheduling files. Among the object files, the load file has to be mentioned first. It contains the data elements characterizing each load. These include origin, destination, number of cars, characteristics of cars, time constraints, and duration of the loading process. References to tours and trucks are also included. A truck file containing information concerning availability, location, capacity, employment, and assignment to drivers and tours is also needed. In addition to loads, data about customers at pickup locations and delivery locations has to be recorded. Addresses, time windows, and additional information (when needed) are also gathered in this file. Data on drivers are stored in another file. Time of employment and vacation schedules are important for various reasons. Coordinates of regions have to be stored as well, together with the zip code and a special alphabetical code for identification of the corresponding region. Finally, the distances between each pair of regions are recorded. These distances (e.g., road miles) can be used for cost calculations. The most important scheduling file is the file for tours. It contains entries on loads and trucks assigned to each tour, and on duration, length, and capacity utilization. Besides the above, the system also offers lists of loads, trucks, and locations of trucks to assist in manual scheduling. The dispatcher may also obtain a list of loads sorted according to a procedure described below. The dispatcher may perform an array of operations on these files: he may insert, search, change, or delete data items as he chooses. He must be able to change or cancel assignments of loads or trucks to tours, and alter the sequence of loading and unloading on the same tour.

4.2. User Interface

The information system is designed to lead the user. This means that the user may choose only between the functions offered by the system's current state of dialogue. Thus, even the untrained user will feel secure with the choice of functions and will be able to perform the desired operation. Data are entered via screen masks. Function calls are initiated by pressing function keys. Data and route plans are displayed on the screen. In addition, forms involved in the business (bills of delivery or driver schedule forms) can be printed easily.

4.3. Routing and Scheduling

4.3.1. Levels of Automation

The system supports three different levels of automation for the routing and scheduling process:

- manual;
- half-automated;
- fully-automated.

Manual routing and scheduling enables the dispatcher to decide which load will be assigned to which vehicle, and to determine the sequence of loading and unloading. The system, however, assists him in meeting certain constraints, such as the capacity of the truck. It ensures loading before unloading and checks that the same load is not put onto several tours. In addition, the dispatcher may consult lists of all loads, all trucks, and all truck locations. The actual state of the scheduling process is displayed on the screen. The dispatcher may shift a load from one route to another and revise previous decisions. The system enables him to interrupt planning whenever he wishes to in order to perform some other task (e.g., accept a new load).

Half-automated scheduling relieves the dispatcher from routing work. Whenever the system offers a new route and schedule, the dispatcher has the opportunity to accept or reject the proposal. If he rejects it, the system tries to find an alternative solution. The dispatcher may still compose schedules manually within this system.

Fully-automated scheduling primarily leads to faster processing. On this level, the system delivers a comprehensive proposal for all routes and schedules. Afterwards, the dispatcher is given the opportunity to alter any part of this schedule according to his preferences.

The half-automated and fully-automated levels offer the full spectrum of operations for manual changes and information retrieval possibilities. The final decision still remains with the user, however, not with the machine. In order to evaluate the system's recommendations more precisely, the dispatcher may receive the list of loads step by step according to previously defined priorities. Having finished the routing and scheduling, the dispatcher can declare the whole schedule to be "fixed". Up to this point, each part of the schedule may be changed, and last-minute loads may be added. Once all tours have been marked fixed, this state of planning will be taken as the basis for the next run. This feature was very positively received by representatives of the transportation industry. It enables carriers to develop local schedules at their branches and then confirm them at their headquarters.

The dispatcher may influence the scheduling process through several parameters. The following parameters are being considered in order to assign a priority to each load:

- distance a load has to be carried;
- the number of cars a load consists of;
- the time elapsed since the load has been waiting for delivery.

Criteria such as distance between the pickup location and the route, or distance between locations of loads on the same tour also serve as parameters. Hence, the dispatcher has the option to create different schedules, from which he can choose the one that best meets his needs.

4.3.2. The Algorithm

The algorithms used for half-automated and fully-automated scheduling have been simplified more and more during the system's development. This is mainly a result of the complexity of time windows which have to be considered. Before each run, the dispatcher determines the current values of the parameters. He may choose integer priorities between 1 and 9, or enter other parameter values as he wishes. A sorting algorithm then assigns a final value (determined by the different priorities) to each load. All loads are regarded strictly in the order established by this process, i.e., the loads are considered in descending order of priority. In the beginning, all loads are assigned the state "new", which will be changed during the planning process to "loaded" and later to "unloaded".

In the first step, the algorithm tries to assign the current load to a route that already exists. The load has to fulfill certain restrictions. It must not violate the capacity of the truck. The pickup location must be "close enough" to the route as it is shaped at that moment. The demand location must be near the destinations of other loads on that route, or on the way to these destinations. Time window constraints must be met. The algorithm then chooses the most "appropriate" route from those fulfilling these requirements. The length of the route may serve as a measure of "appropriateness".

If the algorithm cannot find a suitable tour, in a second step it will attempt to find a truck that has not yet been assigned to any tour. From amongst all trucks of capacity large enough to hold the load, the algorithm chooses the one which is located nearest the origin of the load.

If there is no such truck, the algorithm performs its third step. One of the current tours is declared "fixed" and the truck assigned to that tour is considered empty again. The algorithm chooses the tour whose last demand location is nearest to the origin of the current load. The restrictions mentioned previously have to be satisfied, naturally.

Whenever the dispatcher rejects a proposal, the corresponding load is considered again and the system offers another alternative. However, if there are no alternatives remaining, the dispatcher will be informed of this and the system presents the first and, according to the algorithm, optimal alternative again. The final decision to accept any proposal or to intervene manually remains with the dispatcher.

5. AN EXAMPLE FOR APPLICATION

5.1. Structure of the Company

The system development has been carried out in cooperation with several carriers in Germany and in the U.S. The following remarks may serve to illustrate the field of application and the difficulties connected with implementing an integrated information and control system.

Ryder Systems, Inc. was founded in 1930 and is now the parent company for a group of service-oriented companies which provide highway transportation, distribution, and business services to customers in the continental United States and Canada. It is headquartered in Miami, Florida. Each of the subsidiaries is assigned to one of the following seven divisions:

- Vehicle Leasing and Services Division;
- Distribution Systems Division;
- Truckstop Division;
- Automotive Carrier Division;
- Freight Management Division;
- Insurance Division;
- Aviation Leasing and Services Division.

The Vehicle Leasing and Services Division is the largest subsidiary; the second largest is the Automotive Carrier Division. [There is another trucking company with the same name, which has no relation to Ryder Systems. The other Ryder's parent company is IU International, and it now operates under the name "Ryder/P.I.E."]

5.2. The Automotive Carrier Division

The Automotive Carrier Division of Ryder Systems is the largest automobile carrier in the U.S., followed by Leaseway (with Anchor Motor Freight as a subsidiary). Ryder transports about 30% of all new cars from assembly plants and rail ramps, as well as imported cars to 13,000 dealerships (Leaseway transports about 20%).

The division considers itself an extension of the assembly line, so its objectives are closely related to the objectives of the manufacturers. Another key objective is the reduction of overall transportation expense, primarily through reducing empty vehicle-miles. Ryder Systems has bought five auto carriers since 1968 (which was also the year of the division's founding):

- M & G Convoy (1968);
- Complete Auto Transit (1970);
- Janesville Auto Transport (1977);
- F. J. Boutell Driveaway (1981);
- Commercial Carriers (1982).

It operates in 52 locations all over the U.S. and serves mainly General Motors, Chrysler, Volkswagen of America, and various imports. The percentage of production hauled for various manufacturers by Ryder Systems are:

- General Motors 47%;
- Chrysler 45%;
- Volkswagen of America 44%;
- Imports 22%.

The percentages of Ryder Systems' total volume that these manufacturers represent are:

- General Motors 72%;
- Chrysler 11%;
- Volkswagen of America 2%;
- Others 3%;
- Imports 12%.

The Automotive Carrier Division uses an equipment pool of 4,750 tractors and 5,250 trailers and has 5,750 employees.

5.3. Computerization Issues and System Control

5.3.1. Computerization

The division intends to implement a computer network (in addition to its IBM mainframe) for local data storage and data communication with manufacturers and headquarters. A major obstacle to the easy use of computers is the employment of different mainframes in the various companies Ryder has bought over the years. At present, about 20 terminals are equipped with computers and partially integrated into the network. Three terminals have a second computer for the load make-up program. The load make-up program assigns car types to truck types within a demand area on the basis of a dynamic programming-like heuristic procedure. The assignment of specific cars to specific trucks is done manually.

5.3.2. System Control and Backhaul

The division has already achieved productivity improvements through the development of two-way hauls and by increasing backhaul opportunities. But more than 40% of driven miles are still empty miles. This is a result of:

- competition between the five major carriers;
- competition between terminals;
- slow information flow between terminals and headquarters;
- no common database;
- incomplete computer network.

The assignment of backhauls to returning empty trucks is accomplished via phone by central dispatch in the headquarters. Load make-up is also primarily a manual operation, with some assistance from the computerized procedure mentioned above.

In summary, the systems control process is not well developed at this time. Ryder sees this as an important area in which further advances must be made.

6. MAN-COMPUTER INTERACTION

One of the most significant impediments to user acceptance of computer systems is a feeling of lack of user control over the underlying system. Through more effective man-machine interaction, better acceptance should floow. Part of our research addresses this problem. We ask questions such as: What is the purpose of the dialogue and what persons will use it? Will specialized operators always use the system or will other people use it, too?

There are many different categories of man-computer dialogue. Some are easy to handle, whereas others require months of experience and much skill. Before attempting to determine the dialogue category that should be used, the capabilities and attitudes of the potential operators need to be considered.

The designer of the system has to relate the category of dialogue to the hardware requirements. These requirements affect aspects of the hardware design such as file organization, the telecommunications network, and the type of terminal used. These are also important because they affect the software design (particularly the control programs), the data base design, and the routing algorithms.

Most real-time systems use display-screen-oriented software for the dialogue and for information to be viewed. A second screen for graphical display may be helpful. This should be considered together with techniques to display data for the routing slgorithm and algorithmic results. Recent hardware and software developments may allow for a combination of both kinds of information on one screen.

It is very important that careful attention be devoted to ergonomics (human engineering) and user psychology. There are three design levels that are worthy of attention:

- The functional level. Which functions should be performed by man, and which by the machine?
- The procedural level. Once the functions of the system are determined, what procedures should be employed? What procedures can best help to achieve an efficient and error-free operation?
- The syntactical level. How can terminal operations best be translated into input and output messages? What display formats are best? What response times are necessary? What forms of encoding can enhance the user's capability?

The implementation of such an information system took an enormous amount of time. The system had to offer different levels of automation (manual, half-automated, and fully-automated). This leads to a step-by-step design process. Only the last contained the automated backhaul procedure. Once the dialogue (including the first step of automation) was tentatively designed, it was desirable to implement a prototype and to try it out with live operators. We are now planning an extensive simulation before programming the final version of the first step.

7. EXPERT SYSTEMS FOR ROUTING AHD SCHEDULING

It is often worthwhile to use different approaches to solve a single problem. If the dimension of a complex problem is very small, a combinatorial algorithm may be acceptable. If a large amount of data is involved and decomposition methods cannot be applied, heuristics are often helpful. But if effective heuristics cannot be obtained, or a full, formal problem specification cannot be written, the development of an expert system might be an appropriate approach.

Expert systems offer a new generation of computerized problem representations which mimic the way humans represent problems and solutions. Expert systems attempt to identify, formalize, encode, and use the knowledge of human experts as a basis for a high performance computer program. Expert systems are based on the integration of knowledge and algorithms for interpreting this knowledge. They simulate human expertise by utilizing a sufficiently large and yet narrow, well-defined knowledge-base to produce high quality, extremely useful output efficiently. They also have the ability to justify and explain any decisions and/or conclusions they may present to the user.

As is well-known, it is often impractical to include all possible constraints in an optimization model. There are often constraints which are ambiguous and hard to define. An expert system would start by solving a model which leaves out many constraints, and periodically it would display solutions (or partial solutions) to the user. If at any time a violated constraint is observed, the user (the knowledge engineer) could incorporate new rules into the knowledge-base, and the application of the partly automatic/partly interactive solution procedure would continue. By using this approach instead of trying to include all constraints at the beginning, a large amount of effort can be saved. Also, the smaller model can be more easily structured and the algorithms used may be more efficient.

It is often difficult for the user to identify constraints a priori. Thus a violated constraint can be easily identified by using a knowledge-based system with a modular structure. This expert systems strategy also provides the opportunity to incorporate rare and important expertise of senior members of a transportation company into the system. It also promotes the easy adaptation of an implemented system to slightly different problems. When the expert system is functioning but the user no longer needs or wants its interactive features, the system may be converted into a conventional computer program. In some very complex problem solving situations, the user may still want to direct a heuristic algorithm even when it has been fully implemented. Combined routing and scheduling algorithms are an excellent example; the generation of routes may take advantage of the user's spatial intuition that otherwise could not be obtained.

Vehicle Routing: Methods and Studies
B.L. Golden and A.A. Assad (Editors)

ROUTE PLANNING FOR COAST GUARD SHIPS

Bruce L. Golden

College of Business and Management, University of Maryland, College Park, Maryland 20742

The U.S. Coast Guard is responsible for maintaining approximately 40,000 buoys each of which must be visited, inspected, and serviced at least once a year by ship. In this paper, we examine the problem of generating cost-effective ship routes for the Coast Guard. First, a simplified version of the problem is presented. Then, extensions and generalizations are discussed.

1. INTRODUCTION

The U.S. Coast Guard is responsible for maintaining a large number of aids to navigation. In particular, there are approximately 40,000 buoys each of which must be visited, inspected, and serviced at least once a year by ship. The problem of generating cost-effective ship routes for the Coast Guard is the focus of this paper. Our contribution here is to summarize the efforts of two computer scientists, B. Kuipers and A. Cline, who have studied this problem extensively.

In 1986, Kuipers and Cline received funding from the U.S. Coast Guard to demonstrate an understanding of the problem and to provide a preliminary solution procedure. Their problem description and solution procedure (see [4]) form the basis for this chapter. In 1987, the Coast Guard awarded a second contract to these two researchers, based upon their earlier work, to implement a fully-functional route planning system for the Coast Guard. This system is not yet operational.

2. INFORMAL PROBLEM STATEMENT

The real-world problem faced by the Coast Guard is riddled with complexities. Some of the constraints encountered in practice are as follows:

1. Ships of different types visit, inspect, and service buoys from different home bases;
2. Different types of ships have different capabilities;
3. Travel routes are ship-dependent in the sense that a route may be feasible for ship A and infeasible for ship B due to water depth or exposure;
4. Different buoys may require different ships, due to weight, size, or depth of mooring;
5. Emergencies may require ships to be reassigned or rescheduled on short notice;

6. There are numerous resource limitations such as distance, fuel, time, daylight, storage space, lifting capacity, crew fatigue, depth of water, weather survivability, and so on, which might vary by ship type as well as by base.

The Coast Guard wants to be able to schedule ships conveniently using an automated procedure so as to minimize a measure of overall resource utilization. As a by-product, the Coast Guard would like to be able to use such a route planning system to help determine which ships to acquire in order to meet anticipated, future needs.

The problem involves time-window features, site/vehicle dependencies, and multiple depots and is indeed difficult to solve. With this in mind, Kuipers and Cline identified a fundamental (simplified) version of the Coast Guard routing problem and proposed a two-step heuristic solution procedure. The basic idea is to generate an initial solution by a "greedy" procedure and to improve upon such a solution using the method of "simulated annealing," which has received much attention recently.

3. FORMAL STATEMENT OF FUNDAMENTAL ROUTE PLANNING PROBLEM

The fundamental route planning problem assumes a single base and a single vehicle. N buoys must be visited over time and the least expensive set of trips is sought. The problem has numerous inputs including the following:

p_i -- the position of the ith buoy (latitude, longitude),

r_i -- resources required to service the ith buoy (a vector),

t_i -- time required to service the ith buoy,

p_0 -- position of ship's port (latitude, longitude),

r_0 -- maximum resources haulable (a vector),

t_0 -- maximum time ship can be away from port each day,

T -- the territory in which the ship can sail (a polygon),

s -- ship's cruising speed,

$d(p_i, p_j)$ -- the distance from p_i to p_j obtained by restricting movement to within T.

In order for a day trip to be feasible, several resource-related constraints must be satisfied. First, we must ensure that resource limitations (given by r_0) are not exceeded on a trip. Second, time limits (given by t_0) must not be violated. If we let n be the number of days associated with the <u>buoy tending schedule</u>, then we require that each buoy is visited on exactly one day trip. In addition, we require that each day trip start and end at the ship's port.

Let TIME (d) be the total time spent away from port on day d. The primary objective of the fundamental routing problem is to minimize n. Over all feasible n-day buoy tending schedules, the secondary objective is to minimize

$$\sum_{d=1}^{n} \text{TIME}(d).$$

4. ALGORITHM FOR THE FUNDAMENTAL ROUTE PLANNING PROBLEM

In this section, we briefly sketch the algorithm presented by Kuipers and Cline. The algorithm consists of two key steps. The first step involves finding a reasonable solution very quickly. The second step consists of a procedure that seeks to improve upon the initial solution. It is computationally more expensive than the first step.

The first step seems to be a variant of the well-known nearest-neighbor heuristic. This step is applied repeatedly to schedule one day trip after another. The second step applies simulated annealing to the initial solution, in search of an improved buoy tending schedule.

In the second step, we consider three types of swaps:

1. Buoys i and j are on the same day trip and their positions are exchanged.

2. Buoys i and j are currently scheduled on different day trips, say α and β, respectively. Remove i and j from their current trips and insert them (in the best possible way) into the other day trip. That is, insert i onto day trip β and insert j onto day trip α.

3. Remove buoy i from day trip α and insert it into day trip β (in the best possible way).

Swaps are performed only if they preserve feasibility with respect to time and resources. Specific candidate swaps are chosen randomly and evaluated. Let Δt be the change in total schedule time (i.e., $\sum_{d=1}^{n} \text{TIME}(d)$) that a feasible swap would incur. The following rules then apply:

(a) If $\Delta t \leq 0$, perform the swap.

(b) If $\Delta t > 0$, perform the swap with probability $e^{-k\Delta t/\tau}$.

In the above, k and τ are annealing parameters. The reader interested in learning more about simulated annealing might consult Kirkpatrick et al. [3], Kirkpatrick [2], or Golden and Skiscim [1].

5. THE GENERALIZED ROUTE PLANNING PROBLEM

The algorithm discussed in the previous section addresses the most fundamental of route planning issues. Numerous extensions and generalizations need to be incorporated into the model and handled by the algorithm before the Coast Guard can begin to regard the resulting system as a useful tool for resource evaluation and strategic planning. Some of these complicating issues are discussed below:

1. Multiple ships. There may be different types of ships, with different capabilities. If so, then idle time (and cost) must be taken into account explicitly. Also, buoy/ship type compatibilities need to be checked.

2. Deep and shallow water. Different ships can operate in different water depths or different levels of exposure. This enlarges the list of buoy/ship type compatibilities that need to be examined.

3. Overnight trips. Beyond densely packed harbors and waterways, a ship may be required to spend several days away from port. In such a case, an effective algorithm must keep careful watch over time since buoys cannot be visited during the night. At sundown, the only permissible action for a ship is to travel between buoys and, if necessary, then wait until dawn.

4. Dynamic scheduling. The route planning algorithm should be capable of addressing a variety of situations. Its primary use may be for "pre-planned" routing. However, in response to emergencies or unanticipated weather conditions, the system should be useful in rapidly generating "real-time" adjustments to pre-planned routes.

5. Impact of weather on cruising speed. The height of waves, the speed and direction of winds, and the weather affect cruising speed in a major way. In some cases, it may be possible to estimate cruising speeds s_{ij} between buoys i and j. In general, however, the planner must recognize that travel time is largely stochastic. The planner may respond by using conservative travel speed estimates or by relying more heavily on real-time modifications to pre-planned routes.

6. Man-machine interactions. An ideal route planning system would allow the user to specify an initial set (or subset) of day trips or a number of partial day trips. The two step algorithm would then fill-out the trips and improve upon the suggested solution. In addition, the ideal system would allow the user to propose certain swaps at the end of the automated procedure. Final approval of trips would come from the user and not the machine.

7. Strategic planning. In addition to its role in operational planning, the route planning system should greatly enhance strategic planning capabilities for the Coast Guard. The following are examples of "what if" questions that could be addressed:

(a) What would happen if additional ships were assigned to a given base?

(b) What would happen if a different mix of ships was assigned to a given base?

(c) What would happen if a different set of buoys was assigned to a given base?

6. CLOSING REMARKS

The route planning problem faced by the Coast Guard exhibits many of the real-world complexities that are discussed elsewhere in this book--time windows, different types of vehicles, stochastic travel times, multiple bases, etc. The fact that the Coast Guard routes ships rather than trucks is of importance to us primarily because it underscores the wide applicability of vehicle routing methods to various modes of transport.

ACKNOWLEDGMENTS

We thank Benjamin Kuipers and Alan Cline for bringing this Coast Guard problem to our attention and for several informative telephone conversations.

REFERENCES

[1] Golden, B. and Skiscim, S., "Using Simulated Annealing to Solve Routing and Location Problems," Naval Research Logistics Quarterly, 33(2), 261-279 (1986).

[2] Kirkpatrick, S., "Optimization by Simulated Annealing: Quantitative Studies," Journal of Statistical Physics, 34(5-6), 975-986 (1984).

[3] Kirkpatrick, S., Gelatt, Jr., C.D., and Vecchi, M.P., "Optimization by Simulated Annealing," Science, 220, 671-680 (May 1983).

[4] Kuipers, B. and Cline, A., "Coast Guard Route Planning," Technical Memo, University of Texas, Dept. of Computer Sciences, Austin, Texas 78712 (December 1, 1986).

DEVELOPMENT OF VEHICLE ROUTING SYSTEMS

Vehicle Routing: Methods and Studies
B.L. Golden and A.A. Assad (Editors)

MARKETING A VEHICLE ROUTING PACKAGE

J. M. Hooban

MicroAnalytics, Inc.
2054 N 14th St Ste 307
Arlington, VA 22201

This paper addresses the role of algorithms, special features, support, documentation, and implementation in marketing a vehicle routing and scheduling software package. The discussion addresses implications for research, software design, and market penetration.

1. INTRODUCTION

1.1. Direction of this Paper

This paper examines the process of marketing a vehicle routing and scheduling software package. Topics range from the state of the market to how routing software is sold (i.e., to whom, for what reasons, at what prices, etc.).

We try to maintain as a theme of this discussion the role of algorithms in marketing routing/scheduling software systems. These algorithms may range from venerable Clarke-Wright routines to new and very interesting developments from a variety of sources. Aside from our perception that this focus will interest many readers, we feel that this should be (but often is not) a central issue in the delivery of commercial packages. Packages themselves vary widely in the extent to which they use algorithms, in the sophistication of their algorithms, and in how they make their algorithms available to their users. On the other hand, users of such systems show a wide range of knowledge and awareness in their ability to evaluate the technological aspects of commercial packages.

1.2. Background

I am president of MicroAnalytics Inc. MicroAnalytics publishes several microcomputer software packages that support decision-making in distribution and vehicle operations. Our TRUCKSTOPS package is a general purpose routing and scheduling system, directed primarily at private fleets, that is designed to meet a wide range of user needs for routing and scheduling. The system allows for multiple day operations, multiple time windows, multiple capacities, and special equipment codes. These and other features make TRUCKSTOPS useable without modification by almost any distribution operation. TRUCKSTOPS runs on IBM PCs and compatibles.

We sell TRUCKSTOPS to a wide variety of users. Our customers range from small local delivery operations to companies operating large fleets on a national or regional basis. Our system users also represent most major industrial classifications, including food and beverage distribution, manufacturing, consumer products, leasing, and postal operations. We even know of one situation where the system is used to route doctors on regular visits to medical facilities.

TRUCKSTOPS uses algorithms to generate route schedules. The overall solution procedure for the system is a proprietary heuristic developed by MicroAnalytics. Although it appears in no published source, a knowledgeable reviewer would recognize elements of several major published routing algorithms in the overall design.

We market TRUCKSTOPS based in part on its ability to perform well in reducing distribution costs. Thus, we focus much of our research and development effort on maintaining familiarity with technical developments in the routing and scheduling field, and trying to incorporate such developments, as appropriate, into the system. We also try to educate our users about the nature and value of using a cost minimization type of system. As discussed below, this sometimes runs counter to perceptions among users of what a routing package does.

We price TRUCKSTOPS aggressively, with a standard system available for about $2000. Although we consciously try to meet or exceed the capabilities of all other systems, our price is 10-20% of the price of many competing products. As part of an overall aggressive pricing strategy, we have "unbundled" the TRUCKSTOPS system. In other words, a user can buy anything from the basic system, "off-the-shelf", through a range of enhancements and support options, up to a full "turnkey" installation.

We attribute TRUCKSTOPS' profitability to several factors. We sell a larger number of packages than our competitors. We also sell more systems per customer than other vendors. In addition, we have sales outlets in several countries outside the U.S., as well as through a number of companies in the U.S. Finally, we try to be judicious about our marketing expenses. For example, we do not maintain an outside sales force, relying instead on demonstration products or visits by customers to our offices. Because the system is unbundled and can be installed easily by users or by third parties, we avoid much of the expense associated with keeping large numbers of consulting personnel on our payroll.

We think that our experience with TRUCKSTOPS illustrates that you can publish generic software at a reasonable price for routing and scheduling applications.

2. THE STATE OF ROUTING/SCHEDULING SYSTEM USE

2.1. Size of the Market

The potential market for packaged routing software is sizeable. Depending on the source, estimates run around 100,000-200,000 companies in the U.S. with private vehicle fleets, and perhaps 15,000-20,000 common carriers. We believe these levels are reasonably accurate. In addition, many companies with vehicle fleets operate from multiple terminals. With most routing packages, this means that they will need to purchase more than one copy of the software, if they wish to use it for all their distribution activities. This is reflected in our sales, by the fact that we sell an average of about three copies of our routing system to each buyer.

Unfortunately, potential markets can be quite different from actual markets. A major gap appears to exist between actual and potential in the market for routing systems. Some statistics may help to illustrate this:

- Typical attendance figures at the Distribution Computer Expo trade show held each spring in Chicago are around 1200-1500 individuals, commonly with multiple attendees per company, and this includes vendors and consultants. This show focuses solely on software for distribution managers. The Distribution Computer Expo is promoted partly under the aegis of Handling & Shipping Management, which probably has the largest circulation of any publication in this area. A comparable fall show on the East Coast draws around 700 attendees. The entrance fee for such a show is about $50, which should not pose an impediment to attendance for anyone. This suggests that only 1500-2000 companies have enough interest in distribution software of any kind (not just routing) to spend a few hundred dollars to look at it.
- Directory listings in computer issues of major national publications read by distribution managers typically draw a maximum of only several hundred requests for information, out of circulations of as much as 80,000+. There is usually no charge (not even postage) for such inquiries. In the case of the Handling & Shipping Management software directory (published in May of each year), information requests are made by product category, not by vendor. All requests for information in a category go to all vendors listed in that category. We can thus be reasonably sure that the requests we see from this source reflect the overall level of interest, and not just interest in our specific products. We have found that the Handling & Shipping Management listing draws the maximum level of interest, and that other publications draw less.
- Attendance at the Council of Logistics Management (CLM) annual conference, the "educational" conference for distribution managers, runs about 2500 per year. This is not a software oriented activity, but tends to measure

the level of interest in the "technological" and educational aspects of distribution.

- Published figures on sales of packages indicate that a typical vendor has been marketing a package since 1983-1984, and has sold from 3-4 up to 60-70 packages. This suggests that the total number of packages sold in the last 3-4 years probably does not exceed 1000. MicroAnalytics' TRUCKSTOPS, by this reckoning, would account for about half of this total, with about 500 systems sold to about 200 customers. These totals, by the way, include sales in Canada, Europe, and Australia, as well as in the U.S. It is important to note that these estimates reflect our notion of what a routing system is. For example, the annual CLM survey of distribution software lists a large number of packages under the heading "Traffic Routing & Scheduling." Judging by the names and descriptions (e.g., "freight rate maintenance and audit") of many of these items, they are not routing systems as we use the term. They are definitely not packages we see in competitive situations.

It thus appears that the level of penetration of computerized microcomputer packages has been very limited, although the potential is enormous.

2.2. Level of Interest and Growth

The brighter side of this picture is that growth does appear to be taking place. Our own experience with TRUCKSTOPS suggests that it is reasonable to expect 25-50% increases in unit sales on an annual basis. We also make it a point to examine surveys of the field closely as they become available, and to listen to our customers and potential customers. These indicate that a number of our competitors have seen similar growth patterns.

We also find that the market is now more aware of the existence and potential of routing packages. In the early 1980's, when we began marketing routing systems as a consulting product, we had to orient our demonstrations so as to show the customer what such a system does. Currently, we can generally assume that a customer knows in some sense about routing systems, and we can thus orient presentations to showing the merits of our system versus others. Although it might seem to be preferable to have little or no competition, as in the early years, this is not generally the case. The presence of several systems on the market helps to educate the consumer, and creates market awareness to the benefit of all vendors.

This growing awareness is also reflected in the lead time for users to make a purchase. In the past, it has been quite common for users to take years (literally) to decide to buy a package. At a recent NAWGA (October 1986) conference, one user of a competing system indicated that his company had taken 4 years to decide to buy a routing package. Another user of a different system indicated that his company had taken two years. Our own experience suggests that this was not at all out of the ordinary. Although this still happens, we also find more and more customers who make their decisions within a few months or weeks. We attribute this to the fact that the existence of several packages "legitimizes" the whole "industry" and that the potential user may actually know or have heard of someone else who uses such a system.

An additional factor which has come into play in the marketing of routing systems is the presence of several "major" players in the field. Leaseway may have been the first of these, with their Route-Assist package, but they have been followed into the routing market by Ryder, Navistar (in connection with DART), TRW Systems (as a major investor in Transtech), STSC, and UPS (currently owners of Roadnet). In many cases, routing system sales are subsidiary to some other goal of these organizations. For example, Ryder and Leaseway seem to see vehicle routing systems as a means of increasing the attractiveness of their fleet leases and other services. The major effect from our perspective, however, is again to legitimize the product, and enhance the awareness of potential customers.

It is worth noting that very few companies which sell routing software are solely and strictly distribution software companies. MicroAnalytics has chosen this path, but most other firms seem to be offshoots of, or supported by, some other line of business. For example, several consulting companies offer software in this area, to supplement and enhance their consulting practice. Some products also emanate from professors, or from companies formed by professors, with active university affiliations. While it seems common for software vendors for the broader PC markets to be dedicated to their particular product area, this appears to be the exception among routing systems.

We have some difficulty seeing a clear pattern of survival or development among vendors of routing software. Several products which were available 2-3 years ago have indeed dropped out of the market. However, new products continue to emerge, some mainframe systems are migrating to microcomputers, and nearly all systems continue to develop and add features. Thus, we continue to encounter much of the same major competition as a few years ago, with occasional appearances by other products which seem to show up once or twice and then fall from view. As the comments above on market size indicate, packaged routing systems have barely begun to penetrate their potential market. We believe that a much greater level of use will have to occur before a pattern of survival will emerge.

2.3. Types of Systems

A fundamental problem in discussing the market for routing and scheduling systems is to determine what the term means. To many users, a "routing system" is anything that produces or helps to produce load and sequence information for their vehicles, including "systems" which simply generate reports based on manual input from the dispatcher. Another sizeable group tends to define "routing" as the generation of an optimal path between two points. Most OR/MS professionals, on the other hand, would probably define a routing system as something which solves the (usually multiple) travelling salesman problem. Although we hold the OR/MS view, we have to deal with perceptions which span the entire range.

One result of this situation is that there is no clear set of standards for routing systems. For example, there are no generally recognized test problems which a system is expected to solve, and no basic set of features a system is expected to possess. Thus, no publication that we know of publishes benchmarks of routing systems, and articles that discuss this type of software tend to miss packages altogether, or to list everything that might fall in this category.

Vendors tend to react to the various perceptions in different ways, and thus produce a wide variety of system configurations. We tend to think of these as lying on a spectrum from "analytic" systems (the "high end") to simple "pre-list" systems, with significant offshoots in between. A rough set of categories might be as follows:

- Analytic Systems - Systems which attempt to optimize costs (or some other performance standard) comprise one category of software. A typical system in this area has some mechanism for designing routes so as to optimize total cost, total distance, vehicle utilization, or a similar measure. Such a routine normally does not require the user to specify the loading for vehicles or the delivery sequence, although the ability to choose may be present as an option. Commonly, the user needs to specify basic data describing the problem to the system. This includes stop locations, delivery requirements, time windows and similar information, truck capacities, time factors, costs, etc., and general information such as terminal location, speeds, and so forth. Generally, it is possible to add other information to the problem description, such as special codes, barrier data, or a "true distance" matrix. Once this is done, the system can generate loads and delivery sequences without further intervention from the user.
- Spreadsheet - Systems of this type allow the user to specify loadings and route sequences, and then provide output regarding the impact of such decisions. They include relatively little optimization capability, but are set up to produce multiple displays and reports. This type of system is marked by the need for the user to specify key decisions to the system about which stops go on which vehicles and/or what the delivery sequence for stops should be. For example, it may be necessary for the user to

define zones and assign each stop and truck to a specific zone for the system to operate.

- Pre-List - This type of system really isn't much of a system at all. It consists of a mechanism whereby the user predefines "master routes" which include all stops. When delivery activity is known, the stops are routed according to the vehicle assignment and delivery sequence given by the "master routes". Typical output from this type of system consists of delivery documents in the predefined order. Usually the user has to do considerable shuffling of documents to arrange the routes so that they can actually be driven. We suspect that many reports of automated dispatch systems refer to this type of operation. We are not aware of a specific "package" which is sold on this basis, but we have seen a number of instances where this has been a feature of some other software, such as an order entry system.
- Graphics - Although most systems have some sort of graphics capability to support their routing functions, a recent development seems to be systems that comprise technically sophisticated graphics displays with some sort of routing capability. Typically, such systems present very detailed, map-like displays of the operating area, often requiring special hardware. Two examples are the Transtech system and the Roadshow system. In each case, the publishing company's spokesmen have stated that the key to good routing lies in the dispatcher's ability to see clearly the implications of what he is doing on a detailed map of the dispatch area. Our experience suggests that a graphics capability significantly enhances, but does not replace, the analytic functions of a routing system. Graphics are a powerful means of illustrating what a routing system is doing, and provide a convenient way to check inputs to the system. Concepts and data which are difficult to grasp in text form are immediately apparent in a graphic display. For this reason, graphics are also extremely useful in selling routing systems. By the same token, it is sometimes easy for the user to feel that the graphics component is the major element of the system.
- Artificial Intelligence - At least one system claims to provide a way of incorporating AI concepts into the routing process. Apparently, the idea behind this is that the dispatcher already knows everything that needs to be known to route the vehicles. He just doesn't have the ability to access and organize all this knowledge in a timely manner. The AI type of system basically treats the dispatcher as an expert, and builds an "expert system", similar to those used for medical diagnoses, around his knowledge.

The different types of systems reflect very different philosophies of what a routing and scheduling system should do, and what users actually need from such software. For example, analytic systems seem to assume that computers and software can perform at least some parts of the dispatching function faster or better than humans. Other types of systems tend to assume that the human dispatcher really has all or almost all of the knowledge he needs to do effective dispatching, but lacks the tools to organize this data and focus it on the problem. To some extent, the non-analytic systems also seem to consider dispatching a clerical, rather than a management, function.

2.4. Types of Use and Users

There are some marked distinctions among users of computerized routing systems. When we first began marketing TRUCKSTOPS, we tended to assume that a typical user would be a dispatcher sending trucks out on the road on a daily basis. Although this represents one major element of the market, there are several others. At least at present, we believe that the "daily dispatch" portion of the market does not represent the majority of systems in use, although it is a significant minority.

A number of companies use routing systems as planning or modelling tools, rather than for day-to-day operational management. For example, leasing companies routinely submit bids for contract carriage. This requires them to determine amounts and types of equipment, numbers of drivers, and other factors which will be needed to meet the contract requirements. It has become common for such companies to model these factors using routing software. Since the bids are normally very competitive, the bidder usually looks for software which minimizes overall cost, so the analytic capabilities of the system are quite important.

In other cases, users generate routes only periodically, then update these as they go along, without rerunning the entire problem. For example, following the merger between American Hospital Supply and Travenol, analysts at American Hospital Supply used TRUCKSTOPS to study ways to combine delivery and pickup operations of the two companies in a sensible manner. Another user, a dairy, uses the system to generate fixed routes on a rotating basis for several terminals. Thus, about once a quarter, a routing analyst sets up a new group of routes for each terminal, based on expected volumes during the coming three months. These routes stay in place for the period, with minor adjustments, until it is time to update the structure again.

At the other end of the spectrum, companies like Simmons U.S.A., Dannon Yogurt, and others use the system virtually every day, to generate new sets of routes for actual implementation. Often, these functions are mixed with the planning and modelling functions mentioned above. For example, Simmons has evaluated new vehicle equipment purchases by running TRUCKSTOPS using existing delivery data, but modifying fleet information to reflect prospective equipment changes.

Clear lines between these categories are difficult to draw. We would guess that between half and three-quarters of TRUCKSTOPS users use the system for planning and/or periodic route restructuring, while the remainder generate daily routes. It is important to note, however, that some of these "non-daily" users simply don't dispatch their vehicles on a daily basis.

The type of person using the system also presents some interesting distinctions. We tend to think of the spectrum as being anchored on one end by actual dispatchers, and on the other by analysts or managers. This distinction tends to reflect the level of use. For example, when a system is used every day, the dispatcher is probably the one who uses it. When it is used for planning, the user is probably an analyst or analytically oriented manager. Even in situations where the user is a dispatcher, however, it is often an analyst who sets the system up. In many cases, we have found that an organization will set the system up so that the operator has virtually no discretion about running the routes.

It is also interesting to note that in many cases where the system is used in a daily dispatch environment, it is not the actual dispatcher who operates the computer. More commonly, the person who presses the keys is a clerical person, usually a woman. We believe that this reflects both a lack of basic typing skills in the type of person who becomes a dispatcher, and a perception among such people that anything involving a keyboard is work for women. This tends to remove the dispatcher himself from the process, and reduce his control of the system.

In general, we find that users of routing systems lack familiarity with computers in general and microcomputers in particular. In addition, they often have little or no exposure to analytical tools in general. For example, they are unlikely to have used a spreadsheet. By contrast, users of our facility locations product (usually staff analysts) show much greater familiarity with computers and software tools than users of the routing system (usually distribution managers). We suspect that computer/analytical skills have not been considered part of the "tool kit" of the distribution manager, and that management and educational institutions have only recently begun to correct this.

3. SELLING THE SYSTEM

This section explores how users and vendors look at the decision to buy a routing system. We try to cover the major criteria that go into this decision, and to point out areas in which anomalies exist.

3.1. Packaging and Pricing

Vendors differ greatly in how they package and deliver vehicle routing systems. Prices of systems reflect this, as well as differences in the vendors' perceptions of how to maximize revenue from an emerging and still poorly defined market.

The software itself (i.e., the floppy disks and manuals) costs very little to deliver to a user, assuming that the software actually exists in a deliverable form. Sending out a copy of software probably never costs more than $50.00 for any vendor. Thus, once the programs in a system are written, a rational vendor should price the item so as to garner the maximum contribution to fixed expenses and profits he can get. Depending on the vendor's assessment of the market and the price elasticity of the product, this may imply trying to sell a few systems at a fairly high price, or selling a lot of systems at a fairly low price.

The vendor may also decide that the system must be "bundled" with additional products and services, so that the overall sale price to the user is higher than it would be for a simple software sale. The most common "bundling" technique seems to be the inclusion of direct, on-site support in the package price. We believe system prices vary to a large extent in terms of the amount of such support the user receives with the system. In other words, the difference between a low-priced system and a high-priced system often seems to result from the inclusion of a substantial amount of installation and training time built into the price of the more expensive system. Where installation support is an integral part of the delivered system, it also appears that part of the system cost may come from "customization." Systems may also vary in the inclusion and pricing of special hardware, or in the need for special data files.

We have encountered some systems with prices based on fleet size, number of stops, etc. At least on microcomputers, this seems uncommon, and is decidedly unpopular with users. What seems more common (and acceptable) is to charge for modules or enhancements, as well as for support services. Naturally, where vendor costs actually vary with number of stops or fleet size (e.g., geocoding of stops), user charges also vary.

Pricing reflects both the cost of delivering the system and the vendor's philosophy about selling the system. In general, personnel costs drive prices up very quickly, so that systems which include sizeable amounts of support, customization, and special data gathering tend to have much higher prices than those which do not. For example, a week of on-site support by a qualified analyst can easily cost a vendor over $1000 in direct expense. In addition, the user normally pays not only for the time spent at his site, but also for the time the vendor analyst spends waiting to be sent to his site, and for other "overhead" factors. Since it usually takes a greater amount of sales effort to sell a system at the higher end of the price scale, the user will also pay for on-site sales visits and related expenses. It is quite common for these factors to increase the base rate by a factor of four to five. Thus a week of support can easily cost the user $4000-5000.

The other key pricing factor is the vendor's view of the market. We think many vendors do not trust the market to produce enough sales volume to support them if they price their systems low. They believe that they have a better chance of selling, for example, 20 systems at $30,000 than they have of selling 300 systems at $2000, so they set the price at $30,000. Naturally, this decision largely dictates the strategy and tactics the vendor will then follow in marketing the system. Once this type of pricing decision is made, each individual sale becomes very important, with several implications for the vendor. One of these implications is that each lead should be pursued very intensively, implying significant expenses for direct face-to-face sales efforts. If the vendor expects to survive, these expenses must be passed on to users. In addition, users will expect more for this type of outlay than just a few diskettes, a book and some telephone support. This means that the vendor needs to be able to guarantee the availability of people who can provide extensive support at the user's site. These people cannot be counted on to provide such things as development or telephone support, since they may have to be gone for extended periods of time. Thus the vendor also has to staff to fill these functions as well. All of these costs must be passed on to the user.

The initial pricing/packaging decision thus defines major elements of the vendor's operations. Once the process has started, it is very difficult to change directions. For example, the presence of a large installation support staff means that management will seek ways to ensure that these people are used on billable work. Unless the vendor cuts staff drastically, this means that the system price will remain high. If there is price elasticity, this suggests that unit sales will remain fairly low, reinforcing the notion that the system price must be high for the vendor to survive. The rest of the cycle should be fairly obvious.

We have said nothing here about the algorithms as elements in pricing. This is because we don't believe that they are factors. We are not usually in a position to benchmark different systems ourselves, but we are routinely in contact with people who do, and we try to gather as much information as we can about this aspect of competing products. The overall impression we get is that price has little or no relation to algorithmic performance. In a benchmark conducted by a customer of ours, the top three performers ranged in price from under $1000 to over $100,000. At this writing, one of our major competitors, priced at $30,000-40,000 or more, doesn't provide for optimization of anything other than distance. Another system in the same price range requires that the user manually specify many of the decisions that we would expect an algorithmic system to calculate automatically. Customers of ours who have seen some of the newer graphics-based systems consider them very sophisticated in terms of their graphics technology, but question their ability to produce better routes than other systems. We even believe that there are cases where a vendor has eliminated or greatly simplified the algorithmic portion of his package, with no change in the price.

3.2. Costs and Benefits

Nearly everyone who looks seriously at acquiring a computerized routing system says that they are making their decision based on the costs and benefits of the competing systems. Our experience is that this common statement has widely varying interpretations for different users. In addition, different types of users react very differently to the various features available in routing systems. For example, the majority of our customers are fairly large companies, often in the "Fortune 1000" size range. At least one major competitor, with a base system price 10-20 times higher than ours, seems to sell primarily to small-medium size firms. We believe that one major reason for this is that the more expensive systems tend to be more expensive because they include more on-site installation support than TRUCKSTOPS. Larger companies often have the resources to fulfill this function themselves, while smaller companies often do not. In other words, the smaller company may have no other means of implementing a routing system than to purchase a system which comes with large amounts of consulting support.

3.2.1. Benefit Evaluation

We tend to think of benefits as consisting primarily of the amount of cost reduction achieved by running a system, and we think that this is a fairly common viewpoint among OR/MS people. Our experience suggests that this is not such a common viewpoint among actual users of systems. For larger companies, which may have OR analysts or Industrial Engineers on their staffs, cost reduction tends to be a significant factor. For others, however, things can be quite different. We find that users will routinely bring up some combination of the following:

- Route Results - Typical technical papers on routing systems describe overall results (mileage, cost, etc.) and the technology of the system. A purchaser, however, often shows little concern about the overall reduction in miles or cost. More often, he will see if the routes look "OK", and that the system doesn't do "screwy" things. Generally, the main criterion for OK-ness and screwiness is the current practice of the organization. Normally the lower you go in the organization, the more the criterion for route quality is "how different is it from what I do now?" with more deviation meaning lower quality. This can produce some interesting situations. For example, many dispatchers tend to batch stops geographically, using existing political boundaries as a guideline. For example, there may be a "Virginia Truck" and a "DC Truck" and a "Montgomery County Truck," and so forth. This is a natural consequence of trying to meet the pressures of time in a complex task, since it makes the dispatcher's job easier. A routing system, however, will often put a couple of Maryland stops on the Virginia truck, in order to make efficient use of the capacity of the vehicle. This can offend some dispatchers, partly because it is different, and partly, we think, because it undermines the sense of expertise they have managed to create over the years. One consequence of this seems to be that some systems leave many or even most key decisions in the hands of the dispatcher. This ensures that the results cannot be blamed on the system. In other cases, vendors expend considerable effort to ensure that the algorithm will not do "strange" things. Given the rather arbitrary criteria that a system may be held to, however, virtually any system must provide substantial capabilities for manual overrides in order to be saleable.

- Dispatcher Time/Efficiency - Many companies look at routing as a primarily clerical function. In other words, they view a routing system as a way to get the same job done as before, but in less time. Thus, it may take the dispatcher 4 hours a day to work out the routes by hand. If a computer system allows this to be done in 2 hours, then this is seen as a major benefit. In a sense, these users are conditioned to view a routing system as similar to a payroll system, or other transaction oriented system.
- Optimization - Some companies actually look at how much money they are likely to save if they use a system. As noted earlier, these tend to be larger, more sophisticated entities, who have the capability to make such an evaluation. In such cases, we find that the user will actually keep track of the results, and be in a position to say how much has been saved as a result of the implementation. What is surprising to us is how few such companies there are. All too often, users have little or no idea whether or how much things have changed following implementation of a computerized system. In general, it is fairly difficult to conduct a meaningful evaluation of cost reduction from a routing system. To be truly valid, this would presumably involve running both the automated and manual systems and comparing the results. This is obviously impractical. Instead, companies tend to run the routing system on historical route data, and extrapolate the change to future results. Other evaluations

include prior to current year comparisons and ratios such as cost per cargo unit, cost per stop, etc. Companies performing evaluations between packages also find that they must take into account variations in ability to meet user requirements such as time windows. For example, one system was described to us as producing very good routes, but violating a sizeable proportion of time windows, making the routes unuseable.

3.2.2. Costs

Generally costs fall into the following categories:

- Direct Costs - There is considerable variation in the installed costs of routing systems. Much of this results from the extent to which systems are "bundled" with features or support services. For example, TRUCKSTOPS can be purchased "off the shelf", with no additional features or on-site implementation support, for a price of about $2000. Adding a set of enhancements to the system (e.g., matrix distance input, extra window and capacity fields, etc.) can add another $1000. A graphics enhancement (i.e., multiple route displays, map backgrounds, etc.) can add another $900. Location data files can range from $500 on up (for example, a file of ZIP centroid locations for the U.S. costs $3000, while detailed street files for cities in the entire U.S. would run about $200,000). Implementation support can cost from $2000 to over $100,000, and is available either from MicroAnalytics or from a wide variety of consulting firms. Most vendors, however, follow a somewhat different pattern, in that they include several such features in the base price of the system. For example, a typical "package" will include software, a specified amount of on-site support, setup for data files, and possibly hardware. As a result of pressure from packages such as TRUCKSTOPS and RIG-PC, it appears that some prices are coming down. We have seen one company's price drop from $20,000 to $2000 per package in the course of negotiations for a multi-terminal installation with a major company. We have also noticed restructuring of pricing in several other packages.
- Indirect/Support Costs - It virtually always takes a good bit of time and effort to begin using a routing system effectively. In the area of dispatching systems, much of this effort is simply the result of almost total lack of exposure to computers, or at least to microcomputers. We have, for example, watched in chagrin as a user spent several seconds per key trying to locate individual letters on the PC's keyboard, in order to be able to enter simple (single keystroke) commands to our system. Similarly, much of the time we spend in telephone support involves stepping users through such basic operations as getting a directory listing or copying a file. This can get to the point of having to spell out each command for the user at the other end of the phone

line. We also, of course, encounter problems more directly related to running the system. For example, one common beginning mistake is to enter unloading rates in some form other than the units/hour format the system requires. Examples might include minutes per unit, units per minute, etc. Either of these tends to produce situations where the system calculates that it will take 15-20 days to unload some or all of the stops. Naturally, the results of all this are a bit bizarre, and produce support calls to us, and delays to the user.

3.3. Features

Several different aspects of the user/computer interface seem to have considerable importance in the purchase decision. In general, the points which follow describe features which users demand in a system:

- Command Structure - As with any software, the command structure has to be intelligible and reasonably efficient. With the average routing system user having a relatively low level of "computer literacy", the commands can be a major positive or negative force in the purchase decision. Simple things, such as using a "flat" structure come naturally to mind. Some conflicts also occur, however. For example, TRUCKSTOPS normally uses single letter commands, and displays some sort of promptline on the screen at all times. Generally, the promptline is designed to remind, but not to define. It can be fairly terse, especially when the "flat" command structure goal means that several frequently used commands are available at one "level". Once the user is familiar with the system, this is a very efficient way to operate, but it implies a steeper learning curve.
- Extent of Control - Most successful software provides the user a lot of control over program operations. DBase II, Lotus 1-2-3, and even WordStar are examples of this, and a user can spend a lot of time with any of these and still have much to learn. The same is true of routing systems. The "simpler" systems are usually simplified at the expense of the user's control of the system. Many of the more sophisticated users demand greater flexibility. Conversely, many relatively naive users find this type of feature confusing.
- Reports - Reports are a very tangible aspect of a dispatching system. Virtually any viable system will include reports to show the route schedules and key statistics (e.g., costs, capacity utilization, etc.) for the fleet. Beyond this, each user tends to have his own unique set of needs, and vendors react to this situation in different ways. Some systems provide a plethora of output reports, covering nearly any conceivable aspect of the user's operation. Others provide basic reports common to most users, and also provide a means (e.g., ASCII file output)

for the knowledgeable user to manipulate the output of the system so as to tailor additional reporting mechanisms to his specific needs. For example, TRUCKSTOPS includes a programming interface which allows users to write additional code to manipulate the system's data in whatever way they wish.

- Graphics - Graphics can be a very helpful adjunct to an otherwise sound dispatching system. A graphics capability is also very helpful in selling systems, since it has a lot more impact than text output, and allows the user to see immediately the results the system produces. We suspect that graphics is becoming a sine qua non for such systems, although we believe that it is unwise to base an entire system on this capability. Nearly all commercially available systems include some form of graphics, and users have come to expect that they will have the ability to display route patterns.
- Downloads and Uploads - It is important to be able to pass data from the routing system to another package or computer. It should be borne in mind, however, that it is not usually difficult to do this, given adequate definition of file structures. On the other hand, it is practically impossible for a package to include an interface structure which can be plugged in directly to a generic "host" machine. In other words, downloads and uploads virtually always involve some effort by the user, but should never require very much. This feature can be critical in selling to users who process large numbers of stops on a short turnaround basis. For others, however, it can be totally irrelevant.
- Time Windows - Virtually all commercially viable systems provide some means of confining the vehicle's arrival time to some time window at the stop. Some systems also offer users something called "soft" time windows. We think that there is some merit in this notion, as it tends to reflect some real world circumstances. However, we also think it produces an extra, and largely unnecessary, level of detail for most users, and can be a fairly complex thing for an average user to implement.
- Special Features - Any number of other features may have a bearing on the purchase decision. In general, these are specialized requirements of some group of users. Users who need them often will not buy a system without them, while other users will not even be aware of their existence. For example, TRUCKSTOPS includes features to split large stops into multiple loads, to redispatch vehicles, to route vehicles to variable destinations, and several other items. Although virtually all of these items are standard, probably none of them is used by more than 20% or so of our customers. However, the presence of a particular capability often makes the difference between whether the system is sold or not.

3.4. Distances

Distances sometimes get to be a major source of confusion in selling routing systems. We believe that this situation comes about because many users consider distances to be a system output, while many analysts view them as inputs. Although we tend to line up with the analysts, we also recognize that anyone who wants to sell a system has to address this issue. This becomes particularly important when competing systems start to be marketed based on the accuracy of their distance calculations.

Virtually all routing systems require some form of distance input. A common approach is to approximate distances based on x,y coordinates assigned to each stop. The main alternative to computing distances in this way is to enter them in the form of a matrix. Often the matrix is based on the definition of a network of streets or roads in the delivery area, and application of a shortest path algorithm to obtain the distance between any pair of points. Some vendors seem to make this a major element of their sales approach, by focusing on the accuracy of the shortest path generated distances. For example, we have attended presentations (e.g., the 1986 NAWGA Computer Conference) where this was touted as a major advantage of STSC's TRUCKS system, and have read material from the president of Transtech which made a similar claim. It is interesting to note that the Roadnet system, which was originally heavily oriented towards this distance generation approach, now relies primarily on x,y coordinates (at least according to a Roadnet representative at the 1986 NAWGA Computer Conference).

From a marketing perspective, focusing on the accuracy of distances offers some advantages. Distances are one of the most visible outputs of a routing system, and may be the easiest thing for the user to understand. The idea of finding a shortest path through a street network is often much easier to follow than the notion of minimizing total cost based on vehicle assignment and sequencing decisions. In the same vein, shortest path algorithms are truly optimal (i.e., they really do find the shortest path), while route optimization algorithms are not (i.e., they rely on heuristics). It has been our experience that one of the first things which a user will criticize about a routing system is the distances. Street or road network distances eliminate this objection. We find it difficult to criticize vendors for this focus, although we think it is inappropriate. So many users bring up distances in their "evaluation" of systems that it is virtually impossible to ignore this factor.

The problems with all this, however, are several. First of all, generating the street/road network data base is a major task. At the 1986 NAWGA conference mentioned above, a user of this type of system noted that setting up a city the size of Chicago took about 6 months (he noted, however, that with this experience behind him, he was able to complete Cleveland in only two months). It is also fraught with the potential for error. The most obvious type of error comes from transcribing the data into the computer. Our own experience with user entry of numeric data is that manually entered data will contain a large number of simple typing errors (simple to make, that is, but very difficult to eliminate). Depending on how the matrix data is entered, other errors can also arise. For example, TRUCKSTOPS includes as an option the ability to use a distance matrix specified by the user. One major problem we encounter with this is that users routinely violate the Triangle Inequality in entering distances. For example, a user of ours once entered something like the following (supposedly based on "real" driving experience):

Providence to Asbury Park - 225 miles
Providence to Teterboro - 150 miles
Teterboro to Asbury Park - 60 miles .

Since this tells the computer that it is shorter to go from Providence to Asbury Park via Teterboro than to proceed directly, the results can sometimes be peculiar.

There is no particular reason to believe that the actual users of a routing system in fact know with any degree of precision how far it is from one place to the next. The example noted immediately above was drawn from a table of distances maintained carefully by an operations manager in a major company, specifically for purposes of routing the company's vehicles. When the matrix distance results began to show peculiarities, we took a very detailed, careful look at the whole table. Virtually all distances were rounded to the nearest 5 miles, with many of the longer trips rounded to the nearest 10 or even 25 miles. In addition, many of them were simply wrong, for example, showing two locations five miles apart on the same interstate highway as 20 miles away from one another. When these items were pointed out to the user, he did not disagree with this assessment. In the same vein, we have had user personnel insist that the distance from Akron OH to Columbus OH was over 225 miles, rather than the computer's estimate of about 130, because the driver's log read 227 miles. It turned out that the driver in question had been driving the extra 100 miles in order to do a little "private hauling," using company time and equipment. This kind of incident does nothing to raise our estimation of the geographical knowledge of the average dispatcher. In general, we have found that calculated distances produce distance results that are within 4-5% of actual values about 90-95% of the time. Errors can be mitigated still further by specifying key physical barriers (e.g., lakes, rivers, etc.) for use by the system in calculating distances.

On top of all of the above, there is no research that we have been able to uncover about the effect of more accurate distances on the routing process.

However, there are some avenues that are worth exploring in this regard. First, it is important to note that the main value of accurate distances is that they allow for more accurate driving time estimates. However, there is significant variation in driving times themselves under the best of circumstances. For example, several studies of city driving conditions suggest that time to drive from one point to another in a major city can vary by as much as 50% in a more or less random fashion. This suggests to me that squeezing the last 4% out of the distance estimate on which the time estimate is based is somewhat fruitless.

4. CONCLUSIONS AND FUTURE DIRECTIONS

Several thoughts tend to emerge from all this. In most cases, these items have implications for the future of vehicle routing software marketing. We touch on them briefly here.

- Lack of Standards - As mentioned above, there is no generally accepted definition of what a routing/scheduling system is supposed to do. From a marketing perspective, this means that you can never be quite sure how your product will be perceived by a potential user. Although there is quite a bit of OR/MS literature on the subject, many users have never seen this, and wouldn't understand it in any case. In some cases, they tend to think that the technology is pretty much the same in the different systems ("you guys all use some kind of linear program for this - right?"). In others, they do not know, or do not believe, that the results of their current procedures can be improved. We see some companies beginning to perform quantitative comparisons of the capabilities of different software. We expect that some form of general comparison will evolve. This could take the form of a set of problems which routing systems should be able to solve, and a set of features which they should possess. Publication of results from this type of analysis would be a tremendous service to potential users of such systems. It would be particularly interesting to many users to be able to see in quantitative terms how much difference there is between the various systems available.
- Where the Algorithm Fits In - We continue to think that algorithmic performance is important, and can be used to sell routing systems. However, we think that a major element of the design must be to ensure that the algorithm avoids the types of results which can destroy a user's confidence in the system. This requires a delicate balance between educating users about better ways of doing things, and accommodating their current practice. One element of a generalized evaluation system would have to be a mechanism for assessing the general reasonableness of results.

- Evaluation Capability - In many cases, users simply lack the ability to do a comprehensive evaluation of any computer package. They tend to be very unsophisticated about computers, software in general, and analytic systems in particular. This trend, by the way, does not appear to be limited to distribution software. Several microcomputer publications have noted the same kinds of result for microcomputers in general, as their use spreads to a wider and wider audience. A general trend seems to be to enhance the level of education of distribution managers. For example, many universities offer logistics degrees. In addition, many current managers are actively seeking additional education in this area, often with the help of organizations like CLM or WERC. We have also noticed that companies are changing personnel assignments in some cases to ensure that they have personnel with computer qualifications in responsible positions in logistics.
- Focus on Direct Results - As noted above, some users tend to evaluate packages mainly in terms of how much time they save the dispatcher. This seems to miss the mark widely, since the potential effect of using such a package can be a significant fraction of total operating costs. This seems to result from an inability to perform a benchmark, or from a simple unwillingness to rely on the results. We expect that users will come to view this reduction in clerical burden as a secondary issue, and will come to see the operational improvements possible with a routing system as the key issue.
- Ease of Use - Because of the general unfamiliarity of users with computers, there is a tendency to look very favorably on the package which is easiest to get started, and to confuse the use of the package on a demonstration basis or in the first 2-3 weeks or months of ownership with the long term ramifications of the system. This in many ways reflects the same type of situation in other types of software. For example, many word processors offer a simple, easy to start, WYSIWYG ("what you see is what you get") sort of interface that appeals to beginning users. Our experience is that it takes about six months to exhaust the abilities of such a package for "industrial" use such as producing sizeable documents, etc. On the other hand, it takes longer to learn to use a true production word processor (we still discover features of our package that we didn't know existed). However, you can do almost any job you need done with it. Something similar is true for routing packages. For example, it is simpler to set up a system with only one cost factor than one which has several, but it is very limiting. Similarly, it can be very reassuring to use a system which essentially acts as a scratchpad to keep track of your existing routes. When those routes change (e.g., with the addition

of new stops), however, this type of system must be totally restructured to accommodate the new information. With use of systems extended over more companies and longer times, however, we anticipate that users will become more aware of the longer term possibilities for such systems. They will then seek systems which provide capabilities beyond those they are able to use in the first few months of ownership. It may happen that there will be a two-tiered market for routing systems, much as there is for word processors. The "lower" tier will consist of systems which are easy to use for the unsophisticated user, while the second will be those with the additional power a user comes to expect after using a system for a longer time.

- Pricing - We expect that prices will generally fall. This development will have to be accompanied by changes in the way systems are delivered to users. In other words, it will not be possible for vendors to include implementation support as part of the package price, and still make money selling systems. For many users, such support is not necessary in any case, or can be obtained from sources other than the vendor. In addition, the differences in performance between higher and lower-priced systems bear little or no relation to price. As users become more aware of this, there will be great pressure on the higher priced vendors to reduce their system prices.
- The "Baby Duck" Syndrome - Researchers have found that a baby duck will "imprint" on the first moving object it sees after hatching, and assume that it is the duck's mother. Software users can be similar, in that they will tend to evaluate software by comparison with the first package of a particular type they have seen. In a way, this makes sense, since there really is a cost associated with changing from one package to another. In another, however, it can be frustrating to some vendors, since it reduces the possibility of users switching from one system to another. One advantage that lower-priced software vendors have over their higher-priced counterparts is probably in this area. It seems to take an overwhelming set of advantages to get a user to change packages. Since users can more easily acquire a lower-priced package on a "test" basis, there is a good chance they will become accustomed to the way that package works, and will be disinclined to change. A reverse effect also sometimes comes into play. Some companies have had prior experience with, and abandoned, a mainframe system (often VSPX) sometime in the last 20 years. It is very difficult to sell this type of customer a system, even though you can fairly easily dispose of his objections to the older system.

This paper has been reflective of a particular point of view. We hope that this perspective, in conjunction with others presented in this volume, helps to convey the multi-faceted nature of computerized vehicle routing.

Vehicle Routing: Methods and Studies
B.L. Golden and A.A. Assad (Editors)

CUSTOMIZATION VERSUS A GENERAL PURPOSE CODE FOR ROUTING AND SCHEDULING PROBLEMS: A POINT OF VIEW

Jean-Marc Rousseau

Université de Montréal
Centre de recherche sur les transports

This paper tries to summarize our experience with real life problems involving the routing and scheduling of vehicles. Our experience, coming from a biased sample of these problems, reveals that each problem tends to be different and necessitates particular attention. We suggest that no single algorithm may be appropriate in all, or even in a large number of, situations. Consequently, we feel that it is necessary to develop a general approach that can be adapted by experts to each particular situation. We briefly describe the problems we have dealt with to illustrate our points.

1. THE NEED FOR CUSTOMIZATION: SOME EXPERIENCES

This paper summarizes our experience with real vehicle routing and scheduling problems. Based on our experience with a non-random sample of these problems, we believe that each problem tends to be different and requires particular attention. However, it is difficult to provide such attention in the market place due to the limited knowledge of various professionals in the area of routing and scheduling. We suggest that no single algorithm is appropriate in all situations, or even in a large number of them. Thus, there is a necessity to develop a general approach which can be adapted by experts to each specific situation. We briefly describe problems on which we have worked to exemplify our points.

But first, we tell a story inspired by a real life situation we recently came across.

1.1. A Real Situation

Once upon a time there was a large distribution company. It owned about 200 stores in Central Canada and four depots. It owned a fleet of tractors and trailors but could also use common carriers when appropriate. It could supply its stores only during operating hours and had to observe certain regulations governing the driving time of its drivers. Not all depots carried all goods, and a store's order could be supplied from more than one depot. Split delivery and back orders were allowed, as were multiple trips for the vehicles. The firm allowed itself up to three working days to supply its stores after receiving an order. Depending on the season, the average quantity of goods delivered to each store could vary from 1/5 to 2/3 of a truckload. The trucks could occasionally pick up goods at manufacturers (backhauling) and eventually deliver goods directly from manufacturers to stores if profitable. Obviously the company wanted to reduce its total operating and delivery costs. It also wanted to minimize the inconvenience to stores of receiving multiple deliveries. However the exact costs were as usual ill-defined, and linearly approximated.

Recently, the company decided to award a contract to one of the large consulting firms in computer science, to computerize all its operations (not only the distribution). This large firm, which we will call ABC Inc., is competent in computer science but has no expertise in operations research.

For the distribution problem, the ABC consultants decided to evaluate the packages on the market. Being computer scientists they looked mainly at the specification of the codes available and how easy they were to use. They also obviously looked at cost and ease of implementation of these codes.

When they explained their problem to us, we tried to convince them that they needed a customized solution, not because we wanted to make money, but because we thought their problem was too specific to be handled by the packages on the market. The algorithms in most packages are geographically based, trying to use the same vehicle to serve clients that are geographically close to one another. Obviously, vendors of commercial codes insist that they have extremely good general purpose algorithms based on state-of-the-art techniques that they cannot reveal for proprietary reasons.[1] However, with a limited number of clients per truck, the full packing of the trucks becomes critical if the number of required trucks is to be reduced. In addition, the full use of the multiple-trip options should go beyond merely combining independently generated routes; options for generating routes which may be less than full, but that combine well into multiple trip routes should also be considered (if possible).

The cost of backorders (delivering part of the load with the next delivery) should also be quantified properly, together with costs or penalties for split deliveries. The utilization of common carriers for distant deliveries when there is no opportunity for backhauling goods from a manufacturer in the area should be closely looked at. For all these reasons, we believed that no general algorithm could consider all these factors properly and generate the maximum benefit that a customized approach could give. At best, a general package would give the dispatcher the tools to easily modify its proposed solution. This solution will normally be inadequate or far from optimal, with the added risk that the dispatcher tends to accept it as optimal because it is produced by a computer.

It is difficult, however, for an operations researcher or expert in routing and scheduling algorithms to convince a computer scientist that even when a specification is included in a package, it is far from obvious that it is dealt with properly. The underlying algorithm used by the package may be totally inadequate for the particularities of the problem. It is also far from clear that a general purpose package that fully satisfies other well-known corporations will be appropriate for a problem with different particularities.

When an analyst proposes a customized solution, the proposed technique is obviously not in hand for demonstration. Thus, it is hard to convince others that this technique would produce a solution far superior to what a general purpose code can generate. The cost of merely testing a customized solution is greater than the purchase cost of some of the packages on the market. In this context, a rational M.I.S. person is tempted to say: "Let us try the general purpose code and then we will see if it is appropriate or not." Only blind faith and confidence in the OR people proposing a customized approach can prevail over this type of an apparently rational argument. It is, however, forgotten that nobody may be able to appreciate how far the solutions proposed by a general purpose package are from the best obtainable solution. The dispatcher may also get tired of the package and just not use it without telling anybody. In all probability, nobody will ever want to take the responsibility of declaring a package recently bought inadequate, considering all the costs and effort invested in trying to make it perform reasonably well in a particular context for which the package was not designed to address specifically.[2]

The preceding story has not concluded yet; we are still in the midst of it. It could end in several different ways but it is likely that a general purpose code will be purchased. The moral for the OR community is that operations researchers are not very successful in transmitting to the engineers (or MBA's, computer scientists, etc.) even the most basic notions of OR. Those without OR expertise cling to the idea that LP produces optimal solutions, and that probably every algorithm generated by the OR community also produces optimal solutions in most situations and therefore believe vendors who claim to offer optimal or near optimal algorithms.

To exemplify this point further, we will describe some of our experience with real problems emphasizing the main characteristics of each problem, characteristics which require special attention and customized solutions.

1.2. School Transportation

In Québec, school children must be transported by the School Board if they live more than one mile away from their school. In rural areas, students must be picked up at their front door and in urban areas they are picked up at stops that are generally located less than 1/5 of a mile from their home. At first sight, the problem of picking up groups of students at stops is trivial and most packages should be able to do this properly.

But if we look at the problem more deeply, we realize that the issue of routing buses to pick up students with minimal mileage is minor if considered in relation to the global objective.

The School Boards rent these bus services from private companies. The cost structure is such that the most important cost factor is the number of buses and not the number of miles. The number of buses required is largely a function of the number of school bus routes (picking up students and bringing them to a school) that can be performed by the same bus. This, in turn, is largely influenced by school starting and ending times. We have shown (see Desrosiers et al., 1986) that in a practical context 30% more buses would be required if all schools start at the same time compared to the situation where schools start at times spaced more optimally relative to each other. In this context, what was needed in order of importance was: 1) a good algorithm to choose school times, 2) a good algorithm to construct sequences of tasks with time windows for the buses (there is some flexibility for the arrivals of buses at schools, e.g., 5-20 minutes before school starts), and 3) a good algorithm to construct pick-up routes. Even in this context, the general purpose algorithms used to pick up or deliver goods are not necessarily appropriate because we are transporting people in this case and the shape of the route has special importance. For example, the routes should tend to move in the direction of the schools because parents will complain strongly if the bus passes twice in front of their house or if their children are picked up and subsequently driven away from schools. In addition, there are constraints for equalizing the load between various buses and reducing the total duration of trips to the schools. Seeking to address these issues, we found the algorithms in the literature inadequate for the school bus problem and proposed specially designed algorithms (Chapleau et al., [1], [2]).

This is also a context where we believe that it is necessary to have a detailed network, to locate the students and to evaluate with precision their eligibility for transportation and their walking distance to the bus stop. The network is also needed to evaluate with precision the travel times of the buses on the relatively sparse road network which, in rural areas, is complicated by the presence of rivers, mountains, and lakes. We have seen situations where the south side of a relatively small mountain is about 15 miles away by road from its north side, while they are separated by a linear distance of about 2 miles. The option of using a full detailed network is not always available in commercial codes.

1.3. Postal Delivery Problems

In the context of postal delivery we have looked at several problems. A first problem is concerned with picking up the contents of mail boxes.

Again, at first sight, the problem appears to be trivial. There is practically no capacity constraint on the vehicles, and because the problem is segmented by period of the day, there is no specific time window except that the length of the routes is limited. However, because there is a relatively large number of mail boxes within a relatively small territory the characteristics of the network become extremely important. In addition, union constraints require that the driver pick up the mail only on the right-hand side of the vehicle (the driver is not allowed to cross the street, for security reasons) and U-turns are generally prohibited. Moreover, many mail boxes are located at street corners and the movement of vehicles is limited after picking up such mail boxes as demonstrated in Figure 1. This implies that the distance from mail box x to mail box y depends also on which mail box z has been serviced before mail box x (see Figure 2).

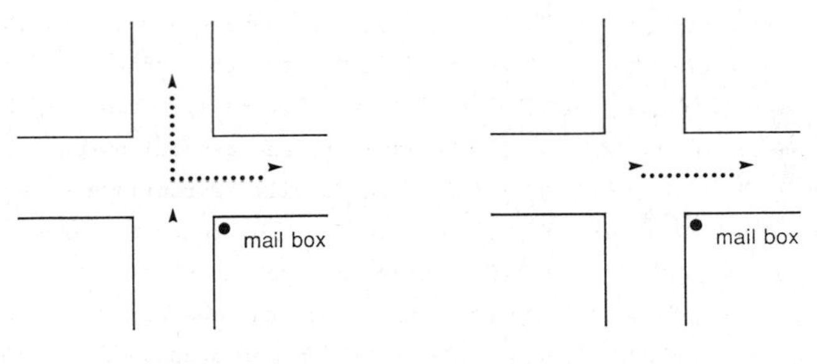

Option 1

The vehicle cannot turn left after picking up at the mail box.

Option 2

The vehicle can only continue straight ahead after picking up at the mail box.

Figure 1: Turn options when a mail box is located at a street corner.

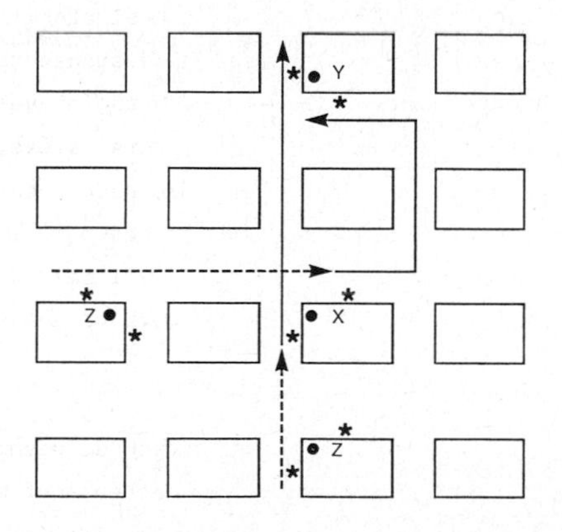

Figure 2: - actual (•) vs artificial (*) mail boxes;
- the distance (—) between two mail boxes depends on which mail box has been services previously (---).

While we readily admit that one-way streets and no left turns are really irrelevant for most delivery problems on a large territory, they clearly play a very important role in this context. In fact, if such network characteristics are ignored in the algorithm, the solution produced must generally be totally resequenced manually. The relative error in the distance between two points is very high if the points are close to each other.

Because the end result must be a set of detailed routes given to truck drivers (including where to turn), the quality of the route proposed must be excellent and hard to improve on manually. Because of this, we also decided to use an exact traveling salesman algorithm to sequence the points on a route using the exact distances calculated on the network. The heuristics used previously were producing nearly optimal solutions but there was some imperfection (crossover) not very important for the total distance but important for the perception of the quality of the routes. In fact we used a version of the traveling salesman algorithm developed by G. Laporte et al. [4] to take into account mail boxes on street corners (these mail boxes are artificially doubled and located on each of the adjacent arcs; the vehicle is required to service only one of these artificial mail boxes; by doing this the distance between two artificial mail boxes is uniquely determined (see, Figure 2)).

It is worth mentioning that several years ago, Canada Post Corporation decided to select a package for this problem and did not listen to our comments that a customized system was needed. They bought several copies of a well-known package, and are currently spending a great deal of effort coding data and trying to make it work. From the previous analysis of the problem, we expect the results to be far from satisfactory.

At Canada Post, we are currently looking at the generation of letter carrier walking routes--a capacitated arc routing problem. For this type of problem there are very few algorithms even published in the literature and, to our knowledge, no package on the market can solve this problem. A paper by Levy and Bodin [5] in this book describes the work of those authors in this area for the U.S. context. The main characteristic of the problem is that most arcs of the network must be visited and the routes include very little deadheading (traveling a given arc without delivering mail). All the deadheading occurs at the beginning and at the end of the route, and at noon time to go back to the depot for lunch. In the urban areas of Canada, transit buses and taxis are used to transport letter carriers to and from their routes. Balancing of work amongst carriers is also a major goal. Without describing the problem in detail, we can say that because of different ways of transporting the letter carriers to and from the routes, and different union and operating constraints, the Canadian and U.S.A. letter-carrier problems may require significantly different algorithms.

Finally, we are also studying a routing and scheduling problem for the priority mail system (24-hour delivery system). Pickup points may have fairly restrictive time windows (sometimes ½ hour), but there are no capacity constraints on the vehicles. All vehicles must reach the airport by a specific deadline time. In this context we have tried to adapt a more general algorithm that we have which takes geography into account, but several tests have convinced us that a specific algorithm must be designed for this context as well.

1.4. Other Problems

We have worked on several other problems that we will not describe in detail here. For example, the transportation of handicapped persons is one of the most complex routing and scheduling problems because it includes multiple capacity constraints (wheel-chair and ambulatory persons), very restrictive time windows on both pickup and drop-off, and constraints to ensure the obvious fact that a handicapped person must first be picked up and dropped off later by the same vehicle. This clearly requires a specially-designed algorithm. Other problems include fuel oil delivery, where the main savings does not come from the routing of vehicles but from the careful selection of customers to be serviced during a given day to prevent stockouts and minimize routing costs. Again the problem requires special treatment not available in a general purpose code.

2. A FRAMEWORK FOR ALGORITHM DEVELOPMENT

In summarizing what we have learned, we would like to emphasize the following points.

Our first point is that in many distribution contexts, routing and especially the minimization of mileage is not necessarily the most important element of the problem. Organizational problems may generate much greater benefits or may need to be solved before other benefits can be realized. However, the ability to efficiently produce good routes for vehicles may be essential for the effectiveness of the overall process. For example, in the school bus problem it is important to produce good routes automatically in order that the rest of the scheduling system be applied effectively to achieve all potential savings. Otherwise, the process of elaborating and manually fixing all routes might be too much of a burden and might prevent the use of the rest of the system.

While the simple redesign of letter-carrier routes can bring only a few percent savings, options for modifying the rules governing the design of these routes could generate sizeable savings. On the other hand, any change in organization would necessitate the total redesign of all letter-carrier walking routes, which is an enormous task. While most savings would come from such a reorganization, the evaluation of the impact and implications of the implementing of such a change would need a good system for designing these routes.

Our second point is that most packages on the market are based on heuristics that place a heavy emphasis on geography that can be partially, if not totally, inadequate for certain types of problems.

Time windows and packing constraints on vehicles can sometimes almost totally destroy the geographical nature of the problem. In other contexts, the shape of the routes is important. We have mentioned the school bus example. Another example is when the quantity to be delivered is unknown and the vehicles may have to come back to their depot for reloading to continue their routes. In this case, the routes should normally start from the farthest points and converge toward the depot so that if stockouts occur they happen, as much as possible, in the neighborhood of the depot.

We should point out, however, that the packages found in the market may be useful on a subset of routing and scheduling problems. They should produce good solutions on problems where the detailed network is not important, the packing aspect is not critical, time constraints on clients are rather wide with possibly a few exceptions, and no other constraints or operational characteristics are of prime importance for the quality of the solution. At first sight, this may cover a large number of problems, but we are sure that once one develops a full understanding of the fundamental nature of each problem, there may not be too many practical problems for which the packages are fully appropriate. Of course, as mentioned previously, the facilities generally included to modify interactively the solution proposed may in all cases be useful to obtain a good solution even if the originally proposed solution is totally inadequate. These features are useful and even necessary in most cases, but too often they also serve as an excuse not to invest more in the generation of an initial solution of high quality.

Finally, we believe that what is needed is the development of an algorithmic shell that could be easily adapted to nearly every context by a routing and scheduling expert. Then a general package could be adapted and have significantly better performance without the necessity of generating a new code each time. Such a shell has been recently proposed by J.-Y. Potvin (Potvin et al., [6], [7]).

Fundamentally, the general heuristic proposed is composed of four basic operators manipulating sets of objects (Figure 3).

In the figure, we can identify three major elements:

a) Objects

The objects have attributes providing information to the operators acting over them. In particular, the major attributes used by the general heuristic are:

i) for the origin : coordinates and number of vehicles available;
ii) for the stops : coordinates, time window and demand;
iii) for the routes : number of stops, length, travel time and total demand;
iv) for the vehicles : speed, capacity and maximal authorized distance or travel time.

b) Basic operators

The objects are manipulated by four basic operators symbolizing distinct phases of the general heuristic. These operators are:

i) INIT : operator for the creation and initialization of new routes;
ii) ADD : operator for the insertion of new stops into existing routes;
iii) MERGE : operator for a two-by-two merging of the routes;
iv) EXCHANGE : operator for the modification of the sequence of stops in the routes.

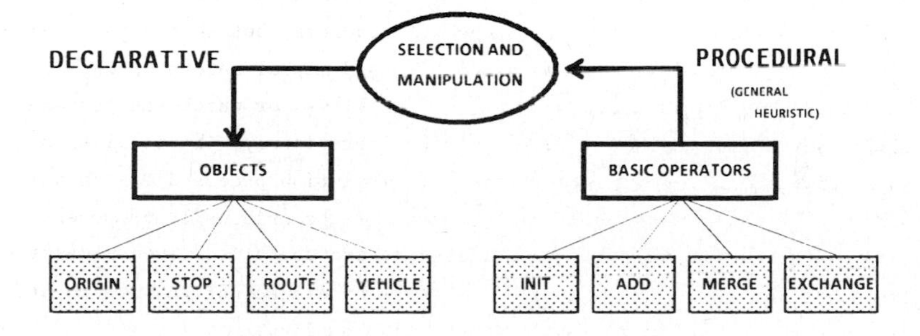

Figure 3: Objects and basic operators.

c) A selection mechanism

Each operator first selects one or more objects and then manipulates these objects. The manipulation process is intrinsically defined in each operator but the selection process is put under user's control by means of selection formulae. These formulae associate values with the objects involved in the problem's definition. Such values, called OMC for Object Measure Coefficient, allow ordering and selection of objects (for example, we can select the object with minimal or maximal OMC). The most usual selection formulae involve real functions of the following types:

i) f: $E_n \rightarrow R$ (ex: the demand at a stop);

ii) f: $E_r \rightarrow R$ (ex: the length of a route);

iii) f: $E_v \rightarrow R$ (ex: the capacity of some vehicle);

iv) f: $E_n x E_n \rightarrow R$ (ex: the distance between two nodes);

v) f: $E_n x E_r \rightarrow R$ (ex: the distance between a node and a route).

where

E_n = the set of nodes in the network;

E_r = the set of routes;

E_v = the set of vehicles.

All operators are independent of each other and can be activated by the user in any desired sequence. The only restriction is obvious: the first operator of the sequence must be INIT (Figure 4). Currently, the user must apply the operators manually but in the near future automatic generation of sequences of properly instantiated operators will be available.

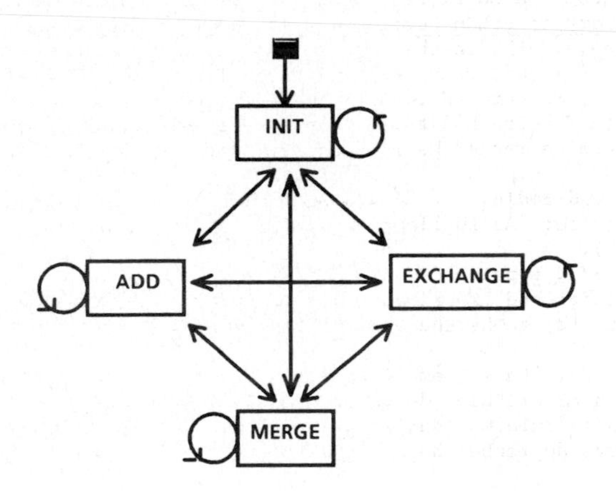

Figure 4: Activation sequences for the basic operators.

In fact, most of the published heuristic algorithms with all their possible variants and mixtures could be easily generated by using appropriate formulae for the selection mechanism and appropriate operators. This general algorithm is at present implemented on a LISP machine and is already helping us in designing new algorithms. We should soon report in the open literature on this recent work. We intend, as part of our development work in this area, to generate the same algorithmic shell on a personal computer so that we can easily design and generate a customized algorithm for each particular situation that might arise.

FOOTNOTES

[1]We have even heard a vendor use the term "linear programming like" algorithm.

[2]I would like to suggest that this human propensity to spend an order of magnitude more effort to salvage a bad decision (rather than admitting the error and starting from scratch) be called the Mirabel syndrome, in honor of the useless airport in the Montréal area whose contribution to the general economy is negative.

REFERENCES

[1] Chapleau, L., Rousseau, L., Ferland, J. A. and Rousseau, J.-M., "Clustering for Routing in Dense Area." European Journal of Operational Research (1985) 20:48-57.

[2] Chapleau, L., Ferland, J. A., Lapalme, G. and Rousseau, J.-M., "A Parallel Insert Method for the Capacitated Arc Routing Problem." Operations Research Letters (1984) 3(2):95-100.

[3] Desrosiers, J., Ferland, J. A., Rousseau, J.-M., Lapalme, G. and Chapleau, L., "TRANSCOL: A Multi-Period School Bus Routing and Scheduling System." In: Delivery of Urban Services, A. J. Swersey and E. J. Ignall (Eds.), Vol. 22, pp. 47-71, in the Series TIMS Studies in the Management Science, Machol, R. E., editor in chief, North-Holland, The Netherlands (1986).

[4] Laporte, G., Mercure, H., and Nobert, Y., "Generalized Travelling Salesman Problem Through n Sets of Nodes: The Asymmetrical Case." Publication #399, Centre de recherche sur les transports, Université de Montréal (1985).

[5] Levy, L., and Bodin, L., "Scheduling the Postal Carriers for the U.S. Postal Service: An Application of Arc Partitioning and Routing," in this book (1988).

[6] Potvin, J.-Y., Lapalme, G., and Rousseau, J.-M., "ALTO: A Computer System for the Design and Experimentation of Routing Algorithms," Publication #525, Centre de recherche sur les transports, Université de Montréal (1987a).

[7] Potvin, J.-Y., "Un système informatique pour le développement et l'expérimentation d'algorithmes de génération de tournées." Thèse de Ph.D., département d'informatique et de recherche opérationnelle, publication #522, Centre de recherche sur les transports, Université de Montréal (1987b).